Applications of the Theory of Groups in Mechanics and Physics

Fundamental Theories of Physics

An International Book Series on The Fundamental Theories of Physics: Their Clarification, Development and Application

Volume 140

Applications of the Theory of Groups in Mechanics and Physics

by

Petre P. Teodorescu

Faculty of Mathematics,
University of Bucharest, Bucharest, Romania

and

Nicolae-Alexandru P. Nicorovici

School of Physics,
The University of Sydney, New South Wales, Australia

KLUWER ACADEMIC PUBLISHERS
DORDRECHT / BOSTON / LONDON

A C.I.P. Catalogue record for this book is available from the Library of Congress.

ISBN 1-4020-2046-5 (HB)
ISBN 1-4020-2047-3 (e-book)

Published by Kluwer Academic Publishers,
P.O. Box 17, 3300 AA Dordrecht, The Netherlands.

Sold and distributed in North, Central and South America
by Kluwer Academic Publishers,
101 Philip Drive, Norwell, MA 02061, U.S.A.

In all other countries, sold and distributed
by Kluwer Academic Publishers,
P.O. Box 322, 3300 AH Dordrecht, The Netherlands.

Printed on acid-free paper

Printed in the Netherlands.

CONTENTS

PREFACE ix
INTRODUCTION xi

1. ELEMENTS OF GENERAL THEORY OF GROUPS 1
 1 Basic notions 1
  1.1 Introduction of the notion of group 1
  1.2 Basic definitions and theorems 4
  1.3 Representations of groups 19
  1.4 The S_3 group 33
 2 Topological groups 40
  2.1 Definitions. Generalities. Lie groups 40
  2.2 Lie algebras. Unitary representations 50
 3 Particular Abelian groups 52
  3.1 The group of real numbers 52
  3.2 The group of discrete translations 54
  3.3 The $\mathsf{SO}(2)$ and C_n groups 58

2. LIE GROUPS 61
 1 The $\mathsf{SO}(3)$ group 61
  1.1 Rotations 61
  1.2 Parametrization of $\mathsf{SO}(3)$ and $\mathsf{O}(3)$ 70
  1.3 Functions defined on $\mathsf{O}(3)$. Infinitesimal generators 74
 2 The $\mathsf{SU}(2)$ group 76
  2.1 Parametrization of $\mathsf{SU}(2)$ 76
  2.2 Functions defined on $\mathsf{SU}(2)$. Infinitesimal generators 82
 3 The $\mathsf{SU}(3)$ and $\mathsf{GL}(n, \mathbb{C})$ groups 89
  3.1 $\mathsf{SU}(3)$ Lie algebra 89

3.2 Infinitesimal generators. Parametrization of SU(3) 102
3.3 The GL(n, ℂ) and SU(n) groups 107
4 The Lorentz group 111
4.1 Lorentz transformations 111
4.2 Parametrization and infinitesimal generators 118

3. SYMMETRY GROUPS OF DIFFERENTIAL EQUATIONS 123
1 Differential operators 123
1.1 The SO(3) and SO(n) groups 123
1.2 The SU(2) and SU(3) groups 127
2 Invariants and differential equations 133
2.1 Preliminary considerations 133
2.2 Invariant differential operators 138
3 Symmetry groups of certain differential equations 149
3.1 Central functions. Characters 149
3.2 The SO(3), SU(2), and SU(3) groups 151
3.3 Direct products of irreducible representations 159
4 Methods of study of certain differential equations 176
4.1 Ordinary differential equations 176
4.2 The linear equivalence method 177
4.3 Partial differential equations 186

4. APPLICATIONS IN MECHANICS 201
1 Classical models of mechanics 201
1.1 Lagrangian formulation of classical mechanics 201
1.2 Hamiltonian formulation of classical mechanics 206
1.3 Invariance of the Lagrange and Hamilton equations 213
1.4 Noether's theorem and its reciprocal 220
2 Symmetry laws and applications 230
2.1 Lie groups with one parameter and with m parameters 230
2.2 The Symplectic and Euclidean groups 235
3 Space-time symmetries. Conservation laws 244
3.1 Particular groups. Noether's theorem 244
3.2 The reciprocal of Noether's theorem 251
3.3 The Hamilton-Jacobi equation for a free particle 259
4 Applications in the theory of vibrations 263
4.1 General considerations 263
4.2 Transformations of normal coordinates 265

5. APPLICATIONS IN THE THEORY OF RELATIVITY AND THEORY OF CLASSICAL FIELDS 279
1 Theory of Special Relativity 279
1.1 Preliminary considerations 279
1.2 Applications in the theory of Special Relativity 289
2 Theory of electromagnetic field 302
2.1 Noether's theorem for the electromagnetic field 302
2.2 Conformal transformations in four dimensions 317
3 Theory of gravitational field 324
3.1 General equations 324
3.2 Conservation laws in the Riemann space 329

6. APPLICATIONS IN QUANTUM MECHANICS AND PHYSICS OF ELEMENTARY PARTICLES 335
1 Non-relativistic quantum mechanics 335
1.1 Invariance properties of quantum systems 335
1.2 The angular momentum. The spin 358
2 Internal symmetries of elementary particles 371
2.1 The isospin and the SU(2) group 371
2.2 The unitary spin and the SU(3) group 383
3 Relativistic quantum mechanics 396
3.1 Basic equations. Symmetry groups 396
3.2 Elementary particle interactions 407

REFERENCES 423

INDEX 431

PREFACE

The notion of group is fundamental in our days, not only in mathematics, but also in classical mechanics, electromagnetism, theory of relativity, quantum mechanics, theory of elementary particles, etc. This notion has developed during a century and this development is connected with the names of great mathematicians as E. Galois, A. L. Cauchy, C. F. Gauss, W. R. Hamilton, C. Jordan, S. Lie, E. Cartan, H. Weyl, E. Wigner, and of many others. In mathematics, as in other sciences, the simple and fertile ideas make their way with difficulty and slowly; however, this long history would have been of a minor interest, had the notion of group remained connected only with rather restricted domains of mathematics, those in which it occurred at the beginning. But at present, groups have invaded almost all mathematical disciplines, mechanics, the largest part of physics, of chemistry, etc. We may say, without exaggeration, that this is the most important idea that occurred in mathematics since the invention of infinitesimal calculus; indeed, the notion of group expresses, in a precise and operational form, the vague and universal ideas of *regularity* and *symmetry*.

The notion of group led to a profound understanding of the character of the laws which govern natural phenomena, permitting to formulate new laws, correcting certain inadequate formulations and providing unitary and non-contradictory formulations for the investigated phenomena. In this way, unitary methods are obtained for solving problems of different physical nature but having common mathematical features; the investigation of these problems in a unitary way permits a better understanding of the special phenomena under consideration.

The present volume is a new edition of a volume published in 1985 ("Aplicaţii ale teoriei grupurilor in mecanică şi fizică", Editura Tehnică, Bucharest, Romania). This new edition contains many improvements concerning the presentation, as well as new topics using an enlarged and updated bibliography. We outline the manner in which the group theory can be applied to the solution and systematization of several problems in the theory of differential equations, in mechanics, and physics. Having in view the great number of published works (as it can be seen in the bibliography), we wish to give the reader a number of preliminary notions and examples which are absolutely necessary for a better understanding of certain works, in different domains, which are based on applications of group theory. Since they are of special importance, being at the same time more accessible, we have considered in detail only the Lie groups, the Lie algebras associated with these groups, and their linear representations.

The work requires as preliminaries only the mathematical knowledge acquired by a student in a technical university. It is addressed to a large audience, to all those interested and compelled to use mathematical methods in various fields of research, like: mechanics, physics, engineering, people involved in research or teaching, as well as students.

P. P. Teodorescu and N.-A. P. Nicorovici

INTRODUCTION

To study a natural phenomenon by means of the methods of group theory, we adopt the following methodology: We consider a phenomenon **F** (mechanical or physical), modelled by a system of differential equations **S**. If the system **S** possesses certain symmetry properties, then there exist certain groups of transformations of coordinates, which act upon a canonical Hilbert space associated with **S** (generally, the space of solutions). The representations, the invariants, and covariants of these groups have certain significances directly connected with the properties of **S**, which can be translated as properties of the phenomenon **F** (as a rule, conservation laws). In this way, the properties of the phenomenon **F** can be enounced and analysed by means of the results of group theory.

We show how the above methodology can be applied to the Lagrangian and Hamiltonian differential systems, to the Hamilton-Jacobi equation and the Lagrange equations for vibrations (in classical mechanics), to the Maxwell equations (in electromagnetism), to the Schrödinger equation (in non-relativistic quantum mechanics), and to the Klein-Gordon and the Dirac equations (in relativistic quantum mechanics). Several symmetry properties of these differential systems are analysed by means of certain local analytical groups, namely: the Symplectic group, the Euclidean group and the non-Euclidean group, the orthogonal group, the unitary group, and the special unitary group. The physical expression of the invariance under group actions of physical entities (for instance, linear momentum, angular momentum, electromagnetic or gravitational field, etc.) is, usually, a conservation law. In particular, the connection between the invariance of the Lagrangian of a physical system under a transformation group and the corresponding conservation laws is expressed by Noether's theorem.

As an example, a mechanical system evolves in the Euclidean space which, in classical mechanics, is assumed to be isotropic and homogeneous, that is invariant with respect to space translations and space rotations, respectively. We may also add the invariance with respect to time translations. In other words, we assume that there is no preferred point in space such that we may set the origin of the coordinates anywhere, there is no preferred direction, therefore the orientation of axes is arbitrary, and we may choose any moment as the origin of time. These invariance properties dictate the analytic form of the Lagrangian, or the Hamiltonian, that are the basic ingredients for any axiomatic theory. In turn, the symmetry properties of the Lagrangian lead to conservation laws; in our case, the conservation of momentum, angular momentum, and energy. Hence, the most common conservation laws in classical mechanics are actually consequences of the geometrical properties of space-time.

Another example comes out from the use of group theory to classify the elementary particles. Thus, the spin quantum number associated to elementary particles has found a mathematical formalization in the theory of the $\mathsf{SU}(2)$ group. By the aid of the same group, Werner Heisenberg (Nobel Prize 1932) introduced the isospin quantum number for elementary particles. In 1961, Murray Gell-Mann (Nobel Prize 1969) and Yuval Ne'eman suggested, independently of each other, a way of classifying the hadrons (elementary particles that take part in strong interactions) known at that time; this classification is based on the $\mathsf{SU}(3)$ group. The most important confirmation of the profoundness of this idea was obtained in 1962 by the experimental discovery of the hadrons Δ and Σ, theoretically predicted by this model. During the same year, the doublet Ξ was also discovered experimentally. Finally, in 1963, the particle Ω^- was found, thus completing a multiplet of 10 hadrons (decuplet) predicted theoretically by the $\mathsf{SU}(3)$ classification. This was the proof that the $\mathsf{SU}(3)$ symmetry is deeply rooted in the world of hadrons, although a symmetry model does not constitute a complete theory. Nevertheless, by the aid of the $\mathsf{SU}(3)$ classification several regularities were discovered concerning the properties of hadrons, like in the case of the Mendeleev table for chemical elements.

In an attempt to model the properties of hadrons, George Zweig and Murray Gell-Mann introduced in 1964 the *quark model*, based also on the $\mathsf{SU}(3)$ symmetry group (now called $\mathsf{SU}(3)$ flavour symmetry group). In the original version of this model it is supposed that the hadrons are composed of several combinations of three fundamental particles and three antiparticles, called quarks and antiquarks, respectively. The most revolutionary aspect of the quark model consists in the fact that these fundamental particles possess fractionary electric charges, unlike the elementary particles whose electric charges are multiples of the electron charge. However, the quark model was used with great success to predict static properties of hadrons. To avoid some difficulties related to the ground states of hadrons, the quark model was changed by adding the *colour* quantum number (*quantum chromodynamics* theory, or QCD, based on the colour symmetry group $\mathsf{SU}(3)_c$). Later, in 1974, the experimental discovery of the J/ψ particle required a new quark (and a new quantum number *charm*), in 1977 the discovery of the Υ (Upsilon) particle necessitated another quark (*bottom*) with a new quantum number attribute, and, in 1995, the sixth quark (*top*) has been discovered at Fermilab. Accordingly, higher symmetry groups have been introduced: $\mathsf{SU}(3)\times\mathsf{SU}(3)$, $\mathsf{SU}(4)$, $\mathsf{SU}(6)$ etc. Even if the free quarks have never been observed, the actual existence of quarks was indirectly demonstrated in 1981 by the experimental discovery of the *gluons* (particles that ensure the binding between the quarks which form a hadron).

In the domain of weak interactions, the introduction of group theory was entailed by the discovery of the parity violation in the β decay (Tsung Dao Lee and Chen Ning Yang, Nobel Prize 1956), which corresponds to the non-invariance of this process under the discrete group of spatial inversions. In 1967, Steven Weinberg and, independently, some months later, Abdus Salam, introduced a unified theory of electromagnetic and weak interactions, a theory based on an elegant mathematical formalism: the $\mathsf{U}(1)\times\mathsf{SU}(2)$ symmetry (Nobel Prize 1979, together with Sheldon Lee Glashow). The outstanding success of this theory was the experimental discovery of the particles $W^{\pm}$ (1982) and Z^0 (1983), predicted theoretically.

All these examples show the importance of the theory of groups as a mathematical method in modelling of natural phenomena. Galois invented the group theory and applied it to algebraic equations. It is the merit of Lie who developed a theory based on the assumption that the symmetries are the primary features of natural phenomena, and they dictate the form of dynamical laws. Lie applied his theory to dynamical systems modelled by systems of ordinary differential equations, and created the symmetry analysis of ordinary differential equations. There are many extensions of Lie's theory due to different authors, but the most important is the group analysis of partial differential equations, developed by Ovsiannikov. From all these contributions and successive extensions of the original theory, we have today a very powerful mathematical tool which allows the investigation of phenomena pertaining to different domains of physics in a unitary way.

Finally, some comments about notations. The set of all $n \times n$ complex matrices $\mathcal{M}(n)$ form a Lie algebra with respect to the internal composition law

$$[\mathbf{a}, \mathbf{b}] \equiv \mathbf{ab} - \mathbf{ba}\,, \quad \mathbf{a}, \mathbf{b} \in \mathcal{M}(n)\,.$$

The corresponding Lie group is the *general linear group in n dimensions*, denoted by $\mathsf{GL}(n, \mathbb{C})$, and defined as the set of linear transformations acting in the n-dimensional linear complex space $\mathbb{C}^n$. To each element in $\mathsf{GL}(n, \mathbb{C})$, we may associate a matrix g such that we may define the group $\mathsf{GL}(n, \mathbb{C})$ in terms of its matrix representation

$$\mathsf{GL}(n, \mathbb{C}) = \{\mathbf{g} \mid \mathbf{g} \in \mathcal{M}(n),\ \det(\mathbf{g}) \neq 0\}\,,$$

the internal composition law being the product of matrices. Many applications of the group theory in physics involve subgroups of $\mathsf{GL}(n, \mathbb{C})$, that are:

- the *special (unimodular) linear group*

$$\mathsf{SL}(n,\mathbb{C}) = \{\mathbf{g} \mid \mathbf{g} \in \mathsf{GL}(n,\mathbb{C}),\ \det(\mathbf{g}) = 1\},$$

- the *unitary group*

$$\mathsf{U}(n) = \{\mathbf{g} \mid \mathbf{g} \in \mathsf{GL}(n,\mathbb{C}),\ \mathbf{g}^{\dagger}\mathbf{g} = \mathbf{e}\},$$

  where the symbol $\dagger$ denotes the Hermitian conjugation, and $\mathbf{e}$ is the identity matrix,

- the *special unitary group*

$$\mathsf{SU}(n) = \{\mathbf{g} \mid \mathbf{g} \in \mathsf{SL}(n,\mathbb{C}),\ \mathbf{g}^{\dagger}\mathbf{g} = \mathbf{e}\} = \mathsf{U}(n) \cap \mathsf{SL}(n,\mathbb{C}),$$

- the *linear real group*

$$\mathsf{GL}(n,\mathbb{R}) = \{\mathbf{g} \mid \mathbf{g} \in \mathsf{GL}(n,\mathbb{C}),\ \mathsf{Im}\,\mathbf{g} = 0\},$$

  where Im means the imaginary part of the argument,

- the *special linear real group*

$$\mathsf{SL}(n,\mathbb{R}) = \{\mathbf{g} \mid \mathbf{g} \in \mathsf{GL}(n,\mathbb{R}),\ \det(\mathbf{g}) = 1\},$$

- the *orthogonal group*

$$\mathsf{O}(n) = \{\mathbf{g} \mid \mathbf{g} \in \mathsf{GL}(n,\mathbb{R}),\ \mathbf{g}^{\mathrm{T}}\mathbf{g} = \mathbf{e}\} = \mathsf{U}(n) \cap \mathsf{GL}(n,\mathbb{R}),$$

  where the superscript T denotes a transposed matrix,

- the *special orthogonal group*

$$\mathsf{SO}(n) = \{\mathbf{g} \mid \mathbf{g} \in \mathsf{SL}(n,\mathbb{R}),\ \mathbf{g}^{\mathrm{T}}\mathbf{g} = \mathbf{e}\} = \mathsf{O}(n) \cap \mathsf{SL}(n,\mathbb{R}).$$

Other groups, like the *Symmetric group* S_n, the *Symplectic group* $\mathsf{Sp}(2n,\mathbb{C})$, the *Euclidean group* $\mathsf{E}(n)$, etc., will be defined in the corresponding chapters.

Note that there are many different notations for some groups (for instance $\mathsf{O}^{+}(n)$ instead of $\mathsf{SO}(n)$), but in this book we used the notations which seems to match better the scheme from above. We also have to point out that, generally we used *Einstein's summation convention* for dummy indices. However, at some places we explicitly mentioned the ranges for the dummy indices to avoid any confusion.

In the examples of applications of group theory we used the fully integrated environment for technical computing *Mathematica*® developed at *Wolfram Research*. Some of the symbolic calculations have been carried out by means of the *Mathematica*® packages *MathLie*™ by G. Baumann, and *YaLie* by J. M. Diaz.

Chapter 1

ELEMENTS OF GENERAL THEORY OF GROUPS

1. Basic notions

1.1 Introduction of the notion of group

1.1.1 Solution of algebraic equations

The group theory appeared from the necessity to find a mathematical tool capable to describe the symmetry properties of the real world. Indeed, the knowledge of symmetry properties of various mathematical or physical objects provides essential information concerning the structure and evolution of these objects.

The first contributions to the field are due to Joseph-Louis Lagrange (1736-1813). In "Considerations on the solution of algebraic equations", published in 1770-1771, he analysed in a critical manner all the known methods to solve by radicals algebraic equations of second, third and fourth degree, and observed that the success of these methods was based on certain properties which do not pertain to equations of fifth or higher degree. In fact, the behaviour of algebraic equations with respect to permutations of roots plays a major rôle in solving such equations by radicals. Lagrange expressed even the idea that the *theory of permutations* represents the "real philosophy of the whole problem", as it was confirmed later by Évariste Galois (1811-1832).

Let us examine the behaviour of a function $f(x_1, x_2, \ldots, x_n)$ with respect to different permutations of its variables. The function may remain unchanged (for instance, in the case of a linear function $f = x_1 + x_2 + \ldots + x_n$), or we may obtain distinct values for each permutation. Thus, in the case of the function $f = x_1 + 2x_2 + \ldots + nx_n$ we have $n!$ possible values. In the general case of an arbitrary function in n variables $f(x_1, x_2, \ldots, x_n)$, Lagrange has shown that the number of distinct values which can be taken by the function, for all possible permutations of variables, is always a divisor of $n!$. For instance, the function

$f = x_1x_2 + x_3x_4$, has only three distinct values for all possible permutations of its $n = 4$ variables, and three is a divisor of $n! = 4! = 24$. Lagrange has also demonstrated that the existence of such functions for $n = 3$ and $n = 4$ explains the possibility to solve by radicals equations of the third and fourth degree.

In 1799 Paolo Ruffini (1765-1822) published a book on the theory of equations entitled "General theory of equations" in which it is shown that the algebraic solution of the general equation of degree greater than four is impossible. Ruffini showed that, for $n = 5$ ($n! = 120$), the number of possible values taken by a function in five variables under all permutations of variables, cannot be three, four, or eight, being thus led to the proof of the impossibility to solve the *quintic equation* (the algebraic equation of fifth degree) by radicals. Between 1803 and 1813, Ruffini published different versions of a general proof of the *impossibility of solving algebraically equations of higher than the fourth degree*. In his work, Ruffini used the group theory but he had to invent the subject for himself, and thus he was the first to introduce the notion of the order of an element, conjugacy, the cycle decomposition of elements of permutation groups, and the notions of primitive and imprimitive.

In 1813, Augustin-Louis Cauchy (1789-1857) had the idea to consider the substitution operations of variables of a function as objects by themselves. He denoted by

$$\begin{pmatrix} a_1 & a_2 & a_3 & \dots \\ b_1 & b_2 & b_3 & \dots \end{pmatrix} \equiv \begin{pmatrix} A \\ B \end{pmatrix}$$

the operation of replacing a_1 by b_1, a_2 by b_2, etc.,

$$\begin{pmatrix} A \\ B \end{pmatrix}^n$$

being the *iterated* of nth order. The composition of these operations obeys to the rule

$$\begin{pmatrix} A \\ B \end{pmatrix} \begin{pmatrix} B \\ C \end{pmatrix} = \begin{pmatrix} A \\ C \end{pmatrix},$$

and the operation denoted by

$$\begin{pmatrix} A \\ A \end{pmatrix}$$

represents the *identical substitution*. By combining all these results, Galois clarified in 1830, in a final form, the problem of solving algebraic equations by radicals; and so the theory of groups was born. The denomination of *group*, introduced by Galois, simplified Cauchy's notation by using a single letter to design a substitution. Also, Galois' group of substitutions led to the abstract notion of *composition law*, and thus enlarging the research field of algebra.

1.1.2 Computations on equivalence classes

The computation on equivalence classes represents another source for the general theory of groups. Thus, studying the problem of finding the solutions in integer numbers of the equation $\varphi(x,y) = n$, where n is an integer, and $\varphi(x,y)$ is the *binary quadratic form* (with integer coefficients)

$$\varphi(x,y) = ax^2 + bxy + cy^2 ,$$

Lagrange observed that this problem does not differ from the similar one concerning the binary quadratic form

$$\varphi'(x',y') = \varphi(\alpha x' + \beta y', \gamma x' + \delta y') ,$$

where $\alpha, \beta, \gamma, \delta$ are integers satisfying the relation $\alpha\delta - \beta\gamma = 1$. From the relations

$$x = \alpha x' + \beta y' , \quad y = \gamma x' + \delta y' ,$$

it follows that

$$x' = \delta x - \beta y , \quad y' = -\gamma x + \alpha y ;$$

therefore, the statements "x and y are integers" and "x' and y' are integers" are equivalent. The quadratic forms $\varphi(x,y)$ and $\varphi'(x',y')$ having the same discriminant $D = b^2 - 4ac$ are called *equivalent quadratic forms*, and the set of all quadratic forms equivalent to a given quadratic form represents a *class of equivalent quadratic forms*.

On the other hand, in the case of quadratic forms in two variables, having the same discriminant, there are various relations, and among them the identity

$$(x^2 + y^2)(x'^2 + y'^2) = (xx' + yy')^2 + (xy' - yx')^2 ,$$

generalized by Lagrange in the form

$$(x^2 - ay^2)(x'^2 - ay'^2) = (xx' - ayy')^2 - a(xy' - yx')^2 .$$

Also, if there are three binary quadratic forms φ_1, φ_2, and φ_3, satisfying the identity $\varphi_1(x,y)\varphi_2(x',y') = \varphi_3(x'',y'')$, where

$$\begin{aligned} x'' &= axx' + bxy' + cyx' + dyy' , \\ y'' &= a'xx' + b'xy' + c'yx' + d'yy' , \end{aligned}$$

the coefficients $a, b, \ldots, d'$ being integers, then the integer solutions x, y of $\varphi_1(x,y) = n_1$ (n_1 integer), and the integer solutions x', y' of $\varphi_2(x',y') = n_2$ (n_2 integer), determine the integer solutions x'', y'' of $\varphi_3(x'',y'') = n_1n_2$. This example justifies the interest presented by the composition of quadratic forms in the study of the equation $\varphi(x,y) = n$, with n integer.

Carl Friedrich Gauss (1777-1855) considered a class of equivalent quadratic forms as a *single object*, denoted by C. He also introduced the notation $C + C'$ for the composition of two classes, and proved that this composition law is what we call nowadays the composition law of a commutative finite group. William Rowan Hamilton (1805-1865) and Arthur Cayley (1821-1916) dealt with some mathematical objects (*quaternions*, respectively *matrices*) as knowing that they from groups, but without using this mathematical concept. In 1877, Julius Wilhelm Richard Dedekind (1831-1916) proved, for the first time in an explicit manner, that the set of classes of binary quadratic forms, or of modulo n integers, form a group. Finally, the first systematic exposition of the theory of groups, more precisely of the permutation groups, is due to Camille Jordan (1838-1922), who elaborated the first study of *infinite groups*.

1.1.3 Links to other branches of science

The important rôle of symmetry laws was also put in evidence in other branches of science: geometry, crystallography, mechanics, physics, chemistry etc., and so the methods and results of the group theory became widely known. Because each domain of applicability has put its specific problems to the theory of groups, the growing number of these domains also exerted an influence, leading to the development of new chapters of this theory. Thus, the modern group theory, unitary concerning its basic notions, can be regarded as being formed by several more or less independent disciplines: the *general theory of groups*, the *theory of finite groups*, the *theory of topological (continuous) groups*, the *theory of groups of discrete transformations*, the *theory of group representations*, and the *theory of group characters*. Developing rapidly, the methods and notions specific to the theory of groups have proved their importance not only in the study of symmetry laws, but also to derive solutions of many problems in various branches of science.

1.2 Basic definitions and theorems

1.2.1 Composition laws

To give the definition of a group, we start with the notion of *set*, considered as a primary notion.

DEFINITION 1.1.1 *Let M be a nonempty set ($M \neq \emptyset$). We call* **internal composition law** *or* **internal algebraic operation** *on M any mapping $f : M \times M \to M$.*

Thus, to any pair of elements $x, y \in M$ one can associate the element $f(x, y) \in M$, called the *compound* of x and y. Generally, one of the signs: $+$, $\cdot$, $*$, $\circ$, $\vee$, $\wedge$, etc. is used to denote an internal composition law. An internal composition law on M, denoted by the sign "$+$", will be called *addition*, and the compound $f(x, y) = x + y$ of two elements, will be called the *sum* of x and y. In a similar manner, an internal composition law on M, denoted by the sign "$\cdot$", will be called *multiplication*, and the compound $f(x, y) = x \cdot y$ will be called the *product* of x and y. In the first case, the composition law has been denoted *additively* and in the second one, *multiplicatively*. We may also consider internal composition laws for which f is a mapping of a subset of the Cartesian product $M \times M$ into M. In the case of a mapping $f : M \times M \to M$ (as it has been considered in the Definition 1.1.1) we say that the composition law is *defined everywhere* and in what follows we will consider only such composition laws omitting the words "defined everywhere". As examples of internal composition laws we mention: the addition and the multiplication on the set of *natural numbers* $\mathbb{N}$ ($\mathbb{N} = \{0, 1, 2, \ldots\}$), the addition, the subtraction, and the multiplication on the set of *integer numbers* $\mathbb{Z}$ (note that the subtraction is not an internal composition law on $\mathbb{N}$, because the difference of two natural numbers is not always a natural number), the addition (composition) of two vectors on the set of vectors, the *union* "$\cup$" and the *intersection* "$\cap$" on the *power set of M* denoted by $\mathcal{P}(M)$ (the set of *all subsets of M, including the empty set $\emptyset$ and the set M itself*).

Generally, if n is a natural number, and $M^n = M \times M \times \ldots \times M$ (the Cartesian product of n factors), then a mapping $f : M^n \to M$ is called an *n-ary operation* on M. Hence, the internal composition law in the Definition 1.1.1 is a *binary operation*. Sometimes, 0-*ary* operations are used; such an operation depends on no arguments at all, and it simply picks out a particular element of the set M.

We also mention the following properties of the internal composition law f defined on the set M:

- the composition law f is *associative* if

$$\forall x, y, z \in M\,, \quad f(f(x, y), z) = f(x, f(y, z))\,;$$

- the composition law f is *commutative* if

$$\forall x, y \in M\,, \quad f(y, x) = f(x, y)\,.$$

We can also write $(x \circ y) \circ z = x \circ (y \circ z) = x \circ y \circ z$ (i.e., we may omit the parentheses), respectively $y \circ x = x \circ y$, where we have denoted the internal composition law by "$\circ$". The addition and the multiplication defined on $\mathbb{N}$ or $\mathbb{Z}$ are associative and commutative; the union and the intersection on $\mathcal{P}(M)$ exhibit the same properties. The subtraction and the division defined on $\mathbb{N}$, $\mathbb{Z}$, $\mathbb{Q}$ (the set of *rational numbers*), or $\mathbb{R}$ (the set of *real numbers*) are nonassociative composition laws.

Let us consider now a set $M \neq \emptyset$, endowed with two internal composition laws, denoted by "$*$" and "$\circ$". Then, the composition law "$*$" is *right* *distributive*, respectively *left distributive*, with respect to the composition law "$\circ$" if $\forall x, y, z \in M$ we have $(x \circ y) * z = (x * z) \circ (y * z)$, respectively $z * (x \circ y) = (z * x) \circ (z * y)$. The composition law "$*$" is *bilaterally distributive* (we will say only distributive, omitting the word "bilaterally") with respect to the composition law "$\circ$" if it is right and left distributive. If the composition law "$\circ$" is commutative and one of the distributivity properties takes place, then the other distributivity property also takes place. For instance, the multiplication is distributive with respect to addition on $\mathbb{N}$ or $\mathbb{Z}$; as well, the intersection of sets is distributive with respect to the union of sets, and the union of sets is distributive with respect to the intersection of sets, on $\mathcal{P}(M)$. If $\forall x, y \in M$, $(x \circ y) * x = x$ and $(x * y) \circ x = x$, then the property of *absorption* takes place. Such a property exists if $M = \mathcal{P}(M)$, "$\circ$" is "$\cup$", and"$*$" is "$\cap$" (or "$\circ$" is "$\cap$", and "$*$" is "$\cup$"), but does not subsist for addition and multiplication on $\mathbb{N}$ or $\mathbb{Z}$.

DEFINITION 1.1.2 *A set $M \neq \emptyset$, endowed with two internal composition laws "$\circ$" and "$*$" satisfying the properties of associativity, commutativity, and absorption is called* **lattice**.

We may say that a lattice is an *ordered triplet* $(M, \circ, *)$; for instance the triplet $(\mathcal{P}(M), \cup, \cap)$.

THEOREM 1.1.1 (THE DUALITY PRINCIPLE) *From every theorem pertaining to a lattice $(M, \circ, *)$ one can deduce another theorem by interchanging "$\circ$" and "$*$".*

An element $x \in M$ is called *idempotent* if $x \circ x = x$. Every element in a lattice is idempotent with respect to both internal composition laws.

DEFINITION 1.1.3 *Let f be an internal composition law (denoted by "$\circ$") defined on a set $M \neq \emptyset$. An element $e \in M$ is called* **neutral element** *for the composition law f if $\forall x \in M$, $f(x, e) = f(e, x) = x$ (or, $x \circ e = e \circ x = x$).*

In other words, the neutral element leaves any element of the set M unchanged when combined with it by f. The neutral element may also be regarded as the value of a 0-*ary* operation.

If an internal composition law is denoted additively, then the neutral element, if it exists, is called the *null element*, and if the internal composition law is denoted multiplicatively, then the neutral element, if it exists, is called the *identity element*. The null element is usually denoted by "0" and the identity element by "1". For instance, the addition and multiplication defined on $\mathbb{N}$ or $\mathbb{Z}$ admit the null element and the identity element, respectively.

THEOREM 1.1.2 *If an internal composition law admits a neutral element, then this element is unique.*

DEFINITION 1.1.4 *Let f be an internal composition law (denoted by "$\circ$") defined on a set $M \neq \emptyset$, having a neutral element e. We say that the element $x' \in M$ is* **symmetric** *to the element $x \in M$ if $f(x, x') = f(x', x) = e$ (or, $x \circ x' = x' \circ x = e$).*

We say that $x \in M$ is a *symmetrizable element* if there exists $x' \in M$ symmetric to x. If the composition law is denoted additively, then the element symmetric to x is called the *opposite* of x (denoted by $-x$), while if the composition law is denoted multiplicatively, then the element symmetric to x is called the *inverse* of x (denoted by x^{-1}). In the last case we say that x is an *invertible* element. For instance, the only symmetrizable element with respect to addition on $\mathbb{N}$ is 0, while with respect to addition on $\mathbb{Z}$, $\mathbb{Q}$, or $\mathbb{R}$, every element has an opposite one ($x + (-x) = 0$), and therefore all the elements in $\mathbb{Z}$, $\mathbb{Q}$, or $\mathbb{R}$ are symmetrizable. Also, only 1 and -1 are invertible with respect to multiplication on $\mathbb{Z}$, while any element $x \neq 0$ is invertible with respect to multiplication on $\mathbb{Q}$, or $\mathbb{R}$ ($xx^{-1} = 1$).

THEOREM 1.1.3 *Let f be an associative internal composition law, with neutral element, defined on $M \neq \emptyset$. If $x \in M$ has a symmetric element $x' \in M$, then x' is unique.*

If an element $x \in M$ is symmetrizable, then $(x')' = x$ (or $(x^{-1})^{-1} = x$). Also, if the elements $x, y \in M$ are symmetrizable, then $[f(x, y)]' = f(y', x')$ (or $(x \circ y)^{-1} = y^{-1} \circ x^{-1}$).

DEFINITION 1.1.5 *Let f be an internal composition law on $M \neq \emptyset$. We call a* **regular** *(or* **cancellable***)* **element** *with respect to the composition law f an element $a \in M$ such that $\forall x, y \in M$, $f(a, x) = f(a, y)$ and $f(x, a) = f(y, a)$ $\Leftrightarrow x = y$.*

THEOREM 1.1.4 *In case of an associative internal composition law f, defined on $M \neq \emptyset$ and with neutral element, any symmetrizable element is also a regular one.*

Any $x \in \mathbb{N}$ is a regular element with respect to addition and any $x \in \mathbb{R} \setminus \{0\}$ is a regular element with respect to multiplication.

DEFINITION 1.1.6 *Let $M \neq \emptyset$ and $\Omega \neq \emptyset$ be two sets. A mapping $\varphi : \Omega \times M \to M$ or $\varphi : M \times \Omega \to M$ is called* **external composition law** *on M with the* **operators domain** Ω.

The element $\varphi(\alpha, x) \in M$, where $\alpha \in \Omega$ and $x \in M$, is called the *composite* of α by x with respect to the external composition law φ. In a left or right multiplicative notation we write αx, respectively $x\alpha$, and in exponential notation x^α. For example, the mapping $\varphi : \mathbb{Z} \times \mathbb{R} \to \mathbb{R}$, defined as

$\varphi(n, x) = nx$, is an external composition law on $\mathbb{R}$ with the operators domain $\mathbb{Z}$. As well, the mapping $f : \mathbb{R} \times \mathbb{R}^n \to \mathbb{R}^n$, defined as $f(\alpha, x) = \alpha x$, where $x \equiv \{x_1, x_2, \ldots, x_n\} \in \mathbb{R}^n$ and $\alpha \in \mathbb{R}$, is an external composition law on $\mathbb{R}^n$ with the operators domain $\mathbb{R}$.

1.2.2 Algebraic structures. Semigroups

DEFINITION 1.1.7 *Let M be a nonempty set. We call* **algebraic structure** *on M any structure determined on M by one or several internal composition laws and by one or several external composition laws (each of them having an operators domain). These composition laws are subjected to certain conditions (for instance, associativity, or commutativity) or are interconnected by some relations (for instance, distributivity of a composition law with respect to another one).*

DEFINITION 1.1.8 *A set $M \neq \emptyset$ endowed with only one internal composition law (denoted by "$\circ$") is called* **groupoid** *and it is denoted by $(M, \circ)$.*

If this composition law is associative, then the groupoid is called *semigroup*. If the internal composition law admits a neutral element, then the semigroup is called *monoid*. In case of a commutative internal composition law the semigroup is called *commutative* or *Abelian* (after Niels Henrik Abel (1802-1829)).

Thus, the sets $\mathbb{N}$ and $\mathbb{Z}$ with an operation of addition, respectively of multiplication, are Abelian monoids, "0" being the null element for addition and "1" the identity element for multiplication. The set of even integers endowed with an operation of multiplication is only an Abelian semigroup (without the identity element). On a set M which has only one element we may define only one internal composition law, such that M is an Abelian monoid, the respective element being also the neutral element. The power set of $M \neq \emptyset$, $\mathcal{P}(M)$, together with the operations of union and intersection, is an Abelian monoid; $\emptyset$ is the neutral element for "$\cup$" and M is the neutral element for "$\cap$".

1.2.3 Groups

Using the notions considered above, we can now introduce the notion of group, for which three equivalent definitions will be given.

DEFINITION 1.1.9 *A set $G \neq \emptyset$ endowed with an internal composition law is called a* **group** *if this internal composition law verifies the axioms of the group:*

G_1) *the composition law is* **associative**;

G_2) *the composition law has a* **neutral element**;

G_3) *any element in G is* **symmetrizable**.

Hence, a monoid for which any element is symmetrizable is a group.

DEFINITION 1.1.10 *A semigroup* $(G, \circ)$ *is called a* **group** *if* $\forall a, b \in G$ *the equations* $a \circ x = b$ *and* $y \circ a = b$ *have a unique solution* $x, y \in G$.

DEFINITION 1.1.11 *A semigroup* $(G, \circ)$ *is called a* **group** *if the internal composition law "*$\circ$*" has a left identity element (i.e.,* $\forall x \in G$, $e_l \circ x = x$*) and any element* $x \in G$ *has a left symmetric element (i.e.,* $\forall x \in G$, $\exists x' \in G$ *such that* $x' \circ x = e_l$*).*

One can thus see that one of the axioms G_2 and G_3 in the Definition 1.1.9 is redundant, but this form of the definition is much more simple. A group G is called a *finite group* if the set G is finite, otherwise G is an *infinite group*. If the internal composition law is commutative, then the group is called *Abelian* (or *commutative*). The *order* of a group G is, by definition, the *cardinality* of the set G (denoted by card G, or $|G|$); if G is finite, then $|G|$ is the number of elements in the set G. It can be shown that $|G| = 1 \Leftrightarrow G = \{e\}$, or $G = \{0\}$; that is, the identity group (multiplicative) and the null group (additive) are the only group structures on a set consisting of only one element. In any group $(G, \circ)$ the relations

$$(x \circ y)^{-1} = y^{-1} \circ x^{-1}, \quad \left(x^{-1}\right)^{-1} = x, \quad \forall x, y \in G,$$

are satisfied; as well, any element of G is a regular element.

It is easy to verify that $(\mathbb{Z}, +)$, $(\mathbb{Q}, +)$, $(\mathbb{R}, +)$, $(\mathbb{C}, +)$, where $\mathbb{C}$ is the set of *complex numbers*, are additive Abelian groups. Also, $(\mathbb{Q}^*, \cdot)$, $(\mathbb{R}^*, \cdot)$, $(\mathbb{C}^*, \cdot)$, where $\mathbb{Q}^* = \mathbb{Q} \setminus \{0\}$, $\mathbb{R}^* = \mathbb{R} \setminus \{0\}$, and $\mathbb{C}^* = \mathbb{C} \setminus \{0\}$, are multiplicative Abelian groups. In contrast, the set of natural numbers endowed with an operation of addition or multiplication does not form a group. As well, the set of even integers endowed with an operation of multiplication does not form a group.

Let $R_n = \{z | z \in \mathbb{C}, z^n = 1\}$ be the set of the complex nth roots of unity, given by

$$z_k = \cos\left(\frac{2k\pi}{n}\right) + i \sin\left(\frac{2k\pi}{n}\right), \quad k = 0, 1, \ldots, n-1.$$

One can easily verify that $(R_n, \cdot)$ forms a multiplicative Abelian group, that is the group of the complex nth roots of unity. In particular, for $n = 2$, $R_2 = \{-1, 1\}$, and for $n = 4$, $R_4 = \{\pm 1, \pm i\}$.

A *transformation* T of a set M is a one-to-one mapping of the elements of M onto themselves. Thus, $\forall x \in M$, it corresponds a unique element $y \in M$, and $\forall y \in M$ there exists one element and only one $x \in M$ whose image with respect to the transformation T is the element y. In other words, the transformation T

is one-to-one if the equation $\mathsf{T}x = y$ has a unique solution $x \in M$, $\forall y \in M$. For every transformation T of the set M onto itself one may define the inverse transformation, denoted by T^{-1}; if the transformation T assigns to an arbitrary element $x \in M$ the element $y \in M$, then the transformation which associates the element x to y is called the *inverse of the transformation* T:

$$\mathsf{T}x = y \iff x = \mathsf{T}^{-1}y\,.$$

The successive application of transformations is associative, and therefore, the set of transformations forms a group.

For instance, let $M = \{1, 2, 3, 4\}$ be a set of numbers, and T the transformation defined by the table

$$\mathsf{T} = \begin{pmatrix} 1\,2\,3\,4 \\ 2\,3\,4\,1 \end{pmatrix},$$

in which 1 is replaced by 2, 2 by 3, 3 by 4, and 4 by 1. The inverse of T is defined by the table

$$\mathsf{T}^{-1} = \begin{pmatrix} 2\,3\,4\,1 \\ 1\,2\,3\,4 \end{pmatrix} = \begin{pmatrix} 1\,2\,3\,4 \\ 4\,1\,2\,3 \end{pmatrix}.$$

Let T_1 and T_2 be two such transformations defined by the tables

$$\mathsf{T}_1 = \begin{pmatrix} 1\,2\,3\,4 \\ 4\,3\,2\,1 \end{pmatrix}, \quad \mathsf{T}_2 = \begin{pmatrix} 1\,2\,3\,4 \\ 2\,1\,3\,4 \end{pmatrix};$$

the corresponding composition law of these transformations has the form

$$\mathsf{T}_1 \circ \mathsf{T}_2 = \begin{pmatrix} 1\,2\,3\,4 \\ 4\,3\,2\,1 \end{pmatrix} \circ \begin{pmatrix} 1\,2\,3\,4 \\ 2\,1\,3\,4 \end{pmatrix} = \begin{pmatrix} 1\,2\,3\,4 \\ 3\,4\,2\,1 \end{pmatrix}.$$

One can see that

$$\mathsf{T} \circ \mathsf{T}^{-1} = \mathsf{T}^{-1} \circ \mathsf{T} = \begin{pmatrix} 1\,2\,3\,4 \\ 1\,2\,3\,4 \end{pmatrix} = \mathsf{E},$$

where we have denoted by E the identity transformation. By taking into account the Definition 1.1.9 it follows that the set of these transformations forms a finite group with $4! = 24$ elements, defined by the tables:

$$\begin{pmatrix} 1\,2\,3\,4 \\ 1\,2\,3\,4 \end{pmatrix}, \begin{pmatrix} 1\,2\,3\,4 \\ 2\,3\,4\,1 \end{pmatrix}, \begin{pmatrix} 1\,2\,3\,4 \\ 3\,4\,1\,2 \end{pmatrix}, \ldots$$

Generally, one can easily establish a criterion to decide if a transformation defined by a table is one-to-one or not; it is necessary and sufficient that every element in the set to appear once and only once in the lower line of the corresponding table. Thus, the transformation defined by the table

$$\begin{pmatrix} 1\,2\,3\,4 \\ 2\,1\,2\,3 \end{pmatrix}$$

is not one-to-one, because by this transformation no element in the set $M = \{1,2,3,4\}$ has as image the number 4. Also, the transformation of the set of natural numbers $\mathbb{N}^* = \mathbb{N} \setminus \{0\}$, defined by the table

$$\begin{pmatrix} 1\ 2\ 3\ 4\ 5\ 6\ \ldots \\ 1\ 1\ 2\ 2\ 3\ 3\ \ldots \end{pmatrix},$$

is not one-to-one. Although for every number n does exist the number $2n$, which is transformed into n, the number $2n$ is not the only one having this property, because n is also the image of $2n-1$.

1.2.4 Homomorphism. Isomorphism

DEFINITION 1.1.12 *Let $(F,*)$ and $(G,\circ)$ be two groups. A mapping $f : F \to G$ is called a* **homomorphism** *(or* **morphism***)* **of groups** *if $\forall x,y \in F$, $f(x*y) = f(x)\circ f(y)$.*

A homomorphism of groups is called an *injective homomorphism* (or a *monomorphism*), respectively a *surjective homomorphism* (or *epimorphism*), if f is an injective mapping, respectively a surjective one. An injective and surjective homomorphism is a *bijective homomorphism.* A homomorphism of groups defined by the mapping $f : G \to G$ is an *endomorphism* of the group G (denoted by $\mathsf{End}(G)$).

THEOREM 1.1.5 *Let $(F,*)$ and $(G,\circ)$ be two groups. If the mapping $f : F \to G$ is a homomorphism of groups, then $f(e_F) = e_G$, e_F and e_G being the identity elements in F and G, respectively.*

That is, the identity elements correspond one to the other in a homomorphism of groups.

THEOREM 1.1.6 *The image of the symmetric of an element by a homomorphism of groups is the symmetric of the image of that element.*

Let $f : F \to G$ be a homomorphism of two groups F and G, assumed to be multiplicative groups. It follows that $f(x^{-1}) = [f(x)]^{-1}$. If $(F,*)$ is a multiplicative group and $(G,+)$ is an additive group, then $f(x^{-1}) = -f(x)$; this result can be easily verified, for instance in the case of the homomorphism of groups defined by the mapping $f : (\mathbb{R}^*,\cdot) \to (\mathbb{R},+)$, where $f(x) = \log_a x$, $a > 0$.

THEOREM 1.1.7 *If $f : G \to G'$ and $g : G' \to G''$ are homomorphisms of groups, then the composition of mappings $g\circ f$ is also a homomorphism of groups ($g\circ f : G \to G''$).*

By denoting the groups multiplicatively, we can write

$$(g\circ f)(xy) = g(f(xy)) = (g\circ f)(x)(g\circ f)(y)\,, \quad \forall x,y \in G.$$

DEFINITION 1.1.13 *A homomorphism of groups $f : F \to G$ is called an* **isomorphism of groups** *if there exists a homomorphism of groups $g : G \to F$, such that $g \circ f = 1_F$ and $f \circ g = 1_G$, where 1_F and 1_G are the identity endomorphisms of the groups F and G, respectively.*

In other words, a one-to-one mapping between the elements of the two groups F and G ($x \leftrightarrow y$, $x \in F$, $y \in G$) is called an isomorphism of groups if from any pair of relations $x_1 \leftrightarrow y_1$ and $x_2 \leftrightarrow y_2$, where $x_1, x_2 \in F$ and $y_1, y_2 \in G$, it results the relation $x_1x_2 \leftrightarrow y_1y_2$. Actually, isomorphic groups are roughly the same except for the names of their elements.

Note that, a group G is homomorphic to a group F if any element $y \in G$ can be assigned to an element $x \in F$ such that, from the relations $y_1 \to x_1$ and $y_2 \to x_2$ it follows $y_1y_2 \to x_1x_2$. Thus, the homomorphism of two groups is different from their isomorphism because a homomorphism does not require a bijective correspondence between the elements of the groups. Therefore, the isomorphism is a particular case of homomorphism. The groups related by an isomorphism are called *isomorphic groups*, and every theorem established for a group G can be automatically applied to all the groups isomorphic to G. Consequently, from the point of view of the group theory, the isomorphic groups are identical.

An isomorphism of groups $f : G \to G$ is called an *automorphism* of G.

THEOREM 1.1.8 *A homomorphism of groups $f : F \to G$ is an isomorphism of groups if and only if it is bijective.*

For instance, the identity mapping $1_G : G \to G$ is an isomorphism of groups, hence it is an automorphism of G. If G is a multiplicative group and $x \in G$, then one can show that the mapping $f_x : G \to G$, defined by $f_x(g) = xgx^{-1}$, $\forall g \in G$, is an automorphism of G, called an *inner automorphism* of G. If the group G is Abelian, then all the inner automorphisms of G coincide with the identity automorphism 1_G of G.

Note that, the mapping $f'_x : G \to G$, defined by $f'_x(g) = x^{-1}gx$, $\forall g \in G$, is also an inner automorphism, and the relation $f'_x = f_{x^{-1}}$ takes place. Also, the composition of two inner automorphisms can be regarded as an internal composition law ($f_x \circ f_y = f_{xy}$) in the set $\widetilde{G}$ of the inner automorphisms of G. It is easy to show that this internal composition law is associative, has a neutral element (the inner automorphism f_e), and $\forall x \in G$, f_x is invertible (i.e., $(f_x)^{-1} = f_{x^{-1}}$). Hence, the set $\widetilde{G}$ is actually a group, that is called the *group of inner automorphisms* of G.

Let $\mathsf{Hom}(F, G) = \{f | f : F \to G, f \text{ homomorphism}\}$, where F and G are additive Abelian groups. In this case, one can prove that $\mathsf{Hom}(F, G)$ is an Abelian group. If $F = G$, then $\mathsf{Hom}(G, G) = \mathsf{End}(G)$ is *the group of endomorphisms* of G (also an Abelian group).

DEFINITION 1.1.14 *In a group G, an element $g \in G$ is called the* **conjugate** *to the element $h \in G$, if $\exists x \in G$ such that $xgx^{-1} = h$.*

It follows that the mapping f_x, previously considered as an inner automorphism of G, defines the conjugate elements. The set of all the elements conjugate to a given element forms a *class*, and the entire group can be decomposed in *classes of conjugate elements* (or *conjugacy classes*). One can easily see that the class of the elements conjugate to the identity element $e \in G$, contains only the element e, because $xex^{-1} = e, \forall x \in G$. In the case of an Abelian group each class contains only one element ($xgx^{-1} = gxx^{-1} = g$).

We point out the following properties concerning conjugate elements:

a) every element $g \in G$ is conjugate to itself ($ggg^{-1} = g$);

b) if g is conjugate to h, then h is also conjugate to g, because

$$xgx^{-1} = h \rightarrow g = x^{-1}hx = (x^{-1})h(x^{-1})^{-1}\,;$$

c) if g is conjugate to h and h is conjugate to f, then g is conjugate to f

$$xgx^{-1} = h\,,\ yhy^{-1} = f \rightarrow y(xgx^{-1})y^{-1} = (yx)g(yx)^{-1} = f\,;$$

d) if g is conjugate to h, then g^{-1} is conjugate to h^{-1}

$$xgx^{-1} = h \rightarrow h^{-1} = (xgx^{-1})^{-1} = xg^{-1}x^{-1}\,.$$

1.2.5 Subgroups

DEFINITION 1.1.15 *Let G be a group. A subset $H \subset G$, $H \neq \emptyset$, is called a* **subgroup** *of G if the internal composition law of G restricted to H is an internal composition law that makes H into a group.*

To check if a nonempty subset of a group G is a subgroup we can use the following two theorems:

THEOREM 1.1.9 *A subset $H \neq \emptyset$ of a multiplicative (or additive) group G is a subgroup in G if and only if the conditions:*

1) $\forall x, y \in H,\, xy \in H$ (or $x + y \in H$),

2) $\forall x \in H,\, x^{-1} \in H$ (or $-x \in H$)

are satisfied.

THEOREM 1.1.10 *A subset $H \neq \emptyset$ of a multiplicative (or additive) group G is a subgroup in G if and only if*

$$\forall x, y \in H\,,\quad xy^{-1} \in H \text{ (or } x - y \in H).$$

The subsets G and $\{e\}$ of a group G are called the *improper subgroups* of G. Any other subgroup of G is a *proper subgroup*.

THEOREM 1.1.11 *Let $f : F \to G$ be a homomorphism of groups. Then:*

1) *$f(F)$ is a subgroup of G, which is called the image of F by f;*

2) *if H is a subgroup of G, then $f^{-1}(H)$ is a subgroup of F.*

In particular, $f^{-1}(e)$, where e is the identity element in G, is a subgroup of F, and it is called the *kernel* of f (denoted by $\mathsf{Ker}\, f$).

DEFINITION 1.1.16 *Let G be a group and X a nonempty subset of G. The intersection of all the subgroups of G, which contain the subset X, is called the* **subgroup of G generated by** *X, and X is called the* **system of generators** *of this subgroup.*

THEOREM 1.1.12 *Let G be a group and $X \subset G$, $X \neq \emptyset$. Then, the subgroup of G generated by X consists of all the finite products of elements in X and of the inverse of these elements.*

This subgroup is the *lowest subgroup* of G (in the sense of the inclusion relations) which contains the subset X. If H_1 and H_2 are subgroups of G, then the subgroup generated by them (denoted by $\{H_1, H_2\}$) is the lowest subgroup of G which contains H_1 and H_2; finite products or finite sums of elements of H_1 and H_2, as G is multiplicative or additive, respectively, form this subgroup. In the case of an Abelian group G, the subgroup $\{H_1, H_2\}$ consists of elements of the form $h_1 h_2$ (or $h_1 + h_2$), $h_1 \in H_1$, $h_2 \in H_2$, and is denoted by $H_1 H_2$ (or $H_1 + H_2$).

DEFINITION 1.1.17 *A group G is called of* **finite type** *(or* **finitely generated***) if there exists a finite set of elements in G which generate the whole group G. A group generated by only one element is called a* **cyclic group** *(or* **monogenic group***).*

For instance, the subgroups of the additive group $\mathbb{Z}$ are cyclic groups, and $\mathbb{Z}$ is a cyclic group generated by 1 or -1.

DEFINITION 1.1.18 *Let G be a group. A subgroup H of G is called an* **invariant** *(or* **normal***)* **subgroup** *if $\forall x \in G$ and $\forall h \in H$, $xhx^{-1} \in H$.*

Therefore, H is an invariant subgroup in G (denoted by $H \lhd G$) if it contains together with any of its elements $h \in H$ all the elements conjugate to h (these are the transformations of h by any inner automorphism of G). The improper groups G and $\{e\}$ are invariant subgroups in any group G. Any subgroup of an Abelian group is an invariant subgroup; indeed, the only inner automorphism of an Abelian group is the identity automorphism. In a group G the set

$$\mathcal{C}(G) = \{x | x \in G \text{ and } \forall y \in G, xy = yx\}$$

is called the **centre** of G; one can easily prove that the centre of a group G is an invariant subgroup in G ($\mathcal{C}(G) \triangleleft G$). It can also be shown that the group of inner automorphisms of G is an invariant subgroup in the group of all automorphisms of G.

THEOREM 1.1.13 *If $f : F \to G$ is a homomorphism of groups and $H \triangleleft G$, then $f^{-1}(H) \triangleleft F$. In particular,* $\mathsf{Ker}\, f \triangleleft F$.

If $\{H_i\}_{i=1}^n$ is a set of subgroups in G, then their intersection $H = \bigcap_{i=1}^n H_i$ is also a subgroup in G. If H_i are invariant subgroups in G, then their intersection is also an invariant subgroup in G.

1.2.6 The factor group

DEFINITION 1.1.19 *Let A and B be two sets, not necessarily distinct. A subset R in the Cartesian product $A \times B$ is called a* **binary relation** *(***relation** *or* **correspondence***) between the two sets A and B.*

The elements associated by this relation are those elements $a \in A$ and $b \in B$ for which $(a, b) \in R \subset A \times B$. If $A = B$, then (a, b) is an ordered pair of elements in A and we say that we have defined a relation in A. Generally, if the relation is denoted by the symbol R, then we write bRa to show that a and b are associated by the relation R. If $A \subset A'$, $B \subset B'$, and $R \subset A \times B$, then $R \subset A' \times B'$. If the relation R is reflexive ($\forall a \in A$, aRa), symmetric ($\forall a \in A, b \in B$, $bRa \Leftrightarrow aRb$), and transitive ($\forall a \in A, b \in B, c \in C$, bRa and $cRb \Leftrightarrow cRa$), then R is called an *equivalence relation*. Two elements associate by an equivalence relation are called *equivalent*. A relation is called *antireflexive* if aRa, $a \in A$, does not exists; as well, a relation is *antisymmetric* if aRb and bRa imply $a = b$, where $a \in A$ and $b \in B$.

Generally, we call *n-ary relation* between n sets A_i, $i = 1, 2, \ldots, n$, which can be distinct or partially coincident or identical, a subset $R \subset A_1 \times A_2 \times \ldots \times A_n$; the elements of an ordered n-tuple $(a_1, a_2, \ldots, a_n)$, where $a_i \in A_i$, are called *R-related* if $(a_1, a_2, \ldots, a_n) \in R$.

DEFINITION 1.1.20 *Let M be a set. A class C of nonempty subsets of M, such that the subsets in C are disjoint, and the union of all subsets in C is M, is called a* **partition** *of M.*

Thus, if C is a partition of M, then any element of M belongs only to one subset in C. We are led to similar results if we establish a partition of M by means of an equivalence relation.

DEFINITION 1.1.21 *Let M be a set, R an equivalence relation in M, and $x \in M$. The set of all elements in M which are R-related to x, denoted by $C_x = \{y | yRx; y \in M\}$, is called the* **equivalence class of x**.

THEOREM 1.1.14 *Let R be an equivalence relation in M. Then, the equivalence classes of all the elements in M form a partition of M. Reciprocally, a partition of M being given, the relation defined by yRx, where x and y belong to the same subset, is an equivalence relation.*

DEFINITION 1.1.22 *The set of equivalence classes, determined by an equivalence relation R in M, is called the* **factor set** *(***quotient set***) of M with respect to R, and is denoted by M/R.*

In regard to groups, because their structure is determined by only one internal composition law, an equivalence relation is compatible to the group structure if it is compatible to the respective composition law.

DEFINITION 1.1.23 *Let G be a group and H a subgroup of G. Then, for $y \in G$ the subset $yH = \{x | x = yh, h \in H\} \subseteq G$ is called the* **left coset** *of H in G. Also, the* **right coset** *of H in G is the subset $Hy = \{x | x = hy, h \in H\} \subseteq G$.*

It can be shown that the right (left) cosets of a subgroup H in G form a partition of G, that is any two right (left) cosets of H in G are disjoint, and the union of all right (left) cosets in G is G (see the Definition 1.1.20).

THEOREM 1.1.15 *Let R be an equivalence relation in the multiplicative group G. If R is left (respectively right) compatible to the internal composition law of the group, then there exists a subgroup $H \subset G$ such that $yRx \Leftrightarrow y^{-1}x \in H$ (respectively $yRx \Leftrightarrow xy^{-1} \in H$). Reciprocally, if H is a subgroup, then the relation R determined by $yRx \Leftrightarrow y^{-1}x \in H$ (respectively $yRx \Leftrightarrow xy^{-1} \in H$) is an equivalence relation in G, left (respectively right) compatible to the composition law of G.*

The cardinality of the set of left (respectively right) equivalence classes is called the *left* (respectively *right*) *index* of the subgroup H.

THEOREM 1.1.16 *Let G be a group and H a subgroup of G. Then, the left index of H is equal to the right index of H, and it is called the index of H in G.*

THEOREM 1.1.17 *An equivalence relation R defined on a group G is compatible with the group structure of G if and only if it is determined by an invariant subgroup of G.*

An equivalence relation determined by an invariant subgroup H of a group G is called a *congruence relation modulo H*, and it is denoted by $x \equiv y (\mathrm{mod}\, H)$. This is the particular case when the subgroup H in Definition 1.1.23 is an invariant subgroup of G, and the right and left cosets of H in G are identical ($H \lhd G$ implies $xHx^{-1} = H$, $\forall x \in G$, and therefore $xH = Hx$). The corresponding equivalence classes are simply called *cosets* of the group G with respect to the subgroup H. The set of equivalence classes determined by the invariant subgroup $H \lhd G$ form a group called the *factor (quotient) group* of G with respect to H, denoted by G/H. If the factor group is finite, then its order is equal to the index of H in G, being denoted by $[G : H]$; if G/H is infinite we say that the index of H in G is ∞. For instance, $\forall p \in \mathbb{N}^*$ the subgroup $p\mathbb{Z} \equiv \{mp | m \in \mathbb{Z}\}$ is an invariant subgroup of the group $(\mathbb{Z}, +)$, and the factor group $\mathbb{Z}_p = \mathbb{Z}/p\mathbb{Z}$ is an Abelian additive group of order p (the order of $\mathbb{Z}_0$ is ∞).

THEOREM 1.1.18 (LAGRANGE) *If H is a subgroup of order n of a finite group G of order N, then $N = n\,i$, where i is the index of H in G.*

Therefore, the order and the index of a subgroup of a finite group G are divisors of the order of G. An immediate consequence is the fact that a finite group G whose order is a prime number has only the trivial subgroups $H = G$ and $H = \{e\}$, so that we can have only $n = N$ or $n = 1$. A group G is called *simple* if it has no invariant subgroups (normal divisors) excepting the trivial subgroups, and *semisimple* if it contains no Abelian invariant subgroups. Hence, a finite group whose order is a prime number is a simple group.

The mapping $p : G \to G/H$ ($H \lhd G$), defined by $p(x) = xH$, is a surjective homomorphism (epimorphism) of groups, called the *canonical homomorphism* or the *canonical surjection* of G onto G/H. The set of elements in G, conjugate with a given element (see Definition 1.1.14), forms an equivalence class, so that one obtains a partition of G, determined by the classes of conjugate elements.

1.2.7 Cyclic groups

DEFINITION 1.1.24 *A group G generated by only one of its elements is called a* **cyclic (monogenic)** *group.*

THEOREM 1.1.19 *Let G be a multiplicative (or additive) group and $x \in G$. Then, $\forall n, m \in \mathbb{Z}$ (or $\forall n, m \in \mathbb{N}^*$):*

1) $x^n x^m = x^{n+m}$ [or $nx + mx = (n+m)x$];

2) $(x^n)^m = x^{nm}$ [or $n(mx) = (nm)x$].

The cyclic subgroups constitute an important example of subgroups. They consist of all positive and negative integer powers of an element $g \in G$. Thus, the set of all integer powers of g, that is $H = \{\ldots, g^{-n}, \ldots, g^{-1}, g^0 = e, g, \ldots, g^n, \ldots\}$, is a subgroup of G, namely the *cyclic subgroup* generated by g.

DEFINITION 1.1.25 *The order of the cyclic subgroup generated by an element g in the group G is called the* **order of the element** g.

Any subgroup of a finite group G is finite; hence, the order of any element of G is finite. The order of an element $g \in G$ is 1 if and only if g is the neutral element of G. As a consequence of Lagrange's theorem (see Theorem 1.1.18), the order of any element of a finite group G is a divisor of the order of the group. The order of the group $\mathbb{Z}$ as well as the order of any of its subgroups, other than the null one (i.e., $\{0\}$), is infinite. If an element $g \in G$ is of finite order n, then n is the smallest nonzero natural number for which $g^n = e$. If G is a finite group of order N, then $\forall g \in G$ we have $g^n = e$, $n \leq N$.

THEOREM 1.1.20 *Any cyclic group is isomorphic to a factor group of $\mathbb{Z}$.*

Consequently, any infinite cyclic group is isomorphic to $\mathbb{Z}$.

1.2.8 Permutation groups

In Chap. 1 Sec. 1.2.3 we introduced the group of transformations of a set M. This is the group of all *permutations* (or *substitutions*) of a set M, and is called the *symmetric* group of M, being denoted by $\mathsf{S}(M)$. In the particular case where $M = \{1, 2, \ldots, n\}$, the symmetric group $\mathsf{S}(M)$ is denoted by S_n and is called the *permutation group of degree* n. The order of S_n is $n!$. One can prove that the groups $\mathsf{S}(M)$ and $\mathsf{S}(M')$ of *equipotent* sets (i.e., which have the same cardinality) M and M' are isomorphic. Therefore, the symmetric group $\mathsf{S}(M)$ of any finite set M with n elements is isomorphic to S_n:

$$\mathsf{S}(M) \simeq \mathsf{S}_n.$$

A permutation of degree n, $\boldsymbol{\sigma} \in \mathsf{S}_n$, is represented in the form

$$\boldsymbol{\sigma} = \begin{pmatrix} 1 & 2 & \ldots & n \\ m_1 & m_2 & \ldots & m_n \end{pmatrix},$$

where each element in the second row is the image of the corresponding element in the first row, under $\boldsymbol{\sigma}$. Thus, the object placed initially in position 1 is moved to position m_1, the object placed initially in position 2 is moved to position m_2, a.s.o. The product of two permutations $\boldsymbol{\sigma}_1$ and $\boldsymbol{\sigma}_2$, denoted by $\boldsymbol{\sigma}_1\boldsymbol{\sigma}_2$, is defined as the permutation obtained by applying first the permutation σ_2 and then the permutation σ_1. The *identity permutation* is

$$\sigma = \begin{pmatrix} 1 \; 2 \; \dots \; n \\ 1 \; 2 \; \dots \; n \end{pmatrix},$$

and the *inverse permutation* of σ can be written in the form

$$\sigma^{-1} = \begin{pmatrix} m_1 & m_2 & \dots & m_n \\ 1 & 2 & \dots & n \end{pmatrix}.$$

Note that, in general, the group S_n is not Abelian.

It can be shown that the symmetries of a rectangle form a permutation group, isomorphic to the group

$$G_4 = \left\{ \begin{pmatrix} 1\,2\,3\,4 \\ 1\,2\,3\,4 \end{pmatrix}, \begin{pmatrix} 1\,2\,3\,4 \\ 2\,1\,4\,3 \end{pmatrix}, \begin{pmatrix} 1\,2\,3\,4 \\ 3\,4\,1\,2 \end{pmatrix}, \begin{pmatrix} 1\,2\,3\,4 \\ 4\,3\,2\,1 \end{pmatrix} \right\},$$

called *Klein's group.*

THEOREM 1.1.21 (CAYLEY) *Any group G (finite or infinite) is isomorphic to a subgroup of the symmetric group* $\mathsf{S}(G)$.

Let G be a finite group of order n. We have $\mathsf{S}(G) \simeq \mathsf{S}_n$ and according to Cayley's theorem, G is isomorphic to a subgroup of $\mathsf{S}(G)$, therefore G is also isomorphic to a subgroup of S_n. It follows that any finite group of order n is isomorphic to a subgroup of S_n.

1.3 Representations of groups

1.3.1 General considerations. Equivalent representations

The theory of group representations is the result of the introduction of calculus and classical algebra as methods of the group theory. In a similar manner to the analytical geometry, which does not provide only methods to solve the problems in geometry by means of calculus but also clarifies many intricate problems of this field, the theory of group representations is not only an auxiliary tool to study the properties of groups; it enables to establish closed connections between the basic notions and fundamental problems characteristic to calculus and group theory. Thus, one can obtain numerical expressions for group properties as well as interpretations of analytical dependencies, in the frame of group theory. In general, the objective of the theory of group representations is to study the homomorphisms of an arbitrary group onto all possible groups of linear operators.

DEFINITION 1.1.26 *Let G be a group and $\mathcal{L}$ a complex linear space. Any mapping* T *by which $\forall g \in G$ it corresponds a* **linear operator** $\mathsf{T}(g)$ *acting in $\mathcal{L}$, such that:*

1) $\mathsf{T}(e) = 1$, *where* 1 *is the identity operator in* $\mathcal{L}$,

2) $\mathsf{T}(g_1 g_2) = \mathsf{T}(g_1)\,\mathsf{T}(g_2)$, $\forall g_1, g_2 \in G$,

is called a **representation of the group** G **in the space** $\mathcal{L}$.

The space $\mathcal{L}$ is called the representation space, and the operators $\mathsf{T}(g)$ are the *operators of the representation*. The conditions 1) and 2) in the Definition 1.1.26 imply $\mathsf{T}(g^{-1})\mathsf{T}(g) = \mathsf{T}(g^{-1}g) = \mathsf{T}(e) = 1$, and analogously $\mathsf{T}(g)\mathsf{T}(g^{-1}) = 1$. Therefore, every operator $\mathsf{T}(g)$ is a bijection of $\mathcal{L}$ onto $\mathcal{L}$, and $\mathsf{T}(g^{-1}) = [\mathsf{T}(g)]^{-1}$. Thus, according to condition 2), the representation in the space $\mathcal{L}$ is a homomorphism of the group G onto the group $G_{\mathcal{L}}$, where $G_{\mathcal{L}}$ is the group of linear operators defined on $\mathcal{L}$, which map $\mathcal{L}$ onto itself in a one-to-one manner.

THEOREM 1.1.22 (CAYLEY) *Every group is homomorphic to a group of transformations mapping a set onto itself.*

If a group G is isomorphic with the group $G_{\mathcal{L}}$, then the representation is called *faithful* (or *pointwise*). A simple group (whose invariant subgroups are only $\{e\}$ and itself) has only faithful representations, because the kernel of the homomorphism consists only of its identity element. If T is a representation of G in $\mathcal{L}$, we also say that the space $\mathcal{L}$ transforms in compliance with this representation.

If the elements of a group G are transformations in an arbitrary space, then the correspondence $g \to \mathsf{T}(g) = g$ provides a representation of G. This representation, which coincides with G, is called the *identical representation*. In other words, the group G is a representation of itself. Of course, there are other representations of G that do not coincide with G. If we associate the identity operator on $\mathcal{L}$ (denoted by 1) to any element $g \in G$, that is $g \to 1$ ($\forall g \in G$), we obtain a representation of G called the *identity representation*.

The dimension of $\mathcal{L}$ is called the *dimension of the representation*, so that a group can have finite- or infinite-dimensional representations. If the space $\mathcal{L}$ is a finite-dimensional linear space ($\dim \mathcal{L} = n$), we may choose a basis $\{\mathbf{e}_i\}_{i=1}^{n}$ in $\mathcal{L}$, and the operators $\mathsf{T}(g)$ may be expressed in this basis as square nonsingular matrices of order n

$$\mathsf{T}(g) \to \mathbf{t}(g) = \begin{bmatrix} t_{11}(g) & \dots & t_{1n}(g) \\ \dots & \dots & \dots \\ t_{n1}(g) & \dots & t_{nn}(g) \end{bmatrix}, \tag{1.1.1}$$

whose elements are defined by the relations

$$\mathsf{T}(g)\,\mathbf{e}_k = \sum_{j=1}^{n} t_{jk}(g)\,\mathbf{e}_j\,, \quad k = 1, 2, \dots, n. \tag{1.1.2}$$

In terms of the matrices $\mathbf{t}(g)$, the conditions 1) and 2) in Definition 1.1.26 become

$$\mathbf{t}(e) = \mathbf{1}\,, \quad \mathbf{t}(g_1 g_2) = \mathbf{t}(g_1)\,\mathbf{t}(g_2)\,, \tag{1.1.3}$$

where $\mathbf{1}$ is the identity matrix, and

$$t_{jk}(g_1 g_2) = \sum_{l=1}^{n} t_{jl}(g_1) t_{lk}(g_2)\,, \quad j,k = 1,2,\ldots,n.$$

The matrix $\mathbf{t}(g)$ is called the *matrix of representation* T and the functions $t_{jk}(g)$ are the *matrix elements* of this representation in the basis $\{\mathbf{e}_i\}$. Reciprocally, let $g \to \mathbf{t}(g)$ be a matrix function of nth order defined on a group G, and let us suppose that this function satisfies the conditions 1) and 2) in the Definition 1.1.26. Now, in the space of complex numbers $\mathbb{C}^n$ we associate to each $g \in G$ an operator $\mathsf{T}(g)$ of matrix $\mathbf{t}(g)$, such that

$$\mathsf{T}(g) : \mathbf{x} \to \mathbf{x}'\,, \tag{1.1.4}$$

where $\mathbf{x} = \{x_1, x_2, \ldots, x_n\}$ and $\mathbf{x}' = \{x'_1, x'_2, \ldots, x'_n\}$ are elements of $\mathbb{C}^n$, and

$$x'_j = \sum_{k=1}^{n} t_{jk}(g)\, x_k\,, \quad j = 1,2,\ldots,n. \tag{1.1.5}$$

According to the Definition 1.1.26, the mapping $g \to \mathsf{T}(g)$ is a representation of G. Consequently, a finite-dimensional representation of a group G can be considered as a matrix function $g \to \mathbf{t}(g)$, satisfying the conditions (1.1.3). In particular, a one-dimensional representation can be assimilated to a numerical function satisfying the same conditions, with the matrix $\mathbf{t}(g)$ reduced to only one element.

Let T be a representation of a group G in the space $\mathcal{L}$ and H a subgroup of G. The set of operators

$$\mathsf{T}|_H = \{\mathsf{T}(h) | h \in H \subset G\}$$

form a representation of H, called the *restriction of the representation* T *to the subgroup* H. The subspace $\mathcal{M} \subset \mathcal{L}$ is called *invariant* with respect to the representation T if it is invariant with respect to all the operators $\mathsf{T}(g)$ of T (that is, the operators $\mathsf{T}(g)$ transform the vectors of $\mathcal{M}$ into themselves). Let us now suppose that the subspace $\mathcal{M} \subset \mathcal{L}$ is invariant with respect to the representation T of G in $\mathcal{L}$. The restrictions of the operators $\mathsf{T}(g)$ to the subspace $\mathcal{M}$, constitute a representation of G in $\mathcal{M}$, called the *restriction of the representation* T *to* $\mathcal{M}$, and denoted by $\mathsf{T}|_{\mathcal{M}}$. On the other hand, the operators $\mathsf{T}(g)$ induce the operators $\widetilde{\mathsf{T}}(g)$ in the quotient space $\widetilde{\mathcal{L}} = \mathcal{L}/\mathcal{M}$ and it is easy to verify that the homomorphism $g \to \widetilde{\mathsf{T}}(g)$ satisfies the conditions required by the definition of a representation. The representation of G in $\widetilde{\mathcal{L}}$, defined by the homomorphism $g \to \widetilde{\mathsf{T}}(g)$, is called the *representation induced by* T *in the quotient space* $\mathcal{L}/\mathcal{M}$, and it is denoted by $\widetilde{\mathsf{T}}$.

A representation T of a group G in a space $\mathcal{L}$ is called *irreducible*, if the representation space $\mathcal{L}$ does not contain nontrivial subspaces invariant with respect to T. Otherwise, the representation is called *reducible*.

THEOREM 1.1.23 *Let* T *be a group representation in a finite-dimensional space* $\mathcal{L}$. *Then, there is a subspace* $\mathcal{M} \subset \mathcal{L}$, $\mathcal{M} \neq \{0\}$, *such that the restriction of* T *to* $\mathcal{M}$ *is irreducible.*

Note that, in the same space $\mathcal{L}$ may exist several representations of the same group G, but not all these representations are essentially different. There are representations which we call *equivalent*. Let S be a representation of G in $\mathcal{L}$, defined by the mapping $g \to \mathsf{S}(g)$. We introduce the operators

$$\mathsf{T}(g) = \mathsf{A}^{-1}\mathsf{S}(g)\mathsf{A}\,, \quad g \in G\,, \tag{1.1.6}$$

where A is an arbitrary operator acting in $\mathcal{L}$. The operators $\mathsf{T}(g)$ satisfy the relations

$$\mathsf{T}(g_1)\mathsf{T}(g_2) = \mathsf{A}^{-1}\mathsf{S}(g_1)\mathsf{A}\mathsf{A}^{-1}\mathsf{S}(g_2)\mathsf{A} = \mathsf{A}^{-1}\mathsf{S}(g_1g_2)\mathsf{A} = \mathsf{T}(g_1g_2)\,,$$

so that the mapping $g \to \mathsf{T}(g)$ defines a new representation T of G in $\mathcal{L}$. The representations S and T correlated by (1.1.6) are called *equivalent representations.*

DEFINITION 1.1.27 *Two representations* S *and* T *of the same group* G, *acting in the spaces* $\mathcal{L}$ *and* $\mathcal{M}$, *respectively, are called* **equivalent**, *if there exists a linear operator* A *which maps bijectively* $\mathcal{L}$ *onto* $\mathcal{M}$, *and satisfies the condition*

$$\mathsf{A}\mathsf{T}(g) = \mathsf{S}(g)\mathsf{A}\,, \quad \forall g \in G. \tag{1.1.7}$$

If $\mathcal{M} = \mathcal{L}$, then (1.1.7) takes the form (1.1.6).

THEOREM 1.1.24 *If the spaces* $\mathcal{L}$ *and* $\mathcal{M}$ *are of finite dimension* $\dim \mathcal{L} = \dim \mathcal{M} = n$, *and the dimension of the representation* T *is* $n_{\mathsf{T}} = n$, *then the equivalence of the representations* S *and* T *implies* $n_{\mathsf{S}} = n$, *and the matrix elements of* S *and* T *are the same for a certain choice of bases in* $\mathcal{L}$ *and* $\mathcal{M}$.

In particular, the equivalent one-dimensional representations are defined by the same numerical function $t(g)$.

THEOREM 1.1.25 *If* T *and* S *are equivalent representations and* T *is irreducible, then* S *is also irreducible.*

1.3.2 Schur's lemmas

LEMMA 1.1.1 (SCHUR) *Let* T *and* S *be two irreducible representations of a group* G*, in the spaces* $\mathcal{L}$ *and* $\mathcal{M}$*, respectively, and let* A *be an operator which maps* $\mathcal{L}$ *onto* $\mathcal{M}$ *and satisfies the condition (1.1.7). Then, either the mapping of* $\mathcal{L}$ *onto* $\mathcal{M}$ *defined by* A *is one-to-one, that is* T *is equivalent to* S*, or* $\mathsf{A} = 0$.

LEMMA 1.1.2 (SCHUR) *Let* T *be a finite irreducible representation of a group* G *in the space* $\mathcal{L}$*. Then, every linear operator* B *acting in* $\mathcal{L}$*, which commutes with all the operators* $\mathsf{T}(g)$*,* $g \in G$*, has the form* $\mathsf{B} = \lambda 1$*, where* λ *is a scalar.*

In other words, if the operator B satisfies the commutation relation

$$\mathsf{BT}(g) = \mathsf{T}(g)\mathsf{B}, \quad \forall g \in G, \tag{1.1.8}$$

then $\mathsf{B} = \lambda 1$.

Let T be an irreducible representation of an Abelian group G in the finite-dimensional space $\mathcal{L}$. Then,

$$\mathsf{T}(g_0)\mathsf{T}(g) = \mathsf{T}(g_0 g) = \mathsf{T}(g g_0) = \mathsf{T}(g)\mathsf{T}(g_0), \quad \forall g_0, g \in G,$$

that is any operator $\mathsf{T}(g_0)$ commutes with all the operators $\mathsf{T}(g)$, and therefore $\mathsf{T}(g_0) = \lambda(g_0)1$. If $\dim \mathcal{L} > 1$, then any subspace of $\mathcal{L}$ is invariant with respect to the operators $\mathsf{T}(g) = \lambda(g)1$, in contradiction to the hypothesis that the representation T is irreducible. It follows that $\dim \mathcal{L} = 1$, and we may assert that *all the irreducible finite-dimensional representations of an Abelian group are one-dimensional.*

In this case of Abelian groups, the one-dimensional irreducible representations are defined by a numerical function $t(g)$, satisfying the conditions

$$t(e) = 1, \quad t(g_1 g_2) = t(g_1)t(g_2). \tag{1.1.9}$$

DEFINITION 1.1.28 *The numerical functions, which are defined on an Abelian group* G *and satisfy the conditions (1.1.9), are called* **characters on** G.

1.3.3 Adjoint representations

DEFINITION 1.1.29 *Let* $\mathcal{L}$ *and* $\mathcal{M}$ *be two linear spaces. We call* **bilinear form** *on the pair* $\mathcal{L}$*,* $\mathcal{M}$ *a numerical function, denoted by* $(\mathbf{x}, \mathbf{y})$*, defined on* $\mathcal{L} \times \mathcal{M}$ *and satisfying the conditions:*

1) $(\alpha\mathbf{x}, \mathbf{y}) = \alpha\,(\mathbf{x}, \mathbf{y})$,

2) $(\mathbf{x}, \alpha\mathbf{y}) = \overline{\alpha}\,(\mathbf{x}, \mathbf{y})$,

3) $(\mathbf{x}_1 + \mathbf{x}_2, \mathbf{y}) = (\mathbf{x}_1, \mathbf{y}) + (\mathbf{x}_2, \mathbf{y})$,

4) $(\mathbf{x}, \mathbf{y}_1 + \mathbf{y}_2) = (\mathbf{x}, \mathbf{y}_1) + (\mathbf{x}, \mathbf{y}_2)$,

$\forall \mathbf{x}, \mathbf{x}_1, \mathbf{x}_2 \in \mathcal{L}$, $\forall \mathbf{y}, \mathbf{y}_1, \mathbf{y}_2 \in \mathcal{M}$, and $\forall \alpha \in \mathbb{C}$. If $\mathcal{L} = \mathcal{M}$ we obtain a bilinear form on $\mathcal{L}$.

Here, the superposed bar on α denotes the complex conjugation.

Let $\mathbf{x} \in \mathcal{L}$ and $\mathbf{y} \in \mathcal{M}$ be two vectors. We say that $\mathbf{x}$ is orthogonal to $\mathbf{y}$ (denoted by $\mathbf{x} \perp \mathbf{y}$) if $(\mathbf{x}, \mathbf{y}) = 0$. For a subspace $\mathcal{E} \subset \mathcal{L}$, the set of vectors $\mathbf{y} \in \mathcal{M}$ orthogonal to all the vectors $\mathbf{x} \in \mathcal{E}$ forms the *orthogonal complement* of $\mathcal{E}$ in $\mathcal{M}$ with respect to $(\mathbf{x}, \mathbf{y})$.

DEFINITION 1.1.30 *We say that the pair $\mathcal{L}$, $\mathcal{M}$ is* **dual** *with respect to the bilinear form $(\mathbf{x}, \mathbf{y})$ if, besides the conditions 1) – 4) in the Definition 1.1.29, the conditions:*

5) *if $(\mathbf{x}_0, \mathbf{y}) = 0$, $\forall \mathbf{y} \in \mathcal{M}$, then $\mathbf{x}_0 = \mathbf{0}$,*

6) *if $(\mathbf{x}, \mathbf{y}_0) = 0$, $\forall \mathbf{x} \in \mathcal{L}$, then $\mathbf{y}_0 = \mathbf{0}$,*

are also satisfied.

If the spaces $\mathcal{L}$ and $\mathcal{M}$ are dual with respect to the bilinear form $(\mathbf{x}, \mathbf{y})$, then they have the same dimension. Furthermore, if $(\mathbf{x}, \mathbf{y})$ is a bilinear form on the pair $\mathcal{L}$, $\mathcal{M}$, then the function $(\mathbf{y}, \mathbf{x}) = \overline{(\mathbf{x}, \mathbf{y})}$ is a bilinear form on the pair $\mathcal{M}$, $\mathcal{L}$.

Let $\mathcal{L}$ and $\mathcal{M}$ be two linear spaces, dual with respect to the bilinear form $(\mathbf{x}, \mathbf{y})$, and let T and S be the representations of a group G in the spaces $\mathcal{L}$ and $\mathcal{M}$, respectively.

DEFINITION 1.1.31 *The representation S is called the* **adjoint of the representation T with respect to** $(\mathbf{x}, \mathbf{y})$ *if the relation*

$$(\mathsf{T}(g)\mathbf{x}, \mathsf{S}(g)\mathbf{y}) = (\mathbf{x}, \mathbf{y}), \quad \forall g \in G, \quad \forall \mathbf{x} \in \mathcal{L}, \quad \forall \mathbf{y} \in \mathcal{M}, \tag{1.1.10}$$

is fulfilled.

We can use the fact that $\mathsf{T}(g)$ is a one-to-one mapping of the space $\mathcal{L}$ onto itself, to replace the vector $\mathbf{x}$ in (1.1.10) by $\mathsf{T}(g^{-1})\mathbf{x}$. Thus, we obtain

$$(\mathsf{T}(g)\mathsf{T}(g^{-1})\mathbf{x}, \mathsf{S}(g)\mathbf{y}) = (\mathsf{T}(g^{-1})\mathbf{x}, \mathbf{y}),$$

or

$$(\mathsf{T}(g^{-1})\mathbf{x}, \mathbf{y}) = (\mathbf{x}, \mathsf{S}(g)\mathbf{y}), \tag{1.1.11}$$

which is equivalent to (1.1.10).

THEOREM 1.1.26 *If the linear spaces $\mathcal{L}$ and $\mathcal{M}$ are dual with respect to the bilinear form $(\mathbf{x}, \mathbf{y})$, T is a representation in $\mathcal{L}$, and there exists a representation in $\mathcal{M}$, adjoint to T with respect to $(\mathbf{x}, \mathbf{y})$, then the representation T is unique.*

THEOREM 1.1.27 *If the representation S is adjoint to the representation T with respect to the bilinear form $(\mathbf{x}, \mathbf{y})$, then the representation T is adjoint to the representation S with respect to the bilinear form $(\mathbf{y}, \mathbf{x}) = \overline{(\mathbf{x}, \mathbf{y})}$.*

THEOREM 1.1.28 *Two finite-dimensional representations T and S, in the spaces $\mathcal{L}$ and $\mathcal{M}$, respectively, are adjoint with respect to a certain bilinear form $(\mathbf{x}, \mathbf{y})$, if and only if their matrix elements are connected by the relations*

$$t_{jk}(g^{-1}) = \overline{s_{kj}(g)}, \quad j, k, = 1, 2, \ldots, n, \quad n = \dim \mathcal{L} = \dim \mathcal{M}\,, \quad (1.1.12)$$

for a certain choice of bases in $\mathcal{L}$ and $\mathcal{M}$.

THEOREM 1.1.29 *Let $\mathcal{L}$ and $\mathcal{M}$ be two finite-dimensional spaces, dual with respect to the bilinear form $(\mathbf{x}, \mathbf{y})$, and let T be a representation of a group G in $\mathcal{L}$. Then, there exists a unique representation S of G in $\mathcal{M}$, adjoint to the representation T.*

THEOREM 1.1.30 *If T and S are finite-dimensional adjoint representations, then T is irreducible if and only if S is irreducible.*

1.3.4 The direct sum of finite-dimensional representations

Let $\mathsf{T} = \{\mathbf{t}(g) | g \in G\}$ and $\mathsf{S} = \{\mathbf{s}(g) | g \in G\}$ be two finite-dimensional representations of a group G, in terms of square matrices of mth and nth order, respectively. Let us now consider the homomorphism

$$g \to \begin{bmatrix} \mathbf{t}(g) & \mathbf{0} \\ \mathbf{0} & \mathbf{s}(g) \end{bmatrix}, \quad g \in G\,. \quad (1.1.13)$$

By taking into account the rules of multiplication of block matrices, we can write

$$\begin{aligned} gh \to \begin{bmatrix} \mathbf{t}(gh) & \mathbf{0} \\ \mathbf{0} & \mathbf{s}(gh) \end{bmatrix} &= \begin{bmatrix} \mathbf{t}(g)\mathbf{t}(h) & \mathbf{0} \\ \mathbf{0} & \mathbf{s}(g)\mathbf{s}(h) \end{bmatrix} \\ &= \begin{bmatrix} \mathbf{t}(g) & \mathbf{0} \\ \mathbf{0} & \mathbf{s}(g) \end{bmatrix} \begin{bmatrix} \mathbf{t}(h) & \mathbf{0} \\ \mathbf{0} & \mathbf{s}(h) \end{bmatrix}, \end{aligned} \quad (1.1.14)$$

where $g, h \in G$. Therefore, the homomorphism (1.1.13) is also a representation of G, called the *direct sum of the representations* T *and* S, and denoted by $\mathsf{T}(g) \oplus \mathsf{S}(g)$. By permuting the block matrices on the main diagonal, we obtain a new representation

$$g \to \begin{bmatrix} \mathbf{s}(g) & \mathbf{0} \\ \mathbf{0} & \mathbf{t}(g) \end{bmatrix}, \quad g \in G\,, \quad (1.1.15)$$

which is equivalent to (1.1.13). To prove the equivalence of the representations (1.1.13) and (1.1.15) we use the constant matrix

$$\mathbf{a} = \begin{bmatrix} \mathbf{0}_{n\times m} & \mathbf{1}_{n\times n} \\ \mathbf{1}_{m\times m} & \mathbf{0}_{m\times n} \end{bmatrix},$$

where we have shown explicitly the dimensions of the null ($\mathbf{0}$) and identity ($\mathbf{1}$) matrices, to form the products

$$\begin{bmatrix} \mathbf{0}_{n\times m} & \mathbf{1}_{n\times n} \\ \mathbf{1}_{m\times m} & \mathbf{0}_{m\times n} \end{bmatrix} \begin{bmatrix} \mathbf{t}(g)_{m\times m} & \mathbf{0}_{m\times n} \\ \mathbf{0}_{n\times m} & \mathbf{s}(g)_{n\times n} \end{bmatrix} = \begin{bmatrix} \mathbf{0}_{n\times m} & \mathbf{s}(g)_{n\times n} \\ \mathbf{t}(g)_{m\times m} & \mathbf{0}_{m\times n} \end{bmatrix} \tag{1.1.16}$$

and

$$\begin{bmatrix} \mathbf{s}(g)_{n\times n} & \mathbf{0}_{n\times m} \\ \mathbf{0}_{m\times n} & \mathbf{t}(g)_{m\times m} \end{bmatrix} \begin{bmatrix} \mathbf{0}_{n\times m} & \mathbf{1}_{n\times n} \\ \mathbf{1}_{m\times m} & \mathbf{0}_{m\times n} \end{bmatrix} = \begin{bmatrix} \mathbf{0}_{n\times m} & \mathbf{s}(g)_{n\times n} \\ \mathbf{t}(g)_{m\times m} & \mathbf{0}_{m\times n} \end{bmatrix}. \tag{1.1.17}$$

The products (1.1.16) and (1.1.17) are equal $\forall g \in G$. Hence, according to the Definition 1.1.27 applied to the matrices of the corresponding operators, the representations (1.1.13) and (1.1.15) are equivalent. It follows that, if we do not mark as different the equivalent representations, then the direct sum of representations is a *commutative operation*. One can easily see that this condition also ensures the associativity of the direct sum. Starting from a certain set of representations $\{\mathsf{T}(g), \mathsf{S}(g), \mathsf{U}(g), \ldots\}$ of a group G we can obtain representations of any degree: $\mathsf{T}(g) \oplus \mathsf{S}(g) \oplus \mathsf{U}(g)$, $\mathsf{T}(g) \oplus \mathsf{T}(g) \oplus \mathsf{T}(g) \oplus \mathsf{T}(g)$ etc.

For instance, the set $G = \{1, -1, i, -i\}$, that is the set of the roots of equation $x^4 = 1$, forms a group with respect to multiplication. If we associate to each number in G its own numerical value (that can be represented as a 1×1 matrix)

$$1 \to [1], \quad -1 \to [-1], \quad i \to [i], \quad -i \to [-i],$$

we obtain the identical representation of G (of first degree, or one-dimensional). Another one-dimensional representation is given by the mapping

$$1 \to [1], \quad -1 \to [-1], \quad i \to [-i], \quad -i \to [i].$$

The direct sum of these two representations will be the representation of the second degree

$$1 \to \begin{bmatrix} 1 & 0 \\ 0 & 1 \end{bmatrix}, \quad -1 \to \begin{bmatrix} -1 & 0 \\ 0 & -1 \end{bmatrix}, \quad i \to \begin{bmatrix} i & 0 \\ 0 & -i \end{bmatrix}, \quad -i \to \begin{bmatrix} -i & 0 \\ 0 & i \end{bmatrix}.$$

We may transform this representation by means of the constant matrix

$$\begin{bmatrix} 1 & i \\ 1 & -i \end{bmatrix},$$

to obtain the equivalent representation

$$1 \to \begin{bmatrix} 1 & 0 \\ 0 & 1 \end{bmatrix}, \quad -1 \to \begin{bmatrix} -1 & 0 \\ 0 & -1 \end{bmatrix}, \quad i \to \begin{bmatrix} 0 & 1 \\ -1 & 0 \end{bmatrix}, \quad -i \to \begin{bmatrix} 0 & -1 \\ 1 & 0 \end{bmatrix},$$

which contains only real matrices.

Let $\mathcal{L}_1, \mathcal{L}_2, \ldots, \mathcal{L}_m$ be m linear spaces, and let $\mathcal{L} = \mathcal{L}_1 \oplus \mathcal{L}_2 \oplus \ldots \oplus \mathcal{L}_m$ be their direct sum. Then, a vector $\mathbf{x} \in \mathcal{L}$ can be uniquely written in the form $\mathbf{x} = \mathbf{x}_1 + \mathbf{x}_2 + \ldots + \mathbf{x}_m$, $\mathbf{x}_k \in \mathcal{L}_k$. We also suppose that a representation T^k of a group G is given in each space $\mathcal{L}_k$. Now, we introduce the linear operator $\mathsf{T}(g)$ acting on $\mathcal{L}$, and defined by the relation

$$\begin{aligned} \mathsf{T}(g)\mathbf{x} &= \mathsf{T}(g)(\mathbf{x}_1 + \mathbf{x}_2 + \ldots + \mathbf{x}_m) \\ &= \mathsf{T}^1(g)\mathbf{x}_1 + \mathsf{T}^2(g)\mathbf{x}_2 + \ldots + \mathsf{T}^m(g)\mathbf{x}_m \,. \end{aligned} \qquad (1.1.18)$$

Thus, we have

$$\mathsf{T}(e) = 1, \quad \mathsf{T}(g_1, g_2) = \mathsf{T}(g_1)\mathsf{T}(g_2), \quad \forall g_1, g_2 \in G\,,$$

so that the mapping $g \to \mathsf{T}(g)$ is a representation of G in $\mathcal{L}$. This representation is the *direct sum of the representations* T^1, T^2, ..., T^m, denoted by $\mathsf{T} = \mathsf{T}^1 \oplus \mathsf{T}^2 \oplus \ldots \oplus \mathsf{T}^m$. It is evident that each space $\mathcal{L}_k$ is a subspace of $\mathcal{L}$, invariant with respect to T, and the restriction of T to $\mathcal{L}_k$ is T^k.

Let us consider the case when the linear spaces $\mathcal{L}_k$, $k = 1, 2, \ldots, m$, have the finite dimension $\dim \mathcal{L}_k = n_k$. Also, let $\{\mathbf{e}_i^k\}_{i=1}^{n_k}$ be a basis in $\mathcal{L}_k$, and $t_{jl}^k(g)$, $j, l = 1, 2, \ldots, n_k$, the matrix elements of the representation T^k in this basis; we denote by $\mathbf{t}^k(g)$ the corresponding matrix. The union of the bases defined in the spaces $\mathcal{L}_k$ form a basis in the space $\mathcal{L}$ (the direct sum of the spaces $\mathcal{L}_k$), so that the relations

$$\mathsf{T}(g)\mathbf{e}_j^k = \mathsf{T}^k(g)\mathbf{e}_j^k = \sum_{i=1}^{n_k} t_{ij}^k(g)\mathbf{e}_i^k \qquad (1.1.19)$$

take place. If we order the bases $\{\mathbf{e}_i^k\}$ with respect to the index k, the matrix of the operator $\mathsf{T}(g)$ in the space $\mathcal{L}$, takes the block diagonal form

$$\mathbf{t}(g) = \begin{bmatrix} \mathbf{t}^1(g) & \mathbf{0} & \ldots & \mathbf{0} \\ \mathbf{0} & \mathbf{t}^2(g) & \ldots & \mathbf{0} \\ . & . & \ldots & . \\ \mathbf{0} & \mathbf{0} & \ldots & \mathbf{t}^m(g) \end{bmatrix}, \qquad (1.1.20)$$

where

$$\mathbf{t}^k(g) = \begin{bmatrix} t_{11}^k(g) & \ldots & t_{1\,n_k}^k(g) \\ . & \ldots & . \\ t_{n_k\,1}^k(g) & \ldots & t_{n_k\,n_k}^k(g) \end{bmatrix}, \quad k = 1, 2, \ldots, m. \qquad (1.1.21)$$

In general, a representation T in a finite-dimensional space $\mathcal{L}$ is called *fully reducible*, or *decomposable*, if T is the direct sum of a finite number of irreducible representations; we say that T is decomposed into a *direct sum of irreducible representations*. A representation T is called a *multiple* of an irreducible T^1 (denoted by $\mathsf{T} = n\mathsf{T}^1$), if T can be decomposed into a direct sum of n representations, which are equivalent to the irreducible representation T^1. When a fully reducible representation T is decomposed into a direct sum of irreducible representations, among which n_1 are equivalent to T^1, n_2 are equivalent to $\mathsf{T}^2, \ldots, n_p$ are equivalent to T^p, then we can write

$$\mathsf{T} = n_1\mathsf{T}^1 \oplus n_2\mathsf{T}^2 \oplus \ldots \oplus n_p\mathsf{T}^p\,, \tag{1.1.22}$$

and we say that the irreducible representation T^k is contained in T with the multiplicity n_k.

THEOREM 1.1.31 *Let T be a representation of a group G in a linear space $\mathcal{L}$ and let $\mathcal{L}_1, \mathcal{L}_2, \ldots$ be a sequence of subspaces of $\mathcal{L}$, satisfying the conditions:*

1) *each subspace $\mathcal{L}_k$ is invariant with respect to T;*

2) *the restrictions T^k of T to $\mathcal{L}_k$ are irreducible.*

Then, one can choose the subspaces $\mathcal{L}_{k_1}, \mathcal{L}_{k_2}, \ldots$ in the sequence $\mathcal{L}_1, \mathcal{L}_2, \ldots$, such that

a) *$\mathcal{L}_{k_1}, \mathcal{L}_{k_2}, \ldots$ are linearly independent;*

b) $\sum_j^{\oplus} \mathcal{L}_{k_j} = \sum_k^{\oplus} \mathcal{L}_k$.

In particular, if $\sum_k^{\oplus} \mathcal{L}_k = \mathcal{L}$, then $\sum_j^{\oplus} \mathcal{L}_{k_j} = \mathcal{L}$.

Let us assume that all the matrices of a certain nth degree representation of a group G have the form

$$\mathbf{t}(g) = \begin{bmatrix} \mathbf{t}^1(g) & \mathbf{t}^2(g) \\ \mathbf{0} & \mathbf{t}^3(g) \end{bmatrix}, \quad g \in G\,, \tag{1.1.23}$$

where $\mathbf{t}^1(g)$ and $\mathbf{t}^3(g)$ are square matrices. For $g, h \in G$, by multiplying the corresponding matrices $\mathbf{t}(g)$ and $\mathbf{t}(h)$ we obtain

$$\begin{aligned} \mathbf{t}(gh) = \mathbf{t}(g)\mathbf{t}(h) &= \begin{bmatrix} \mathbf{t}^1(g)\mathbf{t}^1(h) & \mathbf{t}^1(g)\mathbf{t}^2(h) + \mathbf{t}^2(g)\mathbf{t}^3(h) \\ \mathbf{0} & \mathbf{t}^3(g)\mathbf{t}^3(h) \end{bmatrix} \\ &= \begin{bmatrix} \mathbf{t}^1(gh) & \mathbf{t}^1(g)\mathbf{t}^2(h) + \mathbf{t}^2(g)\mathbf{t}^3(h) \\ \mathbf{0} & \mathbf{t}^3(gh) \end{bmatrix}. \end{aligned}$$

It results that $\mathsf{T}^1(g)$ and $\mathsf{T}^3(g)$ are also representations of G but of a lower degree. The representation $\mathsf{T}(g)$, in terms of the block triangular matrices (1.1.23), is called *triangular representation*, and any representation equivalent to a triangular representation is said to be *reducible*. A representation which is not equivalent to any triangular representation is called *irreducible*.

1.3.5 The regular representation of a finite group

For any finite group G we can introduce a regular representation, constructed by the following method: let $g_1, g_2, \ldots, g_n$ be the elements of G, labelled in an arbitrary order, and let $g_{i_k} = g_i g_k$, $i, k = 1, 2, \ldots, n$. Then, we fix the element g_k and form a square matrix of order n, by setting to 1 the i_kth element in each row i, $i = 1, 2, \ldots, n$, and to 0 all the other elements in the row. We denote this matrix by $\mathbf{r}(g_k)$. The mapping $g_k \to \mathbf{r}(g_k)$, $k = 1, 2, \ldots, n$, defines a representation R called the *regular representation of* G. By changing the initial order of elements in G we obtain a representation equivalent to R; hence, *leaving out equivalent representations, any finite group has only one regular representation.*

It can be shown that the number of distinct (nonequivalent) irreducible representations of a finite group is finite, and equal to the number of classes of conjugate elements of this group. The degrees of the irreducible representations are divisors of the order of the group, and the regular representation is equivalent to the direct sum of all distinct irreducible representations, each of them having a multiplicity equal to its degree.

Let us denote by n the order of a finite group G, by k the number of classes of conjugate elements, and by $n_1, n_2, \ldots, n_k$ the degrees of the irreducible representations of G. By construction, the regular representation of G is of degree n. Because the regular representation is equivalent to the direct sum of n_1 representations equivalent to the first irreducible representation, to which we add n_2 representations equivalent to the second irreducible representation etc., and because the degree of a direct sum of representations is equal to the sum of the degrees of the component representations, the relation

$$n = n_1^2 + n_2^2 + \ldots + n_k^2 \tag{1.1.24}$$

has to be fulfilled. If we associate to each element in G the number 1, we obtain the trivial irreducible representation of degree 1, that is admitted by any group. Hence, by assuming that n_1 in (1.1.24) is the degree of this identity representation, we may rewrite (1.1.24) in the form

$$n = 1 + n_2^2 + \ldots + n_k^2, \tag{1.1.25}$$

where $n_2, \ldots, n_k$ are now the degrees of nontrivial irreducible representations. Taking into account the fact that $n_2, \ldots, n_k$ have to be divisors of n, and determining k from the structure of G, one can sometimes find all the numbers $n_2, \ldots, n_k$ directly from (1.1.25).

1.3.6 The direct product of finite-dimensional representations

Let T^1 and T^2 be two representations of a group G in the spaces $\mathcal{L}_1$ and $\mathcal{L}_2$, respectively, of dimensions $\dim \mathcal{L}_1 = n_1$ and $\dim \mathcal{L}_2 = n_2$, and let $\{\mathbf{e}_i\}$ and

$\{\mathbf{f}_j\}$ be the bases of these two spaces. We consider the space $\mathcal{L} = \mathcal{L}_1 \otimes \mathcal{L}_2$ of dimension $n_1 n_2$ having the basis $\{\mathbf{h}_{ij}\} = \{\mathbf{e}_i \otimes \mathbf{f}_j\}$. Thus, every element $\mathbf{z} \in \mathcal{L}$ will be of the form

$$\mathbf{z} = \sum_{i=1}^{n_1} \sum_{j=1}^{n_2} (\mathbf{e}_i \otimes \mathbf{f}_j) z_{ij} = \sum_{i=1}^{n_1} \sum_{j=1}^{n_2} \mathbf{h}_{ij} z_{ij} . \qquad (1.1.26)$$

The representations T^1 and T^2 satisfy the relations

$$\begin{aligned} \mathsf{T}^1(g)\mathbf{e}_i = \mathbf{e}'_i = \sum_{j=1}^{n_1} t^1_{ji}(g)\mathbf{e}_j , \qquad i = 1, 2, \ldots, n_1 , \\ \mathsf{T}^2(g)\mathbf{f}_i = \mathbf{f}'_i = \sum_{j=1}^{n_2} t^2_{ji}(g)\mathbf{f}_j , \qquad i = 1, 2, \ldots, n_2 ; \end{aligned} \qquad (1.1.27)$$

hence, the components (x_i) and (y_i) of two vectors $\mathbf{x} \in \mathcal{L}_1$ and $\mathbf{y} \in \mathcal{L}_2$ are transformed by the operators of T^1 and T^2 according to

$$\begin{aligned} x'_i = \sum_{j=1}^{n_1} t^1_{ij}(g) x_j , \qquad i = 1, 2, \ldots, n_1 , \\ y'_i = \sum_{j=1}^{n_2} t^2_{ij}(g) y_j , \qquad i = 1, 2, \ldots, n_2 . \end{aligned} \qquad (1.1.28)$$

If the relations (1.1.28) define the action of representations T^1 and T^2 in $\mathcal{L}_1$ and $\mathcal{L}_2$, respectively, then the same relations also define a representation T in $\mathcal{L}$. Indeed, according to (1.1.27) the basis vector $\mathbf{h}_{ij}$ is transformed into a vector $\mathbf{h}'_{ij}$ such that

$$\begin{aligned} \mathsf{T}(g)\mathbf{h}_{ij} &= \mathbf{h}'_{ij} = \mathbf{e}'_i \otimes \mathbf{f}'_j = \mathsf{T}^1(g)\mathbf{e}_i \otimes \mathsf{T}^2(g)\mathbf{f}_j \\ &= \sum_{k=1}^{n_1} t^1_{ki}(g)\mathbf{e}_k \otimes \sum_{l=1}^{n_2} t^2_{lj}(g)\mathbf{f}_l = \sum_{k=1}^{n_1} \sum_{l=1}^{n_2} t^1_{ki}(g) t^2_{lj}(g) (\mathbf{e}_k \otimes \mathbf{f}_l) \\ &= \sum_{k=1}^{n_1} \sum_{l=1}^{n_2} t^1_{ki}(g) t^2_{lj}(g) \mathbf{h}_{kl} = \sum_{k,l} t_{(kl)(ij)}(g) \mathbf{h}_{kl} . \end{aligned} \qquad (1.1.29)$$

Therefore, the transformation of the components $\{z_{ij}\}$ of a vector $\mathbf{z} \in \mathcal{L}$ under the action of T is given by the equations

$$z'_{ij} = \sum_{k,l} t_{(ij)(kl)}(g) z_{kl} = \sum_{k=1}^{n_1} \sum_{l=1}^{n_2} t^1_{ik}(g) t^2_{jl}(g) z_{kl} .$$

Thus, the matrix elements of the transformation T are connected to the matrix elements of the representations T^1 and T^2 by the relations

$$t_{(ij)(kl)}(g) = t^1_{ik}(g) t^2_{jl}(g) \,. \tag{1.1.30}$$

Now, let u and v be two elements in G. From (1.1.30) we have

$$\begin{aligned} t_{(ij)(kl)}(uv) &= t^1_{ik}(uv) t^2_{jl}(uv) = \sum_{s=1}^{n_1} t^1_{is}(u) t^1_{sk}(v) \sum_{p=1}^{n_2} t^2_{jp}(u) t^2_{pl}(v) \\ &= \sum_{s,p} t_{(ij)(sp)}(u) t_{(sp)(kl)}(v) \,; \end{aligned} \tag{1.1.31}$$

hence, the mapping $g \to \mathsf{T}(g)$ is a representation of G in $\mathcal{L}$. This representation is called the *direct product of the representations* T^1 *and* T^2. In general, the direct product of irreducible representations is a reducible representation, that can be decomposed into a direct sum of irreducible representations, when it is fully reducible.

The direct product of a finite number of representations, of finite dimension, can be defined in an analogous manner.

DEFINITION 1.1.32 *A representation* T *of a group* G *in the space* $\mathcal{L} = \mathcal{L}_1 \otimes \mathcal{L}_2 \otimes \ldots \otimes \mathcal{L}_m$, *whose operators* $\mathsf{T}(g)$ *transform the vectors* $\mathbf{x}_1 \otimes \mathbf{x}_2 \otimes \ldots \otimes \mathbf{x}_m$, $\mathbf{x}_k \in \mathcal{L}_k$, $k = 1, 2, \ldots, m$, *according to the relation*

$$\mathsf{T}(g)(\mathbf{x}_1 \otimes \mathbf{x}_2 \otimes \ldots \otimes \mathbf{x}_m) = \mathsf{T}^1(g)\mathbf{x}_1 \otimes \mathsf{T}^2(g)\mathbf{x}_2 \otimes \ldots \otimes \mathsf{T}^m(g)\mathbf{x}_m, \tag{1.1.32}$$

is called the **direct product of the finite-dimensional representations** T^k **of** G **in** $\mathcal{L}_k$, $k = 1, 2, \ldots, m$, *and it is denoted by* $\mathsf{T} = \mathsf{T}^1 \otimes \mathsf{T}^2 \otimes \ldots \otimes \mathsf{T}^m$.

Let $n_1, n_2, \ldots, n_m$ be the dimensions of the spaces $\mathcal{L}_1, \mathcal{L}_2, \ldots, \mathcal{L}_m$, respectively. Also, for $k = 1, 2, \ldots, m$, let $\{\mathbf{e}^k_i\}$ be a basis of $\mathcal{L}_k$, and $t^k_{ij}(g)$ the matrix elements of a finite-dimensional representation T^k in this basis. Then, the matrix elements of the direct product of representations $\mathsf{T} = \mathsf{T}^1 \otimes \mathsf{T}^2 \otimes \ldots \otimes \mathsf{T}^m$ are given by

$$t_{(i_1 i_2 \cdots i_m)(j_1 j_2 \cdots j_m)}(g) = t^1_{i_1 j_1}(g) t^2_{i_2 j_2}(g) \ldots t^m_{i_m j_m}(g) \,. \tag{1.1.33}$$

1.3.7 The characters of finite-dimensional representations

DEFINITION 1.1.33 *The trace of the matrix* $\mathbf{t}(g)$ *associated to an operator* $\mathsf{T}(g)$ *of an n-dimensional representation* T *of a group* G, *that is*

$$\chi_{\mathsf{T}}(g) = \mathrm{Tr}\, \mathbf{t}(g) = t_{11}(g) + t_{22}(g) + \ldots + t_{nn}(g) \,, \tag{1.1.34}$$

where $t_{jk}(g)$ *are the matrix elements of this representation in a given basis, is called the* **character of the representation** T.

On the basis of the definition one can easily show that the character of the identity element gives the dimension of the corresponding representation. Also, the characters are constant on conjugacy classes, and therefore the characters are *class functions*. If the group G is Abelian, and T is one-dimensional, then the characters $\chi_{\mathsf{T}}(g)$ are identical to the characters on G (see Definition 1.1.28).

The characters of finite-dimensional representations exhibit some specific features which play an important rôle in the study of decomposition of representations:

1) the characters of equivalent representations are identical;

2) the characters are constant on each conjugacy class;

3) the characters of two adjoint representations T and S satisfy the relation

$$\chi_{\mathsf{S}}(g) = \overline{\chi_{\mathsf{T}}(g^{-1})}\,;$$

4) the character of a direct sum of a finite number of representations is equal to the sum of characters of these representations;

5) the character of a direct product of a finite number of representations is equal to the product of characters of these representations.

Another important feature of group characters is the orthogonality relation satisfied by characters of different representations. Thus, let G be a finite group of order n, and let $C_1, C_2, \ldots, C_k$ be the classes of conjugate elements of G. We denote by n_i the number of elements in the class C_i. Also, the group G has precisely k irreducible representations $\mathsf{T}^1, \mathsf{T}^2, \ldots, \mathsf{T}^k$, and we denote the characters of these representations by $\chi^{(i)}$, $i = 1, 2, \ldots, k$. In terms of these notations, the orthogonality relation of characters reads

$$\sum_{i=1}^{k} n_i \chi^{(j)}(C_i)\overline{\chi^{(l)}(C_i)} = n\delta_{jl}\,, \qquad (1.1.35)$$

where we have used a conjugacy class as argument of characters rather than a group element, to stress the fact that the characters are actually class functions. A representation of G, if it is fully reducible, can be decomposed into a direct sum of irreducible representations having different multiplicities:

$$\mathsf{T} = m_1\mathsf{T}^1 \oplus m_2\mathsf{T}^2 \oplus \ldots \oplus m_k\mathsf{T}^k\,,$$

where the multiplicity of an irreducible representation is equal to zero, if that representation is not contained in T. Consequently, for each class, the character of T will be the sum of characters of the irreducible representations

$$\chi(C_j) = \sum_{i=1}^{k} m_i \chi^{(i)}(C_j)\,, \quad j = 1, 2, \ldots, k.$$

Now, we can multiply both sides by $n_j \overline{\chi^{(l)}(C_j)}$ and sum over j. Thus, on account of orthogonality relations (1.1.35) we obtain

$$m_l = \frac{1}{n} \sum_{j=1}^{k} n_j \overline{\chi^{(l)}(C_j)} \chi(C_j)\,, \quad l = 1, 2, \ldots, k\,. \tag{1.1.36}$$

Also, if T is the regular representation of G then the multiplicities m_l have to satisfy (1.1.25).

1.4 The S_3 group

1.4.1 Definition. Subgroups. Equivalence classes

To make clearer the implications of the definitions and theorems concerning the theory of finite groups, we present here a detailed analysis of one of the simplest nontrivial finite groups, that is the symmetric group of the set $M = \{1, 2, 3\}$. This is the permutation group of the third degree, denoted by S_3. The order of S_3 is $3! = 6$ (see Chap. 1 Sec. 1.2.8).

A group isomorphic to S_3 is the group of transformations of an equilateral triangle onto itself, called the *symmetry group of the equilateral triangle*. In crystallography, this is one of the point groups of trigonal systems and it is denoted by C_{3v} (Schoenflies notation). Thus, let us consider an equilateral triangle whose vertices are labelled by the numbers 1, 2, 3. It is easy to see that all the transformations of C_{3v}:

1) the identity transformation,
2) the set of rotations of the triangle by $\pm 2\pi/3$ and $\pm 4\pi/3$ about its centre,
3) the set of reflections of the triangle about an axis passing through its centre and one of the vertices,

can be mapped onto the set of 6 permutations

$$\{\sigma_0, \sigma_1, \sigma_2, \sigma_3, \sigma_4, \sigma_5\} \equiv \left\{ \begin{pmatrix} 1\,2\,3 \\ 1\,2\,3 \end{pmatrix}, \begin{pmatrix} 1\,2\,3 \\ 2\,3\,1 \end{pmatrix}, \begin{pmatrix} 1\,2\,3 \\ 3\,1\,2 \end{pmatrix}, \right.$$
$$\left. \begin{pmatrix} 1\,2\,3 \\ 1\,3\,2 \end{pmatrix}, \begin{pmatrix} 1\,2\,3 \\ 3\,2\,1 \end{pmatrix}, \begin{pmatrix} 1\,2\,3 \\ 2\,1\,3 \end{pmatrix} \right\},$$

that are the elements of S_3. Note that, the groups S_3 and C_{3v} are isomorphic ($\mathsf{S}_3 \simeq \mathsf{C}_{3v}$) if we consider only once identical transformations of C_{3v} (for instance, rotations of $+2\pi/3$ and $-4\pi/3$, or of $-2\pi/3$ and $+4\pi/3$ etc.), otherwise they are homomorphic. Hence, the set of considered rotations forms a finite group of sixth order isomorphic to S_3.

The internal composition law of S_3 is the product of permutations, and the *group table* (*Cayley's table*) has the form

	σ_0	σ_1	σ_2	σ_3	σ_4	σ_5
σ_0	σ_0	σ_1	σ_2	σ_3	σ_4	σ_5
σ_1	σ_1	σ_2	σ_0	σ_5	σ_3	σ_4
σ_2	σ_2	σ_0	σ_1	σ_4	σ_5	σ_3
σ_3	σ_3	σ_4	σ_5	σ_0	σ_1	σ_2
σ_4	σ_4	σ_5	σ_3	σ_2	σ_0	σ_1
σ_5	σ_5	σ_3	σ_4	σ_1	σ_2	σ_0

We observe in the Cayley's table that

a) the composition law is associative (for instance, $\sigma_1(\sigma_2\sigma_3) = \sigma_1\sigma_4 = \sigma_3$ and $(\sigma_1\sigma_2)\sigma_3 = \sigma_0\sigma_3 = \sigma_3$);

b) the identity element is σ_0;

c) the inverse of the group elements are:

$$\sigma_1^{-1} = \sigma_2\,,\ \sigma_2^{-1} = \sigma_1\,,\ \sigma_3^{-1} = \sigma_3\,,\ \sigma_4^{-1} = \sigma_4\,,\ \sigma_5^{-1} = \sigma_5\,.$$

Also, it is easy to see that the set $\mathsf{H} = \{\sigma_0, \sigma_1, \sigma_2\}$ is a subgroup of S_3.

To determine the subgroup structure of S_3, we introduce the homomorphism $f : \mathsf{S}_3 \to \{1, -1\}$, where $\{1, -1\}$ is the multiplicative group of the roots of equation $x^2 = 1$, defined by

$$f(\sigma) = \begin{cases} 1, & \sigma = \sigma_0, \sigma_1, \sigma_2, \\ -1, & \sigma = \sigma_3, \sigma_4, \sigma_5. \end{cases} \tag{1.1.37}$$

In other words, we have associated the number 1 to even permutations, and -1 to odd permutations. It is easy to verify that $f(\sigma_i\sigma_j) = f(\sigma_i)f(\sigma_j)$. Also, the mapping f is surjective (or onto) and it follows that we have defined a surjective homomorphism (or epimorphism). Because

$$\mathrm{Ker}\, f = f^{-1}(1) = \{\sigma_0, \sigma_1, \sigma_2\} = \mathsf{H} \neq \{\sigma_0\},$$

the mapping f is not one-to-one (or bijective). On the other hand, $\mathsf{Ker}\, f = \mathsf{H}$ is an invariant subgroup of S_3 (note that, $\{1\}$ is an invariant subgroup of $\{1, -1\}$). Thus, we may infer that the invariant subgroups of S_3 are $\{\sigma_0\}$, H, and S_3 itself. Note that, generally, we may define the homomorphism $f : \mathsf{S}_n \to \{1, -1\}$ for any permutation group S_n. Therefore, one invariant subgroup of S_n is always $\mathsf{Ker}\, f \lhd \mathsf{S}_n$, consisting of the identity element and all the even permutations in S_n.

Other subgroups of S_3 are $\mathsf{G}_1 = \{\sigma_0, \sigma_3\}$, $\mathsf{G}_2 = \{\sigma_0, \sigma_4\}$, and $\mathsf{G}_3 = \{\sigma_0, \sigma_5\}$. These are cyclic groups of the second order, generated by σ_3, σ_4, and σ_5, respectively. Note that, the invariant subgroup H is also a cyclic group of the third order generated by σ_1 or σ_2. The order of each of these subgroups is a divisor of the order of S_3. Any intersection of the subgroups H, G_1, G_2, and G_3 is the subgroup $\{\sigma_0\}$.

By means of Cayley's table one finds out that S_3 has three distinct conjugacy classes

$$
\begin{aligned}
C_0 &= \{\sigma\sigma_0\sigma^{-1} | \sigma \in \mathsf{S}_3\} = \{\sigma_0\}\,, \\
C_1 &= \{\sigma\sigma_1\sigma^{-1} | \sigma \in \mathsf{S}_3\} = \{\sigma_1, \sigma_2\} = C_2\,, \\
C_3 &= \{\sigma\sigma_3\sigma^{-1} | \sigma \in \mathsf{S}_3\} = \{\sigma_3, \sigma_4, \sigma_5\} = C_4 = C_5\,,
\end{aligned}
\tag{1.1.38}
$$

which form a partition $C = \{C_0, C_1, C_2\}$ (see Definition 1.1.20), such that $\mathsf{S}_3 = C_0 \cup C_1 \cup C_2$. From the structure of these conjugacy classes we also deduce that $\mathsf{H} = C_0 \cup C_1$ is an invariant subgroup of S_3 (i.e., $\sigma \mathsf{H} \sigma^{-1} = \mathsf{H}$, $\forall \sigma \in \mathsf{S}_3$).

The conjugates of all the elements in S_3 by σ_4, are

$$
\{\sigma_4\sigma\sigma_4^{-1} | \sigma \in \mathsf{S}_3\} = \{\sigma_0, \sigma_2, \sigma_1, \sigma_5, \sigma_4, \sigma_3\} = \mathsf{S}_3',
$$

where we have shown the order of the conjugated elements. Hence, by this operation we obtain the whole group S_3 (with its elements reordered), and it is easy to show that S_3 and S_3' are isomorphic. In fact, this isomorphism is an inner automorphism induced by σ_4. There are other four inner automorphisms induced, respectively, by σ_1, σ_2, σ_3, and σ_5. In all these inner automorphisms, the elements of $\mathsf{H} = \{\sigma_0, \sigma_1, \sigma_2\}$ are transformed between them; hence, $\mathsf{H} \lhd \mathsf{S}_3$ is left invariant by all the inner automorphisms of S_3 (see Definition 1.1.18).

The left cosets of H are defined as $\sigma \mathsf{H} = \{\sigma\sigma' | \sigma' \in \mathsf{H}\}$, $\forall \sigma \in \mathsf{S}_3$. Similarly, the right cosets of H are $\mathsf{H}\sigma = \{\sigma'\sigma | \sigma' \in \mathsf{H}\}$, $\forall \sigma \in \mathsf{S}_3$. By inspecting Cayley's table one can see that the left and right cosets of H are identical (note that H is an invariant subgroup and $\sigma \mathsf{H} \sigma^{-1} = \mathsf{H}$, or $\sigma \mathsf{H} = \mathsf{H}\sigma$, $\forall \sigma \in \mathsf{S}_3$), and that there are only two distinct cosets (equivalence classes in this case, according to the Theorem 1.1.17):

$$
\begin{aligned}
K_1 &= \mathsf{H}\sigma_0 = \mathsf{H}\sigma_1 = \mathsf{H}\sigma_2 = \{\sigma_0, \sigma_1, \sigma_2\} = C_0 \cup C_1, \\
K_2 &= \mathsf{H}\sigma_3 = \mathsf{H}\sigma_4 = \mathsf{H}\sigma_5 = \{\sigma_3, \sigma_4, \sigma_5\} = C_2,
\end{aligned}
$$

which also form a partition of S_3.

By introducing the product of cosets $(Hx)(Hy) = H(xy), x, y \in S_3$, one can easily see that the set $\{K_1, K_2\}$ forms a group with the internal composition law given by Cayley's table

	K_1	K_2
K_1	K_1	K_2
K_2	K_2	K_1

This is the factor group $S_3/H = \{K_1, K_2\}$. The order of the factor group is the index of H in S_3 (denoted by $[S_3 : H] = 2$), and thus, the order of S_3 is the product of the order and index of H (see Lagrange's Theorem 1.1.18). The factor group S_3/H is isomorphic to the group S_2, that is the group of permutations of the subset $\{1, 2\} \subset \{1, 2, 3\} = M$, so that we obtain a relation between permutation groups of the third and second degree

$$S_3/H = S_2.$$

In relation to the trivial invariant subgroup $\{\sigma_0\}$, of order $n = 1$ and index $[S_3 : \{\sigma_0\}] = 6$, there are six cosets (equivalence classes), which coincide with the elements of S_3. Consequently, the corresponding factor group is

$$S_3/\{\sigma_0\} = S_3.$$

The other trivial invariant subgroup is S_3 itself, and it has the order $n = 6$ and the index $[S_3 : S_3] = 1$. Now, there is only one equivalence class $K_0 = S_3\sigma_0$, which coincides with S_3 itself, and the corresponding factor group is

$$S_3/S_3 = \{K_0\},$$

where K_0 plays the rôle of an identity element ($K_0K_0 = (S_3\sigma_0)(S_3\sigma_0) = S_3(\sigma_0\sigma_0) = S_3\sigma_0 = K_0$).

The centre of S_3

$$\mathcal{C}(S_3) = \{\sigma | \sigma \in S_3 \text{ and } \forall \sigma' \in S_3, \sigma\sigma' = \sigma'\sigma\} = \{\sigma_0\} \tag{1.1.39}$$

is identical to the trivial invariant subgroup $\{\sigma_0\}$.

1.4.2 Representations

The regular representation of S_3 comes out directly from Cayley's table. The method is to consider that the composition of two elements can be expressed as a linear combination of all the elements in the group (see Chap. 1 Sec. 1.3.5). For instance, we write the product $\sigma_3\sigma_1 = \sigma_4$ in the form

$$\sigma_3\sigma_1 = 0 \cdot \sigma_0 + 0 \cdot \sigma_1 + 0 \cdot \sigma_2 + 0 \cdot \sigma_3 + 1 \cdot \sigma_4 + 0 \cdot \sigma_5.$$

In an analogous manner, with the element σ_1 being fixed, we obtain other five relations

$$\begin{aligned}
\sigma_0\sigma_1 &= 0\cdot\sigma_0 + 1\cdot\sigma_1 + 0\cdot\sigma_2 + 0\cdot\sigma_3 + 0\cdot\sigma_4 + 0\cdot\sigma_5,\\
\sigma_1\sigma_1 &= 0\cdot\sigma_0 + 0\cdot\sigma_1 + 1\cdot\sigma_2 + 0\cdot\sigma_3 + 0\cdot\sigma_4 + 0\cdot\sigma_5,\\
\sigma_2\sigma_1 &= 1\cdot\sigma_0 + 0\cdot\sigma_1 + 0\cdot\sigma_2 + 0\cdot\sigma_3 + 0\cdot\sigma_4 + 0\cdot\sigma_5,\\
\sigma_4\sigma_1 &= 0\cdot\sigma_0 + 0\cdot\sigma_1 + 0\cdot\sigma_2 + 0\cdot\sigma_3 + 0\cdot\sigma_4 + 1\cdot\sigma_5,\\
\sigma_5\sigma_1 &= 0\cdot\sigma_0 + 0\cdot\sigma_1 + 0\cdot\sigma_2 + 1\cdot\sigma_3 + 0\cdot\sigma_4 + 0\cdot\sigma_5,
\end{aligned}$$

and associate to σ_1 the matrix of coefficients

$$\sigma_1 \to \mathbf{r}(\sigma_1) = \begin{bmatrix} 0 & 1 & 0 & 0 & 0 & 0\\ 0 & 0 & 1 & 0 & 0 & 0\\ 1 & 0 & 0 & 0 & 0 & 0\\ 0 & 0 & 0 & 0 & 1 & 0\\ 0 & 0 & 0 & 0 & 0 & 1\\ 0 & 0 & 0 & 1 & 0 & 0 \end{bmatrix}. \tag{1.1.40}$$

We continue this process, fixing one group element at a time, to obtain all the other matrices in the regular representation

$$\mathsf{R} = \{\mathbf{r}(\sigma)|\sigma \in \mathsf{S}_3\}. \tag{1.1.41}$$

The characters of a representation have been defined in (1.1.34), and in terms of the conjugacy classes (1.1.38), the characters of the regular representation (1.1.41) can be arranged in the form of a character table

Class	χ	n
C_0	6	1
C_1	0	2
C_2	0	3

where n represents the number of elements in each class.

To find the other representations of S_3, we take into account the fact that all permutation groups can be represented as groups of linear transformations. Thus, let us consider in the real three-dimensional space a point of coordinates (x_1, x_2, x_3) which, under the action of σ_1, is transformed into (x_1', x_2', x_3'), where $x_1' = x_3$, $x_2' = x_1$, and $x_3' = x_2$. This is the linear transformation

$$\begin{aligned}
x_1' &= 0\cdot x_1 + 0\cdot x_2 + 1\cdot x_3\,,\\
x_2' &= 1\cdot x_1 + 0\cdot x_2 + 0\cdot x_3\,,\\
x_3' &= 0\cdot x_1 + 1\cdot x_2 + 0\cdot x_3\,,
\end{aligned}$$

and we denote its matrix by $\mathbf{s}(\sigma_1)$. By applying the same method to all elements of S_3 we are led to the three-dimensional representation

$$\mathbf{s}(\sigma_0) = \begin{bmatrix} 1 & 0 & 0 \\ 0 & 1 & 0 \\ 0 & 0 & 1 \end{bmatrix}, \quad \mathbf{s}(\sigma_1) = \begin{bmatrix} 0 & 0 & 1 \\ 1 & 0 & 0 \\ 0 & 1 & 0 \end{bmatrix}, \quad \mathbf{s}(\sigma_2) = \begin{bmatrix} 0 & 1 & 0 \\ 0 & 0 & 1 \\ 1 & 0 & 0 \end{bmatrix},$$
$$\mathbf{s}(\sigma_3) = \begin{bmatrix} 1 & 0 & 0 \\ 0 & 0 & 1 \\ 0 & 1 & 0 \end{bmatrix}, \quad \mathbf{s}(\sigma_4) = \begin{bmatrix} 0 & 0 & 1 \\ 0 & 1 & 0 \\ 1 & 0 & 0 \end{bmatrix}, \quad \mathbf{s}(\sigma_5) = \begin{bmatrix} 0 & 1 & 0 \\ 1 & 0 & 0 \\ 0 & 0 & 1 \end{bmatrix}. \tag{1.1.42}$$

These matrices verify Cayley's table, the composition of two elements being the product of the corresponding matrices. We may obtain a new representation, equivalent to (1.1.42), in terms of the matrices $\mathbf{t}(\sigma) = \mathbf{A}\mathbf{s}(\sigma)\mathbf{A}^{-1}$, $\sigma \in \mathsf{S}_3$:

$$\begin{aligned}
\mathbf{t}(\sigma_0) &= \begin{bmatrix} 1 & 0 & 0 \\ 0 & 1 & 0 \\ 0 & 0 & 1 \end{bmatrix} = [1] \oplus \begin{bmatrix} 1 & 0 \\ 0 & 1 \end{bmatrix}, \\
\mathbf{t}(\sigma_1) &= \begin{bmatrix} 1 & 0 & 0 \\ 0 & -c & -s \\ 0 & s & -c \end{bmatrix} = [1] \oplus \begin{bmatrix} -c & -s \\ s & -c \end{bmatrix}, \\
\mathbf{t}(\sigma_2) &= \begin{bmatrix} 1 & 0 & 0 \\ 0 & -c & s \\ 0 & -s & -c \end{bmatrix} = [1] \oplus \begin{bmatrix} -c & s \\ -s & -c \end{bmatrix}, \\
\mathbf{t}(\sigma_3) &= \begin{bmatrix} 1 & 0 & 0 \\ 0 & -1 & 0 \\ 0 & 0 & 1 \end{bmatrix} = [1] \oplus \begin{bmatrix} -1 & 0 \\ 0 & 1 \end{bmatrix}, \\
\mathbf{t}(\sigma_4) &= \begin{bmatrix} 1 & 0 & 0 \\ 0 & c & s \\ 0 & s & -c \end{bmatrix} = [1] \oplus \begin{bmatrix} c & s \\ s & -c \end{bmatrix}, \\
\mathbf{t}(\sigma_5) &= \begin{bmatrix} 1 & 0 & 0 \\ 0 & c & -s \\ 0 & -s & -c \end{bmatrix} = [1] \oplus \begin{bmatrix} c & -s \\ -s & -c \end{bmatrix},
\end{aligned} \tag{1.1.43}$$

where $c = \cos(\pi/3) = 1/2$, $s = \sin(\pi/3) = \sqrt{3}/2$, and $\mathbf{A}$ is the orthogonal matrix

$$\mathbf{A} = \begin{bmatrix} 1/\sqrt{3} & 1/\sqrt{3} & 1/\sqrt{3} \\ 0 & -1/\sqrt{2} & 1/\sqrt{2} \\ \sqrt{2}/\sqrt{3} & -1/\sqrt{6} & -1/\sqrt{6} \end{bmatrix}, \quad \det \mathbf{A} = 1, \tag{1.1.44}$$

which determines the similarity transformation that simultaneously reduce the matrices (1.1.42) to a block diagonal form. The last equality in (1.1.43) points out the fact that the representation $\mathsf{T} = \{\mathbf{t}(\sigma)|\sigma \in \mathsf{S}_3\}$ is decomposed into a direct sum of two representations (one being one-dimensional and the other two-dimensional), that is

$$\mathsf{T} = \mathsf{T}_1^s \oplus \mathsf{T}_2,$$

where

$$\mathsf{T}_1^s = \{[1], [1], [1], [1], [1], [1]\}$$

is the trivial irreducible representation of degree 1. Another one-dimensional representation, non-equivalent to T^s, is given by the homomorphism (1.1.37)

$$\mathsf{T}_1^a = \{[1], [1], [1], [-1], [-1], [-1]\}.$$

Here, $[\pm 1]$ stands for a matrix with one single element, and the upper indices s (for symmetric) and a (for antisymmetric) have been introduced to distinguish the two one-dimensional representations.

The group S_3 has three distinct conjugacy classes (1.1.38), that is it has three irreducible representations. The regular representation (1.1.41) is six-dimensional, so that (1.1.25) becomes

$$6 = 1 + n_2^2 + n_3^2, \tag{1.1.45}$$

where n_2 and n_3 are integers, divisors of 6. Hence, the only solutions of (1.1.45) are $\{n_2 = 1, n_3 = 2\}$, or $\{n_2 = 2, n_3 = 1\}$, and we can say that the group S_3 has two distinct irreducible representations of degree $n_1 = n_2 = 1$ (or one-dimensional) and one irreducible representation of degree $n_3 = 2$ (or two-dimensional). Therefore, T_1^s, T_1^a, and T_2 exhausts the irreducible representations of S_3. The character table for these irreducible representations reads

Class	$\chi_s^{(1)}$	$\chi_a^{(1)}$	$\chi^{(2)}$	n
C_0	1	1	2	1
C_1	1	1	-1	2
C_2	1	-1	0	3

where n is the number of elements in each class. Now, we can use the orthogonality relations (1.1.36) to show that the decomposition of the regular representation of S_3 into irreducible representations, has the form

$$\mathsf{R} = \mathsf{T}_1^s \oplus \mathsf{T}_1^a \oplus 2\mathsf{T}_2, \tag{1.1.46}$$

in accordance to (1.1.45), for $n_2 = 1$ and $n_3 = 2$.

The most difficult part in this analysis of S_3 has been to find the matrix $\mathbf{A}$, used to reduce simultaneously all the matrices in (1.1.42) to a block diagonal form, by a similarity transformation. In fact, this is related to the main issue in the theory of group representations, namely the finding of all irreducible representations of a given group, and there are different methods to solve this problem, depending on the individual characteristics of each group. However, in some applications, we do not need the explicit form of the similarity transformation, but only a proof that a certain representation is fully reducible. Thus, in Chap. 4 Sec. 4.2.2 we will present a method, based on transformations that leave invariant an equilateral triangle (in a two-dimensional space), which enables us to find directly the irreducible representation T_2 of S_3, and to construct the block diagonal form (1.1.46) of the regular representation.

2. Topological groups

2.1 Definitions. Generalities. Lie groups

2.1.1 Topological space. Topological group

DEFINITION 1.2.1 *It is called* **topological space** *any set $\mathcal{L}$ comprising a collection of subsets, called* **vicinities**, *that satisfies the axioms:*

1) *every point $\mathbf{x} \in \mathcal{L}$ is also contained in a vicinity $V(\mathbf{x})$;*

2) *every subset including $V(\mathbf{x})$ is also a vicinity of $\mathbf{x}$;*

3) *$V(\mathbf{x})$ is also a vicinity for another point $\mathbf{y} \in \mathcal{L}$, sufficiently close to $\mathbf{x}$.*

An example of topological space is the Euclidean n-dimensional space. In this case, the vicinity of a point $\mathbf{x}$ can be regarded as the set of points inside a sphere of radius R centred at $\mathbf{x}$, that is the subset of points $\mathbf{y}$ satisfying the condition $\|\mathbf{x} - \mathbf{y}\| \leq R$.

A topological space is called *compact* in the limit point sense, if each sequence of its points contains a sequence which converges towards a point of this space, that is each infinite set of points has a limit point. A subset $\mathcal{M} \subset \mathcal{L}$ is called *compact* in $\mathcal{L}$ if any sequence of points in $\mathcal{M}$ contains a sequence converging towards a point in $\mathcal{L}$. If this condition is fulfilled, and the limit point belongs to the subset $\mathcal{M}$, then $\mathcal{M}$ is said to be *compact in itself.* For instance, the real line is not compact, because it contains at least one infinite and non-convergent sequence of points. On the other hand, any finite interval of the real line is compact, because each infinite sequence of its points has a limit point contained in this interval. In the same manner, one can prove that each subset of the Euclidean n-dimensional space, defined by $\|\mathbf{x}\| \leq M$ (M finite) is compact.

DEFINITION 1.2.2 *It is called* **topological group** *(or* **continuous group**) *a group G, that also forms a topological space, the group and topological structures being compatible.*

Let us denote by "$\circ$" the internal composition law in G. The compatibility of the group and topological structures entails that:

1) if $\mathbf{x}, \mathbf{y} \in G$, then for any vicinity W of the product $\mathbf{x} \circ \mathbf{y}$ there exist vicinities X and Y such that, $\mathbf{x} \in X$ and $\mathbf{y} \in Y$ imply $\mathbf{x} \circ \mathbf{y} \in W$;

2) for any vicinity Z of the inverse element $\mathbf{x}^{-1}$ there exists a vicinity X such that $\mathbf{x} \in X$ implies $\mathbf{x}^{-1} \in Z$.

Therefore, the product $\mathbf{x} \circ \mathbf{y}$ is a continuous function of $\mathbf{x}$ and $\mathbf{y}$, and $\mathbf{x}^{-1}$ is a continuous function of $\mathbf{x}$.

A topological group can be *compact* or *noncompact*. For instance, the group of translations on the real line is a noncompact Abelian group. In contrast, the group of rotations about a fixed axis is a compact Abelian group, because any rotation is characterized by an angle θ in the interval $[0, 2\pi)$, which is a compact set. The group of proper rotations in the Euclidean three-dimensional space (denoted by $\mathsf{SO}(3)$) is compact, because the Euler angles $\{\psi, \theta, \varphi\}$, which determine the rotations, are defined on compact sets: $\psi \in [0, 2\pi)$, $\theta \in [0, \pi]$, and $\varphi \in [0, 2\pi)$.

2.1.2 The invariant integral

THEOREM 1.2.1 *If* $\{g_1 = e, g_2, \ldots, g_n\}$ *are the elements of a finite group G, and* $g_k \in G$ *is an arbitrary element, then each element of G is contained once and only once in the set* $Gg_k = \{eg_k = g_k, g_2 g_k, \ldots, g_n g_k\}$.

The same property is exhibited by the set $g_k G = \{g_k, g_k g_2, \ldots, g_k g_n\}$. Actually, Gg_k and $g_k G$ are the right and left cosets of the trivial invariant subgroup $G \lhd G$, and any two right (or left) cosets are either disjoint or exactly the same. Also, the Theorem 1.2.1 expresses the fact that each element of G appears once and only once in each column and each row in Cayley's table. The simplest and at the same time the most important consequence of this theorem is the following: if $f : G \to \mathbb{R}$ is a numerical function on G (i.e., $\{f(e), f(g_2), \ldots, f(g_n)\}$ are numbers such that to each element $g_k \in G$ it corresponds the number $f(g_k)$), then

$$\sum_{i=1}^{n} f(g_i) = \sum_{i=1}^{n} f(g_i g_j) = \sum_{i=1}^{n} f(g_j g_i), \quad \forall g_j \in G. \tag{1.2.1}$$

Indeed, from the Theorem 1.2.1 it follows that each sum in (1.2.1) contains precisely the same terms, but in a different order.

In the case of topological groups, their elements are characterized by parameters which vary continuously in a certain domain, called the *group space*. Each point (set of parameter values) in the group space defines an element in the group and, reciprocally, to each element in the group it corresponds a point in the group space, such that there is a one-to-one correspondence between the group elements and the points in the group space.

Let G be a topological group whose elements are determined by a finite set of n parameters (*finite topological group*), and let $\alpha_1(g), \alpha_2(g), \ldots, \alpha_n(g)$ be the parameters characterizing an element $g \in G$. We suppose that the parameters $\alpha_1(gh), \alpha_2(gh), \ldots, \alpha_n(gh)$, associated to the product $gh \in G$, are at least piecewise functions of class C^1 of the parameters $\alpha_1(g), \alpha_2(g), \ldots, \alpha_n(g)$ and $\alpha_1(h), \alpha_2(h), \ldots, \alpha_n(h)$, associated to the two factors g and h, respectively. We also suppose that the same property holds for the parameters $\alpha_1(g^{-1}), \alpha_2(g^{-1}), \ldots, \alpha_n(g^{-1})$, as functions of the parameters of g. The group space can be a simple or multiple connected domain, or the union of several disjoint domains. The topological group corresponding to this latter case is called a *mixed topological group*, in contrast to the *simple topological group* whose group space is a simple connected domain.

For topological groups, the first sum in (1.2.1) is replaced by an integral of the form

$$\int f(g)dg = \int f(g)\rho(g)d\alpha_1 d\alpha_2 \ldots d\alpha_n \,, \tag{1.2.2}$$

where $\alpha_1, \alpha_2, \ldots, \alpha_n$ are the parameters of the group element g, and the integral extends over the whole group space. The function $\rho(g)$ represents the *density* in the vicinity of g, in the group space. The integral (1.2.2) is called the *Hurwitz integral* or the *invariant integral* in the group space. In this case, the relation (1.2.1) becomes

$$\int f(g)dg = \int f(g_0 g)dg = \int f(g g_0)dg \,, \tag{1.2.3}$$

where g_0 is a fixed element. If the topological group is compact, then the invariant integral exists for any function of bounded variation in the group space, unlike the case of non-compact groups for which this integral does not exist.

2.1.3 Lie groups

Let $\mathbf{x} = (x_1, x_2, \ldots, x_n)$ be a vector in an n-dimensional linear space $\mathcal{L}$, and let

$$x_i' = f_i(x_1, x_2, \ldots, x_n; a_1, a_2, \ldots, a_m) \,, \quad i = 1, 2, \ldots, n, \tag{1.2.4}$$

be a set of transformations depending on m real and continuous independent parameters $\{a_1, a_2, \ldots, a_m\}$. We admit that f_i are analytical functions of the parameters a_j such that, the set of values $a_1 = a_2 = \ldots = a_m = 0$ defines the identity transformation

$$x_i = f_i(x_1, x_2, \ldots, x_n; 0, 0, \ldots, 0)\,, \quad i = 1, 2, \ldots, n\,. \tag{1.2.5}$$

We also suppose that there exists a set of values of the parameters a_j such that

$$x_i = f_i(x'_1, x'_2, \ldots, x'_n; \widetilde{a}_1, \widetilde{a}_2, \ldots, \widetilde{a}_m)\,, \quad i = 1, 2, \ldots, n\,, \tag{1.2.6}$$

that is the inverse of the transformation (1.2.4).

Let us now consider the effect of two successive transformations

$$\begin{aligned} x''_i &= f_i(x'_1, x'_2, \ldots, x'_n; a'_1, a'_2, \ldots, a'_m) \\ &= f_i(f_1(x_1, x_2, \ldots, x_n; a_1, a_2, \ldots, a_m), \ldots; a'_1, a'_2, \ldots, a'_m)\,, \end{aligned}$$

for $i = 1, 2, \ldots, n$, and let us admit that there exists a set of values of the parameters such that

$$x''_i = f_i(x_1, x_2, \ldots, x_n; a''_1, a''_2, \ldots, a''_m)\,, \quad i = 1, 2, \ldots, n\,. \tag{1.2.7}$$

In the frame of these hypotheses, the set of transformations (1.2.4) forms a finite topological group with m parameters, and it is evident that relations of the form

$$a''_j = \varphi_j(a_1, a_2, \ldots, a_m; a'_1, a'_2, \ldots, a'_m)\,, \quad j = 1, 2, \ldots, m \tag{1.2.8}$$

hold. If φ_j are analytical functions of the arguments a_k and a'_l, then we say that the transformations (1.2.4) form a *Lie group*. If m is the smallest number of independent parameters required to characterize the elements of a Lie group, then these parameters are called *essential parameters* and m is the *order* or the *dimension of the group*. In what follows we will assume that the parameters of a considered group are essential.

Let $\mathbf{x} = \mathbf{f}(\mathbf{x}^0; \mathbf{a})$ be a transformation of the form (1.2.4), where $\mathbf{a} = \{a_1, a_2, \ldots, a_m\}$. At the same time, from (1.2.5) it results that $\mathbf{x} = \mathbf{f}(\mathbf{x}; \mathbf{0})$. Now, an infinitesimal displacement of the point $\mathbf{x}$ is a differential displacement that can be obtained in two ways: as the result of two successive transformations, $\mathbf{x} = \mathbf{f}(\mathbf{x}^0; \mathbf{a})$ followed by $\mathbf{x} + d\mathbf{x} = \mathbf{f}(\mathbf{x}; \delta\mathbf{a})$, or directly from $\mathbf{x} + d\mathbf{x} = \mathbf{f}(\mathbf{x}^0; \mathbf{a} + d\mathbf{a})$;. Taking into account (1.2.8) the considered parameters are connected by a relation of the form $\mathbf{a} + d\mathbf{a} = \varphi(\mathbf{a}; \delta\mathbf{a})$. Because we have considered infinitesimal displacements, we may retain only terms of the first order in a Taylor series of $\mathbf{f}$. Thus, we obtain

$$d\mathbf{x} = \sum_{k=1}^{m} \left[\frac{\partial \mathbf{f}(\mathbf{x}; \mathbf{a})}{\partial a_k} \right]_{\mathbf{a}=\mathbf{0}} \delta a_k$$

in the first case, where $\mathbf{a} = \mathbf{0}$ stands for $a_1 = a_2 = \ldots = a_m = 0$, and

$$d\mathbf{x} = \sum_{k=1}^{m} \frac{\partial \mathbf{f}(\mathbf{x}^0; \mathbf{a})}{\partial a_k}\, da_k$$

in the second case. Then, by introducing the notation

$$u_k^i(\mathbf{x}) = \left[\frac{\partial f_i(\mathbf{x};\mathbf{a})}{\partial a_k}\right]_{\mathbf{a}=\mathbf{0}}, \quad i = 1,2,\ldots,n, \quad k = 1,2,\ldots,m, \quad (1.2.9)$$

we can write

$$dx_i = \sum_{k=1}^{m} u_k^i(\mathbf{x})\delta a_k, \quad i = 1,2,\ldots,n. \quad (1.2.10)$$

As the functions φ are analytical, and $\varphi(\mathbf{a};\mathbf{0}) = \mathbf{a}$, we may use a Taylor series expansion to obtain

$$\mathbf{a} + d\mathbf{a} = \mathbf{a} + \sum_{k=1}^{m} \left[\frac{\partial \varphi(\mathbf{a};\mathbf{a}')}{\partial a'_k}\right]_{\mathbf{a}'=\mathbf{0}} \delta a_k. \quad (1.2.11)$$

Hence, da_j is a linear combination of δa_k, that can be written in the form

$$da_j = \sum_{k=1}^{m} \mu_k^j(\mathbf{a})\delta a_k, \quad j = 1,2,\ldots,m, \quad (1.2.12)$$

where

$$\mu_k^j(\mathbf{a}) = \left[\frac{\partial \varphi_j(\mathbf{a};\mathbf{a}')}{\partial a'_k}\right]_{\mathbf{a}'=\mathbf{0}}, \quad j,k = 1,2,\ldots,m. \quad (1.2.13)$$

The inverse relations of (1.2.12) have the form

$$\delta a_k = \sum_{j=1}^{m} \lambda_j^k(\mathbf{a})da_j, \quad k = 1,2,\ldots,m, \quad (1.2.14)$$

where $\boldsymbol{\lambda}\boldsymbol{\mu} = \mathbf{1}$ (or $\sum_{j=1}^{m} \lambda_j^i \mu_k^j = \delta_k^i$, δ_k^i being Kronecker's symbol). Finally, from (1.2.10) and (1.2.14) we derive the differential equations of the Lie group

$$\frac{\partial x_i}{\partial a_j} = \sum_{k=1}^{m} u_k^i(\mathbf{x})\lambda_j^k(\mathbf{a}), \quad i = 1,2,\ldots,n, \quad j = 1,2,\ldots,m. \quad (1.2.15)$$

This system of differential equations must be completely integrable, such that all the conditions

$$\frac{\partial^2 x_i}{\partial a_j \partial a_k} = \frac{\partial^2 x_i}{\partial a_k \partial a_j}$$

have to be identically satisfied. Now, from (1.2.15) we obtain

$$\frac{\partial^2 x_i}{\partial a_l \partial a_j} = \sum_{s=1}^{n} \sum_{k,p=1}^{m} u_p^s \frac{\partial u_k^i}{\partial x_s} \lambda_l^p \lambda_j^k + \sum_{k=1}^{m} u_k^i \frac{\partial \lambda_j^k}{\partial a_l},$$

where we have omitted the arguments of u and λ, and consequently the integrability conditions of (1.2.15) become

$$\sum_{s=1}^{n}\sum_{p,r=1}^{m}\left(u_p^s\frac{\partial u_r^i}{\partial x_s}-u_r^s\frac{\partial u_p^i}{\partial x_s}\right)\lambda_l^p\lambda_j^r+\sum_{r=1}^{m}u_r^i\left(\frac{\partial\lambda_j^r}{\partial a_l}-\frac{\partial\lambda_l^r}{\partial a_j}\right)=0\,.$$

The orthogonality relations satisfied by the quantities λ and μ allow us to rewrite these last equations in the form

$$\sum_{s=1}^{n}\left(u_p^s\frac{\partial u_r^i}{\partial x_s}-u_r^s\frac{\partial u_p^i}{\partial x_s}\right)=\sum_{k=1}^{m}c_{pr}^k(\mathbf{a})u_k^i\,,\qquad(1.2.16)$$
$$i=1,2,\ldots,n,\quad p,r=1,2,\ldots,m,$$

where

$$c_{pr}^k(\mathbf{a})=\sum_{l,j=1}^{m}\left(\frac{\partial\lambda_l^k}{\partial a_j}-\frac{\partial\lambda_j^k}{\partial a_l}\right)\mu_p^l\mu_r^j\,,\quad k,p,r=1,2,\ldots,m.\qquad(1.2.17)$$

The functions u are independent of $\mathbf{a}$ (they only depend on $\mathbf{x}$), and thus, by differentiating (1.2.16) with respect to a_j, we are led to

$$\sum_{k=1}^{m}\frac{\partial c_{pr}^k}{\partial a_j}u_k^i=0\,,\quad i=1,2,\ldots,n\,,\quad j,p,r=1,2,\ldots,m\,.$$

We have assumed that the parameters a_j are essential, hence the functions u_k^i are linearly independent, and it results that $\partial c_{pr}^k/\partial a_j=0$; therefore, the quantities c_{pr}^k are constants. Finally, (1.2.16) and (1.2.17) can be rewritten in the form

$$\begin{gathered}\sum_{s=1}^{n}\left(u_p^s\frac{\partial u_r^i}{\partial x_s}-u_r^s\frac{\partial u_p^i}{\partial x_s}\right)=\sum_{k=1}^{m}c_{pr}^k u_k^i\,,\\ \frac{\partial\lambda_l^k}{\partial a_j}-\frac{\partial\lambda_j^k}{\partial a_l}=\sum_{p,r=1}^{m}c_{pr}^k\lambda_l^p\lambda_j^r\,.\end{gathered}\qquad(1.2.18)$$

Let $F(\mathbf{x})$ be a function of class C^2. The variation of F induced by an infinitesimal transformation of the point $\mathbf{x}$ (corresponding to an infinitesimal displacement) has the form

$$\begin{aligned}dF(\mathbf{x})&=\sum_{i=1}^{n}\frac{\partial F}{\partial x_i}dx_i=\sum_{i=1}^{n}\sum_{j=1}^{m}u_j^i(\mathbf{x})\frac{\partial F}{\partial x_i}\delta a_j\\ &\equiv\sum_{j=1}^{m}\mathsf{X}_jF(\mathbf{x})\,\delta a_j=\sum_{j=1}^{m}\frac{\partial F}{\partial a_j}\delta a_j\,.\end{aligned}\qquad(1.2.19)$$

The operators

$$\mathsf{X}_j = \sum_{i=1}^{n} u_j^i(\mathbf{x}) \frac{\partial}{\partial x_i}, \quad j = 1, 2, \ldots, m, \tag{1.2.20}$$

are called the *infinitesimal generators* of the group of transformations. From (1.2.19) we observe that the infinitesimal generators can also be expressed in the form

$$\mathsf{X}_j = \frac{\partial}{\partial a_j}, \quad j = 1, 2, \ldots, m, \tag{1.2.21}$$

and sometimes it is convenient to define the infinitesimal generators in the form

$$\mathsf{X}_j = K \frac{\partial}{\partial a_j}, \quad j = 1, 2, \ldots, m, \tag{1.2.22}$$

where K is a complex constant.

Now, we consider the case when the product of two infinitesimal generators is applied to the function $F(\mathbf{x})$, that is

$$\begin{aligned} \mathsf{X}_p \mathsf{X}_r F(\mathbf{x}) &= \sum_{s=1}^{n} u_s^p \frac{\partial}{\partial x_s} \sum_{i=1}^{n} u_r^i \frac{\partial}{\partial x_i} F \\ &= \sum_{i,s=1}^{n} u_p^s \frac{\partial u_r^i}{\partial x_s} \frac{\partial F}{\partial x_i} + \sum_{i,s=1}^{n} u_p^s u_r^i \frac{\partial^2 F}{\partial x_s \partial x_i}, \quad p, r = 1, 2, \ldots, m, \end{aligned}$$

where we have omitted the arguments of u and F. By virtue of Schwarz's theorem, which ensures that the mixed derivatives of the second order of a function of class C^2 do not depend on the order of differentiation, and by means of (1.2.18) we obtain

$$\begin{aligned} (\mathsf{X}_p \mathsf{X}_r - \mathsf{X}_r \mathsf{X}_p) F &= \sum_{i,s=1}^{n} \left(u_p^s \frac{\partial u_r^i}{\partial x_s} - u_r^s \frac{\partial u_p^i}{\partial x_s} \right) \frac{\partial F}{\partial x_i} = \sum_{i=1}^{n} \sum_{k=1}^{m} c_{pr}^k u_k^i \frac{\partial F}{\partial x_i} \\ &= \sum_{k=1}^{m} c_{pr}^k \sum_{i=1}^{n} u_k^i \frac{\partial}{\partial x_i} F = \sum_{k=1}^{m} c_{pr}^k \mathsf{X}_k F. \end{aligned}$$

Therefore, the infinitesimal generators of the group of transformations satisfy the relations

$$\mathsf{X}_p \mathsf{X}_r - \mathsf{X}_r \mathsf{X}_p \equiv [\mathsf{X}_p, \mathsf{X}_r] = \sum_{k=1}^{m} c_{pr}^k \mathsf{X}_k \quad p, r = 1, 2, \ldots, m. \tag{1.2.23}$$

The square bracket thus introduced is called *Poisson's bracket* or the *commutator* of the operators X_p and X_r. The quantities c_{pr}^k are called the *structure constants* of the group, and the property of skew-symmetry

$$[\mathsf{X}_p, \mathsf{X}_r] = -[\mathsf{X}_r, \mathsf{X}_p] \tag{1.2.24}$$

is reflected in the structure constants under the form

$$c_{pr}^{k} = -c_{rp}^{k} . \tag{1.2.25}$$

Also, from Jacobi's identity satisfied by Poisson's brackets

$$[\mathsf{X}_i, [\mathsf{X}_j, \mathsf{X}_k]] + [\mathsf{X}_j, [\mathsf{X}_k, \mathsf{X}_i]] + [\mathsf{X}_k, [\mathsf{X}_i, \mathsf{X}_j]] = 0 , \tag{1.2.26}$$

which can easily be verified, we obtain the relation

$$\sum_{l=1}^{m} \left(c_{il}^{s} c_{jk}^{l} + c_{jl}^{s} c_{ki}^{l} + c_{kl}^{s} c_{ij}^{l} \right) = 0 , \quad i, j, k, s = 1, 2, \ldots, m. \tag{1.2.27}$$

The structure constants of every Lie group are uniquely determined by the group structure.

THEOREM 1.2.2 (LIE, I) *If the set of transformations (1.2.4), containing the identity transformation, satisfies the differential equations (1.2.15), then this set forms a group of one-parameter infinitesimal transformations, each of them being defined by the infinitesimal operator*

$$\alpha_1 \mathsf{X}_1 + \alpha_2 \mathsf{X}_2 + \ldots + \alpha_m \mathsf{X}_m , \quad \mathsf{X}_i = \sum_{j=1}^{n} u_i^j \frac{\partial}{\partial x_j} ,$$

where $\alpha_1, \alpha_2, \ldots, \alpha_m$ *are constants.*

The fact that a Lie group with m parameters can be generated by one-parameter groups, whose operators are arbitrary linear combinations of the infinitesimal generators X_i, leads to the theorems:

THEOREM 1.2.3 (LIE, II) *A Lie group can be defined by means of* m *infinitesimal generators* X_i*, that satisfy the relations (1.2.23), where* c_{pr}^{k} *are constants.*

THEOREM 1.2.4 (LIE, III) *Being given a system of* $m^2(m-1)/2$ *constants* $c_{pr}^{k} = -c_{rp}^{k}$*, which satisfy (1.2.27), there exists a Lie group* G *having these constants as structure constants.*

Any other Lie group G' with the same structure constants c_{pr}^{k} is locally isomorphic to G.

A comparison of (1.2.9) and (1.2.13) shows a formal analogy between the functions μ_k^j and u_m^l. In fact, the relations (1.2.8), which connect the parameters of different transformations of the form (1.2.4), define a group in the same manner as (1.2.4). Hence, the relation (1.2.8) can be regarded as a transformation of the parameters $\mathbf{a}$ into $\mathbf{a}''$, the parameters of the transformation being $\mathbf{a}'$.

The group defined by (1.2.8) is called the *parameter group* and it is isomorphic to the group defined by (1.2.4). Consequently, all the theorems previously enounced for the Lie groups also hold for the parameter group, the infinitesimal generators of the latter being of the form (1.2.21) or (1.2.22).

THEOREM 1.2.5 (ADO) *Any Lie group admits a linear representation.*

From this theorem it follows that any Lie group is isomorphic to a group of linear transformations in a linear space, called the *representation space*. Therefore, each element of a Lie group can be associated to a linear operator acting in such a space. If the Lie group is of order m, then this operator depends on m real parameters $a_1, a_2, \ldots, a_m$, and it is denoted as

$$\mathsf{g} = \mathsf{T}(a_1, a_2, \ldots, a_m)\,, \tag{1.2.28}$$

the identity element being $\mathsf{e} = \mathsf{T}(0, 0, \ldots, 0)$. If we introduce a basis in the representation space, assumed to be n-dimensional, then each operator $\mathsf{T}(a_1, a_2, \ldots, a_m)$ can be expressed by a matrix having as elements the functions $t_{ij}(a_1, a_2, \ldots, a_m)$, $i, j = 1, 2, \ldots, n$. The components of a vector in the representation space are transformed by the operator $\mathsf{T}(a_1, a_2, \ldots, a_m)$ according to the relations

$$x'_i = \sum_{j=1}^{n} t_{ij}(a_1, a_2, \ldots, a_m)\, x_j\,, \quad i = 1, 2, \ldots, n\,. \tag{1.2.29}$$

We assume that all the functions $t_{ij}\,(a_1, a_2, \ldots, a_m)$ are of class C^2. In this case, we may derive from (1.2.29) the corresponding infinitesimal transformation

$$dx_i = \sum_{j=1}^{n}\sum_{k=1}^{m}\left(\frac{\partial t_{ij}}{\partial a_k}\right)_{\mathbf{a}=\mathbf{0}} x_j \delta a_k = \sum_{k=1}^{m}\sum_{j=1}^{n}\left(\frac{\partial t_{ij}}{\partial a_k}\right)_{\mathbf{a}=\mathbf{0}} x_j \delta a_k\,,$$

or, in matrix notation,

$$d\mathbf{x} = \sum_{k=1}^{m}\left(\frac{\partial \mathbf{t}}{\partial a_k}\right)_{\mathbf{a}=\mathbf{0}} \mathbf{x}\,\delta a_k\,. \tag{1.2.30}$$

Now, by comparing (1.2.30) and (1.2.19) we find that the *matrices of the infinitesimal generators* have the form

$$\mathbf{X}_k = \left(\frac{\partial \mathbf{t}}{\partial a^k}\right)_{\mathbf{a}=\mathbf{0}}\,, \quad k = 1, 2, \cdots, m\,. \tag{1.2.31}$$

For example, each element of the group of three-dimensional rotations ($\mathsf{O}(3)$) can be represented as an orthogonal matrix of the third order, depending on three parameters. Instead of Euler's angles, we choose these parameters as follows: to each rotation we associate a vector directed to the positive sense of the rotation axis, having the modulus equal to the absolute value of the rotation angle. The projections ξ_1, ξ_2, and ξ_3 of this vector onto the coordinate axes Ox_1, Ox_2, and Ox_3 (mutually orthogonal) will be the essential parameters of the rotation. Thus, according to (1.2.31) the matrix of the infinitesimal generator corresponding to the parameter ξ_1 will be

$$\mathbf{X}_1 = \left.\frac{\partial \mathbf{t}(\xi_1, 0, 0)}{\partial \xi_1}\right|_{\xi_1 = 0} .$$

But, $\mathbf{t}(\xi_1, 0, 0)$ is the matrix which represents a rotation of angle ξ_1 about the axis Ox_1 in a counter-clockwise direction when looking towards the origin, that is

$$\mathbf{t}(\xi_1, 0, 0) = \begin{bmatrix} 1 & 0 & 0 \\ 0 & \cos\xi_1 & -\sin\xi_1 \\ 0 & \sin\xi_1 & \cos\xi_1 \end{bmatrix},$$

so that

$$\mathbf{X}_1 = \begin{bmatrix} 0 & 0 & 0 \\ 0 & 0 & -1 \\ 0 & 1 & 0 \end{bmatrix}. \tag{1.2.32}$$

Analogously, for

$$\mathbf{t}(0, \xi_2, 0) = \begin{bmatrix} \cos\xi_2 & 0 & \sin\xi_2 \\ 0 & 1 & 0 \\ -\sin\xi_2 & 0 & \cos\xi_2 \end{bmatrix}, \quad \mathbf{t}(0, 0, \xi_3) = \begin{bmatrix} \cos\xi_3 & -\sin\xi_3 & 0 \\ \sin\xi_3 & \cos\xi_3 & 0 \\ 0 & 0 & 1 \end{bmatrix},$$

we obtain the infinitesimal generators

$$\mathbf{X}_2 = \begin{bmatrix} 0 & 0 & 1 \\ 0 & 0 & 0 \\ -1 & 0 & 0 \end{bmatrix}, \quad \mathbf{X}_3 = \begin{bmatrix} 0 & -1 & 0 \\ 1 & 0 & 0 \\ 0 & 0 & 0 \end{bmatrix}. \tag{1.2.33}$$

We observe that these infinitesimal generators satisfy the relations

$$\mathbf{X}_k \mathbf{X}_l - \mathbf{X}_l \mathbf{X}_k = \epsilon_{klm} \mathbf{X}_m , \quad k, l = 1, 2, 3, \tag{1.2.34}$$

where we have used *Einstein's summation convention for dummy indices*, and ϵ_{klm} represents *Ricci's symbol*

$$\epsilon_{klm} = \begin{cases} +1 \text{ if } (k, l, m) \in \{(1,2,3), (2,3,1), (3,1,2)\}, \\ -1 \text{ if } (k, l, m) \in \{(1,3,2), (3,2,1), (2,1,3)\}, \\ 0 \text{ otherwise}. \end{cases} \tag{1.2.35}$$

Note that, in this case, ϵ_{klm} are the structure constants of the $\mathsf{O}(3)$ group. The Ricci symbol can also be defined as the triple scalar product of axis unit vectors in a Cartesian right-handed coordinate system

$$\epsilon_{ijk} = \widehat{\mathbf{x}}_i \cdot (\widehat{\mathbf{x}}_j \times \widehat{\mathbf{x}}_k)\,,$$

and it satisfies the relations

$$\begin{aligned} \delta_{ij}\,\epsilon_{ijk} &= 0\,, \\ \epsilon_{ikl}\,\epsilon_{jkl} &= 2\delta_{ij}\,, \\ \epsilon_{ijk}\,\epsilon_{ijk} &= 6\,, \\ \epsilon_{ijk}\,\epsilon_{lmk} &= \delta_{il}\delta_{jm} - \delta_{im}\delta_{jl}\,, \end{aligned} \tag{1.2.36}$$

where δ_{ij} is the Kronecker symbol.

2.2 Lie algebras. Unitary representations

2.2.1 Lie algebras

DEFINITION 1.2.3 *A set L is called an* **algebra** *onto a field of complex numbers* $\mathbb{C}$ *if it has the properties:*

1) *$\forall\lambda \in \mathbb{C}$ and $\forall x \in L$ there is a well-defined element called the* **product** *of λ and x, and denoted by $\lambda x \in L$;*

2) *for each pair $x, y \in L$ there exists a uniquely determined element called the* **sum** *of x and y, denoted by $x + y \in L$;*

3) *there exists an operation which associates to each pair $x, y \in L$ an element called the* product *of x and y, denoted by $xy \in L$;*

4) *the product of elements* 3) *and the multiplication by a scalar* 1) *are* **distributive with respect to addition**, *and the multiplication by a scalar also satisfies the* **associativity** *relations*

$$\lambda(xy) = (\lambda x)y = x(\lambda y)\,, \quad \forall x, y \in L\,, \quad \forall\lambda \in \mathbb{C}\,.$$

DEFINITION 1.2.4 *An algebra L onto a complex field* $\mathbb{C}$ *is called a* **Lie algebra** *if, in addition to the properties* 1) - 4), *it also has the property*

5) *$xy + yx = 0$ and $x(yz) + y(zx) + z(xy) = 0$, $\forall x, y, z \in L$.*

DEFINITION 1.2.5 *It is called* **finite-dimensional Lie algebra**, *a Lie algebra L for which there exists a collection of independent elements $X_1, X_2, \ldots, X_m \in L$, called the basis of the algebra, such that any other element of L is a linear combination of these elements.*

The number m is called the *dimension of the algebra*. From the properties 1) - 4) in the Definition 1.2.3, it results that the product of two basis elements can be expressed in the form

$$X_iX_j = \sum_{k=1}^{m} c_{ij}^k X_k\,, \quad i,j = 1,2,\ldots,m\,. \tag{1.2.37}$$

Also, from the property 5) in the Definition 1.2.4, it follows that the constants $c_{ij}^k \in \mathbb{C}$ satisfy the relations (1.2.25) and (1.2.27). For instance, the algebra with the basis formed by *Pauli's matrices* $\boldsymbol{\sigma}_1$, $\boldsymbol{\sigma}_2$, and $\boldsymbol{\sigma}_3$ (which are the infinitesimal generators of the $\mathsf{SU}(2)$ group), and the product of elements defined by the commutator $\boldsymbol{\sigma}_1\boldsymbol{\sigma}_2 \equiv [\boldsymbol{\sigma}_1, \boldsymbol{\sigma}_2]$, is a Lie algebra. The infinitesimal generators of any Lie group form a Lie algebra, in which the product of two elements is the commutator of infinitesimal generators. The Lie algebra of the infinitesimal generators of a Lie group is completely determined by this group; the inverse is true up to an isomorphism.

A Lie algebra is called *commutative* if the product of any pair of elements x and y, is $xy = 0$.

DEFINITION 1.2.6 *If a Lie algebra L contains a subset $L' \subset L$ which forms a Lie algebra by itself, then L' is called a* **subalgebra** *of the Lie algebra L.*

An essential property of the subalgebra L' is that $\forall x, y \in L'$ we have $xy = [x,y] \in L'$, which is symbolically denoted by $[L', L'] \subset L'$.

DEFINITION 1.2.7 *If I is a subalgebra of a Lie algebra L, such that $\forall x \in I$ and $\forall y \in L$ we have $[x,y] \in I$ (symbolically denoted by $[I, L] \subset I$), then I is called an* **ideal** *of the Lie algebra L.*

There is a close connection between the subgroups of a Lie group and the subalgebras of the corresponding Lie algebra, and also between the invariant subgroups and ideals. Thus, if G is a Lie group and L is the Lie algebra of the infinitesimal generators of G, then the Lie algebra L' of the subgroup $G' \subset G$ is a subalgebra of L, and the algebra of the invariant subgroup $H \lhd G$ is an ideal of L. In analogy to the denominations of *simple group* (it has no invariant subgroups other than itself and $\{e\}$) and *semisimple group* (it has no Abelian invariant subgroup), a Lie algebra is called *simple* if it has no ideal other than itself, and *semisimple* if it has no commutative ideal. The Lie algebra of a simple Lie group is simple and that of a semisimple Lie group is semisimple. Let G be a Lie group of finite dimension m, and L the m-dimensional Lie algebra of the infinitesimal generators of G. If there exist commutative elements in L, then these elements form a *commutative Lie subalgebra*.

DEFINITION 1.2.8 *The maximal commutative subalgebra of a Lie algebra is called the* **Cartan subalgebra**.

Let l be the dimension of the Cartan subalgebra. The elements of the basis of the Cartan subalgebra are denoted by H_i, and it is evident that $[H_i, H_j] = 0$, for $i, j = 1, 2, \ldots, l$. The dimension l of the Cartan subalgebra of the Lie algebra L is called the *rank of the algebra* L or the *rank of the Lie group* G, whose algebra is L.

2.2.2 Unitary representations

Let $\mathcal{H}$ be a Hilbert space. Hence, to every pair of vectors $\mathbf{x}, \mathbf{y} \in \mathcal{H}$ it corresponds a complex number called *scalar product* (or *inner product*), denoted by $< \mathbf{x}, \mathbf{y} >$, that satisfies the conditions:

a) $< \mathbf{x}, \mathbf{y} >= \overline{< \mathbf{x}, \mathbf{y} >}$,

b) $< \mathbf{x}_1 + \mathbf{x}_2, \mathbf{y} >=< \mathbf{x}_1, \mathbf{y} > + < \mathbf{x}_2, \mathbf{y} >$,

c) $< \mathbf{x}, \lambda\mathbf{y} >=< \overline{\lambda}\mathbf{x}, \mathbf{y} >= \lambda < \mathbf{x}, \mathbf{y} >, \quad \forall \lambda \in \mathbb{C}$.

DEFINITION 1.2.9 *A transformation* T *in a Hilbert space* $\mathcal{H}$ *is called* **unitary**, *if* $\forall \mathbf{x}, \mathbf{y} \in \mathcal{H}$ *we have* $< \mathsf{T}\mathbf{x}, \mathsf{T}\mathbf{y} >=< \mathbf{x}, \mathbf{y} >$.

DEFINITION 1.2.10 *Let* S *be a representation of a group* G *in a Hilbert space* $\mathcal{H}$. *If all the operators* $\mathsf{S}(g)$ *of* S *are unitary, then the representation* S *is called* **unitary representation**.

THEOREM 1.2.6 (SCHUR - AUERBACH) *Any representation of a finite group is equivalent to a unitary representation.*

THEOREM 1.2.7 (MASCHKE) *Any reducible representation of a finite group is fully reducible.*

These two theorems are also true in the case of compact groups. The proof is based on the existence of the invariant integral for compact groups. In particular, the Theorem 1.2.7 implies that any unitary representation is fully reducible (i.e., it can be decomposed into a direct sum of irreducible representations). Thus, we are led to the following result of particular importance in the theory of representations:

Any representation of a compact group is equivalent to a unitary representation which, in turn, is fully reducible.

Therefore, in the case of compact groups it is sufficient to study the unitary irreducible representations.

3. Particular Abelian groups

3.1 The group of real numbers

3.1.1 General considerations

A group G is called *Abelian* if the result of the internal composition law does not depend on the order of terms (i.e., $g_i g_j = g_j g_i$, $\forall g_i, g_j \in G$). Let $\mathbf{t}(g_i)$

and $\mathbf{t}(g_j)$ be the matrices associated to the elements $g_i, g_j \in G$, respectively, in a linear representation of G. It follows that $\mathbf{t}(g_i)\mathbf{t}(g_j) = \mathbf{t}(g_j)\mathbf{t}(g_i)$, that is the product of matrices, which form the representation of an Abelian group, is commutative.

Let us suppose that the representation of the Abelian group G is unitary (i.e., all the matrices in the representation are unitary). As it is known, in this case, there exists a unitary constant matrix $\mathbf{u}$, such that all the transformed matrices $\mathbf{ut}(g)\mathbf{u}^{-1}$ are simultaneously reduced to the diagonal form

$$\mathbf{ut}(g)\mathbf{u}^{-1} = \begin{bmatrix} s_1(g) & 0 & \dots & 0 \\ 0 & s_2(g) & \dots & 0 \\ . & . & \dots & . \\ 0 & 0 & \dots & s_n(g) \end{bmatrix}. \tag{1.3.1}$$

Hence, this representation decomposes into a direct sum of n one-dimensional representations $s_i(g), i = 1, 2, \dots, n$. Therefore, any unitary representation of an Abelian group is equivalent to certain sets of one-dimensional representations, and the corresponding similarity transformation is determined by a matrix $\mathbf{u}$ which is also unitary. This result agrees with Schur's lemma 1.1.2, by virtue of which any irreducible representation, of finite dimension, of an Abelian group is one-dimensional. Each of these irreducible representations is defined by a numerical function $s(g)$, called character on G (see Definition 1.1.28), that satisfies the conditions (1.1.9).

3.1.2 Representations of the group of real numbers

For finite Abelian groups, the characters of all the elements in the group are the roots of the unity, because the order of any element is finite. Thus, if the order of an element g is two, that is $g^2 = e$, then $t(g) = \chi(g) = \pm 1$; if $g^3 = e$, then $t(g) = \chi(g) = \exp(2\pi ik/3)$, $k = 1, 2, 3$. Generally, if the order of g is n (i.e., $g^n = e$), then

$$t(g) = \chi(g) = e^{2\pi ik/n}, \quad k = 1, 2, \dots, n. \tag{1.3.2}$$

Let G be an Abelian group of order m. Each element of G forms a conjugacy class by itself, and thus, for $k = m$ in the formula (1.1.24), we obtain $n_1 = n_2 = \dots = n_m = 1$. Therefore, G has m irreducible representations, all being one-dimensional.

If the Abelian group G is also cyclic, then it is generated by an element g of order m. Hence, if we denote by $\chi(g) = \exp(2\pi il/m)$, $l = 1, 2, \dots, m$, then the characters of all the elements in the group are determined by the powers of $\chi(g)$; e.g., $\chi(g^k) = \exp(2\pi ikl/m)$.

Let us consider the group of real numbers $\mathbb{R}$, the composition law being the addition of two numbers. Because $(\mathbb{R}, +)$ is an Abelian compact group, all its irreducible representations are one-dimensional and, at the same time, equivalent to unitary representations. Consequently, according to (1.1.9), these representations will be continuous functions of a real variable, satisfying the functional equation

$$t(x)t(y) = t(x+y)\,, \quad x, y \in \mathbb{R}\,, \tag{1.3.3}$$

and the condition $t(0) = 1$, where 0 is the null element in $(\mathbb{R}, +)$. Also, the numerical functions $t(x)$ have to satisfy the *unitarity condition*

$$|t(x)| = 1\,. \tag{1.3.4}$$

We obtain the solution of (1.3.3) by differentiating it with respect to y, and setting $y = 0$. This gives the form $t(x) = \exp{(ax)}$, where $a = t'(0)$ is in general a complex number. Now, the unitarity condition (1.3.4) becomes $\exp{(ax)}\exp{(\overline{a}x)} = 1$, $\forall x \in \mathbb{R}$, such that $a + \overline{a} = 0$, a being thus a pure imaginary number $a = ib, b \in \mathbb{R}$. Hence, the unitary irreducible representations of the group $(\mathbb{R}, +)$ are given by the functions

$$x \to t(x) = e^{ibx}\,, \quad b \in \mathbb{R}\,. \tag{1.3.5}$$

3.2 The group of discrete translations

3.2.1 Basic results

A linear vector space $\mathcal{T}$ can be regarded as a group of vectors whose inner composition law is the addition of vectors. This group is called *one-dimensional*, if all its vectors are parallel one to the other, *two-dimensional*, if all its vectors are parallel to a given plane, and *three-dimensional*, if the group contains at least three vectors which are not coplanar.

DEFINITION 1.3.1 *A group of vectors is called* **discrete** *if there exists a positive number p such that the modulus of any non-zero vector in the group is greater than p.*

THEOREM 1.3.1 *Let $\mathcal{T}$ be a discrete group of three-dimensional vectors. Then, there exists three vectors $\mathbf{a}_1, \mathbf{a}_2, \mathbf{a}_3 \in \mathcal{T}$ such that any vector $\mathbf{v} \in \mathcal{T}$ can be expressed in the form*

$$\mathbf{v} = m_1\mathbf{a}_1 + m_2\mathbf{a}_2 + m_3\mathbf{a}_3\,, \quad m_1, m_2, m_3 \in \mathbb{Z}\,. \tag{1.3.6}$$

The vectors $\mathbf{a}_1$, $\mathbf{a}_2$, $\mathbf{a}_3$ are called the *basis vectors* of the group $\mathcal{T}$. At the same time, these vectors determine a three-dimensional lattice and are also called *fundamental translation vectors*. The parallelepiped formed by the fundamental translation vectors is called the *basic parallelepiped* or the *unit cell* of the lattice.

THEOREM 1.3.2 *We may change the shape and orientation of a unit cell by choosing one of its edges directed along any vector* $\mathbf{v}_1 \in \mathcal{T}$, *and a face, which contains this edge, being determined by* $\mathbf{b}_1$ *and any other vector* $\mathbf{v}_2 \in \mathcal{T}$. *Any parallelepiped thus constructed, associated to the vectors* $\mathbf{v}_1$ *and* $\mathbf{v}_2$, *and having a minimum volume, is also a unit cell.*

The volumes of all the unit cells of a given lattice are equal. The volume of any parallelepiped formed by three non-coplanar vectors $\mathbf{v}_1, \mathbf{v}_2, \mathbf{v}_3 \in \mathcal{T}$ is always larger than the volume of the unit cell.

A translation $\mathsf{T}(\mathbf{v})$ can be represented by a vector $\mathbf{v}$; therefore, $\mathcal{T}$ is the *group of discrete translations*.

3.2.2 Representations of the group of discrete translations

Let $\mathcal{T}$ be a three-dimensional discrete group, and $\mathbf{a}_1, \mathbf{a}_2, \mathbf{a}_3$ the basis vectors of $\mathcal{T}$. The group $\mathcal{T}$ is Abelian and thus it has only one-dimensional irreducible representations, the representation operators being numbers. We will denote by

$$t(\mathbf{a}_j) = \tau_j\,, \quad j = 1, 2, 3, \tag{1.3.7}$$

the numbers (characters) associated to translations by the basis vectors. Hence, the operator associated to a translation defined by the vector $\mathbf{v} = m_1\mathbf{a}_1 + m_2\mathbf{a}_2 + m_3\mathbf{a}_3$ (with $m_1, m_2, m_3 \in \mathbb{Z}$) will have the form

$$t(\mathbf{v}) = {\tau_1}^{m_1}{\tau_2}^{m_2}{\tau_3}^{m_3}\,. \tag{1.3.8}$$

Consequently, the triplet of numbers τ_1, τ_2, τ_3 determines all the irreducible representations of the discrete translations group $\mathcal{T}$. If all the three numbers τ_1, τ_2, τ_3 have the modulus equal to one, then the representations are unitary.

Generally, instead of the numbers τ_1, τ_2, τ_3 it is more convenient to use a vector $\mathbf{k}$ in the reciprocal lattice. The basis vectors of the reciprocal lattice are defined as the solutions of the equations $\mathbf{b}_p \cdot \mathbf{a}_q = 2\pi\delta_{pq}$, which are

$$\mathbf{b}_p = 2\pi\epsilon_{pqr}\,\frac{\mathbf{a}_q \times \mathbf{a}_r}{(\mathbf{a}_1, \mathbf{a}_2, \mathbf{a}_3)}\,, \quad p = 1, 2, 3, \tag{1.3.9}$$

where ϵ_{pqr} represents Ricci's symbol, and $(\mathbf{a}_1, \mathbf{a}_2, \mathbf{a}_3) = \mathbf{a}_1 \cdot (\mathbf{a}_2 \times \mathbf{a}_3)$ is the volume of the unit cell in the direct lattice. Any linear combination of these vectors, with integer coefficients, is a vector in the reciprocal lattice. Hence, if we assume that the characters of the unitary irreducible representations of $\mathcal{T}$, have the form

$$\tau_j = e^{i\mathbf{k}\cdot\mathbf{a}_j}\,, \quad j = 1, 2, 3\,, \tag{1.3.10}$$

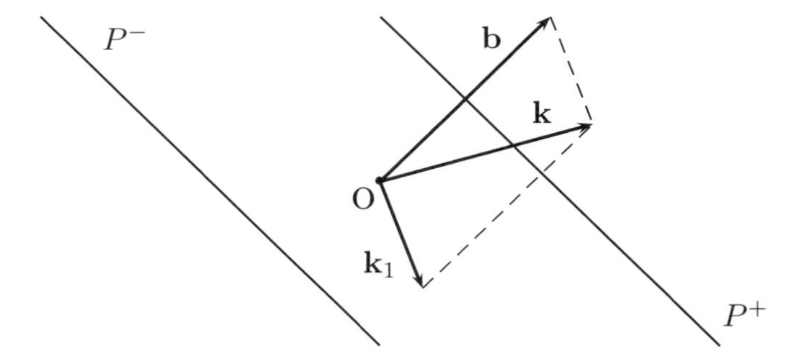

Figure 1.1. The Brillouin zone for one basis vector ($\mathbf{b}$) in the reciprocal lattice.

then

$$\mathbf{k} = \frac{1}{2\pi i}\,(\ln \tau_j)\,\mathbf{b}_j + n_1\mathbf{b}_1 + n_2\mathbf{b}_2 + n_3\mathbf{b}_3\,, \quad n_1, n_2, n_3 \in \mathbb{Z}\,. \tag{1.3.11}$$

Note that $\mathbf{k} = \mathbf{k}_0 + \mathbf{K}$, where $\mathbf{k}_0 = (\ln \tau_j)\mathbf{b}_j/(2\pi i)$ corresponds to what is called "crystal momentum" in solid state physics, and $\mathbf{K} = n_1\mathbf{b}_1 + n_2\mathbf{b}_2 + n_3\mathbf{b}_3$ is an arbitrary vector in the reciprocal lattice. It follows that two vectors $\mathbf{k}_1$ and $\mathbf{k}_2$, which differ by a reciprocal lattice vector, are associated to the same irreducible representation of $\mathcal{T}$. Such vectors are called *equivalent* vectors and are denoted by $\mathbf{k}_1 \equiv \mathbf{k}_2$. In terms of the vectors in the reciprocal lattice, the expression for the characters of a unitary irreducible representations of $\mathcal{T}$ (1.3.8) takes the form

$$t(\mathbf{v}) = e^{i\mathbf{k}\cdot\mathbf{v}}\,, \tag{1.3.12}$$

where

$$\mathbf{k} = \frac{1}{2\pi i}\sum_{j=1}^{3}(\ln \tau_j)\,\mathbf{b}_j + \sum_{j=1}^{3} n_j\mathbf{b}_j\,, \quad n_1, n_2, n_3 \in \mathbb{Z}\,.$$

One can easily see that (1.3.12) represents the three-dimensional generalization of the representation (1.3.5) restricted to $b \in \mathbb{Z}$.

The set of vectors $\mathbf{k}$, whose modulus cannot be reduced by adding a vector of the reciprocal lattice, form the *first Brillouin zone* (or simply the *Brillouin zone*). If the modulus of a vector $\mathbf{k}$ is precisely smaller than the modulus of any other vector equivalent to it, then $\mathbf{k}$ belongs to the Brillouin zone, but when among the vectors equivalent to $\mathbf{k}$ there exists at least one vector having the same modulus as $\mathbf{k}$, then the vector $\mathbf{k}$ points to the boundary of the Brillouin zone. It follows that the interior of the Brillouin zone contains only nonequivalent vectors, while any vector pointing to the boundary of the Brillouin zone is equivalent to one or several vectors also pointing to the boundary of the Brillouin zone.

To determine the Brillouin zone we consider a vector $\mathbf{b}$ in the reciprocal lattice and two parallel planes P^+ and P^-, normal to $\mathbf{b}$, and located at a

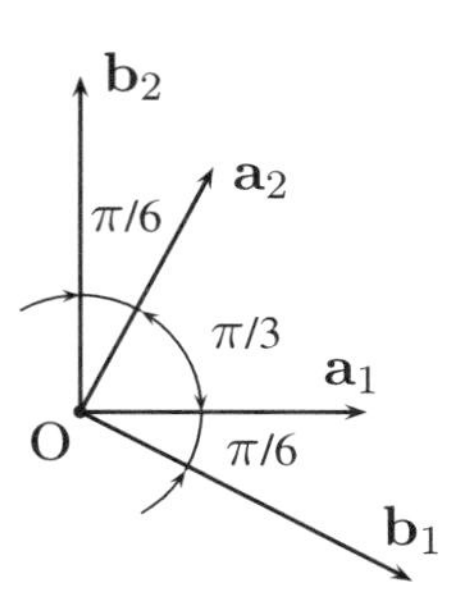

Figure 1.2. The basis vectors of the direct and reciprocal of a hexagonal lattice.

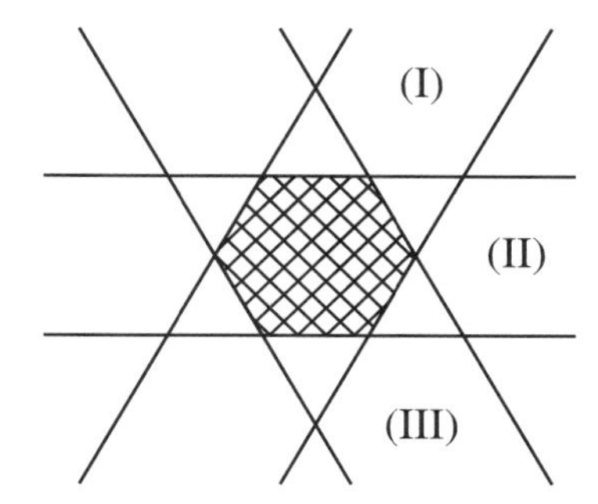

Figure 1.3. The Brillouin zone of a hexagonal lattice.

distance $b/2$ to the origin of the system of coordinates (see Fig. 1.1). It is obvious that for any vector $\mathbf{k}$ pointing to the right side of P^+, there exists an equivalent vector $\mathbf{k}_1 = \mathbf{k} - \mathbf{b}$ of a modulus smaller than $|\mathbf{k}|$. Hence, the vector $\mathbf{k}$ cannot belong to the Brillouin zone, which is included in the domain between the two planes P^+ and P^-. We repeat this procedure choosing instead of $\mathbf{b}$ another vector of the reciprocal lattice, to obtain another domain containing the Brillouin zone. Finally, the Brillouin zone is given by the intersection of all such domains determined by the lattice vectors.

For instance, let us consider a two-dimensional hexagonal lattice formed by the set of vectors

$$\{\mathbf{v}|\mathbf{v} = m_1\mathbf{a}_1 + m_2\mathbf{a}_2, \quad m_1, m_2 \in \mathbb{Z}\},$$

where $|\mathbf{a}_1| = |\mathbf{a}_2|$, and the angle between the basis vectors is $\pi/3$ (see Fig. 1.2). If we denote by $\mathbf{z}$ the unit vector of the cross product $\mathbf{a}_1 \times \mathbf{a}_2$, then the basis vectors of the reciprocal lattice will be

$$\mathbf{b}_1 = 2\pi \frac{\mathbf{a}_2 \times \mathbf{z}}{(\mathbf{a}_1, \mathbf{a}_2, \mathbf{z})}, \qquad \mathbf{b}_2 = 2\pi \frac{\mathbf{z} \times \mathbf{a}_1}{(\mathbf{a}_1, \mathbf{a}_2, \mathbf{z})}.$$

These two vectors are of equal modulus $|\mathbf{b}_1| = |\mathbf{b}_2|$, and the angle between them is $2\pi/3$. Applying the method described above for $\mathbf{b} = \mathbf{b}_1$ we obtain the strip (I) in Fig. 1.3. Then, for $\mathbf{b} = \mathbf{b}_2$ we obtain the strip (II), and finally for $\mathbf{b} = \mathbf{b}_1 + \mathbf{b}_2$ we obtain the strip (III). Any other possible choice $\mathbf{b} = n_1\mathbf{b}_1 + n_2\mathbf{b}_2$ (with $n_1, n_2 \in \mathbb{Z}$) leads to a strip which contains the hexagon marked off in Fig. 1.3. In conclusion, this hexagon represents the Brillouin zone for the discrete translation group of the hexagonal lattice.

In the reciprocal lattice, the first Brillouin zone plays the same rôle as the unit cell for the direct (or real space) lattice[1]. The discrete translation group associated to a lattice is the minimal symmetry group for the pair consisting of the direct and reciprocal lattices. The characters (1.3.12) are symmetrical in $\mathbf{k}$ and $\mathbf{v}$, and we can also regard them as being the characters $t(\mathbf{k}) = \exp(i\mathbf{k}\cdot\mathbf{v})$ for the irreducible representations of the discrete translation group of the reciprocal lattice (i.e., translations in the reciprocal space determined by the vectors $n_1\mathbf{b}_1 + n_2\mathbf{b}_2 + n_3\mathbf{b}_3$, $n_1, n_2, n_3 \in \mathbb{Z}$). Hence, the translational symmetry group of the reciprocal lattice coincides with the translational symmetry group of the of the direct lattice. At the same time, the Brillouin zone, as well as the unit cell, are invariant with respect to all the translations in the symmetry group of the corresponding lattice.

3.3 The $\mathsf{SO}(2)$ and C_n groups

3.3.1 Basic results

Let

$$x' = x\cos\varphi - y\sin\varphi\,,\quad y' = x\sin\varphi + y\cos\varphi\,,$$

be a rotation of an angle φ in the plane Oxy, about the Oz axis. The set of such two-dimensional rotations, determined by all the possible values of the real parameter $\varphi \in [0, 2\pi)$, forms a one-parameter Lie group called $\mathsf{SO}(2)$. If we associate to each element of $\mathsf{SO}(2)$ the matrix

$$\mathbf{t}(\varphi) = \begin{bmatrix} \cos\varphi & -\sin\varphi \\ \sin\varphi & \cos\varphi \end{bmatrix}, \tag{1.3.13}$$

it is obvious that the product of two rotations of angles φ_1 and φ_2 is a rotation of angle $\varphi_1 + \varphi_2$, that is

$$\mathbf{t}(\varphi_2)\mathbf{t}(\varphi_1) = \mathbf{t}(\varphi_1 + \varphi_2)\,.$$

We also have $\mathbf{t}(\varphi_1)\mathbf{t}(\varphi_2) = \mathbf{t}(\varphi_1+\varphi_2)$, so that the considered group is Abelian. The composition of two elements in $\mathsf{SO}(2)$ corresponds to the summation of the values of the parameter φ associated to the two elements. Because the parameter φ varies in a finite interval the group is also compact, and therefore the invariant integral exists for any function of bounded variation in the interval $[0, 2\pi)$. The same $\mathsf{SO}(2)$ group can be used to represent rotations of the planes Oyz and Ozx, about the Ox and Oy axes, respectively.

Together with the groups of discrete translations, the discrete groups of two-dimensional rotations are of particular importance in studies concerning two-dimensional lattices. For instance, in addition to the translational symmetry, a

[1]More precisely, the equivalent of the Brillouin zone in the direct lattice is the Wigner-Seitz cell, which is constructed exactly in the same way as the Brillouin zone, but with respect to the basis vectors $\mathbf{a}_1, \mathbf{a}_2, \mathbf{a}_3$. The Wigner-Seitz cell has the same volume as the unit cell.

hexagonal lattice also shows a rotational symmetry, that is the lattice is invariant with respect to plane rotations of $2\pi n/6$, where $n = 0, 1, \ldots, 5$. The finite set of these rotations forms a cyclic group called C_6.

Generally, the two-dimensional rotations in the Oxy plane, about the Oz axis, by the angles $0, 2\pi/n, 4\pi/n, \ldots, 2(n-1)\pi/n$, for a given $n \in \mathbb{N}^*$, form a cyclic group called C_n. The matrices of these rotations have the form

$$\mathbf{t}\left(\frac{2m\pi}{n}\right) = \begin{bmatrix} \cos\dfrac{2m\pi}{n} & -\sin\dfrac{2m\pi}{n} \\ \sin\dfrac{2m\pi}{n} & \cos\dfrac{2m\pi}{n} \end{bmatrix},$$

and constitute the elements of a finite cyclic group, because

$$\mathbf{t}\left(\frac{2m\pi}{n}\right) = \left[\mathbf{t}\left(\frac{2\pi}{n}\right)\right]^m .$$

If we choose an angle φ_0, which is not a multiple of π, then the transformations

$$[\mathbf{t}(\varphi_0)]^m = \mathbf{t}(m\varphi_0)\,, \quad m \in \mathbb{Z}\,,$$

form an infinite group, because it does not exist an integer number $k \neq 0$ such that $[\mathbf{t}(\varphi_0)]^k$ and the identity element of the group coincide. Although the group is infinite its matrices do not contain any parameter varying continuously. In this case we say that the group is *denumerable*, because there exists a one-to-one correspondence between the group elements and the set of integers $\mathbb{Z}$.

The set of n distinct two-dimensional rotations about the z-axis and the n mirror reflections in planes containing the z-axis, form a discrete group called C_{nv}. The C_n and C_{nv} groups have been used by P. R. McIsaac to classify the electromagnetic modes in waveguides. Recently, K. Sakoda presented an exhaustive study of symmetry properties of band diagrams associated to photonic crystals, based on discrete groups. Also, the discrete symmetries proved to be useful in the band theory of solids.

3.3.2 Representations of $\mathsf{SO}(2)$ and C_n groups

The group of two-dimensional rotations $\mathsf{SO}(2)$ is Abelian, so that all its irreducible representations are one-dimensional. Let $\chi(\varphi)$ be the character of an irreducible representation. The representation being one-dimensional, we may write

$$\chi(\varphi)\chi(\psi) = \chi(\varphi + \psi)\,. \tag{1.3.14}$$

Admitting that the characters are continuous and differentiable, we differentiate (1.3.14) with respect to ψ, and set $\psi = 0$. We thus obtain the differential equation $\chi(\varphi)\chi'(0) = \chi'(\varphi)$, and we look for a solution which also satisfies (1.3.14). Such a solution has the form

$$\chi(\varphi) = e^{\varphi\chi'(0)} . \tag{1.3.15}$$

The number $m = -i\chi'(0)$ is called the *representation index*, and the character of an irreducible representation of index m is

$$\chi_m(\varphi) = e^{im\varphi} , \tag{1.3.16}$$

a form similar to (1.3.5). If the index m is an integer number, then the representation is *one-valued*, otherwise we obtain a *multi-valued* representation. The group $\mathsf{SO}(2)$ has thus a continuum of irreducible representations among which those of real (integer) index are unitary, the other being non-unitary.

The group $\mathsf{SO}(2)$ is also a one-parameter group with the infinitesimal generator

$$\mathsf{X}_m = \frac{d}{d\varphi}\,\chi_m(\varphi)\bigg|_{\varphi=0} = im . \tag{1.3.17}$$

The group C_n is isomorphic to the group of even permutations of n objects, which is an Abelian invariant subgroup of the symmetric group S_n (see Theorem 1.1.21). C_n has n conjugacy classes, and therefore n unitary irreducible representations (one-dimensional) with the characters

$$\chi_m = e^{i2m\pi/n} , \quad m = 0, 1, \ldots, n-1 . \tag{1.3.18}$$

Note that this numbers are the complex nth roots of the unity, which form an Abelian (cyclic) group of order n, also isomorphic to C_n.

Chapter 2

LIE GROUPS

1. The SO(3) group

1.1 Rotations

1.1.1 Matrices

Let $\mathbf{A}$ be a *square matrix* of elements $a_{ik} \in \mathbb{C}$, $i, k = 1, 2, \ldots, n$. The matrix

$$\mathbf{A}^{\dagger} = \overline{\mathbf{A}}^{\mathrm{T}},$$

where $\mathbf{A}^{\mathrm{T}}$ is the *transposed* of $\mathbf{A}$, and $\overline{\mathbf{A}}$ is the *complex conjugate* of $\mathbf{A}$, is called the *adjoint* (or *Hermitian conjugate*) of the matrix $\mathbf{A}$. If $\mathbf{A} = \overline{\mathbf{A}}$ then the matrix $\mathbf{A}$ is *real*, and if $\mathbf{A} = -\overline{\mathbf{A}}$ then the matrix $\mathbf{A}$ is *pure imaginary*. A square matrix $\mathbf{S}$ is called *symmetric* or *skew-symmetric*, if $\mathbf{S} = \mathbf{S}^{\mathrm{T}}$ or $\mathbf{S} = -\mathbf{S}^{\mathrm{T}}$, respectively. A square matrix $\mathbf{H}$ is called *Hermitian* if $\mathbf{H} = \mathbf{H}^{\dagger}$, and *anti-Hermitian* if $\mathbf{H} = -\mathbf{H}^{\dagger}$. Note that, a real and symmetric matrix is also Hermitian.

If a square complex matrix $\mathbf{O}$ satisfies the relation $\mathbf{O}^{\mathrm{T}} = \mathbf{O}^{-1}$, where $\mathbf{O}^{-1}$ is the *inverse* of the matrix $\mathbf{O}$, that is

$$\mathbf{O}\mathbf{O}^{\mathrm{T}} = \mathbf{O}^{\mathrm{T}}\mathbf{O} = \mathbf{E},$$

with $\mathbf{E}$ being the identity matrix, then we say that $\mathbf{O}$ is a *complex orthogonal matrix*. The determinant of a square matrix is equal to the determinant of the transposed matrix such that, the absolute value of the determinant of a complex orthogonal matrix is equal to one. A square complex matrix $\mathbf{U}$ is called *unitary matrix* if $\mathbf{U}^{\dagger} = \mathbf{U}^{-1}$, that is

$$\mathbf{U}\mathbf{U}^{\dagger} = \mathbf{U}^{\dagger}\mathbf{U} = \mathbf{E}.$$

Note that, a unitary ($\mathbf{R}^{\dagger} = \mathbf{R}^{-1}$) and real ($\mathbf{R} = \overline{\mathbf{R}}$) matrix satisfies the relations

$$\mathbf{R}^{\mathrm{T}} = \overline{\mathbf{R}}^{\mathrm{T}} = \mathbf{R}^{\dagger} = \mathbf{R}^{-1},$$

i.e., $\mathbf{R}^{\mathrm{T}} = \mathbf{R}^{-1}$. In this case, we say that $\mathbf{R}$ is a *real orthogonal matrix* (or, simply *orthogonal matrix*).

Let $\mathcal{H}$ be a *Hilbert space*. We may use the *scalar product* introduced in this space to define the adjoint of a matrix $\mathbf{A}$ by the relation

$$< \mathbf{x}, \mathbf{A}\mathbf{y} > = < \mathbf{A}^{\dagger}\mathbf{x}, \mathbf{y} >, \quad \mathbf{x}, \mathbf{y} \in \mathcal{H}.$$

In the case of a real matrix, the same relation defines the transposed matrix:

$$< \mathbf{x}, \mathbf{R}\mathbf{y} > = < \mathbf{R}^{\mathrm{T}}\mathbf{x}, \mathbf{y} >, \quad \mathbf{x}, \mathbf{y} \in \mathcal{H}.$$

Consequently, a unitary matrix $\mathbf{U}$ preserves the scalar product in a complex Hilbert space, that is

$$< \mathbf{U}\mathbf{x}, \mathbf{U}\mathbf{y} > = < \mathbf{U}^{\dagger}\mathbf{U}\mathbf{x}, \mathbf{y} > = < \mathbf{x}, \mathbf{y} >, \quad \forall \mathbf{x}, \mathbf{y} \in \mathcal{H},$$

while an orthogonal matrix $\mathbf{R}$ preserves the scalar product in a real Hilbert space, that is

$$< \mathbf{R}\mathbf{x}, \mathbf{R}\mathbf{y} > = < \mathbf{R}^{\mathrm{T}}\mathbf{R}\mathbf{x}, \mathbf{y} > = < \mathbf{x}, \mathbf{y} >, \quad \forall \mathbf{x}, \mathbf{y} \in \mathcal{H}.$$

THEOREM 2.1.1 *Any unitary matrix* $\mathbf{U}$ *and any Hermitian matrix* $\mathbf{H}$ *can be reduced to the diagonal form by a similarity transformation* $\mathbf{V}^{\dagger}\mathbf{U}\mathbf{V}$, *respectively* $\mathbf{V}^{\dagger}\mathbf{H}\mathbf{V}$, *where* $\mathbf{V}$ *is a unitary matrix.*

In particular, this theorem can also be applied in the case of real orthogonal or real symmetric matrices, but not in the case of complex orthogonal or complex symmetric matrices.

1.1.2 The rotation operator

Let us consider the metric space $\mathcal{E}_3$. The set of the vectors having the origin at a given point O forms a *real linear space* with respect to the addition of vectors and the multiplication of vectors by real numbers. In this set of vectors, we introduce the *scalar product* (or *dot product*) of two vectors defined by the relation

$$< \mathbf{x}, \mathbf{y} > = \|\mathbf{x}\| \, \|\mathbf{y}\| \, \cos(\mathbf{x}, \mathbf{y}), \tag{2.1.1}$$

where $\|.\|$ denotes the *modulus* (or *length*) of the corresponding vector, and $(\mathbf{x}, \mathbf{y})$ is the angle between the directions of $\mathbf{x}$ and $\mathbf{y}$. Being endowed with the scalar product, the set of vectors with a common origin forms the *geometric space* $\mathcal{G}$.

DEFINITION 2.1.1 *It is called* **rotation**, *a mapping of* $\mathcal{G}$ *onto itself, which preserves the length of the vectors and the angle between them.*

Hence, a rotation preserves the scalar product (2.1.1).

THEOREM 2.1.2 *Any rotation can be represented by an orthogonal linear operator* G, *acting in* $\mathcal{G}$.

If we introduce an orthonormal basis $\{\mathbf{f}_1, \mathbf{f}_2, \mathbf{f}_3\} \subset \mathcal{G}$ (i.e., $<\mathbf{f}_i, \mathbf{f}_j> = \delta_{ij}$, $i, j = 1, 2, 3$), then we can associate to the operator G a square matrix $\mathbf{g}$ defined by the relations

$$\mathsf{G}\mathbf{f}_i = g_{ki}\mathbf{f}_k\,, \quad i = 1, 2, 3, \tag{2.1.2}$$

where

$$g_{ki} = <\mathbf{f}_k, \mathsf{G}\mathbf{f}_i>\,, \quad i, k = 1, 2, 3.$$

The linear operator G is orthogonal, hence the matrix $\mathbf{g}$ is also orthogonal, that is $\mathbf{g}\mathbf{g}^{\mathrm{T}} = \mathbf{g}^{\mathrm{T}}\mathbf{g} = \mathbf{e}$, where $\mathbf{e}$ is the identity matrix. This gives nine relations of the form $g_{ij}g_{kj} = \delta_{ik}$, $i, k = 1, 2, 3$, but only six of these relations are independent, so that the nine matrix elements g_{ij} can be expressed as functions of three of them. The elements of the matrix $\mathbf{g}$ also satisfy the relations

$$\begin{aligned}
\epsilon_{\alpha\beta\gamma}\, g_{\alpha i}\, g_{\beta j}\, g_{\gamma k} &= (\det \mathbf{g})\, \epsilon_{ijk}\,, & i, j, k &= 1, 2, 3,\\
\epsilon_{\alpha\beta\gamma}\, g_{\alpha i}\, g_{\beta j} &= (\det \mathbf{g})\, \epsilon_{ijk}\, g_{\gamma k}\,, & i, j, \gamma &= 1, 2, 3,\\
\epsilon_{\alpha\beta\gamma}\, g_{j\beta}\, g_{k\gamma} &= (\det \mathbf{g})\, \epsilon_{ijk}\, g_{i\alpha}\,, & j, k, \alpha &= 1, 2, 3.
\end{aligned} \tag{2.1.3}$$

1.1.3 Proper and improper rotations

Because the matrix of the rotation operator G is orthogonal, for an orthonormal basis in $\mathcal{G}$, and the determinant of a matrix is equal to the determinant of the transposed matrix, we obtain $(\det \mathbf{g})^2 = 1$ that is $\det \mathbf{g} = \pm 1$.

DEFINITION 2.1.2 *The rotations associated to a matrix with* $\det \mathbf{g} = +1$ *are called* **proper rotations**, *while the rotations associated to a matrix with* $\det \mathbf{g} = -1$ *are called* **improper rotations**.

The transformation I, by which any vector $\mathbf{x} \in \mathcal{G}$ becomes its opposite $-\mathbf{x}$, is called *inversion*. An inversion is a type of mirror image (reflection in space) by which a vector $\mathbf{x} \in \mathcal{G}$ is inverted through the point at the origin. The matrix associated to the operator I is $-\mathbf{e}$, therefore the operator I commutes with any rotation operator G. Any *improper rotation* is equal to the product of I and a proper rotation (i.e., a rotation followed by an inversion).

Let $\{\mathbf{f}_1, \mathbf{f}_2, \mathbf{f}_3\} \subset \mathcal{G}$ be an orthonormal basis in $\mathcal{G}$. The cross product $\mathbf{f}_i \times \mathbf{f}_j = \sigma\, \epsilon_{ijk}\, \mathbf{f}_k$ defines the character of the basis; as $\sigma = \pm 1$, the basis is called *right-handed* (or *dextral*), respectively *left-handed* (or *sinistral*).

THEOREM 2.1.3 *A proper rotation does not change the right-handed or left-handed character of a basis, and reciprocally.*

THEOREM 2.1.4 *There always exists a basis in the geometrical space $\mathcal{G}$, such that the matrix associated to a proper rotation* G *takes the particular form*

$$\mathbf{g}^0 = \begin{bmatrix} \cos\varphi_0 & -\sin\varphi_0 & 0 \\ \sin\varphi_0 & \cos\varphi_0 & 0 \\ 0 & 0 & 1 \end{bmatrix}. \tag{2.1.4}$$

The form (2.1.4) of the matrix $\mathbf{g}$ results from the decomposition into the direct sum $\mathcal{G} = \mathcal{M} \oplus \mathcal{N}$, where $\mathcal{M}$ is a two-dimensional subspace and $\mathcal{N}$ is a one-dimensional subspace. Accordingly, the rotation operator decomposes into $\mathsf{G} = \mathsf{R} \oplus \mathsf{E}$, where R is a proper rotation in the two-dimensional subspace $\mathcal{M}$, while E is the identity operator in $\mathcal{N}$. Let $\{\mathbf{f}_1^0, \mathbf{f}_2^0, \mathbf{f}_3^0\} \subset \mathcal{G}$ be an orthonormal basis in $\mathcal{G}$, such that $\{\mathbf{f}_1^0, \mathbf{f}_2^0\}$ is an orthonormal basis in $\mathcal{M}$, and $\{\mathbf{f}_3^0\}$ a basis in $\mathcal{N}$. With these notations, the rotation G will be defined as a *rotation of angle* φ_0 about a *rotation axis* Δ along $\mathbf{f}_3^0$, the *sense of rotation* being determined by the direction of $\mathbf{f}_3^0$.

THEOREM 2.1.5 *If two orthogonal bases are given in $\mathcal{G}$, then they are related by a rotation about an axis Δ, determined by these bases.*

In other words, any two orthogonal bases defined in $\mathcal{G}$ have a common origin O, and there exists an axis Δ passing through O such that a rotation of one basis about Δ, by a certain angle φ, will overlay the corresponding vectors of the two bases. The length of the corresponding vectors in the two bases can be different, but after the rotation they will have the same direction and also the same sense. It is evident that the Theorem 2.1.5 does not hold for non-orthogonal bases.

The two bases being given, we know the rotation matrix $\mathbf{g}$ (respectively its elements g_{ij}), which is associated to the transformation of one basis into another, and by means of this matrix we can determine the rotation axis. We notice that in the particular case of a rotation associated to the matrix (2.1.4), the rotation axis Δ is defined by the unit vector $\mathbf{f}_3^0$, and any vector along Δ has the form $\mathbf{z} = \lambda\mathbf{f}_3^0$, $\lambda \in \mathbb{R}$. From (2.1.4) we also have $\mathbf{g}^0\mathbf{f}_3^0 = \mathbf{f}_3^0$, and this means that we have to look for an eigenvector of $\mathbf{g}$ corresponding to the eigenvalue 1 (i.e., $\mathbf{g}\mathbf{z} = \mathbf{z}$). In fact, the matrix $\mathbf{g}$ is orthogonal and the condition $\det \mathbf{g} = 1$ implies that the most general form of its eigenvalues is $\{1, \exp(i\varphi), \exp(-i\varphi)\}$, with $\varphi \in \mathbb{R}$, such that an orthogonal matrix associated to a proper rotation in three dimensions, always has an eigenvalue equal to 1. The eigenvector corresponding to the eigenvalue 1, which is unaffected by the rotation (i.e., the action of $\mathbf{g}$), thereby defines the axis about which the rotation is taken. From the orthogonality of $\mathbf{g}$ (i.e., $\mathbf{g}^{\mathrm{T}}\mathbf{g} = \mathbf{e}$) we also have $\mathbf{z} = \mathbf{g}^{\mathrm{T}}\mathbf{g}\mathbf{z} = \mathbf{g}^{\mathrm{T}}\mathbf{z}$, or $(\mathbf{g} - \mathbf{g}^{\mathrm{T}})\mathbf{z} = \mathbf{0}$. This last relation represents a homogeneous, linear, algebraic system

$$\left(g_{ij} - g_{ij}^{\mathrm{T}}\right)\zeta_j \equiv \left(g_{ij} - g_{ji}\right)\zeta_j = 0\,, \quad i = 1,2,3, \tag{2.1.5}$$

where the unknowns $\zeta_1, \zeta_2, \zeta_3$ are the components of the vector $\mathbf{z}$. From the system (2.1.5) we can only find the ratios

$$\zeta_1/\zeta_2/\zeta_3 = (g_{32} - g_{23})/(g_{13} - g_{31})/(g_{21} - g_{12}), \tag{2.1.6}$$

which are sufficient to obtain the direction of the vector $\mathbf{z}$. Thus, we may combine (2.1.6) with the normalization condition

$$\mathbf{z} \cdot \mathbf{z} = \zeta_1^2 + \zeta_2^2 + \zeta_3^2 = 1,$$

to determine $\mathbf{z}$ uniquely, and respectively the direction of the rotation axis Δ.

The rotation angle φ is given by the trace of the matrix $\mathbf{g}$. Because the trace of the matrix associated to a linear operator does not depend on the basis, the relations

$$\operatorname{Tr} \mathbf{g} = g_{ii} = \operatorname{Tr} \mathbf{g}^0 = 2 \cos \varphi_0 + 1 \tag{2.1.7}$$

take place. We can thus obtain $\cos \varphi$ and $\sin \varphi$, abstraction of a sign (function of the sense of rotation)

$$-\pi \leq \varphi_0 = \arccos \left[(g_{ii} - 1)/2 \right] \leq \pi, \quad \varphi = \varphi_0 + 2k\pi, \quad k \in \mathbb{Z}. \tag{2.1.8}$$

Finally, to determine the sense of the rotation axis, we introduce the quantities

$$\xi_i = \frac{\varphi_0}{\sin \varphi_0} \eta_i, \quad \eta_i = -\frac{1}{2} \epsilon_{ijk} g_{jk}, \quad i = 1, 2, 3. \tag{2.1.9}$$

The quantities η_i are the components of an axial vector associated to the antisymmetric part of the matrix $\mathbf{g}$. The quantities ξ_i essentially depend on the basis in the space $\mathcal{G}$. Indeed, a transformation defined by an orthogonal matrix $\mathbf{c}$, will bring forth the new basis

$$\mathbf{f}'_i = c_{ki} \mathbf{f}_k, \quad i = 1, 2, 3, \tag{2.1.10}$$

and, according to (2.1.2), the matrix $\mathbf{g}$ will become

$$\mathbf{g}' = \mathbf{c}^{-1} \mathbf{g} \mathbf{c} = \mathbf{c}^{\mathrm{T}} \mathbf{g} \mathbf{c},$$

with elements given by the relations

$$g'_{ij} = c_{ki} g_{kl} c_{lj}, \quad i, j = 1, 2, 3.$$

We thus obtain

$$\begin{aligned} \eta'_i &= -\frac{1}{2} \epsilon_{ijk} g'_{jk} = -\frac{1}{2} \epsilon_{ijk} c_{lj} g_{lm} c_{mk} = -\frac{1}{2} (\det \mathbf{c}) \epsilon_{nlm} g_{lm} c_{ni} \\ &= (\det \mathbf{c}) \eta_n c_{ni} = (\det \mathbf{c}) (\mathbf{c}^{\mathrm{T}})_{in} \eta_n, \quad i = 1, 2, 3, \end{aligned} \tag{2.1.11}$$

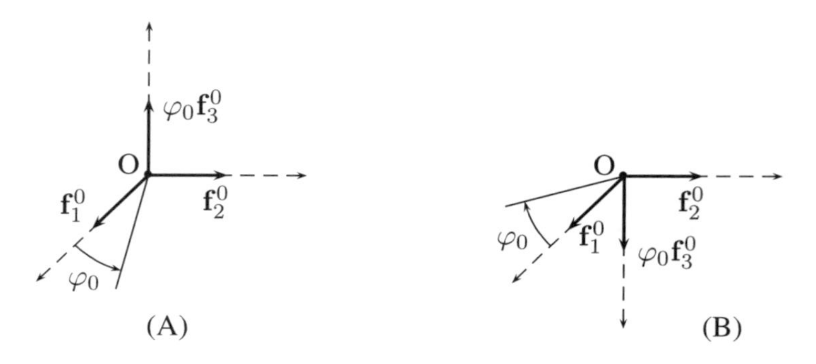

Figure 2.1. The rotations axes for the cases $\varphi_0 > 0$ (A) and $\varphi_0 < 0$ (B). The sense of the rotation axis depends on the sign of φ_0 such that we always have a rotation in the direct sense with respect to the corresponding axis.

where we have used the second relation in (2.1.3). The quantities ξ_i are transformed in the same manner as η_i, because the ratio $\varphi_0/\sin\varphi_0$ is independent of the basis

$$\xi_i' = (\det \mathbf{c})\,(\mathbf{c}^{\mathrm{T}})_{in}\xi_n\,, \quad i = 1, 2, 3. \tag{2.1.12}$$

Hence, the quantities ξ_i form the components of a pseudovector $\boldsymbol{\xi}$. Note that the length of $\boldsymbol{\xi}$ is constant (i.e., it does not depend on the chosen basis), because

$$\xi_i'\xi_i' = (\det \mathbf{c})^2\,(\mathbf{c}^{\mathrm{T}})_{in}(\mathbf{c}^{\mathrm{T}})_{im}\xi_n\xi_m = (\det \mathbf{c})^2\delta_{nm}\xi_n\xi_m = \xi_n\xi_n\,. \tag{2.1.13}$$

Now, we suppose that the matrix $\mathbf{g}$ is given in a right-handed basis $\{\mathbf{f}_1, \mathbf{f}_2, \mathbf{f}_3\}$, the basis $\{\mathbf{f}_1^0, \mathbf{f}_2^0, \mathbf{f}_3^0\}$, in which $\mathbf{g}$ becomes $\mathbf{g}^0$, being also right-handed. From (2.1.4), (2.1.8), and (2.1.9) it results that $\eta_1^0 = \eta_2^0 = 0$, $\eta_3^0 = \sin\varphi_0$, and $\xi_1^0 = \xi_2^0 = 0$, $\xi_3^0 = \varphi_0$. Both bases $\{\mathbf{f}_1, \mathbf{f}_2, \mathbf{f}_3\}$ and $\{\mathbf{f}_1^0, \mathbf{f}_2^0, \mathbf{f}_3^0\}$ are right-handed, hence, when passing from one basis to the other, the quantities ξ_i transform in the same way as the components of a vector (because, $\det \mathbf{c} = 1$ in this case). In the basis $\{\mathbf{f}_1^0, \mathbf{f}_2^0, \mathbf{f}_3^0\}$ the vector $\boldsymbol{\xi}$ is directed along $\mathbf{f}_3^0$ and, regardless of $\varphi_0 > 0$ or $\varphi_0 < 0$, the rotation is of an angle $|\varphi_0|$ about the axis $\varphi_0\mathbf{f}_3^0$, in the direct sense (see Fig. 2.1, A and B).

In the second case, we suppose that the matrix $\mathbf{g}$ is given in a left-handed basis $\{\mathbf{f}_1, \mathbf{f}_2, \mathbf{f}_3\}$. By an inversion $\mathsf{I} = -\mathsf{E}$ we obtain the matrix of the rotation operator in a right-handed base $\{\mathbf{f}_1' = -\mathbf{f}_1, \mathbf{f}_2' = -\mathbf{f}_2, \mathbf{f}_3' = -\mathbf{f}_3\}$, the numbers ξ_i becoming $\xi_i' = -\xi_i$, $i = 1, 2, 3$. Note that the determinant of the matrix associated to an inversion is equal to -1. Then, by knowing the matrix of the rotation operator in the right-handed base $\{\mathbf{f}_1', \mathbf{f}_2', \mathbf{f}_3'\}$, as well as the associated numbers ξ_i', we may further apply the same method as in the first case. In both cases, from (2.1.13) we have

$$\|\boldsymbol{\xi}\|^2 = \xi_i\xi_i = \xi_i^0\xi_i^0 = \varphi_0^2\,. \tag{2.1.14}$$

1.1.4 The $\mathsf{SO}(3)$ and $\mathsf{O}(3)$ groups

The proper rotations of the geometric space $\mathcal{G}$ are elements of the *rotation group* $\mathsf{SO}(3)$. As we have shown, any rotation is determined by a vector $\mathbf{x}$, situated on the rotation axis, and by the rotation angle φ, measured counter-clockwise with respect to the sense of $\mathbf{x}$. Thus, we will denote the rotations by $\mathsf{G}(\varphi, \mathbf{x})$[1]. The product of two rotations $\mathsf{G}(\varphi_2, \mathbf{x}_2)\mathsf{G}(\varphi_1, \mathbf{x}_1)$ is the resultant rotation $\mathsf{G}(\varphi, \mathbf{x})$. This means that an arbitrary vector $\mathbf{r} \in \mathcal{G}$ is transformed into $\mathbf{r}' \in \mathcal{G}$ by $\mathsf{G}(\varphi_1, \mathbf{x}_1)$, and $\mathbf{r}' \in \mathcal{G}$ is transformed into $\mathbf{r}'' \in \mathcal{G}$ by $\mathsf{G}(\varphi_2, \mathbf{x}_2)$, while $\mathbf{r} \in \mathcal{G}$ is directly transformed into $\mathbf{r}'' \in \mathcal{G}$ by $\mathsf{G}(\varphi, \mathbf{x})$.

THEOREM 2.1.6 *The rotations about the same axis commute one with the other.*

By definition, all the rotations $\mathsf{G}(\varphi_1, \mathbf{x}_1)$, conjugated to a fixed arbitrary rotation $\mathsf{G}(\varphi, \mathbf{x})$ have the form

$$\mathsf{G}(\varphi_1, \mathbf{x}_1) = \mathsf{R}\,\mathsf{G}(\varphi, \mathbf{x})\,\mathsf{R}^{-1}\,, \tag{2.1.15}$$

where $\mathsf{R} \in \mathsf{SO}(3)$ represents a rotation[2] that changes a basis $\{\mathbf{f}\} = \{\mathbf{f}_1, \mathbf{f}_2, \mathbf{f}_3\}$ of $\mathcal{G}$ into a new one $\{\mathbf{f}' = \mathsf{R}\mathbf{f}\} \subset \mathcal{G}$. We denote by $\mathbf{g} = [g_{ij}]$ the matrix associated to the rotation $\mathsf{G}(\varphi, \mathbf{x})$ in the basis $\{\mathbf{f}\}$. According to (2.1.2), the action of $\mathsf{G}(\varphi_1, \mathbf{x}_1)$ on the vectors of the new basis can be written in the form

$$\begin{aligned} \mathsf{G}(\varphi_1, \mathbf{x}_1)\mathbf{f}'_j &\equiv \mathsf{R}\,\mathsf{G}(\varphi, \mathbf{x})\,\mathsf{R}^{-1}\mathbf{f}'_j = \mathsf{R}\,\mathsf{G}(\varphi, \mathbf{x})\,\mathbf{f}_j \\ &= \mathsf{R}\,g_{kj}\,\mathbf{f}_k = g_{kj}\,\mathsf{R}\,\mathbf{f}_k = g_{kj}\,\mathbf{f}'_k\,, \end{aligned} \tag{2.1.16}$$

where $j = 1, 2, 3$. Since the matrix associated to a rotation in a basis of $\mathcal{G}$, uniquely determines the rotation axis as well the rotation angle, the relations (2.1.16) show that $\varphi_1 = \varphi$, and the components of $\mathbf{x}_1$ in the basis $\{\mathbf{f}'\}$ are the same as the components of $\mathbf{x}$ in the basis $\{\mathbf{f}\}$. Consequently, by applying the rotation R on the vector $\mathbf{x}$ we obtain the vector $\mathbf{x}_1$ (i.e., $\mathbf{x}_1 = \mathsf{R}\mathbf{x}$), and the relation (2.1.15) becomes

$$\mathsf{R}\,\mathsf{G}(\varphi, \mathbf{x})\,\mathsf{R}^{-1} = \mathsf{G}(\varphi, \mathsf{R}\mathbf{x})\,, \quad \mathsf{R} \in \mathsf{SO}(3)\,. \tag{2.1.17}$$

It follows that a conjugacy class in $\mathsf{SO}(3)$ consists of the rotations of the same angle φ about all possible axes.

It can be shown that the reciprocal of Theorem 2.1.2 is true, therefore, the set of the orthogonal linear operators acting in the real three-dimensional space,

[1] This angle-axis parametrization of G can easily be associated to a *unit quaternion* (see Chap. 2 Sec. 2.1.2).
[2] Note that, here and in what follows we will omit the arguments of some rotation operators when they are not relevant.

and having an associated matrix of determinant 1, forms a group isomorphic to $\mathsf{SO}(3)$, the internal composition law being the usual product of operators.

When discussing a rotation, there are two possible conventions: rotation of an object relative to fixed axes and rotation of the axes. The orthogonal matrix associated to the rotation R, that changes the basis $\{\mathbf{f}\}$ into the new one $\{\mathbf{f}' = \mathsf{R}\mathbf{f}\} \subset \mathcal{G}$, can also be regarded as a transformation of a vector $\mathbf{x} \in \mathcal{G}$, when we rotate the space (not the coordinate axes). Thus, if the basis vectors of the rotated space are expressed, in terms of the initial basis vectors, by relations of the form

$$\mathbf{f}'_j = g_{ij}\mathbf{f}_i\,, \quad j = 1, 2, 3\,, \tag{2.1.18}$$

then, an arbitrary vector $\mathbf{x} \in \mathcal{G}$ will be transformed into a vector $\mathbf{x}' \in \mathcal{G}$, which will have in the basis $\{\mathbf{f}'\}$ the same components as the initial vector $\mathbf{x}$ in the basis $\{\mathbf{f}\}$. Hence, if we have $\mathbf{x} = x_i\mathbf{f}_i$ in the basis $\{\mathbf{f}\}$, then $\mathbf{x}' = x_i\mathbf{f}'_i = x'_i\mathbf{f}_i$, where x'_i are the components of $\mathbf{x}'$ with respect to the initial basis. This gives

$$x_i\mathbf{f}'_i = x_i g_{ki}\mathbf{f}_k = x'_k\mathbf{f}_k\,, \tag{2.1.19}$$

and we find that the coordinates of the vertex of the vector $\mathbf{x} \in \mathcal{G}$ are transformed according to the formulae

$$x'_k = g_{ki}x_i\,, \quad k = 1, 2, 3\,, \tag{2.1.20}$$

or $\mathbf{x}' = \mathbf{g}\mathbf{x}$, in matrix notation. Such rotations are called *active rotations*, and the quantities that transform like the components x_i in (2.1.20) are said to transform *cogrediently* with respect to the basis set $\{\mathbf{f}\}$. We can also have *passive rotations* when the vectors remained fixed, whilst the system of coordinates in $\mathcal{G}$ is rotated[3]. If a passive rotation is performed, we may assume that the basis set is also transformed according to relations of the form

$$\mathbf{f}'_j = g_{ij}\mathbf{f}_i\,, \quad j = 1, 2, 3\,, \tag{2.1.21}$$

but a vector $\mathbf{x} \in \mathcal{G}$ will remain fixed and its vertex coordinates in the new and initial system of coordinates will satisfy the relations

$$\mathbf{x} = x_i\mathbf{f}_i = x'_j\mathbf{f}'_j = x'_j g_{ij}\mathbf{f}_i\,, \tag{2.1.22}$$

which gives

$$x_i = g_{ij}x'_j\,, \quad j = 1, 2, 3\,, \tag{2.1.23}$$

or, in matrix notation,

$$\mathbf{x} = \mathbf{g}\mathbf{x}'\,, \quad \mathbf{x}' = \mathbf{g}^{-1}\mathbf{x} = \mathbf{g}^{\mathrm{T}}\mathbf{x}\,. \tag{2.1.24}$$

[3]Generally, this separation into active and passive transformations exists not only for rotations, but for any other set of transformations acting in a linear space.

Now, the quantities that transform like the components x_i in (2.1.24) are said to transform *contragrediently* with respect to the basis set $\{\mathbf{f}\}$. Hence, the same orthogonal matrix can be used to describe either a rotation of the space ($\mathbf{g}$), or a rotation of the coordinates system ($\mathbf{g}^{\mathrm{T}}$). Thus, the group of the orthogonal matrices, of determinant 1, is isomorphic not only to the group of rotations in the space $\mathcal{G}$ (passive rotations), but also to the group of rotations of the space $\mathcal{G}$ itself (active rotations). Because of this isomorphism these two rotation groups are both called *rotation group*.

It is worth-while to mention here the apparently difference with respect to the nomenclature used in tensor calculus, where a vector whose components transform cogrediently is called *contravariant*, while a vector whose components transform contragrediently is called *covariant*. The reason for this difference in nomenclature resides in the form of the relations which define the transformation of the basis vectors. Thus, if the transformation of the basis vectors is determined by a matrix $\mathbf{t}$ in the form

$$\mathbf{f}_i' = \mathbf{t}_{ij}\mathbf{f}_j\,, \quad j = 1, 2, 3\,,$$

and the components of a vector are transformed according to

$$\mathbf{x}' = \mathbf{t}^{\mathrm{T}}\mathbf{x}\,,$$

then we say, in tensor calculus, that $\mathbf{x}$ is a *contravariant vector*. By inspecting (2.1.18) and (2.1.20), one can easily see that $\mathbf{t} = \mathbf{g}^{\mathrm{T}}$ and, consequently, the transformation matrix for a *cogrediently vector* is $\mathbf{t}^{\mathrm{T}} = \mathbf{g}$. Also, for the same transformation of the basis vectors, we say, in tensor calculus, that $\mathbf{x}$ is a *covariant vector* if its components are changed according to

$$\mathbf{x}' = \left(\mathbf{t}^{\mathrm{T}}\right)^{-1}\mathbf{x}\,.$$

Now, in terms of the matrix $\mathbf{g}$, the transformation matrix of the components of $\mathbf{x}$ becomes

$$\left(\mathbf{t}^{\mathrm{T}}\right)^{-1} = \mathbf{g}^{-1} = \mathbf{g}^{\mathrm{T}}\,,$$

which corresponds to the definition of a *contragrediently vector* (2.1.24).

If we add the improper rotations, that is all the products of the form IG ($\mathsf{G} \in \mathsf{SO}(3)$), to the elements of $\mathsf{SO}(3)$, we obtain the group $\mathsf{O}(3)$. The inversion operator $\mathsf{I} = -\mathsf{E}$ commutes with the rotations such that the elements of $\mathsf{O}(3)$ satisfy the relations

$$\mathsf{G}_1\mathsf{I}\mathsf{G}_2 = \mathsf{I}\mathsf{G}_1\mathsf{G}_2\,, \quad (\mathsf{I}\mathsf{G}_1)(\mathsf{I}\mathsf{G}_2) = \mathsf{G}_1\mathsf{G}_2\,, \quad (\mathsf{I}\mathsf{G})(\mathsf{I}\mathsf{G}^{-1}) = \mathsf{E}\,.$$

To decompose the group $\mathsf{O}(3)$ in conjugacy classes, we substitute R into (2.1.17) by an improper rotation $\mathsf{H} = \mathsf{IR}$ ($\mathsf{R} \in \mathsf{SO}(3)$). Thus, we have

$$\mathsf{H}\,\mathsf{G}(\varphi, \mathbf{v})\,\mathsf{H}^{-1} = \mathsf{I}\,\mathsf{R}\,\mathsf{G}(\varphi, \mathbf{v})\,\mathsf{R}^{-1}\mathsf{I} = \mathsf{G}(\varphi, \mathsf{R}\mathbf{v}) = \mathsf{G}(\varphi, -\mathsf{H}\mathbf{v})\,. \qquad (2.1.25)$$

In the same way, we deduce the relations

$$\mathsf{H}\,\mathsf{S}(\varphi,\mathbf{v})\,\mathsf{H}^{-1}=\mathsf{S}(\varphi,-\mathsf{H}\mathbf{v})\,,\quad \mathsf{R}\,\mathsf{S}(\varphi,\mathbf{v})\,\mathsf{R}^{-1}=\mathsf{S}(\varphi,\mathsf{R}\mathbf{v})\,,$$

where $\mathsf{S}(\varphi,\mathbf{v})=\mathsf{I}\mathsf{G}(\varphi,\mathbf{v})$. It results that the set of all rotations of the same angle, as well as the set of all products inversion-rotation, of the same angle, form conjugacy classes of $\mathsf{O}(3)$.

In general, any displacement in the three-dimensional space can be described by a vector function $\mathbf{a}(\mathbf{r})$, which defines a displacement $\mathbf{a}$ of the vertex of the radius vector $\mathbf{r}$. The function $\mathbf{a}(\mathbf{r})$ must preserve the distances between any two points of the space. The total displacement resulting from two successive displacements is defined as the product of displacements, such that the set of all displacements of this type forms a group, called the *Euclidean group* $\mathsf{E}(3)$. Any element of the Euclidean group can be represented as the product of a rotation, or inversion-rotation, about an arbitrary point O, and a translation[4].

Any subgroup of the rotation group $\mathsf{O}(3)$ is called *point group* (also called *Crystal Class*, in crystallography). The transformations of a point group consist of rotations about an axis, reflections in a plane, inversions about a centre, or sequential rotation and inversion. The point groups which do not contain inversion-rotations are called *point groups of the first kind*, all the other point groups being of the second kind. If a point group C contains a rotation about an axis Δ, then Δ is called the axis of C. An example of point group is the cyclic group of rotations by angles of $2\pi/n$, $n\in\mathbb{Z}^*$, about an axis (see Chap. 1 Sec. 3.3).

1.2 Parametrization of $\mathsf{SO}(3)$ and $\mathsf{O}(3)$

1.2.1 Eulerian parametrization of $\mathsf{SO}(3)$

The simplest parametrization of the elements of the rotation group $\mathsf{SO}(3)$ is the *Eulerian parametrization* in terms of Euler's angles

$$0\le\varphi_0<2\pi\,,\quad 0\le\theta\le\pi\,,\quad 0\le\psi_0<2\pi\,.$$

Note that, the angles

$$\varphi=\varphi_0+2k_1\pi\,,\quad \theta\,,\quad \psi=\psi_0+2k_2\pi\,,\quad k_1,k_2\in\mathbb{Z}\,,$$

define the same rotation as φ_0, θ, ψ_0; hence, the elements of the matrix $\mathbf{g}$, associated to the rotation operator $\mathsf{G}\in\mathsf{SO}(3)$ in a given basis, satisfy the periodicity relations

$$g_{ij}(\varphi,\theta,\psi)=g_{ij}(\varphi+2k_1\pi,\theta,\psi+2k_2\pi)\,,\quad i,j=1,2,3\,,$$

[4]Details about $\mathsf{E}(3)$ will be discussed in Chap. 4 Sec. 2.2.2.

for $0 < \theta < \pi$. If $\theta = 0$ only the sum $\varphi + \psi$ is determined, while for $\theta = \pi$ only the difference $\varphi - \psi$ is determined.

In terms of Euler's angles any rotation of $\mathsf{SO}(3)$ is considered to be a product of three rotations:

- a rotation of angle φ about the Oz axis, which transforms the system of coordinates $Oxyz$ into $Ox'y'z'$ ($z' = z$),
- a rotation of angle θ about the Ox' axis, which transforms the system of coordinates $Ox'y'z'$ into $Ox''y''z''$ ($x'' = x'$),
- a rotation of angle ψ about the Oz'' axis.

Therefore, in matrix notation, we have

$$\mathbf{g}(\varphi, \theta, \psi) = \mathbf{g}(0, 0, \psi)\, \mathbf{g}(0, \theta, 0)\, \mathbf{g}(\varphi, 0, 0)$$

$$= \begin{bmatrix} \cos\varphi & -\sin\varphi & 0 \\ \sin\varphi & \cos\varphi & 0 \\ 0 & 0 & 1 \end{bmatrix} \begin{bmatrix} 1 & 0 & 0 \\ 0 & \cos\theta & -\sin\theta \\ 0 & \sin\theta & \cos\theta \end{bmatrix} \begin{bmatrix} \cos\psi & -\sin\psi & 0 \\ \sin\psi & \cos\psi & 0 \\ 0 & 0 & 1 \end{bmatrix} \qquad (2.1.26)$$

$$= \begin{bmatrix} \cos\varphi\cos\psi - \cos\theta\sin\varphi\sin\psi & -\cos\theta\cos\psi\sin\varphi - \cos\varphi\sin\psi & \sin\theta\sin\varphi \\ \cos\psi\sin\varphi + \cos\theta\cos\varphi\sin\psi & \cos\theta\cos\varphi\cos\psi - \sin\varphi\sin\psi & -\sin\theta\cos\varphi \\ \sin\theta\sin\psi & \cos\psi\sin\theta & \cos\theta \end{bmatrix}.$$

One can easily verify that, for $\theta = 0$, the matrix elements g_{ij} are periodic functions of $\varphi + \psi$, while for $\theta = \pi$, they are periodic functions of $\varphi - \psi$.

1.2.2 Exponential (canonical) parametrization of $\mathsf{SO}(3)$

Another parametrization of the elements of the rotation group $\mathsf{SO}(3)$ is the *exponential* (or *canonical*) *parametrization*[5]. In this parametrization the elements of the matrix $\mathbf{g}$, associated to the rotation operator $\mathsf{G} \in \mathsf{SO}(3)$ in a given basis, are expressed in terms of the quantities ξ_i, $i = 1, 2, 3$, defined in (2.1.9). Hence, we have to find the inverse of the relations (2.1.9).

As a first step, we multiply the quantities ξ_i by Ricci's symbol ϵ_{irs}, and sum with respect to the index i, to obtain (see (1.2.36))

$$\begin{aligned} \xi_i \epsilon_{irs} &= -\frac{1}{2}\frac{\varphi_0}{\sin\varphi_0}\,\epsilon_{irs}\,\epsilon_{ijk}\,g_{jk} = -\frac{1}{2}\frac{\varphi_0}{\sin\varphi_0}\,(\delta_{rj}\delta_{sk} - \delta_{rk}\delta_{sj})\,g_{jk} \\ &= -\frac{1}{2}\frac{\varphi_0}{\sin\varphi_0}\,(g_{rs} - g_{sr}), \end{aligned}$$

that is

$$g_{rs} - g_{sr} = -2\,\frac{\sin\varphi_0}{\varphi_0}\,\xi_i\,\epsilon_{irs}\,, \qquad r, s = 1, 2, 3. \qquad (2.1.27)$$

[5]The reasons for the name given to this parametrization will become clear in Chap. 2 Sec. 2.2.3.

Then, we consider a transformation of the basis of $\mathcal{G}$, defined by an orthogonal matrix $\mathbf{c}$, that transforms the matrix $\mathbf{g}^0$ (see (2.1.4)) into the matrix $\mathbf{g}$ according to the relation $\mathbf{g} = \mathbf{c}^{-1}\mathbf{g}^0\mathbf{c} = \mathbf{c}^{\mathrm{T}}\mathbf{g}^0\mathbf{c}$. Detailed, we can write

$$g_{ij} = c_{ki}\, g^0_{kr}\, c_{rj}\,, \quad i,j = 1,2,3, \tag{2.1.28}$$

hence

$$g_{ij} + g_{ji} = 2\cos\varphi_0\,\delta_{ij} + 2(1-\cos\varphi_0)\,c_{3i}\,c_{3j}\,. \tag{2.1.29}$$

From (2.1.12), and with $\xi^0_1 = \xi^0_2 = 0$, $\xi^0_3 = \varphi_0$, we find

$$\xi_j = (\det\mathbf{c})\,c_{ij}\,\xi^0_i = (\det\mathbf{c})\,c_{3j}\,\varphi_0\,, \tag{2.1.30}$$

and

$$c_{3i}\,c_{3j} = \left(\frac{1}{\det\mathbf{c}}\right)^2 \frac{\xi_i\,\xi_j}{\varphi_0^2} = \frac{\xi_i\,\xi_j}{\varphi_0^2}\,.$$

Finally, by summing (2.1.27) and (2.1.29) we obtain the elements of the matrix $\mathbf{g}$ in terms of the quantities ξ_i, in a given basis,

$$g_{ij} = \cos\varphi_0\,\delta_{ij} + \frac{1-\cos\varphi_0}{\varphi_0^2}\,\xi_i\,\xi_j - \frac{\sin\varphi_0}{\varphi_0}\,\epsilon_{ijk}\,\xi_k\,, \tag{2.1.31}$$

for $i,j = 1,2,3$. We may also employ the relation $\varphi_0^2 = \xi_l\xi_l$ from (2.1.14), to rewrite (2.1.31) in the form

$$g_{ij} = \cos\sqrt{\xi_l\xi_l}\,\delta_{ij} + \frac{1-\cos\sqrt{\xi_l\xi_l}}{\xi_l\xi_l}\,\xi_i\,\xi_j - \frac{\sin\sqrt{\xi_l\xi_l}}{\sqrt{\xi_l\xi_l}}\,\epsilon_{ijk}\,\xi_k\,. \tag{2.1.32}$$

The quantities ξ_i completely determine the matrix $\mathbf{g}$ associated to a proper rotation G, and are called the *exponential parameters* of the $\mathsf{SO}(3)$ group. The exponential parameters also satisfy the relation (2.1.14)

$$\|\boldsymbol{\xi}\|^2 = \xi_i\xi_i = \varphi_0^2\,,$$

where $\boldsymbol{\xi} = (\xi_1,\xi_2,\xi_3)$, and $\varphi_0 \in [-\pi,\pi]$. Consequently, the group space is the volume of a three-dimensional sphere of radius π. Note that the identity element of $\mathsf{SO}(3)$ is determined by the exponential parameters $\xi_1 = \xi_2 = \xi_3 = 0$, and is associated to the identity matrix ($g_{ij} = \delta_{ij}$).

A simple example will illustrate the geometrical meaning of (2.1.31) and (2.1.32). Thus, in Fig. 2.2 (A) and (B) an active rotation of an angle φ_0 about the axis Δ has been applied to the vector $\mathbf{x} \equiv \overrightarrow{OM} = (x_1,x_2,x_3)$, and this rotation changed the vector $\mathbf{x}$ into $\mathbf{x}' \equiv \overrightarrow{OM'} = (x'_1,x'_2,x'_3)$. According to (2.1.20), the components of $\mathbf{x}'$ are given by the formulae

$$x'_i = g_{ij}x_j\,, \quad i = 1,2,3\,, \tag{2.1.33}$$

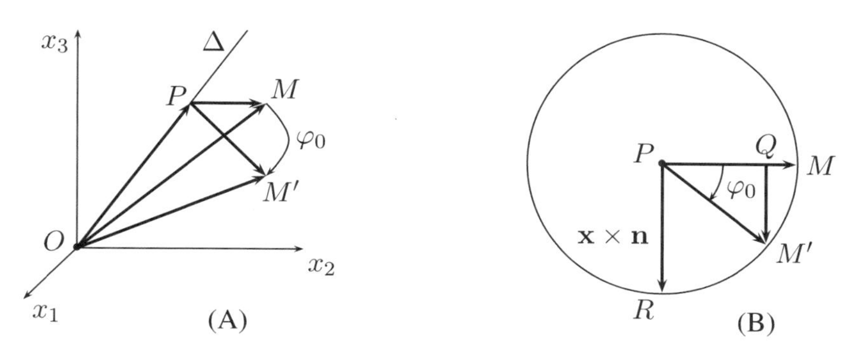

Figure 2.2. The vector $\overrightarrow{OM}$ is transformed into $\overrightarrow{OM}'$ by an active rotation of an angle φ_0 about the axis Δ, in a clockwise sense when looking towards the origin along Δ.

and, substituting (2.1.31) into (2.1.33) we obtain

$$x_i' = \cos\varphi_0\, x_i + \frac{1-\cos\varphi_0}{\varphi_0^2}\,\xi_i\,\xi_j\,x_j - \frac{\sin\varphi_0}{\varphi_0}\,\epsilon_{ijk}\,\xi_k\,x_j\,. \tag{2.1.34}$$

In vector notation (2.1.34) reads

$$\mathbf{x}' = \mathbf{x}\,\cos\varphi_0 + \mathbf{n}(\mathbf{n}\cdot\mathbf{x})\,(1-\cos\varphi_0) + (\mathbf{x}\times\mathbf{n})\,\sin\varphi_0\,, \tag{2.1.35}$$

where $n_i = \xi_i/\varphi_0$.

On the other hand, the points P, M, and M' in Fig. 2.2 (A) determine a plane perpendicular to the axis Δ, such that

$$\mathbf{x}' \equiv \overrightarrow{OM}' = \overrightarrow{OP} + \overrightarrow{PQ} + \overrightarrow{QM}'\,. \tag{2.1.36}$$

A rotation preserves the length of the vectors such that $\|\mathbf{x}'\| = \|\mathbf{x}\| = x$, and if we assume that $\mathbf{n}$ is the unit vector along the axis Δ, then we have

$$\overrightarrow{OP} = (\mathbf{n}\cdot\mathbf{x})\mathbf{n} = (\mathbf{n}\cdot\mathbf{x}')\mathbf{n}\,.$$

Also $\overrightarrow{PM} = \mathbf{n}\times(\mathbf{x}\times\mathbf{n}) = \mathbf{x} - (\mathbf{n}\cdot\mathbf{x})\mathbf{n}$, $\|\overrightarrow{PM}\| = \|\overrightarrow{PM}'\|$, and on account of $\overrightarrow{QM}'$ being perpendicular to $\overrightarrow{PM}$ we obtain

$$\overrightarrow{PQ} = \overrightarrow{PM}'\,\cos\varphi_0 = [\mathbf{x} - (\mathbf{n}\cdot\mathbf{x})\mathbf{n}]\,\cos\varphi_0\,.$$

Finally, with $\|\overrightarrow{PR}\| = \|\overrightarrow{PM}'\|$, and

$$\overrightarrow{QM}' = \overrightarrow{PR}\,\sin\varphi_0 = (\mathbf{x}\times\mathbf{n})\,\sin\varphi_0\,,$$

we are lead to the well-known formula for the rotation of a rigid body

$$\mathbf{x}' = \mathbf{x}\,\cos\varphi_0 + \mathbf{n}(\mathbf{n}\cdot\mathbf{x})\,(1-\cos\varphi_0) + (\mathbf{x}\times\mathbf{n})\,\sin\varphi_0\,, \tag{2.1.37}$$

that is identical to (2.1.35). Therefore, if a set of parameters $\boldsymbol{\xi} = \{\xi_1, \xi_2, \xi_3\}$ is given, then the corresponding matrix $\mathbf{g}(\xi_1, \xi_2, \xi_3)$ represents a rotation of an angle $|\varphi_0| = \sqrt{\xi_i \xi_i}$, about an axis Δ determined by the unit vector $\mathbf{n} = \boldsymbol{\xi}/|\varphi_0|$. Also, according to the conclusions from the proof of Theorem 2.1.5, regardless of $\varphi_0 > 0$ or $\varphi_0 < 0$, the rotation is of an angle $|\varphi_0|$ about the axis Δ, in the direct sense.

1.2.3 Parametrization of $\mathsf{O}(3)$

We have shown that the $\mathsf{O}(3)$ group comprises the subgroup of proper rotations $\mathsf{SO}(3)$, having $\det \mathbf{g} = 1$, and the set of improper rotations, having $\det \mathbf{g} = -1$, which does not form a group. As in the case of proper rotations, a parametrization of improper rotations can be made in terms of three continuous parameters α_1, α_2, and α_3. Consequently, the parametrization of $\mathsf{O}(3)$ is obtained by adding a fourth parameter $\alpha_4 = \det \mathbf{g}$.

1.3 Functions defined on $\mathsf{O}(3)$. Infinitesimal generators

1.3.1 Functions defined on $\mathsf{O}(3)$. The invariant integral

DEFINITION 2.1.3 *Let G be a topological group. A mapping that associates to each element* $\mathrm{g} \in G$ *a unique value* $F(\mathrm{g})$ *is called* **uniform** *(or* **single-valued***)* **function** *on G.*

In the case of the parametrized $\mathsf{O}(3)$ group, the value $F(\mathrm{g})$ is given by a numerical function $f(\alpha_1, \alpha_2, \alpha_3, \alpha_4)$, which has the periodicity property

$$f(\xi_i^0, \alpha_4) = f(\xi_i^0 + 2k\pi\xi_i^0/\varphi_0, \alpha_4)\,, \quad k \in \mathbb{Z}\,, \tag{2.1.38}$$

in the exponential parametrization, or

$$f(\varphi, \theta, \psi, \alpha_4) = f(\varphi + 2k_1\pi, \theta, \psi + 2k_2\pi, \alpha_4)\,, \quad k_1, k_2 \in \mathbb{Z}\,, \tag{2.1.39}$$

for $0 < \theta < \pi$, in the Eulerian parametrization. Without the periodicity property (2.1.38) the function f is a *multiple-valued function* on the group, because the same rotation $\mathsf{G} \in \mathsf{O}(3)$ corresponds to an infinite set of parameter values $\xi_i = \xi_i^0 + 2k\pi\xi_i^0/\varphi_0$.

DEFINITION 2.1.4 *The differential of the function $F(\mathsf{G})$, $\mathsf{G} \in \mathsf{O}(3)$, is given by the formula*

$$dF(\mathsf{G}) = \frac{\partial f}{\partial \alpha_i}\, d\alpha_i\,. \tag{2.1.40}$$

The expression of the differential of $F(\mathsf{G})$ has to be independent of the chosen system of parameters.

Thus, to a continuous and uniform function $F(\mathsf{G})$ on $\mathsf{O}(3)$ there corresponds a continuous and periodic function f. Note that, the group of proper rotations $\mathsf{SO}(3)$ is a continuous and connected group (i.e., a simple topological group), and represents a typical example of a Lie group. Improper rotations contain an inversion, which is a discrete transformation, and therefore $\mathsf{O}(3)$ is a continuous, but not connected group (i.e, a mixed topological group). Nevertheless, $\mathsf{SO}(3)$ and $\mathsf{O}(3)$ are compact Lie groups, such that the invariant integral exists for both these groups.

In the Eulerian parametrization, the invariant integral on $\mathsf{O}(3)$ is defined by the relation

$$\int F(\mathsf{G})\,d\mathsf{G} = \frac{1}{8\pi^2}\int_0^{2\pi}\int_0^{\pi}\int_0^{2\pi} f(\varphi,\theta,\psi)\,\sin\theta\,d\varphi\,d\theta\,d\psi\,. \tag{2.1.41}$$

For a continuous and uniform function $F(\mathsf{G})$ on $\mathsf{O}(3)$ we can write

$$\int F(\mathsf{G})\,d\mathsf{G} = \int F(\mathsf{G}\mathsf{G}_0)\,d\mathsf{G} = \int F(\mathsf{G}_0\mathsf{G})\,d\mathsf{G}\,, \tag{2.1.42}$$

where G_0 is an arbitrary, fixed element of $\mathsf{O}(3)$. If we associate the parameters $\{\varphi,\theta,\psi\}$ to the rotation G, and the parameters $\{\varphi_0,\theta_0,\psi_)\}$ to the rotation G_0, then the rotation $\mathsf{G}\mathsf{G}_0$ will be determined by the parameters $\{\varphi',\theta',\psi'\}$ that can be obtained by means of the matrix (2.1.26). Consequently, the invariance property (2.1.42) becomes

$$\frac{1}{8\pi^2}\int_0^{2\pi}\int_0^{\pi}\int_0^{2\pi} f(\varphi,\theta,\psi)\,\sin\theta\,d\varphi\,d\theta\,d\psi$$
$$= \frac{1}{8\pi^2}\int_0^{2\pi}\int_0^{\pi}\int_0^{2\pi} h(\varphi,\theta,\psi)\,\sin\theta\,d\varphi\,d\theta\,d\psi\,,$$

where $h(\varphi,\theta,\psi) = f(\varphi',\theta',\psi')$. One can easily prove that

$$\int d\mathsf{G} = 1\,.$$

1.3.2 Infinitesimal generators

Let us consider the function $F(\mathsf{G}) = \mathsf{G}$ defined on $\mathsf{SO}(3)$. According to (2.1.32) and (2.1.26), the rotations G are continuous functions of the group parameters. The *infinitesimal generators* of $\mathsf{SO}(3)$, corresponding to the basis in which the matrix g has been determined, will be given by the formulae (1.2.31), that is

$$\mathsf{A}_i = \left[\frac{\partial\mathsf{G}(\xi_1,\xi_2,\xi_3)}{\partial\xi_i}\right]_{\xi_1=\xi_2=\xi_3=0}\,,\quad i=1,2,3\,. \tag{2.1.43}$$

The matrices of these operators are obtained by differentiating the matrix **g** with respect to ξ_i and setting $\xi_1 = \xi_2 = \xi_3 = 0$. We may also introduce the operators

$$\mathsf{H}_k = i\mathsf{A}_k\,, \quad k = 1, 2, 3, \tag{2.1.44}$$

called *infinitesimal Hermitian generators* of $\mathsf{SO}(3)$, because their matrices are Hermitian.

The matrices of the infinitesimal generators of $\mathsf{SO}(3)$ have been shown in (1.2.32) and (1.2.33). Note that, according to (1.2.34) the infinitesimal Hermitian generators will verify the relations

$$\mathsf{H}_j\mathsf{H}_k - \mathsf{H}_k\mathsf{H}_j = i\,\epsilon_{jkl}\,\mathsf{H}_l\,, \quad j, k = 1, 2, 3.$$

2. The $\mathsf{SU}(2)$ group

2.1 Parametrization of $\mathsf{SU}(2)$

2.1.1 Generalities

Let $\mathcal{U}^2$ be a two-dimensional unitary space and U a unitary operator acting in $\mathcal{U}^2$. The matrix $\mathbf{u}$ associated to the operator U, in a given basis, satisfies the relations $\mathbf{u}\mathbf{u}^\dagger = \mathbf{u}^\dagger\mathbf{u} = \mathbf{e}$, where $\mathbf{e}$ is the identity matrix. Hence, $|\det \mathbf{u}|^2 = 1$, and $\det \mathbf{u} = \exp(i\varphi)$ with $\varphi \in \mathbb{R}$. If $\det \mathbf{u} = 1$, then the operator U is said to be a *unimodular unitary* operator.

DEFINITION 2.2.1 *The set of unimodular unitary operators acting in $\mathcal{U}^2$, with the multiplication of operators as internal composition law, forms a group called the* **two-dimensional unimodular** *(***special***)* **unitary group**, *denoted by* $\mathsf{SU}(2)$.

The elements of $\mathsf{SU}(2)$ can be characterized by three continuous parameters. Also, a representation of $\mathsf{SU}(2)$ in the complex two-dimensional space $\mathbb{C}^2$ consists of unitary complex matrices, having the determinant equal to 1.

2.1.2 Cayley-Klein parametrization

Let $\{\mathbf{f}\}$ be an orthonormal basis in the two-dimensional unitary space $\mathcal{U}^2$. Also, let

$$\mathbf{u} = \begin{bmatrix} a & b \\ c & d \end{bmatrix}, \quad a, b, c, d \in \mathbb{C}\,, \tag{2.2.1}$$

be the matrix associated to a unitary operator U, in this basis. Then, the Hermitian conjugate of $\mathbf{u}$ is

$$\mathbf{u}^\dagger = \begin{bmatrix} \overline{a} & \overline{c} \\ \overline{b} & \overline{d} \end{bmatrix},$$

such that the unitarity condition $\mathbf{u}\mathbf{u}^\dagger = \mathbf{u}^\dagger\mathbf{u} = \mathbf{e}$, and the condition $\det \mathbf{u} = 1$, lead to the relations

$$|a|^2 + |b|^2 = 1\,, \quad |c|^2 + |d|^2 = 1\,,$$

$$a\bar{c} + b\bar{d} = 0\,, \quad \bar{a}c + \bar{b}d = 0\,, \tag{2.2.2}$$
$$ad - bc = 1\,.$$

Note that only four of these relations are independent. Now, from relation sequences of the form

$$\bar{b} = \bar{b}(ad - bc) = a(\bar{b}d) - |b|^2 c = a(-\bar{a}c) - |b|^2 c = -c(|a|^2 + |b|^2) = -c\,,$$

we obtain $\bar{b} = -c$, and analogously $\bar{d} = a$. Thus, a unitary unimodular matrix $\mathbf{u}$ has the form

$$\mathbf{u} = \begin{bmatrix} a & b \\ -\bar{b} & \bar{a} \end{bmatrix}, \quad |a|^2 + |b|^2 = 1\,. \tag{2.2.3}$$

The complex numbers a and b are called the *Cayley-Klein parameters* of the $\mathsf{SU}(2)$ group and, as any other parametrization, they depend on the chosen basis in the unitary space $\mathcal{U}^2$.

The parameters a and b can also be represented in the form

$$a = \alpha_0 + i\alpha_3\,, \quad b = \alpha_2 + i\alpha_1\,, \tag{2.2.4}$$

where $\alpha_0, \alpha_1, \alpha_2, \alpha_3 \in \mathbb{R}$ satisfy the relation

$$\alpha_0^2 + \alpha_1^2 + \alpha_2^2 + \alpha_3^2 = 1\,. \tag{2.2.5}$$

The real numbers α_j ($j = 0, 1, 2, 3$) are called the *real Cayley-Klein parameters* of the $\mathsf{SU}(2)$ group[6]. In terms of the real Cayley-Klein parameters the elements of $\mathsf{SU}(2)$ can be written in the form

$$\mathbf{u} = \alpha_0 \mathbf{e} + i\left(\alpha_1 \boldsymbol{\sigma}_1 + \alpha_2 \boldsymbol{\sigma}_2 + \alpha_3 \boldsymbol{\sigma}_3\right), \tag{2.2.6}$$

where $\mathbf{e}$ is the two-by-two identity matrix, and

$$\boldsymbol{\sigma}_1 = \begin{bmatrix} 0 & 1 \\ 1 & 0 \end{bmatrix}, \quad \boldsymbol{\sigma}_2 = \begin{bmatrix} 0 & -i \\ i & 0 \end{bmatrix}, \quad \boldsymbol{\sigma}_3 = \begin{bmatrix} 1 & 0 \\ 0 & -1 \end{bmatrix} \tag{2.2.7}$$

are the Pauli spin matrices.

From (2.2.5) it also follows that we can associate to the elements of $\mathsf{SU}(2)$ the points on the surface of a hypersphere of unit radius, in the four-dimensional space $\mathcal{E}_4$, called *three-sphere*. Whereas the surface of a sphere in $\mathcal{E}_3$ has dimension two and is therefore a two-dimensional manifold, the surface of a three-sphere has dimension three and is a three-dimensional manifold.

---

[6] We may also choose as independent parameters the real numbers $\alpha_1, \alpha_2, \alpha_3$ and $\operatorname{sgn}\alpha_0 = \alpha_0/|\alpha_0|$, the quantity α_0 being given by (2.2.5).

Also, there exists a one-to-one correspondence between the points on a three-sphere and the *unit quaternions*. The field of *quaternions* belongs to the more general class of *hypercomplex numbers*, introduced by Hamilton. They form a non-Abelian group of order eight (with multiplication as the group operation) known as the *quaternion group*. A quaternion can be written in the form

$$q = q_0 + q_1 i + q_2 j + q_3 k \,, \tag{2.2.8}$$

where the quantities i, j, and k satisfy Hamilton's rules: $i^2 = j^2 = k^2 = -1$, $ij = -ji = k$, $jk = -kj = i$, $ki = -ik = j$. The conjugate of a quaternion is

$$\overline{q} = q_0 - q_1 i - q_2 j - q_3 k \,.$$

Quaternions can be interpreted as comprising a scalar (q_1) and a vector ($\mathbf{q}$) by writing $q = (q_1, \mathbf{q})$, where $\mathbf{q} = (q_1, q_2, q_3)$, and in this notation, the multiplication of two quaternions is given by

$$p * q = (p_1 q_1 - \mathbf{p} \cdot \mathbf{q},\; p_1 \mathbf{q} + q_1 \mathbf{p} + \mathbf{p} \times \mathbf{q}) \,,$$

where $p = (p_1, \mathbf{p})$. Note that the quaternion multiplication is not commutative.

A rotation of an angle φ about the unit vector $\mathbf{n} = (n_1, n_2, n_3)$ can be represented in the form of a unit quaternion

$$q = \left(\cos \frac{\varphi}{2},\, \mathbf{n} \sin \frac{\varphi}{2} \right) , \tag{2.2.9}$$

having the norm $\|q\|^2 \equiv q * \overline{q} = 1$. The quaternions can also be represented as complex two-by-two matrices, the basis of this set of matrices being closely related to the Pauli spin matrices. The unit quaternions correspond to special unitary two-by-two matrices, and form a Lie group isomorphic to $\mathsf{SU}(2)$. It is important to note that the unit quaternions q and $-q$ represent the same rotation such that the group of unit quaternions, as well as $\mathsf{SU}(2)$, is the double cover of the group $\mathsf{SO}(3)$. Hence, the group $\mathsf{SU}(2)$ is homomorphic to the group of proper rotations $\mathsf{SO}(3)$.

2.1.3 Exponential parametrization

From (2.2.3) and (2.2.4) we obtain the trace of the matrix $\mathbf{u}$ in the form

$$\mathsf{Tr}\, \mathbf{u} = a + \overline{a} = 2\alpha_0 \,. \tag{2.2.10}$$

The trace of the matrix associated to an operator does not depend on the basis, hence α_0 is an invariant with respect to the choice of the basis $\{\mathbf{f}\}$. If $\{\mathbf{f}\}$ is the basis in which the matrix $\mathbf{u}$, associated to the unimodular unitary operator $\mathsf{U} \in \mathsf{SU}(2)$, is diagonal, then according to (2.2.3) we have $b = \overline{b} = 0$ and

$|a|^2 = 1$, respectively $a = \exp(i\psi)$ with $\psi \in \mathbb{R}$. Therefore, the matrix $\mathbf{u}$ takes the form

$$\mathbf{u} = \begin{bmatrix} e^{i\psi} & 0 \\ 0 & e^{-i\psi} \end{bmatrix} \tag{2.2.11}$$

and, by denoting $\psi = \widetilde{\varphi}/2$, the relation (2.2.10) leads to

$$\alpha_0 = \cos(\widetilde{\varphi}/2)\,. \tag{2.2.12}$$

Starting with an arbitrary solution of (2.2.12) we introduce the quantities

$$\widetilde{\xi}_i = 2\alpha_i \frac{\widetilde{\varphi}/2}{\sin(\widetilde{\varphi}/2)}\,, \qquad i = 1, 2, 3, \tag{2.2.13}$$

that are defined for all values of the parameters α_i ($i = 1, 2, 3$), excepting the case $\alpha_1 = \alpha_2 = \alpha_3 = 0$, when $\alpha_0 = \pm 1$ and $\sin(\widetilde{\varphi}/2) = 0$. In this latter case the quantities $\widetilde{\xi}_i$ are obtained as a limiting process for $\widetilde{\varphi} \to 0$. The system of parameters $\widetilde{\xi}_i$, called *exponential parameters*, characterizes the operators U in the basis $\{\mathbf{f}\}$, and consequently they can replace the real Cayley-Klein parameters.

To find the inverse of (2.2.13) we observe that

$$\begin{aligned}\widetilde{\xi}_i\widetilde{\xi}_i &= \alpha_i\alpha_i \left[\frac{\widetilde{\varphi}}{\sin(\widetilde{\varphi}/2)}\right]^2 = (1 - \alpha_0^2)\left[\frac{\widetilde{\varphi}}{\sin(\widetilde{\varphi}/2)}\right]^2 \\ &= [1 - \cos^2(\widetilde{\varphi}/2)]\left[\frac{\widetilde{\varphi}}{\sin(\widetilde{\varphi}/2)}\right]^2 = \widetilde{\varphi}^2\,,\end{aligned}$$

that is

$$|\widetilde{\varphi}| = \sqrt{\widetilde{\xi}_i\widetilde{\xi}_i}\,. \tag{2.2.14}$$

Thus, in terms of the exponential parameters, the real Cayley-Klein parameters have the expressions

$$\alpha_0 = \cos\left(\sqrt{\widetilde{\xi}_j\widetilde{\xi}_j}/2\right), \quad \alpha_i = \frac{\sin\left(\sqrt{\widetilde{\xi}_j\widetilde{\xi}_j}/2\right)}{\sqrt{\widetilde{\xi}_k\widetilde{\xi}_k}}\,\widetilde{\xi}_i\,, \quad i = 1, 2, 3. \tag{2.2.15}$$

To a given system of exponential parameters there corresponds a well defined operator $\mathsf{U} \in \mathsf{SU}(2)$, while reciprocally, to a given operator U there corresponds an infinite number of systems of exponential parameters. Indeed, the equation (2.2.12) has an infinity of solutions $\widetilde{\varphi}$, such that the most general form of the solutions $\widetilde{\varphi}$, which also satisfy (2.2.13) and (2.2.14), is

$$\begin{aligned}\widetilde{\varphi} &= \widetilde{\varphi}_0 + 4n\pi\,, && 0 < \widetilde{\varphi}_0 < 2\pi\,, && n \in \mathbb{Z}\,, \\ \widetilde{\varphi}' &= \widetilde{\varphi}'_0 + 4n'\pi\,, && -2\pi < \widetilde{\varphi}'_0 < 0\,, && n' \in \mathbb{Z}\,.\end{aligned} \tag{2.2.16}$$

In both cases of (2.2.16) we obtain the sequence of exponential parameters

$$\widetilde{\xi}_i^{(k)} = \widetilde{\xi}_i^{(0)} \left(1 + \frac{4k\pi}{\sqrt{\widetilde{\xi}_j^{(0)} \widetilde{\xi}_j^{(0)}}} \right) , \quad i = 1, 2, 3, \quad k \in \mathbb{Z} , \tag{2.2.17}$$

and, for a given system of exponential parameters $\{\widetilde{\xi}_i^{(0)}\}$, all the systems $\{\widetilde{\xi}_i^{(k)}\}$ are associated to the same unimodular unitary operator U. From (2.2.17) it also follows that

$$\widetilde{\xi}_i^{(k)} \widetilde{\xi}_i^{(k)} = \left(\sqrt{\widetilde{\xi}_j^{(0)} \widetilde{\xi}_j^{(0)}} + 4k\pi \right)^2 , k \in \mathbb{Z} ,$$

where $\sqrt{\widetilde{\xi}_j^{(0)} \widetilde{\xi}_j^{(0)}} = \widetilde{\varphi}_0 < 2\pi$. In the cases $\alpha_1 = \alpha_2 = \alpha_3 = 0$, $\alpha_0 = \pm 1$, when $\cos(\widetilde{\varphi}/2) = \pm 1$ and $\sin(\widetilde{\varphi}/2) = 0$, by a limiting process we obtain $\mathbf{u} = \mathbf{e}$ and $\mathbf{u} = -\mathbf{e}$, respectively.

The Cayley-Klein parametrization leads to a matrix $\mathbf{u}$ of the form

$$\mathbf{u} = \alpha_0 \begin{bmatrix} 1 & 0 \\ 0 & 1 \end{bmatrix} + i \begin{bmatrix} \alpha_3 & \alpha_1 - i\alpha_2 \\ \alpha_1 + i\alpha_2 & -\alpha_3 \end{bmatrix} , \tag{2.2.18}$$

which, in terms of the exponential parameters, reads

$$\mathbf{u} = \begin{bmatrix} 1 & 0 \\ 0 & 1 \end{bmatrix} \cos(\sqrt{\widetilde{\xi}_j \widetilde{\xi}_j}/2) + i \begin{bmatrix} \widetilde{\xi}_3 & \widetilde{\xi}_1 - i\widetilde{\xi}_2 \\ \widetilde{\xi}_1 + i\widetilde{\xi}_2 & -\widetilde{\xi}_3 \end{bmatrix} \frac{\sin(\sqrt{\widetilde{\xi}_j \widetilde{\xi}_j}/2)}{\sqrt{\widetilde{\xi}_k \widetilde{\xi}_k}} . \tag{2.2.19}$$

As a function of the exponential parameters defined in (2.2.17), the matrix $\mathbf{u}$ satisfies the periodicity relations

$$\mathbf{u}\left(\widetilde{\xi}_1^{(k)}, \widetilde{\xi}_2^{(k)}, \widetilde{\xi}_3^{(k)} \right) = \mathbf{u}\left(\widetilde{\xi}_1^{(0)}, \widetilde{\xi}_2^{(0)}, \widetilde{\xi}_3^{(0)} \right) . \tag{2.2.20}$$

By means of the exponential parametrization of the $\mathsf{SO}(3)$ and $\mathsf{SU}(2)$ groups, one can show that to each operator $\mathsf{G} \in \mathsf{SO}(3)$ there correspond the operators $\pm\mathsf{U} \in \mathsf{SU}(2)$. This is another way to prove the following theorem (see also the end of Chap. 2 Sec. 2.1.2):

THEOREM 2.2.1 *The* $\mathsf{SU}(2)$ *group is homomorphic to the* $\mathsf{SO}(3)$ *group.*

2.1.4 Eulerian parametrization

The elements of the matrix $\mathbf{u}$, defined in (2.2.3), satisfy the relation

$$\det \mathbf{u} = |a|^2 + |b|^2 = 1 ,$$

which implies $|a| \leq 1$ and $|b| \leq 1$, such that we may assume that

$$|a| = \cos(\theta/2)\,, \quad |b| = \sin(\theta/2)\,, \tag{2.2.21}$$

where $0 \leq \theta \leq \pi$ is uniquely determined by $|a|$ and $|b|$. For the *Eulerian parametrization* of $\mathsf{SU}(2)$, the other two angles (φ and ψ) are introduced as functions of the arguments of the complex numbers a and b. Thus, we may choose

$$\arg a = -(\varphi + \psi)/2\,, \quad \arg b = -(\varphi - \psi + \pi)/2\,, \tag{2.2.22}$$

that gives

$$\begin{aligned} \varphi &= -\arg a - \arg b - \pi/2\,, \\ \psi &= -\arg a + \arg b + \pi/2\,. \end{aligned} \tag{2.2.23}$$

Because $\arg a \in [0, 2\pi)$ and $\arg b \in [0, 2\pi)$, the domains of the parameters φ and ψ are, respectively $\varphi \in (-9\pi/2, -\pi/2]$ and $\psi \in [-3\pi/2, 5\pi/2)$, φ and ψ being thus defined abstraction of a multiple of 4π. With these notations, the matrix $\mathbf{u}$ takes the form

$$\mathbf{u} = \begin{bmatrix} \cos\frac{\theta}{2}\, e^{-i(\psi+\varphi)/2} & -i\,\sin\frac{\theta}{2}\, e^{i(\psi-\varphi)/2} \\ -i\,\sin\frac{\theta}{2}\, e^{-i(\psi-\varphi)/2} & \cos\frac{\theta}{2}\, e^{i(\psi+\varphi)/2} \end{bmatrix}. \tag{2.2.24}$$

If the matrix elements a and b are given, then there exists an infinite set of systems of angles $\{\varphi, \theta, \psi\}$ corresponding to the same operator $\mathsf{U} \in \mathsf{SU}(2)$. Reciprocally, a given system of angles $\{\varphi, \theta, \psi\}$ uniquely determines the matrix elements a and b, as well the operator U.

Because there exists a one-to-one correspondence between the Cayley-Klein parameters and the elements of $\mathsf{SU}(2)$, we may use the relations between the Cayley-Klein parameters and Euler's angles to find a certain variation domain of these angles, such that the correspondence between Euler's angles and the elements of $\mathsf{SU}(2)$ be also one-to-one. Thus, from (2.2.18) and (2.2.24) we obtain

$$\begin{aligned} \cos\tfrac{\theta}{2}\, e^{-i(\psi+\varphi)/2} &= \alpha_0 + i\alpha_3\,, \\ -i\,\sin\tfrac{\theta}{2}\, e^{i(\psi-\varphi)/2} &= \alpha_2 + i\alpha_1\,, \end{aligned}$$

hence

$$\begin{aligned} \alpha_0 &= \cos\tfrac{\theta}{2}\,\cos\tfrac{\psi+\varphi}{2}\,, & \alpha_1 &= -\sin\tfrac{\theta}{2}\,\cos\tfrac{\psi-\varphi}{2}\,, \\ \alpha_2 &= \sin\tfrac{\theta}{2}\,\sin\tfrac{\psi-\varphi}{2}\,, & \alpha_3 &= -\cos\tfrac{\theta}{2}\,\sin\tfrac{\psi+\varphi}{2}\,. \end{aligned} \tag{2.2.25}$$

Consequently, two systems of Euler's angles $\{\varphi, \theta, \psi\}$ and $\{\varphi + 4\pi, \theta, \psi + 4\pi\}$ lead to the same system of Cayley-Klein parameters $\{\alpha_0, \alpha_1, \alpha_2, \alpha_3\}$. The

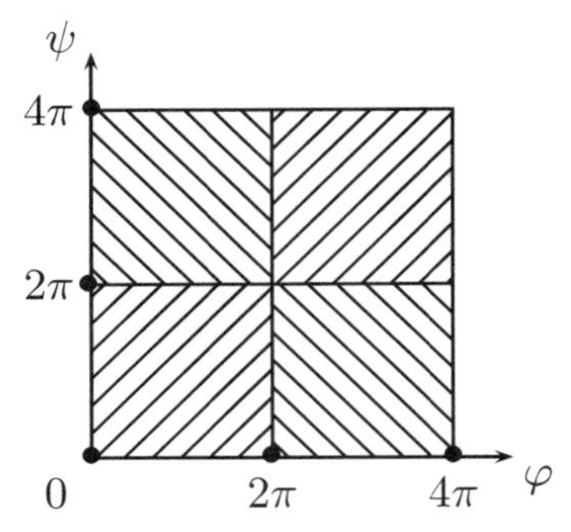

Figure 2.3. The domains defining a one-to-one correspondence between the elements of SU(2) and Euler's angles.

regions identically hatched in Fig. 2.3 correspond to the same operator U, and we have thus a one-to-one correspondence between the elements of $\mathsf{SU}(2)$ and Euler's angles if, for instance, we restrict the variation domain of $\{\varphi, \theta, \psi\}$ to

$$\varphi \in [0, 4\pi)\,, \quad \theta \in [0, \pi]\,, \quad \psi \in [0, 2\pi]\,.$$

2.2 Functions defined on $\mathsf{SU}(2)$. Infinitesimal generators

2.2.1 Functions defined on $\mathsf{SU}(2)$

Any function $\widetilde{F}$ defined on a group generates a corresponding function $\widetilde{f}$ defined on the group space. Thus, in terms of the Cayley-Klein parametrization we have

$$\mathsf{U} \to \widetilde{F}(\mathsf{U}) \to \widetilde{f}(\alpha_0/|\alpha_0|, \alpha_1, \alpha_2, \alpha_3)\,, \tag{2.2.26}$$

where $\mathsf{U} \in \mathsf{SU}(2)$, and α_1, α_2, and α_3 are independent continuous parameters. In terms of the exponential parametrization, the same function $\widetilde{F}$ will generate a different function $\widetilde{g}$, that is

$$\mathsf{U} \to \widetilde{F}(\mathsf{U}) \to \widetilde{g}(\widetilde{\xi}_1, \widetilde{\xi}_2, \widetilde{\xi}_3)\,. \tag{2.2.27}$$

If $\widetilde{F}$ is a uniformly continuous function on $\mathsf{SU}(2)$, then $\widetilde{g}$ is periodic, because to the same operator U there correspond all the systems of parameters $\{\widetilde{\xi}_1^{(k)}, \widetilde{\xi}_2^{(k)}, \widetilde{\xi}_3^{(k)}\}$ defined in (2.2.17), and we can write

$$\widetilde{g}\left(\widetilde{\xi}_1^{(k)}, \widetilde{\xi}_2^{(k)}, \widetilde{\xi}_3^{(k)}\right) = \widetilde{g}\left(\widetilde{\xi}_1^{(k')}, \widetilde{\xi}_2^{(k')}, \widetilde{\xi}_3^{(k')}\right)\,, \quad k, k' \in \mathbb{Z}\,. \tag{2.2.28}$$

The one-to-one correspondence between the pairs $\pm\mathsf{U} \in \mathsf{SU}(2)$ and $\mathsf{G} \in \mathsf{SO}(3)$ shows that functions defined on $\mathsf{SO}(3)$ generate functions on $\mathsf{SU}(2)$, and vice versa. If a function $F(\mathsf{G})$ defined on $\mathsf{SO}(3)$ is a uniformly continuous function, then it will generate a uniformly continuous function $F(\mathsf{U})$ on $\mathsf{SU}(2)$. Indeed, to the elements $\pm\mathsf{U} \in \mathsf{SU}(2)$ there corresponds the element $\mathsf{G} \in \mathsf{SO}(3)$, to which we associate $F(\mathsf{G})$. In turn, the function $F(\mathsf{G})$ generates the function $\widetilde{F}(\mathsf{U})$, according to the schema

$$\left.\begin{matrix} \mathsf{U} \\ \\ -\mathsf{U} \end{matrix}\right\} \rightarrow \mathsf{G} \rightarrow F(\mathsf{G}) \rightarrow \widetilde{F}(\mathsf{U})\,. \tag{2.2.29}$$

The function $\widetilde{F}(\mathsf{U})$ thus obtained is an even function because it is associated simultaneously to U and $-\mathsf{U}$. If $F(\mathsf{G})$ is continuous, then $\widetilde{F}(\mathsf{U})$ is also continuous. Conversely, if $\widetilde{F}(\mathsf{U})$ is a uniformly continuous function on $\mathsf{SU}(2)$, then it generates a function $F(\mathsf{G})$ on $\mathsf{SO}(3)$, according to the schema

$$\mathsf{G} \rightarrow \left\{\begin{matrix} \mathsf{U} \rightarrow \widetilde{F}(\mathsf{U}) \\ \\ -\mathsf{U} \rightarrow \widetilde{F}(-\mathsf{U}) \end{matrix}\right\} \rightarrow F(\mathsf{G})\,, \tag{2.2.30}$$

and $F(\mathsf{G})$ is uniformly continuous only if $\widetilde{F}(\mathsf{U})$ is even and uniformly continuous. Therefore, a uniformly continuous function $\widetilde{F}(\mathsf{U})$ generates a multiple-valued function $F(\mathsf{G})$.

2.2.2 The invariant integral

Let $\{\gamma\} = \{\gamma_1, \gamma_2, \gamma_3\} \in \Gamma$ be a parametrization of $\mathsf{SU}(2)$, where Γ is the group space. We associate the parameters $\{\gamma\}$ and $\{\gamma'\}$, respectively to the elements $\mathsf{U}, \mathsf{U}' \in \mathsf{SU}(2)$, such that $\widetilde{\gamma}_i = \widetilde{\gamma}_i(\gamma, \gamma')$, with $i = 1, 2, 3$, are the parameters of the product $\widetilde{\mathsf{U}} = \mathsf{U}\mathsf{U}'$. Because $\mathsf{SU}(2)$ is a Lie group $\widetilde{\gamma}_i$ are analytical functions of both arguments $\{\gamma\}$ and $\{\gamma'\}$, and there exist inverse relations of the form $\gamma'_i = \gamma'_i(\gamma, \widetilde{\gamma})$ for the parameters of $\mathsf{U}' = \mathsf{U}^{-1}\widetilde{\mathsf{U}}$. The functions $\widetilde{\gamma}_i$ are invertible with respect to the arguments γ'_i, if the necessary condition

$$D(\gamma', \widetilde{\gamma}) \equiv \det\left[\frac{\partial(\gamma')}{\partial(\widetilde{\gamma})}\right] \neq 0 \tag{2.2.31}$$

is satisfied $\forall\{\gamma\}$. In particular, the condition (2.2.31) has to be satisfied by the parameters $\{\gamma_0\}$ associated to the identity element of the group.

Let $\widetilde{F}(\mathsf{U})$ be a uniformly continuous numerical function on $\mathsf{SU}(2)$, to which there corresponds the function $\widetilde{f}(\gamma_1, \gamma_2, \gamma_3)$. The Hurwitz integral of this function has the form

$$J \equiv N \int_\Gamma \widetilde{f}(\gamma_1, \gamma_2, \gamma_3)\left[D(\gamma', \widetilde{\gamma})\right]_{\gamma'=\gamma_0} d\gamma_1\, d\gamma_2\, d\gamma_3 = \int_{\mathsf{SU}(2)} \widetilde{F}(\mathsf{U})\, d\mathsf{U}\,. \tag{2.2.32}$$

The equality in (2.2.32) is obtained for $\gamma_i = \gamma_i(\mathsf{U})$, $\gamma_i' = \gamma_i(\mathsf{U}')$, $\widetilde{\gamma}_i = \gamma_i(\mathsf{U}\mathsf{U}')$, $d\gamma(\mathsf{U}) = d\gamma_1\, d\gamma_2\, d\gamma_3$, and

$$d\mathsf{U} = N\, \left[D(\gamma', \widetilde{\gamma})\right]_{\gamma'=\gamma_0} d\gamma(\mathsf{U})\,.$$

The factor N is determined by the condition

$$\int_{\mathsf{SU}(2)} d\mathsf{U} = 1\,,$$

that, in terms of the parameters γ_i, reads

$$\frac{1}{N} = \int_{\Gamma} \left[D(\gamma', \widetilde{\gamma})\right]_{\gamma'=\gamma_0} d\gamma_1\, d\gamma_2\, d\gamma_3\,.$$

It can be shown that the value of the Hurwitz integral J does not depend on the parametrization, and that the Hurwitz integral is invariant, that is it satisfies the relation

$$\int_{\mathsf{SU}(2)} \widetilde{F}(\mathsf{U}\mathsf{U}_0)\, d\mathsf{U} = \int_{\mathsf{SU}(2)} \widetilde{F}(\mathsf{U}_0\mathsf{U})\, d\mathsf{U} = \int_{\mathsf{SU}(2)} \widetilde{F}(\mathsf{U})\, d\mathsf{U}\,, \tag{2.2.33}$$

where $\mathsf{U}_0 \in \mathsf{SU}(2)$ is an arbitrary fixed element.

Let us consider the Cayley-Klein parametrization, and suppose that the systems of parameters $\{\alpha_0, \alpha_1, \alpha_2, \alpha_3\}$, $\{\alpha_0', \alpha_1', \alpha_2', \alpha_3'\}$, and $\{\widetilde{\alpha}_0, \widetilde{\alpha}_1, \widetilde{\alpha}_2, \widetilde{\alpha}_3\}$, are associated to the operators U, U', and $\widetilde{\mathsf{U}}$, respectively. Note that the parameters of each system of parameters satisfy the condition (2.2.5). Let us also suppose that $\widetilde{\mathsf{U}} = \mathsf{U}\mathsf{U}'$. The inverse relation $\mathsf{U}' = \mathsf{U}^{-1}\widetilde{\mathsf{U}}$ corresponds to the matrix relation $\mathbf{u}' = \mathbf{u}^{\dagger}\widetilde{\mathbf{u}}$, and, by means of expression (2.2.18) and condition (2.2.5), applied to each operator, we obtain the explicit dependence of $\{\alpha'\}$ on $\{\alpha\}$ and $\{\widetilde{\alpha}\}$ in the form

$$\begin{aligned}
\alpha_1' &= -\ \widetilde{\alpha}_2\alpha_3 + \widetilde{\alpha}_1|\alpha_0| + \alpha_2\widetilde{\alpha}_3 - \alpha_1|\widetilde{\alpha}_0|\,,\\
\alpha_2' &= \quad \widetilde{\alpha}_1\alpha_3 + \widetilde{\alpha}_2|\alpha_0| - \alpha_1\widetilde{\alpha}_3 - \alpha_2|\widetilde{\alpha}_0|\,,\\
\alpha_3' &= -\ \widetilde{\alpha}_1\alpha_2 + \widetilde{\alpha}_3|\alpha_0| + \alpha_1\widetilde{\alpha}_2 - \alpha_3|\widetilde{\alpha}_0|\,,
\end{aligned} \tag{2.2.34}$$

where

$$\begin{aligned}
|\alpha_0| &= \sqrt{1 - \alpha_1^2 - \alpha_2^2 - \alpha_3^2}\,,\\
|\widetilde{\alpha}_0| &= \sqrt{1 - \widetilde{\alpha}_1^2 - \widetilde{\alpha}_2^2 - \widetilde{\alpha}_3^2}\,.
\end{aligned}$$

With the expressions (2.2.34) we construct the functional determinant (2.2.31), which takes the form

$$D(\alpha',\widetilde{\alpha}) = \det\begin{bmatrix} \alpha_1\dfrac{\widetilde{\alpha}_1}{|\widetilde{\alpha}_0|}+|\alpha_0| & \alpha_1\dfrac{\widetilde{\alpha}_2}{|\widetilde{\alpha}_0|}-\alpha_3 & \alpha_1\dfrac{\widetilde{\alpha}_3}{|\widetilde{\alpha}_0|}+\alpha_2 \\ \alpha_2\dfrac{\widetilde{\alpha}_1}{|\widetilde{\alpha}_0|}+\alpha_3 & \alpha_2\dfrac{\widetilde{\alpha}_2}{|\widetilde{\alpha}_0|}+|\alpha_0| & \alpha_2\dfrac{\widetilde{\alpha}_3}{|\widetilde{\alpha}_0|}-\alpha_1 \\ \alpha_3\dfrac{\widetilde{\alpha}_1}{|\widetilde{\alpha}_0|}-\alpha_2 & \alpha_3\dfrac{\widetilde{\alpha}_2}{|\widetilde{\alpha}_0|}+\alpha_1 & \alpha_3\dfrac{\widetilde{\alpha}_3}{|\widetilde{\alpha}_0|}+|\alpha_0| \end{bmatrix}$$

$$= |\alpha_0| + \frac{\alpha_1\widetilde{\alpha}_1+\alpha_2\widetilde{\alpha}_2+\alpha_3\widetilde{\alpha}_3}{|\widetilde{\alpha}_0|}\,. \tag{2.2.35}$$

The setting $\gamma' = \gamma_0$ in (2.2.32) means that $\mathsf{U}' = \mathsf{E}$, and therefore $\mathsf{U}' = \mathsf{U}^{-1}\widetilde{\mathsf{U}}$ is satisfied only if $\widetilde{\mathsf{U}} = \mathsf{U}$, that is $\widetilde{\boldsymbol{\alpha}} = \boldsymbol{\alpha}$, where $\widetilde{\boldsymbol{\alpha}} = \{\widetilde{\alpha}_1,\widetilde{\alpha}_2,\widetilde{\alpha}_3\}$ and $\boldsymbol{\alpha} = \{\alpha_1,\alpha_2,\alpha_3\}$. Consequently, we have

$$\big[D(\alpha',\widetilde{\alpha})\big]_{\boldsymbol{\alpha}'=\mathbf{0}} = \big[D(\alpha',\widetilde{\alpha})\big]_{\widetilde{\boldsymbol{\alpha}}=\boldsymbol{\alpha}} = \frac{1}{|\alpha_0|}\,. \tag{2.2.36}$$

Finally, with the function $\widetilde{f}$ defined in (2.2.26), and taking into account that $\alpha_0/|\alpha_0| = \pm 1$, we obtain the form of the invariant integral in terms of the Cayley-Klein parameters

$$\int_{\mathsf{SU}(2)} \widetilde{F}(\mathsf{U})\,d\mathsf{U} \tag{2.2.37}$$

$$= N\int_\Delta \Big[\widetilde{f}(1,\alpha_1,\alpha_2,\alpha_3)+\widetilde{f}(-1,\alpha_1,\alpha_2,\alpha_3)\Big]\frac{1}{|\alpha_0|}\,d\alpha_1\,d\alpha_2\,d\alpha_3\,,$$

where the group space Δ is the volume of the three-dimensional sphere $0 \le r \le 1$, $r^2 = \sum_{i=1}^3 \alpha_i\alpha_i$, $|\alpha_0| = \sqrt{1-r^2}$. Hence, the normalization factor is

$$\frac{1}{N} = 2\int_\Delta \frac{1}{|\alpha_0|}\,d\alpha_1\,d\alpha_2\,d\alpha_3 = 8\pi\int_0^1 \frac{r^2dr}{\sqrt{1-r^2}} = 2\pi^2\,. \tag{2.2.38}$$

By the change of variables defined in (2.2.15) we obtain the invariant integral in terms of the exponential parameters

$$\int_{\mathsf{SU}(2)} \widetilde{F}(\mathsf{U})\,d\mathsf{U} = N\int_{\widetilde{\varphi}\in[0,2\pi]} \widetilde{f}(\widetilde{\xi}_1,\widetilde{\xi}_2,\widetilde{\xi}_3)\,\frac{\sin^2(\widetilde{\varphi}/2)}{2\widetilde{\varphi}^2}\,d\widetilde{\xi}_1\,d\widetilde{\xi}_1\,d\widetilde{\xi}_1$$

$$= \frac{N}{4}\int_{\widetilde{\xi}_k\widetilde{\xi}_k\in[0,4\pi^2]} \widetilde{f}(\widetilde{\xi}_1,\widetilde{\xi}_2,\widetilde{\xi}_3)\,\frac{1-\cos\big(\sqrt{\widetilde{\xi}_i\widetilde{\xi}_i}\big)}{2\widetilde{\xi}_j\widetilde{\xi}_j}\,d\widetilde{\xi}_1\,d\widetilde{\xi}_1\,d\widetilde{\xi}_1\,,$$

where $N = 1/2\pi^2$. In the same manner, by means of relations (2.2.25) we obtain the invariant integral for the Eulerian parametrization

$$\int_{\mathsf{SU}(2)} \widetilde{F}(\mathsf{U})\, d\mathsf{U} = \frac{N}{8} \int_0^{4\pi} \int_0^{\pi} \int_0^{2\pi} \widetilde{f}(\varphi, \theta, \psi)\, \sin\theta\, d\varphi\, d\theta\, d\psi\,, \tag{2.2.39}$$

with $N = 1/2\pi^2$. Note that the normalization factor N has the same value independent of the parametrization.

Let us now consider the set of uniformly continuous, numerical functions $\widetilde{F}(\mathsf{U})$, for which there exists the Hurwitz integral of the squared modulus, denoted by

$$\left\|\widetilde{F}(\mathsf{U})\right\|^2 = \int_{\mathsf{SU}(2)} \left|\widetilde{F}(\mathsf{U})\right|^2 d\mathsf{U}\,, \tag{2.2.40}$$

the integral being a Lebesgue integral. If we define in this set the addition of functions and multiplication by scalars, we obtain an infinite-dimensional linear space. Furthermore, if we associate to each pair of functions the scalar product

$$\left\langle \widetilde{F}(\mathsf{U}), \widetilde{G}(\mathsf{U}) \right\rangle = \int_{\mathsf{SU}(2)} \overline{\widetilde{F}(\mathsf{U})}\, \widetilde{G}(\mathsf{U})\, d\mathsf{U}\,, \tag{2.2.41}$$

then the space becomes unitary. This space is complete in the sense of the general definition of completeness, that is every sequence of functions is strongly convergent in norm and has a limit belonging to the set. Consequently, we obtain a Hilbert space of functions defined on $\mathsf{SU}(2)$, which is analogous to the Hilbert space $\mathcal{L}^2$ of functions defined on a segment or a domain in the three-dimensional space.

2.2.3 Infinitesimal generators

Let $\{\mathbf{f}\}$ be an orthonormal basis in the two-dimensional unitary space $\mathcal{U}^2$. In the exponential parametrization, associated to the basis $\{\mathbf{f}\}$, the infinitesimal generators of $\mathsf{SU}(2)$ are given by the formulae

$$\mathsf{A}_j = \left[\frac{\partial \mathsf{U}(\widetilde{\xi}_1, \widetilde{\xi}_2, \widetilde{\xi}_3)}{\partial \widetilde{\xi}_j} \right]_{\widetilde{\xi}_1 = \widetilde{\xi}_2 = \widetilde{\xi}_3 = 0}, \quad j = 1, 2, 3\,, \tag{2.2.42}$$

while the matrices associated to the operators A_j in the basis $\{\mathbf{f}\}$, are obtained by differentiating the matrix $\mathbf{u}$ in (2.2.19) with respect to $\widetilde{\xi}_j$, and setting $\widetilde{\xi}_1 = \widetilde{\xi}_2 = \widetilde{\xi}_3 = 0$. Thus, we obtain the matrices

$$\mathbf{a}_1 = \frac{i}{2} \begin{bmatrix} 0 & 1 \\ 1 & 0 \end{bmatrix}, \quad \mathbf{a}_2 = \frac{i}{2} \begin{bmatrix} 0 & -i \\ i & 0 \end{bmatrix}, \quad \mathbf{a}_3 = \frac{i}{2} \begin{bmatrix} 1 & 0 \\ 0 & -1 \end{bmatrix}, \tag{2.2.43}$$

which are proportional to the Pauli spin matrices (2.2.7). To the Hermitian operators $\mathsf{H}_k = i\mathsf{A}_k$, $k = 1, 2, 3$, there corresponds the Hermitian matrices

$$\mathbf{h}_1 = -\frac{1}{2}\begin{bmatrix} 0 & 1 \\ 1 & 0 \end{bmatrix}, \quad \mathbf{h}_2 = -\frac{1}{2}\begin{bmatrix} 0 & -i \\ i & 0 \end{bmatrix}, \quad \mathbf{h}_3 = -\frac{1}{2}\begin{bmatrix} 1 & 0 \\ 0 & -1 \end{bmatrix}. \qquad (2.2.44)$$

In terms of the matrices $\mathbf{a}_k$, we have the relation

$$i\begin{bmatrix} \widetilde{\xi}_3 & \widetilde{\xi}_1 - i\widetilde{\xi}_2 \\ \widetilde{\xi}_1 + i\widetilde{\xi}_2 & -\widetilde{\xi}_3 \end{bmatrix} = 2\,\widetilde{\xi}_k\,\mathbf{a}_k$$

such that, from (2.2.19) we may infer the general form of the operators U in the exponential parametrization, in terms of the infinitesimal generators

$$\mathsf{U}(\widetilde{\xi}_1, \widetilde{\xi}_2, \widetilde{\xi}_3) = \cos\left(\sqrt{\widetilde{\xi}_i\widetilde{\xi}_i}/2\right)\mathsf{E} + \frac{\sin\left(\sqrt{\widetilde{\xi}_i\widetilde{\xi}_i}/2\right)}{\sqrt{\widetilde{\xi}_j\widetilde{\xi}_j}/2}\,\widetilde{\xi}_k\mathsf{A}_k\,, \qquad (2.2.45)$$

where E is the identity operator of $\mathsf{SU}(2)$. The infinitesimal generators satisfy the relations

$$\begin{aligned} &\mathsf{A}_k\mathsf{A}_l - \mathsf{A}_l\mathsf{A}_k = -\,\epsilon_{klm}\,\mathsf{A}_m\,, \qquad && \mathsf{H}_k\mathsf{H}_l - \mathsf{H}_l\mathsf{H}_k = -i\,\epsilon_{klm}\,\mathsf{H}_m\,, \\ &\mathsf{A}_k\mathsf{A}_l + \mathsf{A}_l\mathsf{A}_k = -\,\tfrac{1}{2}\,\delta_{kl}\,\mathsf{E}\,, && \mathsf{H}_k\mathsf{H}_l + \mathsf{H}_l\mathsf{H}_k = \tfrac{1}{2}\,\delta_{kl}\,\mathsf{E}\,. \end{aligned} \qquad (2.2.46)$$

THEOREM 2.2.2 *In the exponential parametrization, the elements of* $\mathsf{SU}(2)$ *have the form*

$$\mathsf{U}(\widetilde{\xi}_1, \widetilde{\xi}_2, \widetilde{\xi}_3) = e^{\widetilde{\xi}_k\,\mathsf{A}_k}\,. \qquad (2.2.47)$$

To prove this theorem we introduce the notations $\mathsf{B} = \widetilde{\xi}_k\,\mathsf{A}_k$, and $\omega = \sqrt{\widetilde{\xi}_i\widetilde{\xi}_i}/2$. One thus obtains

$$\mathsf{B}^2 = \widetilde{\xi}_i\,\mathsf{A}_i\,\widetilde{\xi}_j\,\mathsf{A}_j = \frac{1}{2}\,\widetilde{\xi}_i\widetilde{\xi}_j(\mathsf{A}_i\mathsf{A}_j + \mathsf{A}_j\mathsf{A}_i) = -\,\frac{1}{4}\,\widetilde{\xi}_i\widetilde{\xi}_i\,\mathsf{E} = -\omega^2\mathsf{E}\,,$$

such that

$$\mathsf{B}^{2p} = (-\omega^2)^p\,\mathsf{E}\,, \quad \mathsf{B}^{2p+1} = (-\omega^2)^p\,\mathsf{B}\,, \quad p = 0, 1, 2, \ldots$$

It follows that

$$e^{\mathsf{B}} = \sum_{p=0}^{\infty}\frac{1}{(2p)!}\,\mathsf{B}^{2p} + \sum_{p=0}^{\infty}\frac{1}{(2p+1)!}\,\mathsf{B}^{2p+1}$$

$$
\begin{aligned}
&= \mathsf{E} \sum_{p=0}^{\infty} (-1)^p \frac{\omega^{2p}}{(2p)!} + \mathsf{B} \sum_{p=0}^{\infty} (-1)^p \frac{\omega^{2p}}{(2p+1)!} \\
&= \mathsf{E} \sum_{p=0}^{\infty} (-1)^p \frac{\omega^{2p}}{(2p)!} + \frac{\mathsf{B}}{\omega} \sum_{p=0}^{\infty} (-1)^p \frac{\omega^{2p+1}}{(2p+1)!} \\
&= (\cos \omega)\, \mathsf{E} + \frac{\sin \omega}{\omega} \mathsf{B} \,,
\end{aligned}
$$

where the last expression is identical to (2.2.45). We have to mention here that, the form (2.2.47) of the operators U is the reason for the name *exponential parametrization*.

2.2.4 General properties of $\mathsf{SO}(3)$ and $\mathsf{SU}(2)$ representations

Let $\mathsf{T}(\mathsf{G})$ and $\widetilde{\mathsf{T}}(\mathsf{U})$ be representations of $\mathsf{SO}(3)$ and $\mathsf{SU}(2)$, respectively. The representations of Lie groups are uniformly continuous operator functions defined on the group, which satisfy the basic conditions of Definition 1.1.26. Thus, in our case we have

$$
\begin{aligned}
\mathsf{T}(\mathsf{G}_1)\, \mathsf{T}(\mathsf{G}_2) &= \mathsf{T}(\mathsf{G}_1\, \mathsf{G}_2)\,, \quad \mathsf{G}_1, \mathsf{G}_2 \in \mathsf{SO}(3)\,, \\
\widetilde{\mathsf{T}}(\mathsf{U}_1)\, \widetilde{\mathsf{T}}(\mathsf{U}_2) &= \widetilde{\mathsf{T}}(\mathsf{U}_1\, \mathsf{U}_2)\,, \quad \mathsf{U}_1, \mathsf{U}_2 \in \mathsf{SU}(2)\,.
\end{aligned}
$$

In the exponential parametrization, the operator function $\mathsf{T}(\mathsf{G})$ is associated to the function $\mathsf{T}(\xi_1, \xi_2, \xi_3)$, that is periodic with respect to the parameters ξ_k and satisfies the relation

$$
\mathsf{T}(\xi_1', \xi_2', \xi_3')\, \mathsf{T}(\xi_1'', \xi_2'', \xi_3'') = \mathsf{T}(\xi_1, \xi_2, \xi_3)\,,
$$

where $\xi_k = \xi_k(\boldsymbol{\xi}', \boldsymbol{\xi}'')$, $k = 1, 2, 3$.

Let $\widetilde{\mathsf{T}}(\mathsf{U})$ be an irreducible representation of $\mathsf{SU}(2)$. If $\mathsf{U} = \mathsf{E} \in \mathsf{SU}(2)$, then $\widetilde{\mathsf{T}}(\mathsf{E})$ is the identity operator of the representation. From $(-\mathsf{E})\mathsf{U} = \mathsf{U}(-\mathsf{E})$, it follows that $\widetilde{\mathsf{T}}(-\mathsf{E})\, \widetilde{\mathsf{T}}(\mathsf{U}) = \widetilde{\mathsf{T}}(\mathsf{U})\, \widetilde{\mathsf{T}}(-\mathsf{E})$, such that the operator $\widetilde{\mathsf{T}}(-\mathsf{E})$ commutes with all the operators in the representation. We have assumed that the representation is irreducible, therefore, according to Schur's lemma, we have

$$
\widetilde{\mathsf{T}}(-\mathsf{E}) = \lambda\, \widetilde{\mathsf{T}}(\mathsf{E})\,. \tag{2.2.48}
$$

Then, from the relations $(-\mathsf{E})(-\mathsf{E}) = \mathsf{E}$ and $\widetilde{\mathsf{T}}(-\mathsf{E})\, \widetilde{\mathsf{T}}(-\mathsf{E}) = \widetilde{\mathsf{T}}(\mathsf{E})$, we obtain $\lambda^2 = 1$, respectively $\lambda = \pm 1$. Consequently, the equation (2.2.48) shows that the representation is even if $\lambda = 1$, and odd if $\lambda = -1$. For instance, if $\lambda = 1$, from the identity $-\mathsf{U} = (-\mathsf{E})\mathsf{U}$, it results

$$
\widetilde{\mathsf{T}}(-\mathsf{U}) = \widetilde{\mathsf{T}}(-\mathsf{E})\, \widetilde{\mathsf{T}}(\mathsf{U}) = \widetilde{\mathsf{T}}(\mathsf{E})\, \widetilde{\mathsf{T}}(\mathsf{U}) = \widetilde{\mathsf{T}}(\mathsf{U})\,, \quad \forall \mathsf{U} \in \mathsf{SU}(2)\,.
$$

Hence, the irreducible representations of SU(2) are endowed with parity. In general, the reducible representations of SU(2) have no parity, because they can be decomposed into direct sums of irreducible representations, which can have different parities.

THEOREM 2.2.3 *Any irreducible representation of* SO(3) *generates an even irreducible representation of* SU(2).

DEFINITION 2.2.2 *A representation is continuous if the corresponding operator function is continuous.*

THEOREM 2.2.4 *Any continuous finite-dimensional representation of* SU(2) *is equivalent to a unitary representation.*

SU(2) and SO(3) are compact groups such that, according to Schur-Auerbach theorem and Maschke's theorem applied to topological groups (see Chap. 1 Sec. 2.2.2), their representations are equivalent to unitary representations which, in turn, are fully reducible.

3. The SU(3) and GL$(n, \mathbb{C})$ groups

3.1 SU(3) Lie algebra

3.1.1 Definition. Fundamental representations

Let us consider the three-dimensional linear complex space $\mathbb{C}^3$, consisting of column vectors $\mathbf{z} \in \mathbb{C}^3$,

$$\mathbf{z} \equiv \begin{bmatrix} z_1 \\ z_2 \\ z_3 \end{bmatrix} = z_1\mathbf{e}^1 + z_2\mathbf{e}^2 + z_3\mathbf{e}^3, , \quad z_1, z_2, z_3 \in \mathbb{C}\,, \tag{2.3.1}$$

where $\{\mathbf{e}^1, \mathbf{e}^2, \mathbf{e}^3\}$ is a basis in $\mathbb{C}^3$.

DEFINITION 2.3.1 *The set of unimodular unitary operators defined on* $\mathbb{C}^3$, *endowed with a composition law of operators, forms the* **three-dimensional unimodular** (**special**) **unitary group**, *denoted by* SU(3).

An element U $\in$ SU(3) is a unimodular unitary linear transformation in $\mathbb{C}^3$, such that we have the relation

$$\mathbf{z} \xrightarrow{\mathsf{U}} \mathbf{z}' = \mathbf{u}\,\mathbf{z}\,, \tag{2.3.2}$$

or

$$z_i \xrightarrow{\mathsf{U}} z_i' = u_i^j\, z_j\,, \quad i = 1, 2, 3, \tag{2.3.3}$$

where $\mathbf{u}$ is a unimodular ($\det \mathbf{u} = 1$) unitary ($\mathbf{u}\mathbf{u}^\dagger = \mathbf{u}^\dagger\mathbf{u} = \mathbf{e}$) matrix. The set of the matrices $\mathbf{u}$, associated to the elements of SU(3), form a three-dimensional representation of SU(3), called the *fundamental representation* (generally, denoted by 3). The complex numbers z_1, z_2, and z_3, which are transformed according to (2.3.3), form a *covariant triplet* with respect to the SU(3) transformations. The Hermitian conjugate of the vector $\mathbf{z}$ is the row vector

$$\mathbf{z}^\dagger \equiv [\overline{z_1},\, \overline{z_2},\, \overline{z_3}] \;, \tag{2.3.4}$$

where $\overline{z_i}$ is the complex conjugate of z_i. From (2.3.3) we obtain

$$\overline{z'_i} = \overline{u_i^j}\, \overline{z_j} = (\mathbf{u}^\dagger)^i_j\, z^j \,, \quad i = 1, 2, 3, \tag{2.3.5}$$

where we have introduced the notation $z^j = \overline{z_j}$. Thus, one can write

$$z^i \xrightarrow{\mathsf{U}} z'^i = z^j\, (\mathbf{u}^\dagger)^i_j = z^j\, (\mathbf{u}^{-1})^i_j \,, \quad i = 1, 2, 3, \tag{2.3.6}$$

or, in matrix form

$$\mathbf{z}^\dagger \xrightarrow{\mathsf{U}} \mathbf{z}^{\dagger\prime} = \mathbf{z}^\dagger\, \mathbf{u}^\dagger = \mathbf{z}^\dagger\, \mathbf{u}^{-1} \,. \tag{2.3.7}$$

The complex numbers z^1, z^2, and z^3, that are transformed according to (2.3.6), form a *contravariant triplet* with respect to the SU(3) transformations. Note that the transformations of the covariant triplet $\{z_i\}$ are defined by the matrices $\mathbf{u}$, while the transformations of the contravariant triplet $\{z^j\}$ are essentially determined by the matrices $\overline{\mathbf{u}}$. Consequently, there exist two three-dimensional representations of SU(3), called *fundamental representations*. These two representations are not equivalent because a matrix $\mathbf{u}$ cannot be interrelated with its complex conjugate $\overline{\mathbf{u}}$ by a similarity transformation (i.e., it does not exist a constant complex matrix $\mathbf{c}$ such that $\overline{\mathbf{u}} = \mathbf{c}\mathbf{u}\mathbf{c}^{-1}$). The representation formed by the matrices $\mathbf{u}^\dagger = \overline{\mathbf{u}}^{\mathrm{T}}$ is generally denoted by $\overline{3}$, and is called the *contragredient* of the representation 3.

From the relations (2.3.3) and (2.3.6) it follows that the SU(3) transformations leave invariant the quadratic form $\mathbf{z}^\dagger\mathbf{z} = z^1 z_1 + z^2 z_2 + z^3 z_3$.

3.1.2 SU(3) Lie algebra

It can be shown that the SU(3) group is a compact semisimple Lie group. Because it is isomorphic to the group of unimodular unitary transformations in $\mathbb{C}^3$, the elements of SU(3) are represented by complex three-by-three matrices. These matrices are determined by 9 complex numbers or 18 real numbers. The unitarity condition contains 9 relations, while the condition of unimodularity adds one more relation. Hence, the elements of SU(3) are determined by 8 independent real parameters, and it follows that SU(3) is a Lie group of order eight. The infinitesimal generators of SU(3) are obtained by applying the general method detailed in Chap. 1 Sec. 2.1.3, and they form the basis of a Lie algebra in which the product of two elements is their commutator.

A Lie group and its associated Lie algebra can be represented as a matrix Lie group and a matrix Lie algebra, respectively. Then, the relation between the group elements and the elements of the associated algebra is provided by the following theorem:

THEOREM 2.3.1 *Let L be a Lie algebra of matrices. If we associate to each matrix $\mathbf{X} \in L$ the matrix exponential*

$$\mathbf{u} = e^{\mathbf{X}} = \sum_{n=0}^{\infty} \frac{1}{n!} \mathbf{X}^n = \mathbf{E} + \frac{1}{1!} \mathbf{X} + \frac{1}{2!} \mathbf{X}^2 + \dots ,$$

where $\mathbf{E}$ is the identity matrix, then the set of the matrices $\mathbf{u}$ form a Lie group with respect to the usual multiplication of matrices.

Thus, one can write the matrices $\mathbf{u} \in \mathsf{SU}(3)$ in the form

$$\mathbf{u} = e^{i\varepsilon\mathbf{X}} , \tag{2.3.8}$$

where $\varepsilon \in \mathbb{R}$ is an arbitrary parameter, and impose the condition of unitarity

$$\mathbf{u}\,\mathbf{u}^{\dagger} = e^{i\varepsilon(\mathbf{X}-\mathbf{X}^{\dagger})} = \mathbf{E} .$$

Hence, $\mathbf{X}$ must be a Hermitian three-by-three matrix.

Note that, a Lie group completely determines the structure of the associated Lie algebra, but conversely, only the local structure of a Lie group (i.e., the structure in the vicinity of the identity) is completely determined by its associated Lie algebra. Therefore, one has to impose the condition of unimodularity of $\mathbf{u}$ in the vicinity of the identity, that is by assuming that ε is a small number. Thus, we have

$$\det \mathbf{u} = \det \left[\mathbf{E} + \varepsilon\mathbf{X} + \mathcal{O}(\varepsilon^2)\right] = 1 + \varepsilon \operatorname{Tr} \mathbf{X} + \mathcal{O}(\varepsilon^2) ,$$

and $\mathbf{u}$ is unimodular if $\operatorname{Tr} \mathbf{X} = 0$. Consequently, one may infer that the matrix $\mathbf{X}$ has to be a Hermitian three-by-three matrix of null trace. This matrix can be decomposed into a sum of matrices

$$\mathbf{X} = \sum_{i=1}^{3} \sum_{j=1}^{3} \varepsilon^i_j \mathbf{X}^j_i , \tag{2.3.9}$$

with $\varepsilon^i_j \in \mathbb{C}$ satisfying the relations $\overline{\varepsilon^i_j} = \varepsilon^j_i$, and the matrices $\mathbf{X}^j_i$ being of the form

$$\left(\mathbf{X}^j_i\right)_{\mu\nu} = \delta_{i\mu}\delta_{j\nu} - \frac{1}{3}\, \delta_{ij}\delta_{\mu\nu} . \tag{2.3.10}$$

Explicitly, the matrices $\mathbf{X}_i^j$ are

$$\mathbf{X}_1^1 = \begin{bmatrix} \frac{2}{3} & 0 & 0 \\ 0 & -\frac{1}{3} & 0 \\ 0 & 0 & -\frac{1}{3} \end{bmatrix}, \quad \mathbf{X}_2^2 = \begin{bmatrix} -\frac{1}{3} & 0 & 0 \\ 0 & \frac{2}{3} & 0 \\ 0 & 0 & -\frac{1}{3} \end{bmatrix}, \quad \mathbf{X}_3^3 = \begin{bmatrix} -\frac{1}{3} & 0 & 0 \\ 0 & -\frac{1}{3} & 0 \\ 0 & 0 & \frac{2}{3} \end{bmatrix},$$

$$\mathbf{X}_1^2 = (\mathbf{X}_2^1)^\dagger = \begin{bmatrix} 0 & 1 & 0 \\ 0 & 0 & 0 \\ 0 & 0 & 0 \end{bmatrix}, \mathbf{X}_1^3 = (\mathbf{X}_3^1)^\dagger = \begin{bmatrix} 0 & 0 & 1 \\ 0 & 0 & 0 \\ 0 & 0 & 0 \end{bmatrix}, \mathbf{X}_2^3 = (\mathbf{X}_3^2)^\dagger = \begin{bmatrix} 0 & 0 & 0 \\ 0 & 0 & 1 \\ 0 & 0 & 0 \end{bmatrix}. \tag{2.3.11}$$

Note that only eight of the matrices in (2.3.11) are independent, because

$$\mathbf{X}_1^1 + \mathbf{X}_2^2 + \mathbf{X}_3^3 = \mathbf{0}.$$

Also, the matrices $\mathbf{X}_i^j$ satisfy the commutation relations

$$\left[\mathbf{X}_i^j, \mathbf{X}_k^l\right] = \delta_{jk}\mathbf{X}_i^l - \delta_{il}\mathbf{X}_k^j, \tag{2.3.12}$$

such that they form a basis of the Lie algebra associated to $\mathsf{SU}(3)$. This basis has been introduced by Okubo and it is called the *Okubo basis*. Another basis, introduced by Gell-Mann, has the form

$$\mathbf{F}_1 = \tfrac{1}{2}\begin{bmatrix} 0 & 1 & 0 \\ 1 & 0 & 0 \\ 0 & 0 & 0 \end{bmatrix}, \mathbf{F}_2 = \tfrac{1}{2}\begin{bmatrix} 0 & -i & 0 \\ i & 0 & 0 \\ 0 & 0 & 0 \end{bmatrix}, \mathbf{F}_3 = \tfrac{1}{2}\begin{bmatrix} 1 & 0 & 0 \\ 0 & -1 & 0 \\ 0 & 0 & 0 \end{bmatrix}, \quad \mathbf{F}_4 = \tfrac{1}{2}\begin{bmatrix} 0 & 0 & 1 \\ 0 & 0 & 0 \\ 1 & 0 & 0 \end{bmatrix},$$

$$\mathbf{F}_5 = \tfrac{1}{2}\begin{bmatrix} 0 & 0 & -i \\ 0 & 0 & 0 \\ i & 0 & 0 \end{bmatrix}, \quad \mathbf{F}_6 = \tfrac{1}{2}\begin{bmatrix} 0 & 0 & 0 \\ 0 & 0 & 1 \\ 0 & 1 & 0 \end{bmatrix}, \mathbf{F}_7 = \tfrac{1}{2}\begin{bmatrix} 0 & 0 & 0 \\ 0 & 0 & -i \\ 0 & i & 0 \end{bmatrix}, \mathbf{F}_8 = \tfrac{1}{2\sqrt{3}}\begin{bmatrix} 1 & 0 & 0 \\ 0 & 1 & 0 \\ 0 & 0 & -2 \end{bmatrix}, \tag{2.3.13}$$

and the connection between these two bases is given by the relations

$$\begin{aligned} \mathbf{X}_1^1 &= \mathbf{F}_3 + \tfrac{1}{\sqrt{3}}\mathbf{F}_8, \quad \mathbf{X}_2^2 = \tfrac{1}{\sqrt{3}}\mathbf{F}_8 - \mathbf{F}_3, \quad \mathbf{X}_3^3 = -\tfrac{2}{\sqrt{3}}\mathbf{F}_8, \\ \mathbf{X}_1^2 &= \mathbf{F}_1 + i\,\mathbf{F}_2, \qquad \mathbf{X}_1^3 = \mathbf{F}_4 + i\,\mathbf{F}_5, \qquad \mathbf{X}_2^3 = \mathbf{F}_6 + i\,\mathbf{F}_7. \end{aligned} \tag{2.3.14}$$

In terms of the Gell-Mann basis the matrix $\mathbf{X}$ from (2.3.9) becomes

$$\mathbf{X} = \sum_{k=1}^{8} \mu^k \mathbf{F}_k, \tag{2.3.15}$$

where $\mu^k \in \mathbb{R}$ are independent parameters which completely characterize this matrix. The matrices $\mathbf{F}_j$ satisfy the commutation relations

$$[\mathbf{F}_j, \mathbf{F}_k] = i\, f_{jkl}\,\mathbf{F}_l, \tag{2.3.16}$$

where the quantities f_{jkl} are completely antisymmetric, and represent the structure constants of the Lie algebra. The non-zero structure constants are

$$\begin{aligned} f_{123} &= 2\,f_{147} = 2\,f_{246} = 2\,f_{257} = 2\,f_{345} = -2\,f_{156} \\ &= -2\,f_{367} = \frac{2}{\sqrt{3}}\,f_{458} = \frac{2}{\sqrt{3}}\,f_{678} = 1\,. \end{aligned} \tag{2.3.17}$$

Note that, in (2.3.17) each quantity has to be considered for even and odd permutations of the corresponding set of indexes. For instance, $f_{123} = \epsilon_{123}$, $f_{147} = f_{471} = f_{714} = -f_{174} = -f_{417} = -f_{741} = 1/2$, etc.

In terms of the Okubo and Gell-Mann bases, the elements of $\mathsf{SU}(3)$ take a form similar with the form of $\mathsf{SU}(2)$ elements given by (2.2.47), that is

$$\mathbf{u} = e^{\,i\varepsilon^k_j \mathbf{X}^j_k} = e^{\,i\mu^k \mathbf{F}_k}\,. \tag{2.3.18}$$

When the parameters ε^k_j or μ^k are small numbers, the $\mathsf{SU}(3)$ transformations represented by the matrices $\mathbf{u}$ become *infinitesimal transformations*, that are expressed by the relations

$$\mathbf{u} = \mathbf{E} + i\,\varepsilon^k_j \mathbf{X}^j_k = \mathbf{E} + i\,\mu^k \mathbf{F}_k\,. \tag{2.3.19}$$

Hence, the matrices $\mathbf{X}^j_k$, as well as $\mathbf{F}_k$, are matrix representations of the infinitesimal generators of $\mathsf{SU}(3)$ in $\mathbb{C}^3$.

From the commutation relations (2.3.12) and (2.3.16) it follows that the Cartan subalgebra of the Lie algebra associated to the $\mathsf{SU}(3)$ group is two-dimensional (see Definition 1.2.8). Therefore, the $\mathsf{SU}(3)$ group is of rank two.

3.1.3 Root diagrams

To solve the problem of classification of semisimple, compact Lie algebras, Cartan sought the elements B of the algebra, that satisfy the equation

$$[\mathsf{A}, \mathsf{B}] = r(\mathsf{A}, \mathsf{B})\,\mathsf{B}\,, \tag{2.3.20}$$

where A is a fixed element. If X_k ($k = 1, 2, \dots, n$) is a basis of the Lie algebra, then one can write

$$\mathsf{A} = a^k \mathsf{X}_k\,, \quad \mathsf{B} = x^k \mathsf{X}_k\,,$$

such that (2.3.20) becomes a linear algebraic system

$$\sum_{j=1}^{n} \left[\sum_{k=1}^{n} a^k c^i_{kj} - r(\mathsf{A}, \mathsf{B})\,\delta^i_j \right] x^j = 0\,, \quad k = 1, 2, \dots, n, \tag{2.3.21}$$

where c^i_{kj} are the structures constants of the algebra, defined in (1.2.37). The numbers $r(\mathsf{A}, \mathsf{B})$ are the roots of the secular equation

$$\det \left[\sum_{k=1}^{n} a^k c^i_{kj} - r(\mathsf{A}, \mathsf{B})\,\delta^i_j \right] = 0\,. \tag{2.3.22}$$

Cartan proved that, by a suitable choice of the fixed element A, such that the secular equation (2.3.22) has a maximum number of distinct roots, the only degenerate root is zero, and its degeneracy order is equal to the rank of the group (see Definition 1.2.8). The elements B corresponding to $r(\mathsf{A},\mathsf{B}) = 0$, are linearly independent and commute between them; they form the basis of the Cartan subalgebra of the considered Lie algebra. The elements of this basis are denoted by H_i $(i = 1, 2, \ldots, l)$, l being the rank of the algebra (and also the rank of the group). Because the elements H_i correspond to the roots $r(\mathsf{A},\mathsf{H}_i) = 0$, it follows that $[\mathsf{A},\mathsf{H}_i] = 0$ $(i = 1, 2, \ldots, l)$; hence, the algebra element A belongs to the Cartan subalgebra, and it can be expressed in the form

$$\mathsf{A} = \lambda^i \mathsf{H}_i \,.$$

Let us now consider that the basis of an n-dimensional Lie algebra is formed by the elements H_i $(i = 1, 2, \ldots, l)$, and $n - l$ independent elements denoted by E_α. Such a basis is called the *Cartan basis* of the algebra. It can be shown that any root $r(\mathsf{A},\mathsf{E}_\alpha)$ can be represented as a vector $\mathbf{r}(\alpha)$ in an l-dimensional space, and that, in the case of semisimple Lie algebras, the roots $\mathbf{r}(\alpha)$ have the following properties:

1 If $\mathbf{r}(\alpha)$ is a root, then $\mathbf{r}(-\alpha) = -\mathbf{r}(\alpha)$ is also a root. The basis element associated to the root $\mathbf{r}(-\alpha)$ is denoted by $\mathsf{E}_{-\alpha}$, such that the $n - l$ basis elements E_α can be labelled as $\mathsf{E}_{\pm\alpha}$ with $\alpha = 1, 2, \ldots, (n-l)/2$. We also have $\mathsf{E}_{-\alpha} = \mathsf{E}_\alpha^\dagger$.

2 If $\mathbf{r}(\alpha)$ and $\mathbf{r}(\beta)$ are roots, then $2\mathbf{r}(\alpha)\cdot\mathbf{r}(\beta)/\|\mathbf{r}(\alpha)\|^2$, where $\mathbf{r}(\alpha)\cdot\mathbf{r}(\beta)$ represents the dot product of the vectors in an Euclidean metric, is an integer. The integer numbers thus obtained are called *Cartan's numbers.*

3 If $\mathbf{r}(\alpha)$ and $\mathbf{r}(\beta)$ are roots, then $\mathbf{r}(\beta) - 2[\mathbf{r}(\alpha)\cdot\mathbf{r}(\beta)/\|\mathbf{r}(\alpha)\|^2]\mathbf{r}(\alpha)$ is also a root.

4 The roots satisfy the relation

$$\sum_{\alpha=1}^{(n-l)/2} r_i(\alpha) r_j(\alpha) = \delta_{ij}\,, \quad i, j = 1, 2, \ldots, l\,.$$

The commutation relations of the elements in the Cartan basis can be written in the form

$$[\mathsf{H}_i,\mathsf{H}_j] = 0\,, \quad [\mathsf{H}_i,\mathsf{E}_\alpha] = r_i(\alpha)\mathsf{E}_\alpha\,, \quad [\mathsf{E}_\alpha,\mathsf{E}_{-\alpha}] = r_i(\alpha)\mathsf{H}_i\,,$$

$$[\mathsf{E}_\alpha,\mathsf{E}_\beta] = \begin{cases} N_{\alpha\beta}\mathsf{E}_{\alpha+\beta} & \text{if } \mathbf{r}(\alpha)+\mathbf{r}(\beta) \text{ is a nonzero root}\,, \\ 0 & \text{otherwise}\,, \end{cases} \tag{2.3.23}$$

where $i, j = 1, 2, \ldots, l$ and $\alpha, \beta = 1, 2, \ldots, (n-l)/2$. The coefficients $N_{\alpha\beta}$ are given by

$$N_{\alpha\beta} = \pm\sqrt{\frac{(m+1)n}{2}}\,\|\mathbf{r}(\alpha)\|\,, \tag{2.3.24}$$

where m and n are nonzero numbers in the sequence of roots

$$\mathbf{r}(\beta)-m\mathbf{r}(\alpha),\ \mathbf{r}(\beta)-(m-1)\mathbf{r}(\alpha),\ \ldots,\mathbf{r}(\beta),\ \ldots,\ \mathbf{r}(\beta)+n\mathbf{r}(\alpha)\,. \tag{2.3.25}$$

The sign in (2.3.24) has to be chosen such that

$$N_{\alpha\beta} = -N_{\beta\alpha} = -N_{-\alpha,-\beta} = N_{\beta,-\alpha-\beta} = N_{-\alpha-\beta,\alpha}\,. \tag{2.3.26}$$

We will now concentrate on the Lie algebra of rank $l = 2$, whose roots can be represented as two-dimensional vectors. From the root property 2) it follows that

$$\frac{2\mathbf{r}(\alpha)\cdot\mathbf{r}(\beta)}{\|\mathbf{r}(\alpha)\|^2} = m\,,\qquad \frac{2\mathbf{r}(\alpha)\cdot\mathbf{r}(\beta)}{\|\mathbf{r}(\beta)\|^2} = n\,,\qquad m,n\in\mathbb{N}\,, \tag{2.3.27}$$

such that

$$\mathbf{r}(\alpha)\cdot\mathbf{r}(\beta) = \frac{m}{2}\,\|\mathbf{r}(\alpha)\|^2 = \frac{n}{2}\,\|\mathbf{r}(\beta)\|^2 = \pm\frac{1}{2}\sqrt{mn}\,\|\mathbf{r}(\alpha)\|\,\|\mathbf{r}(\beta)\|\,.$$

This means that the angle between two given roots is determined by the relation

$$\cos^2\theta = \frac{mn}{4}\,, \tag{2.3.28}$$

while the ratio of the magnitudes of these roots is

$$\frac{\|\mathbf{r}(\alpha)\|}{\|\mathbf{r}(\beta)\|} = \sqrt{\frac{n}{m}}\,. \tag{2.3.29}$$

Also, from the root property 4) one obtains the normalization condition

$$\sum_{\alpha=1}^{n/2-1} r_i(\alpha)r_i(\alpha) = 1\,,\quad i = 1,2. \tag{2.3.30}$$

The relation (2.3.28) imposes the condition $mn \le 4$, and we thus obtain the following table

mn	0	1	2	3	4
θ	$\pi/2$	$\pi/3,\ 2\pi/3$	$\pi/4,\ 3\pi/4$	$\pi/6,\ 5\pi/6$	$0,\ \pi$
$\|\mathbf{r}(\alpha)\|/\|\mathbf{r}(\beta)\|$	0	1	$\sqrt{2}$	$\sqrt{3}$	1, 2

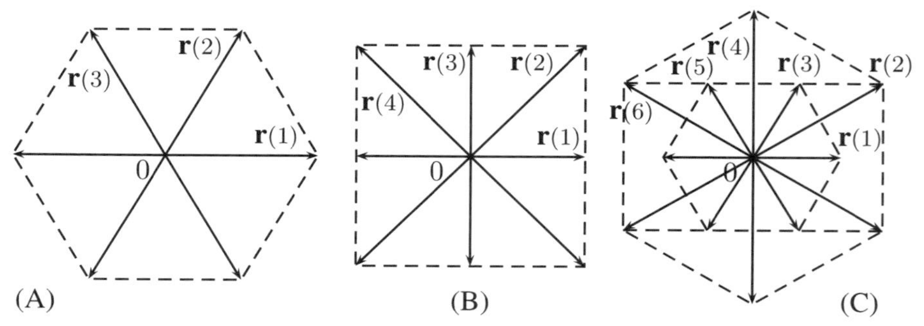

Figure 2.4. The root diagrams of $\mathsf{SU}(3)$, B_2, and G_2. The ratio of different roots (2.3.29) and the normalization condition (2.3.30) determine the length of the root vectors: $1/\sqrt{3}$ (A), $\{1/\sqrt{3}, 1/\sqrt{6}\}$ (B), and $\{1/2, 1/(2\sqrt{3})\}$ (C). Note that only the roots $\mathbf{r}(\alpha)$ for $\alpha \geq 1$ are marked, and that each diagram contains two zero roots.

Note that, each root $\mathbf{r}(\alpha)$ is always accompanied by its opposite $-\mathbf{r}(\alpha)$ such that the angles $\pi - \theta$ do not provide results different than those corresponding to $0 \leq \theta < \pi$.

The angles $\theta = 0$ and $\theta = \pi/2$ represent degenerate cases of the Lie algebra of rank two, while the remaining three angles (i.e., $\theta = \pi/3, \pi/4, \pi/6$) determine the *root diagrams* associated, respectively, to the *unimodular unitary group* $\mathsf{SU}(3)$, the *five-dimensional orthogonal group* B_2, and the *exceptional group* G_2 (subgroup of the seven-dimensional orthogonal group). These three root diagrams are shown in Fig. 2.4. Note that each diagram contains two zero roots. The root diagrams provide information about the coefficients $N_{\alpha,\beta}$ defined in (2.3.24), which exist only if $\mathbf{r}(\alpha)+\mathbf{r}(\beta)$ is a nonzero root. For instance, in the case of the $\mathsf{SU}(3)$ group one can deduce from Fig. 2.4(A) that $N_{12} = 0$, while $N_{13} \neq 0$. As a matter of fact, the nonzero coefficients $N_{\alpha,\beta}$ of the $\mathsf{SU}(3)$ group are

$$N_{13} = -N_{1,-2} = -N_{2,-1} = N_{2,-3} = N_{3,-2} = -N_{-1,-3} = \frac{1}{\sqrt{6}}\,. \quad (2.3.31)$$

These values, together with the root diagram in Fig. 2.4(A), completely determine the commutation relations (2.3.23) satisfied by the elements in the Cartan basis of the Lie algebra associated to the $\mathsf{SU}(3)$ group.

3.1.4 Weight diagrams

Let us consider an n-dimensional representation of the Lie algebra associated to the $\mathsf{SU}(3)$ group. In this representation, the elements $\{\mathsf{H}_1, \mathsf{H}_2, \mathsf{E}_{\pm\alpha}\}$ of the Cartan basis, are square matrices $\{\mathbf{H}_1, \mathbf{H}_2, \mathbf{E}_{\pm\alpha}\}$ of order n.

THEOREM 2.3.2 *Two operators* A *and* B *commute* ($[\mathsf{A}, \mathsf{B}] = 0$) *if and only if there exists a complete orthonormal basis of eigenfunctions shared by the two operators.*

The operators H_1 and H_2 commute and, according to the Theorem 2.3.2, there exists an eigenvector ψ that is common to $\mathbf{H}_1$ and $\mathbf{H}_2$. Thus, we have

$$\mathbf{H}_1\psi = m_1\psi\,, \quad \mathbf{H}_2\psi = m_2\psi\,. \tag{2.3.32}$$

We may assume that the pair (m_1, m_2) forms a vector in the two-dimensional space in which we have represented the roots. The vector $\mathbf{m} \equiv (m_1, m_2)$ is called the *weight* of the eigenvector ψ. A weight is called *simple* if it corresponds to only one eigenvector.

THEOREM 2.3.3 *A finite-dimensional representation of a semisimple Lie algebra has at least one weight.*

From (2.3.23) and (2.3.32) we obtain

$$\begin{aligned} &\mathbf{H}_1(\mathbf{H}_i\psi) = \mathbf{H}_i(\mathbf{H}_1\psi) = m_1(\mathbf{H}_i\psi)\,, \\ &\mathbf{H}_2(\mathbf{H}_i\psi) = m_2(\mathbf{H}_i\psi)\,, \qquad\qquad i = 1, 2, \end{aligned} \tag{2.3.33}$$

and

$$\begin{aligned} \mathbf{H}_1(\mathbf{E}_{\pm\alpha}\psi) &= \mathbf{E}_{\pm\alpha}(\mathbf{H}_1\psi) \pm r_1(\alpha)\mathbf{E}_{\pm\alpha}\psi = [m_1 \pm r_1(\alpha)](\mathbf{E}_{\pm\alpha}\psi)\,, \\ \mathbf{H}_2(\mathbf{E}_{\pm\alpha}\psi) &= [m_2 \pm r_2(\alpha)](\mathbf{E}_{\pm\alpha}\psi)\,, \quad \alpha = 1, 2, 3, \end{aligned} \tag{2.3.34}$$

such that the eigenvector $\mathbf{H}_i\psi$ has the weight $\mathbf{m}$, while the eigenvector $\mathbf{E}_{\pm\alpha}\psi$ has the weight $\mathbf{m} \pm \mathbf{r}(\alpha)$, if $\mathbf{E}_\alpha\psi$, respectively $\mathbf{E}_{-\alpha}\psi$, do not vanish.

If two operators commute, then there exists a basis such that both the matrices associated to these operators are diagonal. Thus, in the representation for which the matrices $\mathbf{H}_1$ and $\mathbf{H}_2$ are diagonal, the common eigenvectors can be written in the form

$$\mathbf{v}_1 = \begin{bmatrix} 1 \\ 0 \\ 0 \\ \vdots \\ 0 \end{bmatrix}, \quad \mathbf{v}_2 = \begin{bmatrix} 0 \\ 1 \\ 0 \\ \vdots \\ 0 \end{bmatrix}, \quad \mathbf{v}_3 = \begin{bmatrix} 0 \\ 0 \\ 1 \\ \vdots \\ 0 \end{bmatrix}, \quad \ldots \tag{2.3.35}$$

If the operator E_α transforms the eigenvector $\mathbf{v}_1$ into an eigenvector having the same weight as $\mathbf{v}_2$, then the matrix associated to the operator E_α is

$$\mathbf{E}_\alpha = k_1 \begin{bmatrix} 0 & 0 & \ldots \\ 1 & 0 & \ldots \\ . & . & \ldots \\ . & . & \ldots \end{bmatrix}, \tag{2.3.36}$$

where k_1 is a positive real constant. In turn, the operator $\mathsf{E}_{-\alpha}$ will transform the eigenvector $\mathbf{v}_2$ into an eigenvector having the same weight as $\mathbf{v}_1$. Hence, the operator $\mathsf{E}_{-\alpha}$ will be represented by the matrix

$$\mathbf{E}_{-\alpha} = k_2 \begin{bmatrix} 0 & 1 & \dots \\ 0 & 0 & \dots \\ . & . & \dots \\ . & . & \dots \end{bmatrix}, \tag{2.3.37}$$

where k_2 is a positive real constant. If, in both cases, we choose the constant factor to be $\sqrt{k_1 k_2}$, then we have $\mathsf{E}_{-\alpha} = \mathsf{E}_{\alpha}^{\dagger}$.

From (2.3.34) it follows that applying the operator E_{α} to an eigenvector of $\mathbf{H}_i$, the weight of the eigenvector is changed by $\mathbf{r}(\alpha)$. The operators H_1 and H_2 are represented by n-dimensional matrices, so that they have at most n distinct eigenvalues, and therefore the number of possible weights is finite. It results that there exists an eigenvector of weight $\mathbf{M}$, denoted by $\boldsymbol{\psi}(\mathbf{M})$, such that $\mathbf{E}_{\alpha}\boldsymbol{\psi}(\mathbf{M}) = \mathbf{0}$. We suppose that the eigenvector $\boldsymbol{\psi}(\mathbf{M})$ is normalized in terms of a scalar product

$$\langle \boldsymbol{\psi}(\mathbf{M}), \boldsymbol{\psi}(\mathbf{M}) \rangle \equiv \boldsymbol{\psi}^{\dagger}(\mathbf{M}) \boldsymbol{\psi}(\mathbf{M}) = 1\,, \tag{2.3.38}$$

such that by applying the operator $\mathsf{E}_{-\alpha}$ we obtain

$$\mathbf{E}_{-\alpha}\boldsymbol{\psi}(\mathbf{M}) = k\, \boldsymbol{\psi}(\mathbf{M} - \mathbf{r}(\alpha))\,, \tag{2.3.39}$$

where k is a normalization constant for the vector $\boldsymbol{\psi}(\mathbf{M} - \mathbf{r}(\alpha))$. Because $[\mathbf{E}_{\alpha}, \mathbf{E}_{-\alpha}] = r_i(\alpha)\mathbf{H}_i = \mathbf{r}(\alpha) \cdot \mathbf{H}$, where $\mathbf{H} = (\mathbf{H}_1, \mathbf{H}_2)$ is a vector of matrices, and $\mathbf{E}_{\alpha}\boldsymbol{\psi}(\mathbf{M}) = \mathbf{0}$, it results

$$\begin{aligned} &\langle \boldsymbol{\psi}(\mathbf{M}), [\mathbf{E}_{\alpha}, \mathbf{E}_{-\alpha}]\boldsymbol{\psi}(\mathbf{M}) \rangle = \langle \boldsymbol{\psi}(\mathbf{M}), \mathbf{E}_{\alpha}\mathbf{E}_{-\alpha}\boldsymbol{\psi}(\mathbf{M}) \rangle \\ &= \left\langle \boldsymbol{\psi}(\mathbf{M}), \mathbf{E}_{-\alpha}^{\dagger}\mathbf{E}_{-\alpha}\boldsymbol{\psi}(\mathbf{M}) \right\rangle = \langle \mathbf{E}_{-\alpha}\boldsymbol{\psi}(\mathbf{M}), \mathbf{E}_{-\alpha}\boldsymbol{\psi}(\mathbf{M}) \rangle \\ &= |k|^2 \langle \boldsymbol{\psi}(\mathbf{M} - \mathbf{r}(\alpha)), \boldsymbol{\psi}(\mathbf{M} - \mathbf{r}(\alpha)) \rangle = |k|^2\,, \end{aligned}$$

and

$$\begin{aligned} &\langle \boldsymbol{\psi}(\mathbf{M}), [\mathbf{E}_{\alpha}, \mathbf{E}_{-\alpha}]\boldsymbol{\psi}(\mathbf{M}) \rangle = \langle \boldsymbol{\psi}(\mathbf{M}), [\mathbf{r}(\alpha) \cdot \mathbf{H}]\boldsymbol{\psi}(\mathbf{M}) \rangle \\ &= [\mathbf{r}(\alpha) \cdot \mathbf{M}] \langle \boldsymbol{\psi}(\mathbf{M}), \boldsymbol{\psi}(\mathbf{M}) \rangle = \mathbf{r}(\alpha) \cdot \mathbf{M}\,. \end{aligned}$$

Consequently, the normalization constant in (2.3.39) is given by

$$|k|^2 = \mathbf{r}(\alpha) \cdot \mathbf{M}\,. \tag{2.3.40}$$

To find the relations between the roots and weights, we consider equations of the form

$$\begin{aligned}\mathbf{E}_\alpha \psi(\mathbf{m}) &= c_{\mathbf{m}} \psi(\mathbf{m} + \mathbf{r}(\alpha)),\\ \mathbf{E}_{-\alpha} \psi(\mathbf{m}) &= d_{\mathbf{m}} \psi(\mathbf{m} - \mathbf{r}(\alpha)).\end{aligned}$$

By applying successively the operators E_α, respectively $\mathsf{E}_{-\alpha}$, we obtain the nonnegative integers j and j', such that

$$\mathbf{E}_\alpha \psi(\mathbf{m} + j\mathbf{r}(\alpha)) = \mathbf{0}, \quad \mathbf{E}_{-\alpha} \psi(\mathbf{m} - j'\mathbf{r}(\alpha)) = \mathbf{0},$$

that is

$$c_{\mathbf{m}+j\mathbf{r}(\alpha)} = 0, \quad d_{\mathbf{m}-j'\mathbf{r}(\alpha)} = 0. \tag{2.3.41}$$

Then, we use the relation

$$[\mathbf{E}_\alpha, \mathbf{E}_{-\alpha}]\psi(\mathbf{m}) = [\mathbf{r}(\alpha) \cdot \mathbf{H}]\psi(\mathbf{m}) = [\mathbf{r}(\alpha) \cdot \mathbf{m}]\psi(\mathbf{m}),$$

which can also be written in the form

$$\begin{aligned}[\mathbf{E}_\alpha, \mathbf{E}_{-\alpha}]\psi(\mathbf{m}) &= \mathbf{E}_\alpha d_{\mathbf{m}} \psi(\mathbf{m} - \mathbf{r}(\alpha))\\ &- \mathbf{E}_{-\alpha} c_{\mathbf{m}} \psi(\mathbf{m} + \mathbf{r}(\alpha))[\mathbf{r}(\alpha) \cdot \mathbf{m}]\psi(\mathbf{m}),\end{aligned}$$

that is

$$d_{\mathbf{m}} c_{\mathbf{m}-\mathbf{r}(\alpha)} \psi(\mathbf{m}) - c_{\mathbf{m}} d_{\mathbf{m}+\mathbf{r}(\alpha)} \psi(\mathbf{m}) = [\mathbf{r}(\alpha) \cdot \mathbf{m}]\psi(\mathbf{m}).$$

We thus find the relation

$$d_{\mathbf{m}+\mathbf{r}(\alpha)} c_{\mathbf{m}} - d_{\mathbf{m}} c_{\mathbf{m}-\mathbf{r}(\alpha)} = -\mathbf{r}(\alpha) \cdot \mathbf{m},$$

in which we substitute successively the weight $\mathbf{m}$ by

$$\mathbf{m} + j\mathbf{r}(\alpha), \mathbf{m} + (j-1)\mathbf{r}(\alpha), \ldots, \mathbf{m} - j'\mathbf{r}(\alpha),$$

and sum up all the obtained relations. The result of this procedure is the equation

$$0 = -(j + j' + 1)[\mathbf{r}(\alpha) \cdot \mathbf{m}] + \frac{1}{2}(j + j' + 1)(j' - j)\|\mathbf{r}(\alpha)\|^2;$$

hence

$$2\,\frac{\mathbf{r}(\alpha) \cdot \mathbf{m}}{\|\mathbf{r}(\alpha)\|^2} = j' - j, \tag{2.3.42}$$

where j and j' are integers. Note that

$$\mathbf{m} - 2\,\frac{\mathbf{r}(\alpha) \cdot \mathbf{m}}{\|\mathbf{r}(\alpha)\|^2}\,\mathbf{r}(\alpha) = \mathbf{m} - (j' - j)\,\mathbf{r}(\alpha) \tag{2.3.43}$$

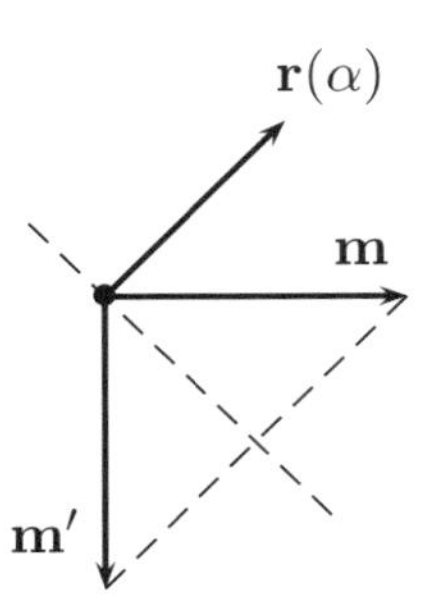

Figure 2.5. Equivalent weights correlated by a Weyl reflection.

is a weight, because it is contained in the sequence $\mathbf{m}-\jmath\mathbf{r}(\alpha),\ldots,\mathbf{m}+\jmath\mathbf{r}(\alpha)$. A consequence of (2.3.43) is that we can pass from one weight $\mathbf{m}$ to another weight $\mathbf{m}'$, by a reflection of $\mathbf{m}$ about an axis perpendicular to a root $\mathbf{r}(\alpha)$ (see Fig. 2.5). Such reflections are called *Weyl reflections*, and the weights correlated by these reflections are *equivalent weights*.

One can see in Fig. 2.4(A) that, for $\alpha \geq 1$, the $\mathsf{SU}(3)$ group has the roots

$$\mathbf{r}(1) = \left(\frac{1}{\sqrt{3}}, 0\right), \quad \mathbf{r}(2) = \left(\frac{1}{2\sqrt{3}}, \frac{1}{2}\right), \quad \mathbf{r}(3) = \left(-\frac{1}{2\sqrt{3}}, \frac{1}{2}\right),$$

such that we have

$$\frac{2\,\mathbf{r}(1)}{\|\mathbf{r}(1)\|^2} = (2\sqrt{3}, 0), \quad \frac{2\,\mathbf{r}(2)}{\|\mathbf{r}(2)\|^2} = (\sqrt{3}, 3), \quad \frac{2\,\mathbf{r}(3)}{\|\mathbf{r}(3)\|^2} = (-\sqrt{3}, 3).$$

Thus, according to (2.3.42), the expressions $2\sqrt{3}m_1$ and $\pm\sqrt{3}m_1 + 3m_2$ are integer numbers, and it follows that the linear combinations $2\sqrt{3}m_1 \pm 6m_2$ are also integers. We may choose $2\sqrt{3}m_1 = \lambda + \mu$ and $6m_2 = \lambda - \mu$, where $\lambda, \mu \in \mathbb{N}$, which lead us to the general expression of the weights for the $\mathsf{SU}(3)$ group

$$\mathbf{m}(\lambda, \mu) = \lambda\left(\frac{\sqrt{3}}{6}, \frac{1}{6}\right) + \mu\left(\frac{\sqrt{3}}{6}, -\frac{1}{6}\right). \tag{2.3.44}$$

DEFINITION 2.3.2 *We say that the weight* $\mathbf{m} = (m_1, m_2, \ldots, m_n)$ *is* **greater than** *the weight* $\mathbf{m}' = (m_1', m_2', \ldots, m_n')$ *if the first nonzero difference (in lexicographical order) is* $m_k - m_k' > 0$.

DEFINITION 2.3.3 *The maximal weight in a set of equivalent weights is called the* **dominant weight**.

THEOREM 2.3.4 *An irreducible representation is uniquely determined by its dominant weight.*

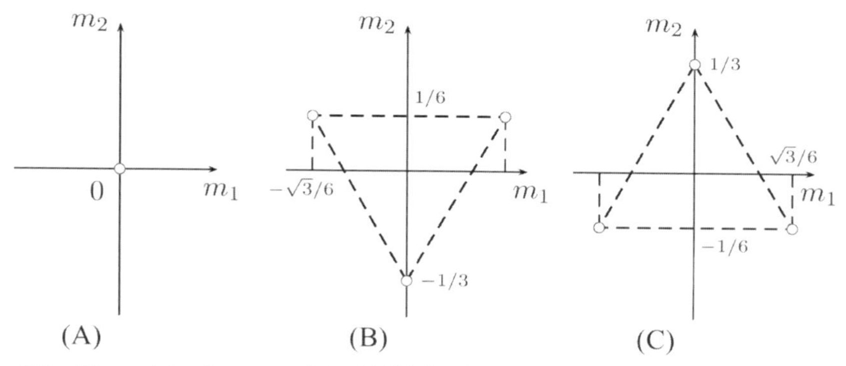

Figure 2.6. The weight diagrams of the $\mathsf{SU}(3)$ irreducible representations $\mathsf{D}(0,0) = 1$ (A), $\mathsf{D}(1,0) = 3$ (B), and $\mathsf{D}(0,1) = \overline{3}$ (C).

It can be shown that the weight (2.3.44) is the dominant weight in an irreducible representation of the $\mathsf{SU}(3)$ group, and that the dimension of that irreducible representation is given by the formula

$$N(\lambda, \mu) = \frac{1}{2}(\lambda + 1)(\mu + 1)(\lambda + \mu + 2). \tag{2.3.45}$$

Thus, for each set of values $\{\lambda, \mu\}$ we obtain the dominant weight in an irreducible representation of $\mathsf{SU}(3)$. All the weights in a given irreducible representation are equivalent, such that starting with the dominant weight we get the complete weight diagram, for the corresponding irreducible representation, by Weyl reflections. Note that the weights determine the diagonal form of matrices $\mathbf{H}_1$ and $\mathbf{H}_2$, and therefore the total number of weights in a given representation is equal to the dimension of the representation $N(\lambda, \mu)$. Furthermore, the explicit forms of $\mathbf{H}_1$ and $\mathbf{H}_2$ enable the finding of the matrices $\mathbf{E}_{\pm\alpha}$ $(\alpha = 1, 2, 3)$, that is the complete set of matrices in the Cartan basis. We may also infer that the irreducible representations of $\mathsf{SU}(3)$ are indexed in terms of a pair of integers, and we will denote them by $\mathsf{D}(\lambda, \mu)$. The notation in terms of the dimension of an irreducible representation is also used (see Chap. 2 Sec. 3.1.1, where we denoted the fundamental representations of $\mathsf{SU}(3)$ by 3 and $\overline{3}$, notations that correspond now to $\mathsf{D}(1,0)$ and $\mathsf{D}(0,1)$, respectively).

For instance, in Fig. 2.6 (A) we show the (trivial) one-dimensional representation $\mathsf{D}(0,0)$ which has only one weight $\mathbf{m} = (0,0)$. The figures 2.6, (B) and (C), display the fundamental representations $\mathsf{D}(1,0) = 3$ and $\mathsf{D}(0,1) = \overline{3}$, defined in Chap. 2 Sec. 3.1.1. Some irreducible representations of higher dimensions are shown in Fig. 2.7 (A), (B), and (C). Note that the weight diagram of the eight-dimensional representation $\mathsf{D}(1,1)$ is identical to the root diagram in Fig. 2.5 (A). The dimension of this irreducible representation is equal to the number of group parameters, and it is called the *regular representation of the* $\mathsf{SU}(3)$ *group*.

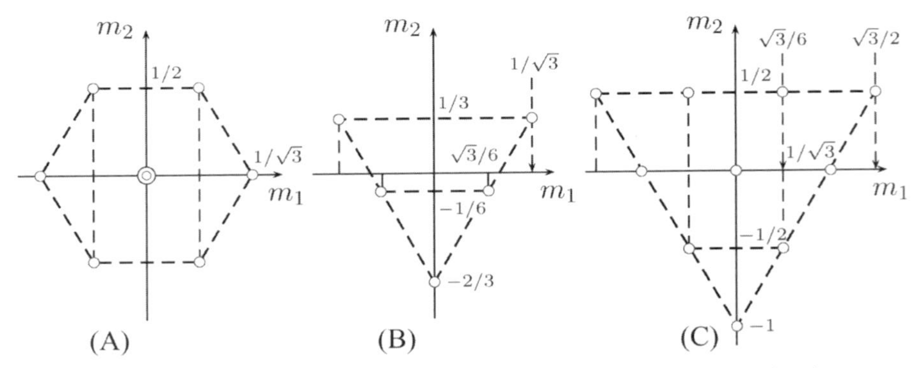

Figure 2.7. The weight diagrams of the SU(3) irreducible representations $\mathsf{D}(1,1) = 8$ (A), $\mathsf{D}(2,0) = 6$ (B), and $\mathsf{D}(3,0) = 10$ (C). Note that the diagram of $\mathsf{D}(1,1)$ contains two zero weights.

3.2 Infinitesimal generators. Parametrization of SU(3)

3.2.1 SU(3) infinitesimal generators in the Cartan basis

The operators in the Cartan basis of the Lie algebra associated to SU(3) are also infinitesimal generators of the SU(3) group, and the weight diagrams provide a general method to find the matrices corresponding to these operators in any representation. Thus, the fundamental representation $\mathsf{D}(1,0)$ of the SU(3) group is determined by the weights $(\sqrt{3}/6, 1/6)$, $(-\sqrt{3}/6, 1/6)$, and $(0, -1/3)$ (see Fig. 2.6 (B)), and let us consider that, in this representation, the eigenvectors of $\mathbf{H}_1$ and $\mathbf{H}_2$ have the form

$$\mathbf{v}_1 = \begin{bmatrix} 1 \\ 0 \\ 0 \end{bmatrix}, \quad \mathbf{v}_2 = \begin{bmatrix} 0 \\ 1 \\ 0 \end{bmatrix}, \quad \mathbf{v}_3 = \begin{bmatrix} 0 \\ 0 \\ 1 \end{bmatrix}. \tag{2.3.46}$$

Then, it follows that the operators H_1 and H_2 will be represented by the diagonal matrices

$$\mathbf{H}_1 = \frac{\sqrt{3}}{6} \begin{bmatrix} 1 & 0 & 0 \\ 0 & -1 & 0 \\ 0 & 0 & 0 \end{bmatrix}, \quad \mathbf{H}_2 = \frac{1}{6} \begin{bmatrix} 1 & 0 & 0 \\ 0 & 1 & 0 \\ 0 & 0 & -2 \end{bmatrix}. \tag{2.3.47}$$

According to (2.3.34), the action of the operators $\mathsf{E}_{\pm 1}$ on the vectors $\mathbf{v}_1$, $\mathbf{v}_2$, and $\mathbf{v}_3$, results in a change of the weights of these vectors into

$$\mathbf{m} \pm \mathbf{r}(1) = \mathbf{m} \pm \left(\frac{1}{\sqrt{3}}, 0 \right).$$

Note that, if the obtained weight does not belong to the diagram in Fig. 2.6 (B), then the result is zero. Consequently, we have the relations

$$\mathbf{E}_1 \mathbf{v}_1 = \mathbf{0}, \quad \mathbf{E}_1 \mathbf{v}_2 = k\, \mathbf{v}_1, \quad \mathbf{E}_1 \mathbf{v}_3 = \mathbf{0},$$

$$\mathbf{E}_{-1}\mathbf{v}_1 = k\,\mathbf{v}_2\,, \quad \mathbf{E}_{-1}\mathbf{v}_2 = \mathbf{0}\,, \quad \mathbf{E}_{-1}\mathbf{v}_3 = \mathbf{0}\,,$$

with the normalization constant k given by (2.3.40)

$$k^2 = \frac{\sqrt{3}}{3}\,\frac{\sqrt{3}}{6} = \frac{1}{6}\,.$$

Finally, we find that, in the fundamental representation $\mathsf{D}(1,0)$, the matrices of the operators $\mathsf{E}_{\pm\alpha}$ are

$$\mathbf{E}_1 = \frac{\pm 1}{\sqrt{6}}\begin{bmatrix} 0 & 1 & 0 \\ 0 & 0 & 0 \\ 0 & 0 & 0 \end{bmatrix}, \; \mathbf{E}_2 = \frac{\pm 1}{\sqrt{6}}\begin{bmatrix} 0 & 0 & 1 \\ 0 & 0 & 0 \\ 0 & 0 & 0 \end{bmatrix}, \; \mathbf{E}_3 = \frac{\pm 1}{\sqrt{6}}\begin{bmatrix} 0 & 0 & 0 \\ 0 & 0 & 1 \\ 0 & 0 & 0 \end{bmatrix},$$
$$\mathbf{E}_{-1} = \frac{\pm 1}{\sqrt{6}}\begin{bmatrix} 0 & 0 & 0 \\ 1 & 0 & 0 \\ 0 & 0 & 0 \end{bmatrix}, \; \mathbf{E}_{-2} = \frac{\pm 1}{\sqrt{6}}\begin{bmatrix} 0 & 0 & 0 \\ 0 & 0 & 0 \\ 1 & 0 & 0 \end{bmatrix}, \; \mathbf{E}_{-3} = \frac{\pm 1}{\sqrt{6}}\begin{bmatrix} 0 & 0 & 0 \\ 0 & 0 & 0 \\ 0 & 1 & 0 \end{bmatrix}. \quad (2.3.48)$$

The sign is determined by commutation relations of the form

$$[\mathbf{E}_1, \mathbf{E}_3] = N_{13}\,\mathbf{E}_2 = \frac{1}{\sqrt{6}}\,\mathbf{E}_2\,,$$

derived from (2.3.23) and (2.3.31), and it follows that we have to choose the factor $+1/\sqrt{6}$ for all the operators $\mathsf{E}_{\pm\alpha}$, $\alpha = 1, 2, 3$. The matrices of the infinitesimal generators in the Cartan basis (2.3.47), (2.3.48) are connected to the corresponding matrices in the Okubo basis (2.3.11), and the Gell-Mann basis (2.3.13), by relations of the form

$$\begin{aligned} \mathbf{H}_1 &= \tfrac{1}{2\sqrt{3}}\left(\mathbf{X}^1_1 - \mathbf{X}^2_2\right) = \tfrac{1}{\sqrt{3}}\,\mathbf{F}_3\,, \\ \mathbf{H}_2 &= \tfrac{1}{\sqrt{6}}\left(\mathbf{X}^1_1 + \mathbf{X}^2_2 - 2\,\mathbf{X}^3_3\right) = \tfrac{1}{\sqrt{3}}\,\mathbf{F}_8\,, \end{aligned} \quad (2.3.49)$$

$$\begin{aligned} \mathbf{E}_1 &= \tfrac{1}{\sqrt{6}}\,\mathbf{X}^1_2 = \tfrac{1}{\sqrt{6}}\,(\mathbf{F}_1 + i\,\mathbf{F}_2)\,, \quad & \mathbf{E}_{-1} &= \tfrac{1}{\sqrt{6}}\,\mathbf{X}^2_1 = \tfrac{1}{\sqrt{6}}\,(\mathbf{F}_1 - i\,\mathbf{F}_2)\,, \\ \mathbf{E}_2 &= \tfrac{1}{\sqrt{6}}\,\mathbf{X}^1_3 = \tfrac{1}{\sqrt{6}}\,(\mathbf{F}_4 + i\,\mathbf{F}_5)\,, \quad & \mathbf{E}_{-2} &= \tfrac{1}{\sqrt{6}}\,\mathbf{X}^3_1 = \tfrac{1}{\sqrt{6}}\,(\mathbf{F}_4 - i\,\mathbf{F}_5)\,, \\ \mathbf{E}_3 &= \tfrac{1}{\sqrt{6}}\,\mathbf{X}^2_3 = \tfrac{1}{\sqrt{6}}\,(\mathbf{F}_6 + i\,\mathbf{F}_7)\,, \quad & \mathbf{E}_{-3} &= \tfrac{1}{\sqrt{6}}\,\mathbf{X}^3_2 = \tfrac{1}{\sqrt{6}}\,(\mathbf{F}_6 - i\,\mathbf{F}_7)\,. \end{aligned} \quad (2.3.50)$$

Also, the weight diagram in Fig. 2.6 (C) leads us to the matrices of the infinitesimal generators, in the Cartan basis, for the fundamental representation $\mathsf{D}(0,1)$

$$\begin{aligned} \mathbf{H}'_i &= \mathbf{H}_i\,, \quad i = 1, 2, \\ \mathbf{E}'_{\pm\alpha} &= -\left(\mathbf{E}_{\pm\alpha}\right)^{\mathrm{T}}\,, \quad \alpha = 1, 2, 3. \end{aligned} \quad (2.3.51)$$

Generally, by this method one can find the matrices of the $\mathsf{SU}(3)$ infinitesimal generators in the Cartan basis of any representation, after finding the corresponding weight diagram. Then, the matrices of the infinitesimal generators in the Okubo and Gell-Mann bases are obtained by inverting the relations (2.3.49) and (2.3.50).

3.2.2 Parametrization of SU(3)

There are several parametrizations of the $\mathsf{SU}(3)$ group. Thus, in the Okubo basis we introduce the complex quantities ε^i_j which satisfy the relation $\overline{\varepsilon^i_j} = \varepsilon^j_i$, such that only nine real numbers are independent. Because the matrices of the infinitesimal generators have to satisfy the relation $\mathbf{X}^1_1 + \mathbf{X}^2_2 + \mathbf{X}^3_3 = \mathbf{0}$, there remain only eight independent real parameters. Another parametrization is that given by Gell-Mann's representation (2.3.15), in terms of the set of eight independent real parameters μ^k. Also, by means of (2.3.49) and (2.3.50) we obtain the parametrizations of the $\mathsf{SU}(3)$ group in the Cartan basis.

The choice of the matrices of infinitesimal generators in the form $\boldsymbol{\lambda}_k = 2\,\mathbf{F}_k$, $k = 1, 2, \ldots, 8$, is convenient, because the matrices $\boldsymbol{\lambda}_k$ represent the $\mathsf{SU}(3)$ counterpart of the Pauli matrices $\boldsymbol{\sigma}_i$, $i = 1, 2, 3$.

The Pauli matrices satisfy the relations

$$[\boldsymbol{\sigma}_j, \boldsymbol{\sigma}_k] = 2\,i\,\epsilon_{jkl}\,\boldsymbol{\sigma}_l\,, \quad \mathsf{Tr}\,(\boldsymbol{\sigma}_j\boldsymbol{\sigma}_k) = 2\,\delta_{jk}\,, \quad j,k = 1,2,3, \tag{2.3.52}$$

and can be used to form the Cartan basis for the $\mathsf{SU}(2)$ group

$$\begin{aligned} \mathbf{H} &= \frac{1}{2}\boldsymbol{\sigma}_3 = \frac{1}{2}\begin{bmatrix} 1 & 0 \\ 0 & -1 \end{bmatrix}, \\ \mathbf{E}_+ &= \frac{1}{2}(\boldsymbol{\sigma}_1 + i\,\boldsymbol{\sigma}_2) = \begin{bmatrix} 0 & 1 \\ 0 & 0 \end{bmatrix}, \\ \mathbf{E}_- &= \frac{1}{2}(\boldsymbol{\sigma}_1 - i\,\boldsymbol{\sigma}_2) = \begin{bmatrix} 0 & 0 \\ 1 & 0 \end{bmatrix}. \end{aligned} \tag{2.3.53}$$

Also, from the commutation relations (2.3.52) we deduce

$$[\mathbf{H}, \mathbf{E}_\pm] = \pm\,\mathbf{E}_\pm\,, \quad [\mathbf{E}_+, \mathbf{E}_-] = 2\,\mathbf{H}\,. \tag{2.3.54}$$

The matrices $\boldsymbol{\lambda}_k$ satisfy the commutation relations

$$[\boldsymbol{\lambda}_j, \boldsymbol{\lambda}_k] = 2\,i\,f_{jkl}\,\boldsymbol{\lambda}_l\,, \quad \mathsf{Tr}\,(\boldsymbol{\lambda}_j\boldsymbol{\lambda}_k) = 2\,\delta_{jk}\,, \quad j,k = 1,2,\ldots,8, \tag{2.3.55}$$

and, for the $\mathsf{SU}(3)$ group, we introduce the operators

$$\begin{aligned} \mathsf{T}_\pm &= (\lambda_1 \pm i\,\lambda_2)/2 = \sqrt{6}\,\mathsf{E}_{\pm 1}\,, \\ \mathsf{V}_\pm &= (\lambda_4 \pm i\,\lambda_5)/2 = \sqrt{6}\,\mathsf{E}_{\pm 2}\,, \\ \mathsf{U}_\pm &= (\lambda_6 \pm i\,\lambda_7)/2 = \sqrt{6}\,\mathsf{E}_{\pm 3}\,, \end{aligned} \tag{2.3.56}$$

and $\mathsf{T}_3 = \lambda_3/2 = \sqrt{3}\,\mathsf{H}_1$. It is easy to verify that the operators $\{\mathsf{T}_\pm, \mathsf{T}_3\}$ satisfy commutation relations similar to (2.3.54), that is

$$[\mathbf{T}_3, \mathbf{T}_\pm] = \pm\,\mathbf{T}_\pm\,, \quad [\mathbf{T}_+, \mathbf{T}_-] = 2\,\mathbf{T}_3\,. \tag{2.3.57}$$

The operators $\{\mathsf{T}_\pm, \mathsf{T}_3\}$ are the infinitesimal generators of a subgroup $\mathsf{SU}(2) \subset \mathsf{SU}(3)$. Analogously, we observe that the operators $\mathsf{V}_\pm$ and

$$\mathsf{V}_3 = \frac{1}{2}\,[\mathsf{V}_+, \mathsf{V}_-] = \frac{\sqrt{3}}{2}\left(\mathsf{H}_1 + \sqrt{3}\,\mathsf{H}_2\right) = \frac{1}{4}\left(\lambda_3 + \sqrt{3}\,\lambda_8\right)\,,$$

as well the operators $\mathsf{U}_\pm$ and

$$\mathsf{U}_3 = \frac{1}{2}\,[\mathsf{U}_+, \mathsf{U}_-] = -\frac{\sqrt{3}}{2}\left(\mathsf{H}_1 - \sqrt{3}\,\mathsf{H}_2\right) = -\frac{1}{4}\left(\lambda_3 - \sqrt{3}\,\lambda_8\right)\,,$$

form the bases for two other $\mathsf{SU}(2)$ type subgroups of the $\mathsf{SU}(3)$ group. Hence, the $\mathsf{SU}(3)$ group contains three different $\mathsf{SU}(2)$ subgroups.

In the general case of the $\mathsf{SU}(n)$ group, the problem of labelling the representations is solved by using the canonical factorization

$$\mathsf{SU}(n) \supset \mathsf{U}(1) \times \mathsf{SU}(n-1)\,,$$

where $\mathsf{SU}(n-1)$ is a subgroup of $\mathsf{SU}(n)$, and $\mathsf{U}(1)$ is an one-parameter Abelian subgroup, generated by a linear combination of the $n-1$ operators H_i in the Cartan basis of $\mathsf{SU}(n)$. Note that, each of the infinitesimal generators of $\mathsf{SU}(n-1)$ has to commute with the infinitesimal generator of $\mathsf{U}(1)$.

In our case, the $\mathsf{SU}(3)$ group has three distinct $\mathsf{SU}(2)$ subgroups, and therefore three possible decompositions of the form

$$\mathsf{SU}(3) \supset \mathsf{U}(1) \times \mathsf{SU}(2)\,.$$

To determine the infinitesimal generator of the $\mathsf{U}(1)$ group we consider the linear combination

$$\mathsf{H}_0 = \alpha\,\mathsf{H}_1 + \beta\,\mathsf{H}_2\,,$$

where α and β are real constants. If the $\mathsf{SU}(2)$ subgroup has the infinitesimal generators $\mathsf{T}_\pm$ and T_3, then the operator H_0 has to satisfy the commutation relations

$$[\mathsf{H}_0, \mathsf{T}_\pm] = [\mathsf{H}_0, \mathsf{T}_3] = 0\,.$$

Replacing the infinitesimal generators by their matrices we find $\alpha = 0$ and $\mathsf{H}_0 = \beta\,\mathsf{H}_2$, where β remains an arbitrary real constant. One can follow the same procedure to find the form of the operator H_0 for the other two canonical decompositions of $\mathsf{SU}(3)$ in terms of the $\mathsf{SU}(2)$ subgroups with infinitesimal generators $\{\mathsf{V}_\pm, \mathsf{V}_3\}$, and $\{\mathsf{U}_\pm, \mathsf{U}_3\}$, respectively. We thus obtain all the three canonical decompositions of the $\mathsf{SU}(3)$ group

$$\mathsf{SU}(3) \supset \mathsf{U}(1)_k \times \mathsf{SU}(2)_k \,, \quad k = \mathsf{T}, \mathsf{V}, \mathsf{U} \,, \tag{2.3.58}$$

having the infinitesimal generators

$$\begin{aligned}
&\mathsf{U}(1)_\mathsf{T} &&: \quad \mathsf{H}_\mathsf{T} = \beta\,\mathsf{H}_2 = \frac{\beta}{2\sqrt{3}}\,\boldsymbol{\lambda}_8 \,, \\
&\mathsf{SU}(2)_\mathsf{T} &&: \quad \{\mathsf{T}_\pm, \mathsf{T}_3\} \,, \\
&\mathsf{U}(1)_\mathsf{V} &&: \quad \mathsf{H}_\mathsf{V} = -\beta\left(\sqrt{3}\,\mathsf{H}_1 - \mathsf{H}_2\right) = \frac{-\beta}{2\sqrt{3}}\left(\sqrt{3}\,\boldsymbol{\lambda}_3 - \boldsymbol{\lambda}_8\right), \\
&\mathsf{SU}(2)_\mathsf{V} &&: \quad \{\mathsf{V}_\pm, \mathsf{V}_3\} \,, \\
&\mathsf{U}(1)_\mathsf{U} &&: \quad \mathsf{H}_\mathsf{U} = \beta\left(\sqrt{3}\,\mathsf{H}_1 + \mathsf{H}_2\right) = \frac{\beta}{2\sqrt{3}}\left(\sqrt{3}\,\boldsymbol{\lambda}_3 + \boldsymbol{\lambda}_8\right), \\
&\mathsf{SU}(2)_\mathsf{U} &&: \quad \{\mathsf{U}_\pm, \mathsf{U}_3\} \,.
\end{aligned} \tag{2.3.59}$$

Finally, if we parametrize the vectors $\mathbf{z} \in \mathbb{C}^3$ in the form

$$\mathbf{z} = \begin{bmatrix} e^{i\varphi_1}\,\cos\theta \\ e^{i\varphi_2}\,\sin\theta\,\cos\psi \\ e^{i\varphi_3}\,\sin\theta\,\sin\psi \end{bmatrix}, \qquad \begin{array}{l} 0 \le \varphi_i < 2\pi, \quad i = 1,2,3, \\ 0 \le \theta, \psi \le \dfrac{\pi}{2}, \end{array} \tag{2.3.60}$$

the relation $\mathbf{z}^\dagger\mathbf{z} = z^1 z_1 + z^2 z_2 + z^3 z_3 = 1$ takes place. This quadratic form is preserved by the $\mathsf{SU}(3)$ transformations. Indeed, according to (2.3.3) and (2.3.6), for an $\mathsf{SU}(3)$ transformation $\mathbf{z}' = \mathbf{u}\mathbf{z}$ we obtain $\mathbf{z}'^\dagger\mathbf{z}' = \mathbf{z}^\dagger\mathbf{z} = 1$. It can be shown that, in the Gell-Mann basis, the matrices associated to the $\mathsf{SU}(3)$ operators for *finite transformations* can be parametrized as

$$\mathbf{u}(\boldsymbol{\alpha}) = \mathbf{T}_2\,\mathbf{T}_3\,\mathbf{T}_2' \,, \tag{2.3.61}$$

where

$$\begin{aligned}
\mathbf{T}_2 &= e^{-i\alpha_1\mathbf{F}_3}\,e^{-i\alpha_2\mathbf{F}_2}\,e^{-i\alpha_3\mathbf{F}_3} \,, \\
\mathbf{T}_3 &= e^{-i(2/\sqrt{3})\alpha_8\mathbf{F}_8}\,e^{-2i\alpha_7\mathbf{F}_7} \,, \\
\mathbf{T}_2' &= e^{-i\alpha_4\mathbf{F}_3}\,e^{-i\alpha_5\mathbf{F}_2}\,e^{-i\alpha_6\mathbf{F}_3} \,.
\end{aligned} \tag{2.3.62}$$

By means of the series expansion of the exponentials, it can be shown that there exist a solution $\{\alpha_4, \alpha_5, \alpha_6\}$ in a minimum range $\alpha_4, \alpha_6 \in [0, 4\pi]$ and $\alpha_5 \in [0, \pi]$, such that

$$\mathbf{T}_2'\mathbf{z} = \begin{bmatrix} 0 \\ e^{i\varphi_3}\cos\psi' \\ e^{i\varphi_3}\sin\psi' \end{bmatrix}, \tag{2.3.63}$$

with $0 \leq \psi' \leq \pi/2$. If

$$\begin{aligned} &\exp\left[(2/3)i\alpha_8\right] = \exp\left(-i\varphi_3\right), && \alpha_8 \in [0, 3\pi], \\ &\alpha_7 = \frac{\pi}{2} - \psi', && \alpha_7 \in \left[0, \frac{\pi}{2}\right], \end{aligned} \tag{2.3.64}$$

then we obtain

$$\mathbf{T}_3\, \mathbf{T}_2'\, \mathbf{z} = \begin{bmatrix} 0 \\ 0 \\ 1 \end{bmatrix}.$$

The matrix $\mathbf{T}_2$ in (2.3.61) is the most general unitary matrix of the form

$$\begin{bmatrix} u_{11} & u_{12} & 0 \\ u_{21} & u_{22} & 0 \\ 0 & 0 & 1 \end{bmatrix}.$$

In this parametrization, the invariant integral on the $\mathsf{SU}(3)$ group has the measure

$$\begin{aligned} \mathrm{d}g = {} & \sin\alpha_2\, \sin\alpha_5\, \sin\left(2\alpha_7\right)\, \sin^2\alpha_7 \\ & \times \frac{\mathrm{d}\alpha_1}{4\pi}\, \frac{\mathrm{d}\alpha_2}{2}\, \frac{\mathrm{d}\alpha_3}{4\pi}\, \frac{\mathrm{d}\alpha_4}{4\pi}\, \frac{\mathrm{d}\alpha_5}{2}\, \frac{\mathrm{d}\alpha_6}{4\pi}\, 2\mathrm{d}\alpha_7\, \frac{\mathrm{d}\alpha_8}{3\pi}, \end{aligned} \tag{2.3.65}$$

such that, with $\alpha_1, \alpha_3, \alpha_4, \alpha_6 \in [0, 4\pi]$, $\alpha_2, \alpha_5 \in [0, \pi]$, $\alpha_7 \in [0, \pi/2]$, and $\alpha_8 \in [0, 3\pi]$, we have

$$\int_{\mathsf{SU}(3)} \mathrm{d}g = 1.$$

3.3 The $\mathsf{GL}(n, \mathbb{C})$ and $\mathsf{SU}(n)$ groups

3.3.1 The $\mathsf{GL}(n, \mathbb{C})$ group

Let $\mathcal{M}(n)$ be the set of $n \times n$ complex matrices. If we introduce in this set an internal operation, defined by the commutator

$$[\mathbf{a}, \mathbf{b}] \equiv \mathbf{ab} - \mathbf{ba}, \quad \mathbf{a}, \mathbf{b} \in \mathcal{M}(n), \tag{2.3.66}$$

then it is easy to verify that

$$\begin{aligned} [\mathbf{a}, \mathbf{b}] &= -[\mathbf{b}, \mathbf{a}], \\ [\mathbf{a}, [\mathbf{b}, \mathbf{c}]] + [\mathbf{b}, [\mathbf{c}, \mathbf{a}]] &+ [\mathbf{c}, [\mathbf{a}, \mathbf{b}]] = \mathbf{0}; \end{aligned} \tag{2.3.67}$$

therefore, the set $\mathcal{M}(n)$ forms a Lie algebra with respect to the commutation operation.

DEFINITION 2.3.4 *The set of nonsingular $n \times n$ complex matrices, endowed with an internal composition law defined as the product of matrices, is called the* **general linear group of nth order**, *and it is denoted by*

$$\mathsf{GL}(n, \mathbb{C}) = \{\mathbf{g} \mid \mathbf{g} \in \mathcal{M}(n),\ \det(\mathbf{g}) \neq 0\}\,.$$

THEOREM 2.3.5 *The Lie algebra of the* $\mathsf{GL}(n, \mathbb{C})$ *group is isomorphic to the Lie algebra* $\mathcal{M}(n)$.

The infinitesimal generators of the $\mathsf{GL}(n, \mathbb{C})$ group are the real matrices $\mathbf{X}_{ij} \in \mathcal{M}(n)$, with $i, j = 1, 2, \ldots, n$, having all the elements equal to zero excepting the element (ij) which is equal to one, that is

$$(\mathbf{X}_{ij})_{kl} = \delta_{ik}\,\delta_{jl}\,.$$

These infinitesimal generators satisfy the commutation relations

$$[\mathbf{X}_{ij}, \mathbf{X}_{kl}] = \delta_{jk}\,\mathbf{X}_{il} - \delta_{il}\,\mathbf{X}_{kj}\,, \tag{2.3.68}$$

and form the basis of the Lie algebra of the $\mathsf{GL}(n, \mathbb{C})$ group, the elements of the algebra being of the form

$$\mathbf{x} = \sum_{i=1}^{n}\sum_{j=1}^{n} \alpha_{ij}\,\mathbf{X}_{ij}\,, \quad \alpha_{ij} \in \mathbb{C}\,.$$

We mention that the commutation relations (2.3.68) are also satisfied by the differential operators

$$\mathsf{X}_{ij} = x_i\,\frac{\partial}{\partial x_j}\,, \quad x_i, x_j \in \mathbb{C}\,, \quad i, j = 1, 2, \ldots, n, \tag{2.3.69}$$

which are the infinitesimal generators of the $\mathsf{GL}(n, \mathbb{C})$ group, in a representation in terms of differential operators.

THEOREM 2.3.6 *The exponential of the* $\mathsf{GL}(n, \mathbb{C})$ *group is defined in terms of the series expansion*

$$e^{t\mathbf{x}} = \sum_{k=0}^{\infty} \frac{1}{k!}\,(t\mathbf{x})^k\,, \quad t \in \mathbb{R}\,, \quad \mathbf{x} \in \mathcal{M}(n)\,,$$

and has the following properties:

a) *the mapping $t \to \exp(t\mathbf{x})$ is an homomorphism of $\mathbb{R}$ into* $\mathsf{GL}(n, \mathbb{C})$*;*

b) *the mapping* $\mathbf{x} \rightarrow \exp(t\mathbf{x})$ *is an isomorphism of a vicinity* $U \subset \mathcal{M}(n)$, *of the zero matrix, into a vicinity* $V \subset \mathsf{GL}(n, \mathbb{C})$, *of the identity element;*

c) *the property* b) *also holds for a subalgebra of* $\mathcal{M}(n)$ *and the subgroup of* $\mathsf{GL}(n, \mathbb{C})$ *to which this subalgebra is associated.*

On the basis of this theorem one can prove that

$$\begin{aligned} e^{\mathbf{x}+\mathbf{y}} &= e^{\mathbf{x}} e^{\mathbf{y}} , \\ e^{\mathbf{g}\,\mathbf{x}\,\mathbf{g}^{-1}} &= \mathbf{g}\, e^{\mathbf{x}} \mathbf{g}^{-1} , \end{aligned} \tag{2.3.70}$$

where $\mathbf{x}, \mathbf{y} \in \mathcal{M}(n)$, $[\mathbf{x}, \mathbf{y}] = \mathbf{0}$, and $\mathbf{g} \in \mathsf{GL}(n, \mathbb{C})$.

THEOREM 2.3.7 *If* λ_i, $i = 1, 2, \ldots, n$, *are the eigenvalues of the matrix* $\mathbf{x} \in \mathcal{M}(n)$, *degenerated or not, then the eigenvalues of the exponential* $\exp(\mathbf{x})$ *are* $\exp(\lambda_i)$, $i = 1, 2, \ldots, n$.

A consequence of this theorem is the relation

$$\det(e^{\mathbf{x}}) = e^{\mathsf{Tr}\,\mathbf{x}} , \quad \mathbf{x} \in \mathcal{M}(n) . \tag{2.3.71}$$

Among the Lie subgroups of the $\mathsf{GL}(n, \mathbb{C})$ group we mention the following:

1 the *special (unimodular) linear group*

$$\mathsf{SL}(n, \mathbb{C}) = \{\mathbf{g} \mid \mathbf{g} \in \mathsf{GL}(n, \mathbb{C}),\ \det(\mathbf{g}) = 1\} ,$$

2 the *unitary group*

$$\mathsf{U}(n) = \{\mathbf{g} \mid \mathbf{g} \in \mathsf{GL}(n, \mathbb{C}),\ \mathbf{g}^{\dagger}\mathbf{g} = \mathbf{e}\} ,$$

3 the *special unitary group*

$$\begin{aligned} \mathsf{SU}(n) &= \{\mathbf{g} \mid \mathbf{g} \in \mathsf{SL}(n, \mathbb{C}),\ \mathbf{g}^{\dagger}\mathbf{g} = \mathbf{e}\} \\ &= \{\mathbf{g} \mid \mathbf{g} \in \mathsf{U}(n) , \det(\mathbf{g}) = 1\} = \mathsf{U}(n) \cap \mathsf{SL}(n, \mathbb{C}) , \end{aligned}$$

4 the *linear real group*

$$\mathsf{GL}(n, \mathbb{R}) = \{\mathbf{g} \mid \mathbf{g} \in \mathsf{GL}(n, \mathbb{C}),\ \mathsf{Im}(\mathbf{g}) = 0\} ,$$

5 the *special linear real group*

$$\mathsf{SL}(n, \mathbb{R}) = \{\mathbf{g} \mid \mathbf{g} \in \mathsf{GL}(n, \mathbb{R}) , \det(\mathbf{g}) = 1\} ,$$

6 the *orthogonal group*

$$\mathsf{O}(n) = \{\mathbf{g} \mid \mathbf{g} \in \mathsf{GL}(n, \mathbb{R}),\, \mathbf{g}^{\mathrm{T}}\mathbf{g} = \mathbf{e}\} = \mathsf{U}(n) \cap \mathsf{GL}(n, \mathbb{R})\,,$$

7 the *special orthogonal group*

$$\begin{aligned}\mathsf{SO}(n) &= \{\mathbf{g} \mid \mathbf{g} \in \mathsf{SL}(n, \mathbb{R}),\, \mathbf{g}^{\mathrm{T}}\mathbf{g} = \mathbf{e}\} \\ &= \{\mathbf{g} \mid \mathbf{g} \in \mathsf{O}(n, \mathbb{R}),\, \det(\mathbf{g}) = 1\} = \mathsf{O}(n) \cap \mathsf{SL}(n, \mathbb{R})\,.\end{aligned}$$

It can be shown that all these Lie groups are compact.

3.3.2 The SU(n) group

DEFINITION 2.3.5 *A mapping $f : \mathcal{E} \to \mathcal{L}$, where $\mathcal{E}$ and $\mathcal{L}$ are topological spaces, is called* **homeomorphism** *or* **topological mapping** *if f is one-to-one, and the mappings f and f^{-1} are continuous.*

DEFINITION 2.3.6 *Two topological spaces $\mathcal{E}$ and $\mathcal{L}$ are called* **homeomorphic** *if there exists a homeomorphism between $\mathcal{E}$ and $\mathcal{L}$.*

Note that, in the case of the existence of a topological group structure, the homeomorphism entails the homomorphism.

THEOREM 2.3.8 *The $\mathsf{U}(n)$ and $\mathsf{U}(1)\times\mathsf{SU}(n)$ groups are homeomorphic (and also homomorphic).*

If we consider a mapping which associates to each matrix $\mathbf{u} \in \mathsf{U}(n)$ the pair $\{\alpha, \mathbf{v}\} \in \mathsf{U}(1) \times \mathsf{SU}(n)$, such that

$$\alpha = \det(\mathbf{u})\,, \quad v_{1k} = \frac{u_{1k}}{\alpha}\,, \; v_{jk} = u_{jk}\,, \quad k = 1, 2, \ldots, n,\; j = 2, 3, \ldots, n,$$

then it is easy to verify that this mapping is a homeomorphism.

In the case of the SU(n) group the Theorem 2.3.6 becomes

THEOREM 2.3.9 *Any matrix $\mathbf{u} \in \mathsf{U}(n)$ can be represented in the form $\mathbf{u} = \exp(i\,\mathbf{h})$, where $\mathbf{h}$ is a $n \times n$ Hermitian matrix. Reciprocally, any matrix $\exp(i\,\mathbf{h})$, where $\mathbf{h}$ is Hermitian, is a unitary matrix.*

On the basis of the Theorem 2.3.1 and the relation (2.3.71), we deduce that the elements of the $\mathsf{SU}(n)$ group can be represented in the form of the matrices $\mathbf{v} = \exp(i\,\mathbf{x})$, where $\mathbf{x}$ is a Hermitian matrix of zero trace. In fact, this general result justifies the form (2.3.8) used for the elements of the $\mathsf{SU}(3)$ group in Chap. 2 Sec. 3.1.2. It also makes clear the relation between a Lie group and its associated Lie algebra, each element in the group being expressed as an exponential of a certain algebra element which, in turn, is a linear combination of the infinitesimal generators (see also Theorem 2.2.2).

THEOREM 2.3.10 *Let* H *be a subgroup of the* $\mathsf{SU}(n)$ *group, formed by the matrices*

$$\mathbf{h} = \begin{bmatrix} 1 & 0 \\ 0 & \mathbf{g} \end{bmatrix}, \quad \mathbf{h} \in \mathsf{H},$$

where $\mathbf{g}$ *is a matrix of order* $n-1$. *Then, the subgroup* H *is isomorphic to the* $\mathsf{SU}(n-1)$ *group, and the factor set* $\mathsf{SU}(n)/\mathsf{H}$ *is homeomorphic to the unit hypersphere in the* $(2n-1)$*-dimensional space.*

Also, one can prove that it always exists a decomposition of the form

$$\mathsf{SU}(n) \supset \mathsf{U}(1) \times \mathsf{SU}(n-1),$$

called the *canonical decomposition*. This decomposition leads to the canonical chain of subgroups

$$\mathsf{SU}(n) \supset \mathsf{SU}(n-1) \supset \ldots \supset \mathsf{SU}(3) \supset \mathsf{SU}(2).$$

4. The Lorentz group

4.1 Lorentz transformations

4.1.1 Special Lorentz transformations

Let us consider a four-dimensional vector space. The scalar product of two vectors, $\mathbf{x}$ and $\mathbf{y}$, in this space is defined by the relation

$$\langle \mathbf{x}, \mathbf{y} \rangle = g_{\mu\nu} x^{\mu} y^{\nu}, \tag{2.4.1}$$

where $g_{\mu\nu}$ is the *metric tensor*, which determines the geometric properties of the space, and x^{μ} and y^{ν} are the components of the vectors $\mathbf{x}$ and $\mathbf{y}$, respectively. If there exists an orthonormal basis in this space such that

$$g_{\mu\nu} = \delta_{\mu\nu}, \quad \mu, \nu = 1, 2, 3, 4,$$

then the space is called *Euclidean* (the space $\mathcal{E}_4$). On the other hand, if there exists an orthonormal basis such that

$$g_{\mu\mu} = -1, \quad \mu = 1, 2, 3, \quad g_{44} = 1,$$

all the other components of the metric tensor being equal to zero, then we obtain the *Minkowskian* space (the space $\mathcal{M}_4$).

Let $\mathbf{\Lambda}$ be the matrix of a linear transformation in $\mathcal{M}_4$. Then, the norm of the transformed vector

$$\mathbf{x}' = \mathbf{\Lambda}\mathbf{x} \tag{2.4.2}$$

can be expressed in the form

$$\|\mathbf{x}'\|^2 = g_{\mu\nu} x'^{\mu} x'^{\nu} = \mathbf{x}'^{\mathrm{T}} \mathbf{G} \mathbf{x}' = (\mathbf{\Lambda}\mathbf{x})^{\mathrm{T}} \mathbf{G} (\mathbf{\Lambda}\mathbf{x}) = \mathbf{x}^{\mathrm{T}} (\mathbf{\Lambda}^{\mathrm{T}} \mathbf{G} \mathbf{\Lambda}) \mathbf{x}, \tag{2.4.3}$$

where

$$\mathbf{G} = [g_{\mu\nu}] = \begin{bmatrix} -1 & 0 & 0 & 0 \\ 0 & -1 & 0 & 0 \\ 0 & 0 & -1 & 0 \\ 0 & 0 & 0 & 1 \end{bmatrix} . \tag{2.4.4}$$

The norm of the vector $\mathbf{x}$ is

$$\|\mathbf{x}\|^2 = \langle \mathbf{x}, \mathbf{x} \rangle = \mathbf{x}^{\mathrm{T}} \mathbf{G} \mathbf{x} ,$$

such that the linear transformations that preserve the norm of the vectors in $\mathcal{M}_4$ (or, equally well, the distance between any two points in the Minkovskian space), that is $\|\mathbf{x}'\|^2 = \|\mathbf{x}\|^2$, satisfy the relation

$$\mathbf{G} = \mathbf{\Lambda}^{\mathrm{T}} \mathbf{G} \mathbf{\Lambda} . \tag{2.4.5}$$

If we assume that the transformation (2.4.2) leaves unchanged the coordinates x^2 and x^3, then the matrix $\mathbf{\Lambda}$ has the general form

$$\mathbf{\Lambda}_{14} = \begin{bmatrix} \lambda_1^1 & 0 & 0 & \lambda_4^1 \\ 0 & 1 & 0 & 0 \\ 0 & 0 & 1 & 0 \\ \lambda_1^4 & 0 & 0 & \lambda_4^4 \end{bmatrix} . \tag{2.4.6}$$

Now, if the transformation defined by the matrix $\mathbf{\Lambda}_{14}$ preserves the norm of the vectors in $\mathcal{M}_4$, then this matrix has to satisfy the condition (2.4.5), and we thus obtain the equations

$$\begin{aligned} 1 - (\lambda_1^1)^2 + (\lambda_1^4)^2 &= 0 , \\ \lambda_1^1 \lambda_4^1 - \lambda_1^4 \lambda_4^4 &= 0 , \\ 1 + (\lambda_4^1)^2 - (\lambda_4^4)^2 &= 0 . \end{aligned} \tag{2.4.7}$$

We consider the square of the second equation in (2.4.7) and substitute $(\lambda_1^1)^2$ and $(\lambda_4^4)^2$ from the first and the last equation, respectively. Thus, we have

$$(\lambda_4^1)^2 = (\lambda_1^4)^2 = (\lambda_4^4)^2 - 1 = (\lambda_1^1)^2 - 1 .$$

We choose the matrix elements λ_1^1 and λ_4^4 to be positive, which means that the transformation defined by $\mathbf{\Lambda}_{14}$ does not change the orientation of the axes, and introduce the notations[7]

$$\lambda_1^1 = \lambda_4^4 \equiv \alpha > 0 , \quad \lambda_4^1 = \lambda_1^4 \equiv -\alpha\beta .$$

[7]The minus sign introduced in the second set of notations is directly related to the physical meaning of these transformations, in the theory of Special Relativity.

Finally, by substituting λ_1^1 and λ_1^4 into the first equation in (2.4.7), we obtain

$$\alpha = 1/\sqrt{1-\beta^2}\,,$$

and the matrix (2.4.6) in the form

$$\mathbf{\Lambda}_{14}(\beta) = \begin{bmatrix} 1/\sqrt{1-\beta^2} & 0 & 0 & -\beta/\sqrt{1-\beta^2} \\ 0 & 1 & 0 & 0 \\ 0 & 0 & 1 & 0 \\ -\beta/\sqrt{1-\beta^2} & 0 & 0 & 1/\sqrt{1-\beta^2} \end{bmatrix}. \tag{2.4.8}$$

The linear transformations in the space $\mathcal{M}_4$ which are determined by a matrix of the form (2.4.8) are called *special Lorentz transformations*. These transformations depend on a single real parameter β, and form a one-parameter Lie group, which is a subgroup of the *Lorentz group*. It is easy to verify that the set of matrices $\mathbf{\Lambda}_{14}(\beta)$ contains an identity transformation defined by $\beta = 0$

$$\mathbf{\Lambda}_{14}(0) = \mathbf{E}$$

and for each transformation defined by β there exists an inverse transformation obtained by changing the sign of β, respectively

$$\mathbf{\Lambda}_{14}(\beta)\mathbf{\Lambda}_{14}(-\beta) = \mathbf{E}\,.$$

The product of two special Lorentz transformations

$$\mathbf{\Lambda}_{14}(\beta_1)\mathbf{\Lambda}_{14}(\beta_2) = \mathbf{\Lambda}_{14}(\beta_3)\,,$$

leads us to the relation

$$\beta_3 = \frac{\beta_1 + \beta_2}{1 + \beta_1\beta_2}\,. \tag{2.4.9}$$

In the theory of Special Relativity, the transformations of the form (2.4.8) determine the relations between the space-time coordinates of an event in two reference frames, in a uniform motion one with respect to the other, along the x^1-axis. If the relative velocity of these two reference frames is v, then the parameter of the corresponding special Lorentz transformation is $\beta = v/c$, where c is the velocity of light in vacuum. The special Lorentz transformations also indicate that one cannot have a relative velocity greater than c, as no special Lorentz transformation to real coordinate systems exists when $\beta > 1$. If we combine two special Lorentz transformations of parameters $\beta_1 = v_1/c$ and $\beta_2 = v_2/c$, then (2.4.9) shows that $\beta_3 \leq 1$, $\forall \beta_1, \beta_2 \in [0,1]$, and the total velocity thus obtained is again less than c. In fact, by substituting each β in

(2.4.9) by its expression in terms of velocity we are lead to the *Einstein addition law for velocities*

$$v_3 = \frac{v_1 + v_2}{1 + \dfrac{v_1 v_2}{c^2}}\,, \tag{2.4.10}$$

which appears here as a direct consequence of the invariance of the *Minkowskian* space $\mathcal{M}_4$ with respect to the special Lorentz transformations.

Note that, the determinant of the matrix (2.4.8) is $\det \mathbf{\Lambda}_{14}(\beta) = 1$, hence the special Lorentz transformations are orthogonal transformations of $\mathcal{M}_4$. In the same manner, one obtains the matrices $\mathbf{\Lambda}_{24}$ and $\mathbf{\Lambda}_{34}$ for the transformations that leave unchanged the coordinates $\{x^1, x^3\}$ and $\{x^1, x^2\}$, respectively.

4.1.2 Inhomogeneous Lorentz transformations

In the preceding section we have analysed the group of special Lorentz transformations, i.e., transformations of $\mathcal{M}_4$ that affect only one of the spatial coordinates (x^1, x^2, or x^3) and the time coordinate (x^4). This is a group of transformations determined by a particular form of the matrices that satisfy the condition (2.4.5), but in general, we have the following definition:

DEFINITION 2.4.1 *The group of transformations which preserve the metric tensor in $\mathcal{M}_4$ is called the* **Lorentz group**, *and is denoted by* L.

Consequently, the Lorentz transformations are characterized by the relation (2.4.5). A definition equivalent to Definition 2.4.1 can be obtained on the basis of (2.4.3) and the condition $\|\mathbf{x}'\|^2 = \|\mathbf{x}\|^2$:

DEFINITION 2.4.2 *The group of transformations which preserve the norm of the vectors in $\mathcal{M}_4$ is called the* **Lorentz group**.

The Lorentz transformations also leave unchanged the infinitesimal distances (the arc elements), defined by the formula

$$ds^2 = g_{\mu\nu} dx^\mu dx^\nu\,, \tag{2.4.11}$$

with $g_{\mu\nu}$ defined in (2.4.4), hence the relation

$$ds'^2 = ds^2 \tag{2.4.12}$$

takes place. One can show that, from (2.4.12) results the linearity of the Lorentz transformations. Indeed, a general transformation in $\mathcal{M}_4$ is of the form

$$x'^\mu = f_\mu(\mathbf{x})\,, \quad \mu = 1, 2, 3, 4, \tag{2.4.13}$$

where $\mathbf{x} \equiv (x^1, x^2, x^3, x^4)$. The transformation (2.4.13) is invertible if the respective Jacobian does not vanish, that is

$$\frac{D(\mathbf{x}')}{D(\mathbf{x})} \equiv \det\left[\frac{\partial x'^\mu}{\partial x^\nu}\right] \neq 0\,. \tag{2.4.14}$$

In terms of the notations

$$\lambda^{\mu}_{\nu} \equiv \frac{\partial f_{\mu}}{\partial x^{\nu}} = \frac{\partial x'^{\mu}}{\partial x^{\nu}} ,$$

the condition (2.4.14) can be rewritten in the form

$$\det\,[\lambda^{\mu}_{\nu}] \neq 0 . \tag{2.4.15}$$

Because

$$dx'^{\mu} = \lambda^{\mu}_{\nu} dx^{\nu} ,$$

we obtain

$$ds'^2 = g_{\mu\nu} dx'^{\mu} dx'^{\nu} = g_{\mu\nu} \lambda^{\mu}_{\alpha} \lambda^{\nu}_{\beta} dx^{\alpha} dx^{\beta}$$

and, by taking into account (2.4.11) and (2.4.12), it follows that

$$g_{\alpha\beta} = g_{\mu\nu} \lambda^{\mu}_{\alpha} \lambda^{\nu}_{\beta} = g_{\beta\alpha} . \tag{2.4.16}$$

We differentiate (2.4.16) with respect to x^{ρ} and consider the relations

$$\begin{aligned}
g_{\mu\nu} \lambda^{\mu}_{\alpha} \frac{\partial \lambda^{\nu}_{\beta}}{\partial x^{\rho}} + g_{\mu\nu} \lambda^{\nu}_{\beta} \frac{\partial \lambda^{\mu}_{\alpha}}{\partial x^{\rho}} &= 0 , \\
g_{\mu\nu} \lambda^{\mu}_{\beta} \frac{\partial \lambda^{\nu}_{\rho}}{\partial x^{\alpha}} + g_{\mu\nu} \lambda^{\nu}_{\rho} \frac{\partial \lambda^{\mu}_{\beta}}{\partial x^{\alpha}} &= 0 , \\
g_{\mu\nu} \lambda^{\mu}_{\rho} \frac{\partial \lambda^{\nu}_{\alpha}}{\partial x^{\beta}} + g_{\mu\nu} \lambda^{\nu}_{\alpha} \frac{\partial \lambda^{\mu}_{\rho}}{\partial x^{\beta}} &= 0 ,
\end{aligned} \tag{2.4.17}$$

corresponding to cyclic permutations of the free indexes α, β, and ρ. We assume that the functions f_{μ} are of class C^2, such that

$$\frac{\partial \lambda^{\mu}_{\alpha}}{\partial x^{\beta}} = \frac{\partial^2 f_{\mu}}{\partial x^{\alpha}\, \partial x^{\beta}} = \frac{\partial^2 f_{\mu}}{\partial x^{\beta}\, \partial x^{\alpha}} = \frac{\partial \lambda^{\mu}_{\beta}}{\partial x^{\alpha}} .$$

Then, taking into account these relations and the fact that the metric tensor $g_{\mu\nu}$ is symmetric, we add the last two equations in (2.4.17) and subtract the first one. The result is the homogeneous linear system

$$\lambda^{\mu}_{\rho} \left(g_{\mu\nu} \frac{\partial \lambda^{\nu}_{\alpha}}{\partial x^{\beta}} \right) = 0 ,$$

whose determinant (i.e., $\det\,[\lambda^{\mu}_{\rho}]$) does not vanish (see (2.4.15)), and therefore

$$g_{\mu\nu} \frac{\partial \lambda^{\nu}_{\alpha}}{\partial x^{\beta}} = 0 .$$

The determinant of this new homogeneous linear system is also different from zero ($\det\,[g_{\mu\nu}] \neq 0$), hence the only solution is the trivial one

$$\frac{\partial \lambda^{\nu}_{\alpha}}{\partial x^{\beta}} = 0 , \quad \forall\, \nu, \alpha, \beta .$$

It follows that the quantities λ^{μ}_{ν} are independent of $\mathbf{x}$, and the general form of the Lorentz transformations is thus given by the linear equations

$$f_{\mu}(\mathbf{x}) = \lambda^{\mu}_{\nu} x^{\nu} + \gamma^{\mu}, \quad \gamma^{\mu} = const., \quad \mu = 1,2,3,4. \tag{2.4.18}$$

Note that the matrix $\mathbf{\Lambda} = [\lambda^{\mu}_{\nu}]$ has to satisfy the condition (2.4.5).

The set of inhomogeneous linear transformations (2.4.18) forms a noncommutative group called the *inhomogeneous Lorentz group* or the *Poincaré group*. In terms of the matrices $\mathbf{\Lambda}$ and the four-vectors $\gamma = (\gamma^{\mu})$, the elements of this group are denoted by the symbol $(\mathbf{\Lambda}, \gamma)$, and the internal composition law of the group can be written as

$$(\mathbf{\Lambda}_1, \gamma_1)(\mathbf{\Lambda}_2, \gamma_2) = (\mathbf{\Lambda}_1\mathbf{\Lambda}_2, \gamma_1 + \mathbf{\Lambda}_1\gamma_2). \tag{2.4.19}$$

One can also represent the elements of the Poincaré group in terms of the matrices of fifth order

$$\begin{bmatrix} \mathbf{\Lambda} & \gamma \\ \mathbf{0} & \mathbf{1} \end{bmatrix}, \tag{2.4.20}$$

associated to operators acting in the space of column vectors

$$\begin{bmatrix} x^1 \\ x^2 \\ x^3 \\ x^4 \\ 1 \end{bmatrix}. \tag{2.4.21}$$

We will concentrate now on the Lorentz group, formed by the homogeneous Lorentz transformations ($\gamma^{\mu} = 0$), which is a subgroup of the Poincaré group.

From (2.4.5) it results that $\det \mathbf{G} = (\det \mathbf{G})(\det \mathbf{\Lambda})^2$, hence

$$\Delta \equiv \det \mathbf{\Lambda} = \pm 1. \tag{2.4.22}$$

Since the quadratic form $g_{\mu\nu}x^{\mu}x^{\nu}$ is left unchanged by Lorentz's transformations, that is the matrix $\mathbf{\Lambda}$ satisfies the condition (2.4.5), we have

$$(\lambda^4_4)^2 = 1 + \sum_{k=1}^{3} (\lambda^k_4)^2 \geq 1,$$

and consequently,

$$\lambda^4_4 \in (-\infty, -1] \cup [1, \infty). \tag{2.4.23}$$

The restrictions (2.4.22) and (2.4.23) lead to a partition of the Lorentz group into classes. We denote by L_+ the subset of transformations for which $\Delta = 1$ (*proper Lorentz transformations*), by L_- the subset of transformations for which $\Delta = -1$ (*improper Lorentz transformations*), by $\mathsf{L}^{\uparrow}_+$ the subset of transformations for which $\Delta = 1$ and $\lambda^4_4 \geq 1$ (*orthochronous proper transformations*), by $\mathsf{L}^{\downarrow}_+$ the subset of transformations for which $\Delta = 1$ and $\lambda^4_4 \leq -1$ (*non-orthochronous proper transformations*), by $\mathsf{L}_-^\uparrow$ the subset of transformations for which $\Delta = -1$ and $\lambda_4^4 \geq 1$ (*orthochronous improper transformations*), and by $\mathsf{L}_-^\downarrow$ the subset of transformations for which $\Delta = -1$ and $\lambda_4^4 \leq -1$ (*non-orthochronous improper transformations*). These notations and definitions show that one can partition the Lorentz group L into four subsets according to the diagram

$$\begin{array}{c} \mathsf{L} \\ \Delta = \pm 1 \\ |\lambda_4^4| \geq 1 \end{array} \rightarrow \left\{ \begin{array}{l} \begin{array}{c} \mathsf{L}_+ \\ \Delta = 1 \end{array} \rightarrow \left\{ \begin{array}{ll} \mathsf{L}_+^\uparrow & \lambda_4^4 \geq 1 \\ & \\ \mathsf{L}_+^\downarrow & \lambda_4^4 \leq -1 \end{array} \right. \\ \\ \begin{array}{c} \mathsf{L}_- \\ \Delta = -1 \end{array} \rightarrow \left\{ \begin{array}{ll} \mathsf{L}_-^\uparrow & \lambda_4^4 \geq 1 \\ & \\ \mathsf{L}_-^\downarrow & \lambda_4^4 \leq -1 \end{array} \right. \end{array} \right. .$$

The proper Lorentz transformations L_+ form an invariant subgroup of L, called the *subgroup of proper rotations in $\mathcal{M}_4$*. The Lorentz group also contains the *three-dimensional inversion*, which is defined by a matrix identical to $\mathbf{G}$, and, evidently, satisfies the condition (2.4.5). It is thus obvious that the group of three-dimensional rotations $\mathsf{SO}(3)$ is a subgroup of L. The factor group L/L_+ contains only two elements $\{\mathbf{E}, \mathbf{G}\}$. The improper Lorentz transformations do not form a group.

The restriction (2.4.23) enables a partition of the Lorentz group into two subsets, namely $\mathsf{L}^\uparrow = \mathsf{L}_+^\uparrow \cup \mathsf{L}_-^\uparrow$ and $\mathsf{L}^\downarrow = \mathsf{L}_+^\downarrow \cup \mathsf{L}_-^\downarrow$, corresponding to orthochronous and non-orthochronous transformations, respectively. The orthochronous transformations $\mathsf{L}^\uparrow$ form an invariant subgroup of L, while the non-orthochronous transformations $\mathsf{L}^\downarrow$ do not form a subgroup. The factor group $\mathsf{L}/\mathsf{L}^\uparrow$ consists of the elements $\{\mathbf{E}, -\mathbf{E}\}$.

THEOREM 2.4.1 *The group L_+ is a simple group.*

A consequence of this theorem is the fact that the group L_+ does not admit homomorphic (i.e., non-isomorphic) representations. Indeed, if a homomorphic representation could exist, then to the identity transformation in the representation group has to correspond, in L_+, the kernel of the homomorphism which is an invariant subgroup of L_+ (see Theorem 1.1.13). But, L_+ is a simple group, such that it has no invariant subgroups, except the trivial subgroups, therefore all the representations of L_+ are isomorphic to L_+.

It can be shown, that there exists a two-to-one mapping between the elements of $\mathsf{SL}(2, \mathbb{C})$ and the elements of L_+ such that, to each element $\mathbf{\Lambda} \in \mathsf{L}_+$ there correspond two elements $\pm\mathbf{v} \in \mathsf{SL}(2, \mathbb{C})$, and to each element $\mathbf{v} \in \mathsf{SL}(2, \mathbb{C})$ there corresponds one element $\mathbf{\Lambda} \in \mathsf{L}_+$. In other words, to each transformation of L_+ there correspond two linear transformations of $\mathsf{SL}(2, \mathbb{C})$, the coefficients of these transformations differing only by their sign. Therefore, we have the following theorem:

THEOREM 2.4.2 *The group* $\mathsf{SL}(2,\mathbb{C})$ *is homomorphic to the group* L_+.

The kernel of the homomorphism is the set $\{\mathbf{e}, -\mathbf{e}\}$ which forms the invariant subgroup $\mathsf{H} \lhd \mathsf{SL}(2,\mathbb{C})$. The subgroup H is also the centre of $\mathsf{SL}(2,\mathbb{C})$. Now, the factor group $\mathsf{SL}(2,\mathbb{C})/\mathsf{H}$ is formed by the cosets

$$\mathbf{v}\mathsf{H} = \mathsf{H}\mathbf{v} = \{\mathbf{v}'\mathbf{v}|\mathbf{v}' \in \mathsf{H}\} = \{\mathbf{v}, -\mathbf{v}\}, \quad \forall\, \mathbf{v} \in \mathsf{SL}(2,\mathbb{C}),$$

and it follows that L_+ is isomorphic to the factor group $\mathsf{SL}(2,\mathbb{C})/\mathsf{H}$.

4.2 Parametrization and infinitesimal generators

4.2.1 Parametrization of L_+. The invariant integral

The parametrization of the Lorentz group L_+ can be obtained by using its relation with the group $\mathsf{SL}(2,\mathbb{C})$ derived from the Theorem 2.4.2. The group $\mathsf{SL}(2,\mathbb{C})$ is isomorphic to the group of unimodular, linear transformations, acting in the complex plane $\mathbb{C}^2$. These transformations change a complex vector $\mathbf{x} = (\xi, \eta) \in \mathbb{C}^2$ into $\mathbf{x}' = (\xi', \eta') \in \mathbb{C}^2$ according to the relations

$$\begin{aligned} \xi' &= a\,\xi + b\,\eta\,, \\ \eta' &= c\,\xi + d\,\eta\,, \end{aligned} \tag{2.4.24}$$

where the coefficients $a, b, c, d \in \mathbb{C}$ satisfy the equation

$$ad - bc = 1\,. \tag{2.4.25}$$

Therefore, the elements of $\mathsf{SL}(2,\mathbb{C})$ are determined by three complex parameters or by six real parameters, all independent. Consequently, the elements of the Lorentz group also depend on six real parameters.

The transformation (2.4.24) is defined by the complex matrix

$$\mathbf{v} = \begin{bmatrix} a & b \\ c & d \end{bmatrix}, \quad a, b, c, d \in \mathbb{C}\,, \tag{2.4.26}$$

having $\det \mathbf{v} = 1$. The internal composition law onto $\mathsf{SL}(2,\mathbb{C})$ is the usual product of matrices, hence for the product $\mathbf{v}' = \mathbf{v}_0\mathbf{v}$ we obtain

$$\begin{aligned} a' &= a_0 a + b_0 c\,, \qquad & b' &= a_0 b + b_0 d\,, \\ c' &= c_0 a + d_0 c\,, \qquad & d' &= c_0 b + d_0 d\,. \end{aligned} \tag{2.4.27}$$

By means of (2.4.25), we may regard the matrix element a as a function of six real parameters $b_1, b_2, c_1, c_2, d_1, d_2$, defined by the relations

$$b = b_1 + i\,b_2\,, \quad c = c_1 + i\,c_2\,, \quad d = d_1 + i\,d_2\,. \tag{2.4.28}$$

Then, the identity element of the group corresponds to $b_1 = b_2 = c_1 = c_2 = d_2 = 0$ and $d_1 = 1$. Indeed, we have $b = 0$ and $c = 0$, such that (2.4.25) becomes $ad = 1$, that is $\operatorname{Re} a = 1$, $\operatorname{Im} a = 0$, and therefore

$$\mathbf{v}_{b_1=b_2=c_1=c_2=d_2=0,\, d_1=1} = \mathbf{e}\,.$$

To determine the invariant integral on the group we use the method from Chap. 2 Sec. 2.2.2. Thus, we have to calculate the functional determinants

$$\begin{aligned} J_1 &= \left.\frac{D(b_1', b_2', c_1', c_2', d_1', d_2')}{D(b_{01}, b_{02}, c_{01}, c_{02}, d_{01}, d_{02})}\right|_{\mathbf{v}_0=\mathbf{e}}, \\ J_2 &= \left.\frac{D(b_1', b_2', c_1', c_2', d_1', d_2')}{D(b_1, b_2, c_1, c_2, d_1, d_2)}\right|_{\mathbf{v}=\mathbf{e}}, \end{aligned} \tag{2.4.29}$$

corresponding to the products $\mathbf{v}_0\mathbf{v}$ and $\mathbf{v}\mathbf{v}_0$. Note that the functional determinants (2.4.29) are actually the inverse of the functional determinants used in Chap. 2 Sec. 2.2.2, but this only means that we have to use $1/J_1$ and $1/J_2$ in the expression of the invariant integral. Now, to evaluate J_1 and J_2 from (2.4.29), we use the following theorem

THEOREM 2.4.3 *Let $w_s = u_s + i\, v_s$, $s = 1, 2, \ldots, k$, be analytical functions of the complex variables $z_s = x_s + i\, y_s$. Then, the functional determinant of the functions $u_1, v_1, \ldots, u_k, v_k$ with respect to the variables $x_1, y_1, \ldots, x_k, y_k$ is equal to the squared modulus of the functional determinant of the functions $w_1, \ldots, w_k$ with respect to the variables $z_1, \ldots, z_k$.*

Hence, for J_1 we have to calculate the functional determinant

$$J_1' = \left.\frac{D(b', c', d')}{D(b_0, c_0, d_0)}\right|_{a_0=d_0=1,\, b_0=c_0=0}.$$

Firstly, we differentiate the relation $a_0 d_0 - b_0 c_0 = 1$, assuming that a_0 depends on b_0, c_0, and d_0:

$$-c_0 + d_0 \frac{\partial a_0}{\partial b_0} = 0\,, \quad -b_0 + d_0 \frac{\partial a_0}{\partial c_0} = 0\,, \quad a_0 + d_0 \frac{\partial a_0}{\partial d_0} = 0\,.$$

Analogously, from (2.4.27) we obtain

$$\frac{\partial b'}{\partial b_0} = d + b \frac{\partial a_0}{\partial b_0}\,, \quad \frac{\partial b'}{\partial c_0} = b \frac{\partial a_0}{\partial c_0}\,, \quad \frac{\partial b'}{\partial d_0} = b \frac{\partial a_0}{\partial d_0}\,,$$

and

$$\frac{\partial c'}{\partial b_0} = \frac{\partial d'}{\partial b_0} = 0\,, \quad \frac{\partial c'}{\partial c_0} = a\,, \quad \frac{\partial d'}{\partial c_0} = b\,, \quad \frac{\partial c'}{\partial d_0} = c\,, \quad \frac{\partial d'}{\partial d_0} = d\,,$$

such that

$$J_1' = \left(d + b\,\frac{\partial a_0}{\partial b_0}\right)(ad - bc)\Big|_{a_0=d_0=1,\, b_0=c_0=0}$$
$$= \left(d + b\,\frac{\partial a_0}{\partial b_0}\right)\Big|_{a_0=d_0=1,\, b_0=c_0=0} = \left(d + b\,\frac{c_0}{d_0}\right)\Big|_{d_0=1,\, c_0=0} = d\,.$$

Finally, according to Theorem 2.4.3, we have

$$J_1 = \left|J_1'\right|^2 = |d|^2 = d_1^2 + d_2^2\,.$$

In the same manner we obtain $J_2 = |d|^2$. The functional determinants in (2.4.29) have the same value, such that we can express the invariant integral on L_+ in the form

$$\int_{\mathsf{L}_+} f(\mathbf{v})\,\mathrm{d}\mathbf{v} = \int_V f(b_1, b_2, c_1, c_2, d_1, d_2)\,\frac{1}{d_1^2 + d_2^2}\,\mathrm{d}b_1\,\mathrm{d}b_2\,\mathrm{d}c_1\,\mathrm{d}c_2\,\mathrm{d}d_1\,\mathrm{d}d_2\,,$$

where the integration domain V is the whole six-dimensional space.

4.2.2 Infinitesimal generators

In the exponential parametrization, the elements of the group of rotations in the three-dimensional space $\mathbb{R}^3$ can be factorized as a product of three rotations, as it has been shown in Chap. 1 Sec. 2.1.3 and Chap. 2 Sec. 1.2.2. In a similar manner, we can factorize the elements of the Lorentz group into a product of the form

$$\mathbf{\Lambda} = \mathbf{R}_{123}(\alpha_1, \alpha_2, \alpha_3)\mathbf{\Lambda}_{14}(\beta_1)\mathbf{\Lambda}_{24}(\beta_2)\mathbf{\Lambda}_{34}(\beta_3)\,,$$

where $\mathbf{R}_{123}$ is a rotation matrix in the subspace $\mathbb{R}^3 \subset \mathcal{M}_4$, while $\mathbf{\Lambda}_{p4}$, $p = 1, 2, 3$, are matrices of the type (2.4.6). We also factorize the rotation matrix as a product of three matrices

$$\mathbf{R}_{123}(\alpha_1, \alpha_2, \alpha_3) = \mathbf{R}_{12}(\alpha_3)\,\mathbf{R}_{23}(\alpha_1)\,\mathbf{R}_{31}(\alpha_2)\,,$$

where

$$\mathbf{R}_{12}(\alpha) = \begin{bmatrix} \cos\alpha & -\sin\alpha & 0 & 0 \\ \sin\alpha & \cos\alpha & 0 & 0 \\ 0 & 0 & 1 & 0 \\ 0 & 0 & 0 & 1 \end{bmatrix}, \quad \mathbf{R}_{23}(\alpha) = \begin{bmatrix} 1 & 0 & 0 & 0 \\ 0 & \cos\alpha & -\sin\alpha & 0 \\ 0 & \sin\alpha & \cos\alpha & 0 \\ 0 & 0 & 0 & 1 \end{bmatrix},$$

$$\mathbf{R}_{31}(\alpha) = \begin{bmatrix} \cos\alpha & 0 & \sin\alpha & 0 \\ 0 & 1 & 0 & 0 \\ -\sin\alpha & 0 & \cos\alpha & 0 \\ 0 & 0 & 0 & 1 \end{bmatrix}.$$

In this parametrization, the matrices of the infinitesimal generators of the Lorentz group are

$$\mathbf{J}_k = \left.\frac{\partial \boldsymbol{\Lambda}}{\partial \beta_k}\right|_{\boldsymbol{\alpha}=\boldsymbol{\beta}=\mathbf{0}}, \qquad \mathbf{I}_k = \left.\frac{\partial \boldsymbol{\Lambda}}{\partial \alpha_k}\right|_{\boldsymbol{\alpha}=\boldsymbol{\beta}=\mathbf{0}}, \qquad k = 1, 2, 3,$$

where $\boldsymbol{\alpha} \equiv (\alpha_1, \alpha_2, \alpha_3)$ and $\boldsymbol{\beta} \equiv (\beta_1, \beta_2, \beta_3)$, have the form

$$\mathbf{J}_1 = \begin{bmatrix} 0 & 0 & 0 & -1 \\ 0 & 0 & 0 & 0 \\ 0 & 0 & 0 & 0 \\ -1 & 0 & 0 & 0 \end{bmatrix}, \; \mathbf{J}_2 = \begin{bmatrix} 0 & 0 & 0 & 0 \\ 0 & 0 & 0 & -1 \\ 0 & 0 & 0 & 0 \\ 0 & -1 & 0 & 0 \end{bmatrix}, \; \mathbf{J}_3 = \begin{bmatrix} 0 & 0 & 0 & 0 \\ 0 & 0 & 0 & 0 \\ 0 & 0 & 0 & -1 \\ 0 & 0 & -1 & 0 \end{bmatrix},$$

$$\mathbf{I}_1 = \begin{bmatrix} 0 & 0 & 0 & 0 \\ 0 & 0 & -1 & 0 \\ 0 & 1 & 0 & 0 \\ 0 & 0 & 0 & 0 \end{bmatrix}, \; \mathbf{I}_2 = \begin{bmatrix} 0 & 0 & 1 & 0 \\ 0 & 0 & 0 & 0 \\ -1 & 0 & 0 & 0 \\ 0 & 0 & 0 & 0 \end{bmatrix}, \; \mathbf{I}_3 = \begin{bmatrix} 0 & -1 & 0 & 0 \\ 1 & 0 & 0 & 0 \\ 0 & 0 & 0 & 0 \\ 0 & 0 & 0 & 0 \end{bmatrix},$$

and satisfy the commutation relations

$$\begin{aligned} [\mathbf{J}_i, \mathbf{J}_j] &= -\epsilon_{ijk}\, \mathbf{I}_k\,, \\ [\mathbf{I}_i, \mathbf{I}_j] &= \epsilon_{ijk}\, \mathbf{I}_k\,, \\ [\mathbf{I}_i, \mathbf{J}_j] &= \epsilon_{ijk}\, \mathbf{J}_k\,, \qquad i, j = 1, 2, 3. \end{aligned} \tag{2.4.30}$$

In some applications, it is more convenient to use the linear combinations of infinitesimal generators

$$\mathsf{A}_k = \frac{1}{2}\left(\mathsf{I}_k + i\,\mathsf{J}_k\right), \quad \mathsf{B}_k = \frac{1}{2}\left(\mathsf{I}_k - i\,\mathsf{J}_k\right), \quad k = 1, 2, 3, \tag{2.4.31}$$

which satisfy the commutation relations

$$\begin{aligned} [\mathsf{A}_j, \mathsf{A}_k] &= \epsilon_{jkl}\, \mathsf{A}_l\,, \\ [\mathsf{B}_j, \mathsf{B}_k] &= \epsilon_{jkl}\, \mathsf{B}_l\,, \\ [\mathsf{A}_j, \mathsf{B}_k] &= 0\,, \qquad j, k = 1, 2, 3. \end{aligned} \tag{2.4.32}$$

Chapter 3

SYMMETRY GROUPS OF DIFFERENTIAL EQUATIONS

1. Differential operators

1.1 The $\mathsf{SO}(3)$ and $\mathsf{SO}(n)$ groups

1.1.1 The SO(3) group

Let us consider the set of functions $f(\theta, \varphi)$, defined on the surface of a sphere in $\mathcal{E}_3$. The subset of square integrable uniform functions forms an infinite dimensional *linear space*, denoted by $\mathcal{L}^2$. This linear space becomes a *unitary space* if we introduce a scalar product defined by the relation

$$\langle f, g\rangle = \int \overline{f(\theta, \varphi)}\, g(\theta, \varphi)\, d\Omega\,, \tag{3.1.1}$$

where $d\Omega = \sin\theta\, d\theta\, d\varphi$ is the element of solid angle. The space $\mathcal{L}^2$ is complete, therefore it is also a *Hilbert space*. To a rotation G of $\mathcal{E}_3$ there corresponds an operator $\mathsf{T}(\mathsf{G})$ acting in $\mathcal{L}^2$, which is unitary, linear, and generates a representation of $\mathsf{SO}(3)$, defined by the formula

$$\mathsf{T}(\mathsf{G}) f(\theta, \varphi) = f(\mathsf{G}^{-1}(\theta, \varphi)) = f(\theta', \varphi')\,.$$

Let G_1 be a rotation of $\mathcal{E}_3$ by an angle ξ_1 about the axis Ox_1. Then, according to the Theorem 2.2.2, in terms of the exponential parameters $\xi_1 \neq 0, \xi_2 = \xi_3 = 0$, we have the relation

$$\mathsf{T}(\mathsf{G}_1) f(\theta, \varphi) = e^{\xi_1 \mathsf{A}_1} f(\theta, \varphi) = f(\theta', \varphi')\,, \tag{3.1.2}$$

where θ' and φ' are functions of ξ_1, and

$$\theta'(\xi_1)\bigg|_{\xi_1=0} = \theta\,, \quad \varphi'(\xi_1)\bigg|_{\xi_1=0} = \varphi\,.$$

By differentiating (3.1.2) with respect to ξ_1, at $\xi_1 = 0$, we have

$$\mathsf{A}_1 f(\theta,\varphi) = \left(\frac{\partial f}{\partial \theta'}\frac{d\theta'}{d\xi_1} + \frac{\partial f}{\partial \varphi'}\frac{d\varphi'}{d\xi_1}\right)_{\xi_1=0},$$

and, because

$$\left(\frac{\partial f}{\partial \theta'}\right)_{\xi_1=0} = \frac{\partial f}{\partial \theta}, \quad \left(\frac{\partial f}{\partial \varphi'}\right)_{\xi_1=0} = \frac{\partial f}{\partial \varphi},$$

we may associate to the infinitesimal generator A_1 a differential operator of the form[1]

$$\mathsf{A}_1 = \left(\frac{d\theta'}{d\xi_1}\right)_{\xi_1=0}\frac{\partial}{\partial \theta} + \left(\frac{d\varphi'}{d\xi_1}\right)_{\xi_1=0}\frac{\partial}{\partial \varphi}.$$

In the same manner we find the expressions of the differential operators A_2 and A_3. Consequently, the differential operators associated to the infinitesimal generators of $\mathsf{SO}(3)$ are given by

$$\mathsf{A}_k = \left(\frac{d\theta'}{d\xi_k}\right)_{\boldsymbol{\xi}=\mathbf{0}}\frac{\partial}{\partial \theta} + \left(\frac{d\varphi'}{d\xi_k}\right)_{\boldsymbol{\xi}=\mathbf{0}}\frac{\partial}{\partial \varphi}, \quad k = 1,2,3, \tag{3.1.3}$$

where $\boldsymbol{\xi} \equiv (\xi_1, \xi_2, \xi_3)$.

To determine the coefficients in (3.1.3) we perform a rotation of $\mathcal{E}_3$ by an angle ξ_1 about the axis Ox_1. Note this is an active rotation and therefore, the coordinates will be changed according to a rotation of angle $-\xi_1$ that is expressed in the form

$$\begin{aligned}
x_1' &= \quad x_1\,,\\
x_2' &= \quad x_2 \cos\xi_1 + x_3 \sin\xi_1\,,\\
x_3' &= -\, x_2 \sin\xi_1 + x_3 \cos\xi_1\,.
\end{aligned}$$

If x_1, x_2, and x_3 are the coordinates of a point on the unit sphere in $\mathcal{E}_3$, then we may use the spherical coordinates $x_1 = \sin\theta\cos\varphi$, $x_2 = \sin\theta\sin\varphi$, $x_3 = \cos\theta$, to obtain the relations

$$\begin{aligned}
\sin\theta' \cos\varphi' &= \quad \sin\theta\cos\varphi\,,\\
\sin\theta' \sin\varphi' &= \quad \sin\theta\sin\varphi\cos\xi_1 + \cos\theta\sin\xi_1\,,\\
\cos\theta' &= -\, \sin\theta\sin\varphi\sin\xi_1 + \cos\theta\cos\xi_1\,.
\end{aligned} \tag{3.1.4}$$

[1]Note that, for the sake of simplicity, we use the same notation for the infinitesimal generators and the corresponding differential operators.

In (3.1.4), θ and φ are independent of ξ_1 and, by differentiating the last equation with respect to ξ_1 and setting $\xi_1 = 0$, we have

$$\left(\frac{d\theta'}{d\xi_1}\right)_{\xi_1=0} = \sin\varphi\,.$$

Then, by differentiating the second equation in (3.1.4) with respect to ξ_1 and setting $\xi_1 = 0$ we get

$$\left(\frac{d\varphi'}{d\xi_1}\right)_{\xi_1=0} = \cot\theta\,\cos\varphi\,,$$

and it results that

$$\mathsf{A}_1 = \sin\varphi\,\frac{\partial}{\partial\theta} + \cot\theta\,\cos\varphi\,\frac{\partial}{\partial\varphi}\,. \tag{3.1.5}$$

Analogously, by considering rotations of angles ξ_2 and ξ_3 about the axes Ox_2 and Ox_3, respectively, we also obtain

$$\begin{aligned}\mathsf{A}_2 &= -\ cos\varphi\,\frac{\partial}{\partial\theta} + \cot\theta\,\sin\varphi\,\frac{\partial}{\partial\varphi}\,,\\ \mathsf{A}_3 &= -\ \frac{\partial}{\partial\varphi}\,.\end{aligned} \tag{3.1.6}$$

The operators (3.1.5) and (3.1.6) represent the differential form of the infinitesimal generators of $\mathsf{SO}(3)$. Likewise their counterparts, that have been defined in terms of the matrices (1.2.32) and (1.2.33), the differential operators (3.1.5) and (3.1.6) satisfy the commutation relations (1.2.34). The *Hermitian differential operators*

$$\mathsf{H}_k = i\,\mathsf{A}_k\,, \quad k = 1, 2, 3, \tag{3.1.7}$$

correspond to the Hermitian infinitesimal generators (2.1.44). In applications, one also uses the linear combinations of Hermitian operators

$$\begin{aligned}\mathsf{H}_+ &\equiv \mathsf{H}_1 + i\,\mathsf{H}_2 = e^{i\varphi}\left(\frac{\partial}{\partial\theta} + i\,\cot\theta\,\frac{\partial}{\partial\varphi}\right),\\ \mathsf{H}_- &\equiv \mathsf{H}_1 - i\,\mathsf{H}_2 = e^{-i\varphi}\left(-\frac{\partial}{\partial\theta} + i\,\cot\theta\,\frac{\partial}{\partial\varphi}\right),\\ \mathsf{H}_3 &= -i\,\frac{\partial}{\partial\varphi}\,,\end{aligned} \tag{3.1.8}$$

which are the differential forms of the operators in the *Cartan basis associated to the Lie algebra of* $\mathsf{O}(3)$. It is easy to show that this algebra is of rank one, and H_3 forms the basis of the Cartan subalgebra.

1.1.2 The SO(n) group

We denote by $\mathbf{x} \equiv (x_1, x_2, \ldots, x_n)$ the vectors in the real Euclidean space $\mathcal{E}_n$, their length being $\|\mathbf{x}\| = r$. In terms of the hyperspherical coordinates $r, \theta_1, \theta_2, \ldots, \theta_{n-1}$, the Cartesian coordinates of $\mathbf{x}$ read

$$\begin{aligned} x_1 &= r \sin\theta_{n-1} \ldots \sin\theta_2 \sin\theta_1 \,, \\ x_2 &= r \sin\theta_{n-1} \ldots \sin\theta_2 \cos\theta_1 \,, \\ &\ldots\ldots\ldots\ldots\ldots\ldots\ldots\ldots \\ x_{n-1} &= r \sin\theta_{n-1} \cos\theta_{n-2} \,, \\ x_n &= r \cos\theta_{n-1} \,, \end{aligned} \tag{3.1.9}$$

where

$$0 \le r < \infty \,, \quad 0 \le \theta_1 < 2\pi \,, \quad 0 \le \theta_k \le \pi \,, \quad k = 2, 3, \ldots, n-1.$$

For a fixed r, the point of coordinates $(r, \theta_1, \theta_2, \ldots, \theta_{n-1})$ is situated on the surface of the hypersphere of radius r, so that the angles $\theta_1, \theta_2, \ldots, \theta_{n-1}$ can be regarded as coordinates on the unit hypersphere S_{n-1}. The integration element on S_{n-1} (that is equivalent to the element of solid angle in $\mathcal{E}_3$) has the form

$$d\omega = \frac{\Gamma(n/2)}{2\pi^{n/2}} \sin^{n-2}\theta_{n-1} \ldots \sin\theta_2 \, d\theta_1 \, d\theta_2 \ldots d\theta_{n-1} \,, \tag{3.1.10}$$

where Γ is the Euler gamma function.

By a *rotation of* $\mathcal{E}_n$ we intend a linear transformation $\mathbf{g}$ of this space, which does not change neither the orientation of axes, nor the distance of an arbitrary point to the origin of coordinates. The set of all these rotations forms the SO(n) group.

We denote by $\mathbf{g}_{jk}(\alpha)$ the rotation of angle α in the plane (x_j, x_k), and analogously $\mathbf{g}_k(\alpha) = \mathbf{g}_{k+1,k}(\alpha)$. Then, the following theorem holds

THEOREM 3.1.1 *Any rotation of* SO(n) *can be expressed in the form* $\mathbf{g} = \mathbf{g}^{(n-1)} \ldots \mathbf{g}^{(2)}\mathbf{g}^{(1)}$, *where* $\mathbf{g}^{(k)} = \mathbf{g}_1(\theta_1^k)\, \mathbf{g}_2(\theta_2^k) \ldots \mathbf{g}_k(\theta_k^k)$.

Here, the angles θ_j^k, $k = 1, 2, \ldots, n-1$, $j = 1, 2, \ldots, k$, represent *Euler's angles* for the rotation g, and are defined on the intervals

$$0 \le \theta_1^k < 2\pi \,, \quad 0 \le \theta_j^k \le \pi \,, \quad j = 2, 3, \ldots, k.$$

In this parametrization, the integration element of the invariant integral over the SO(n) group has the form

$$dg = A_n \prod_{k=1}^{n-1} \prod_{j=1}^{k} \sin^{j-1}\theta_j^k \, d\theta_j^k \,, \quad A_n = \prod_{k=1}^{n} \frac{\Gamma(k/2)}{2\pi^{k/2}} \,. \tag{3.1.11}$$

We denote by $\mathcal{L}^2(S_{n-1})$ the space of square integrable functions $f(\omega)$, $\omega \equiv (\theta_1, \theta_2, \ldots, \theta_{n-1})$, defined on the unit hypersphere S_{n-1}. To each element $g \in \mathsf{SO}(n)$ we associate an operator $\mathsf{T}(g)$ whose action in the space $\mathcal{L}^2(S_{n-1})$ is defined by the formula

$$\mathsf{T}(g)\, f(\omega) = f(g^{-1}\omega)\,. \tag{3.1.12}$$

The set of all operators $\mathsf{T}(g)$, $g \in \mathsf{SO}(n)$, forms a representation of $\mathsf{SO}(n)$. In Cartesian coordinates, a rotation of angle α in the plane (x_j, x_k) transforms a function $f(\mathbf{x}) \in \mathcal{L}^2(S_{n-1})$ into

$$\begin{aligned}\mathsf{T}[\mathbf{g}_{jk}(\alpha)]\, f(\mathbf{x}) &= f[\mathbf{g}_{jk}(-\alpha)\mathbf{x}] \\ &= f(x_1, x_2, \ldots, x_j\, \cos\alpha + x_k\, \sin\alpha, \ldots, -x_j\, \sin\alpha + x_k\, \cos\alpha, \ldots, x_n)\,.\end{aligned}$$

By differentiating with respect to α, at $\alpha = 0$, we obtain the differential form of the operators in the representation T

$$\mathsf{A}_{jk} \equiv \left.\frac{d\mathsf{T}[\mathbf{g}_{jk}(\alpha)]}{d\alpha}\right|_{\alpha=0} = x_k\, \frac{\partial}{\partial x_j} - x_j\, \frac{\partial}{\partial x_k}\,. \tag{3.1.13}$$

These operators are skew-symmetric ($\mathsf{A}_{jk} = -\mathsf{A}_{kj}$), and satisfy the commutation relations

$$[\mathsf{A}_{jk}, \mathsf{A}_{pq}] = \delta_{kp}\mathsf{A}_{jk}\,, \quad j < k\,, \quad p < q.$$

1.2 The $\mathsf{SU}(2)$ and $\mathsf{SU}(3)$ groups

1.2.1 The $\mathsf{SU}(2)$ group

Let $\mathcal{F}$ be the space of uniform functions defined on the $\mathsf{SU}(2)$ group. We introduce the scalar product (2.2.41) to obtain a Hilbert space. Then, to each transformation $\mathsf{U} \in \mathsf{SU}(2)$, acting in the two-dimensional complex space $\mathbb{C}^2$, we associate the operator $\mathsf{T}(\mathsf{U})$ defined by the relation

$$\mathsf{T}(\mathsf{U}')\, f(\mathsf{U}) = f(\mathsf{U}\mathsf{U}')\,, \tag{3.1.14}$$

where $f \in \mathcal{F}$. The set of all operators $\mathsf{T}(\mathsf{U})$, $\mathsf{U} \in \mathsf{SU}(2)$, constitutes a representation of $\mathsf{SU}(2)$ in the Hilbert space $\mathcal{F}$. If we denote by $\mathsf{U}'' = \mathsf{U}\mathsf{U}'$, then in the Eulerian parametrization (see Chap. 2 Sec. 2.1.4), the equation (3.1.14) can be written in the form

$$\mathsf{T}(\varphi', \theta', \psi')\, f(\varphi, \theta, \psi) = f(\varphi'', \theta'', \psi'')\,, \tag{3.1.15}$$

where the connection between the arguments $\{\varphi, \theta, \psi\}$ and $\{\varphi'', \theta'', \psi''\}$ follows from the relation $\mathsf{U}'' = \mathsf{U}\mathsf{U}'$. The infinitesimal generators of the representation are uniquely defined in terms of the exponential parameters, which we associate to the same basis as for Euler's parameters,

$$\mathsf{A}_k \equiv \left.\frac{\partial\mathsf{T}(\varphi', \theta', \psi')}{\partial\xi_k}\right|_{\boldsymbol{\xi}=\mathbf{0}}\,, \qquad k = 1, 2, 3, \tag{3.1.16}$$

with $\boldsymbol{\xi} \equiv (\xi_1, \xi_2, \xi_3)$. Now, taking into account that the arguments $\{\varphi, \theta, \psi\}$ are independent of $\{\xi_1, \xi_2, \xi_3\}$, we differentiate (3.1.15) with respect to ξ_k and set $\boldsymbol{\xi} = \mathbf{0}$. We thus obtain

$$\mathsf{A}_k f(\varphi, \theta, \psi) = \left.\frac{\partial f(\varphi'', \theta'', \psi'')}{\partial \xi_k}\right|_{\boldsymbol{\xi}=\mathbf{0}} = \left.\frac{\partial f(\varphi'', \theta'', \psi'')}{\partial \varphi''}\frac{\partial \varphi''}{\partial \xi_k}\right|_{\boldsymbol{\xi}=\mathbf{0}} + \left.\frac{\partial f(\varphi'', \theta'', \psi'')}{\partial \theta''}\frac{\partial \theta''}{\partial \xi_k}\right|_{\boldsymbol{\xi}=\mathbf{0}} + \left.\frac{\partial f(\varphi'', \theta'', \psi'')}{\partial \psi''}\frac{\partial \psi''}{\partial \xi_k}\right|_{\boldsymbol{\xi}=\mathbf{0}}. \tag{3.1.17}$$

If $\boldsymbol{\xi} \to \mathbf{0}$, then $\mathsf{U}' \to \mathsf{E}$ such that $\varphi'' \to \varphi$, $\theta'' \to \theta$, and $\psi'' \to \psi$, and the infinitesimal generators A_k can be expressed as differential operators on $\mathcal{F}$, having the form

$$\mathsf{A}_k = \left.\frac{\partial \varphi''}{\partial \xi_k}\right|_{\boldsymbol{\xi}=\mathbf{0}} \frac{\partial}{\partial \varphi} + \left.\frac{\partial \theta''}{\partial \xi_k}\right|_{\boldsymbol{\xi}=\mathbf{0}} \frac{\partial}{\partial \theta} + \left.\frac{\partial \psi''}{\partial \xi_k}\right|_{\boldsymbol{\xi}=\mathbf{0}} \frac{\partial}{\partial \psi}. \tag{3.1.18}$$

In terms of the associated matrices, the relation $\mathsf{U}'' = \mathsf{U}\mathsf{U}'$ reads

$$\mathbf{u}(\varphi'', \theta'', \psi'') = \mathbf{u}(\varphi, \theta, \psi)\mathbf{u}(\varphi', \theta', \psi'), \tag{3.1.19}$$

where the matrices $\mathbf{u}$ are of the form (2.2.24). We differentiate (3.1.19) with respect to ξ_k at $\boldsymbol{\xi} = \mathbf{0}$. Thus, on the basis of the definition (2.2.42), we get

$$\left.\frac{\partial \mathbf{u}(\varphi'', \theta'', \psi'')}{\partial \xi_k}\right|_{\boldsymbol{\xi}=\mathbf{0}} = \mathbf{u}(\varphi, \theta, \psi)\, \mathbf{a}_k, \quad k = 1, 2, 3, \tag{3.1.20}$$

the matrices $\mathbf{a}_k$ of the infinitesimal generators being given in (2.2.43). Because

$$\left.\frac{\partial \mathbf{u}(\varphi'', \theta'', \psi'')}{\partial \xi_k}\right|_{\boldsymbol{\xi}=\mathbf{0}} = \left.\frac{\partial \mathbf{u}(\varphi'', \theta'', \psi'')}{\partial \varphi''}\frac{\partial \varphi''}{\partial \xi_k}\right|_{\boldsymbol{\xi}=\mathbf{0}} + \ldots$$
$$= \frac{\partial \mathbf{u}(\varphi, \theta, \psi)}{\partial \varphi}\left(\frac{\partial \varphi''}{\partial \xi_k}\right)_{\boldsymbol{\xi}=\mathbf{0}} + \frac{\partial \mathbf{u}(\varphi, \theta, \psi)}{\partial \theta}\left(\frac{\partial \theta''}{\partial \xi_k}\right)_{\boldsymbol{\xi}=\mathbf{0}} + \frac{\partial \mathbf{u}(\varphi, \theta, \psi)}{\partial \psi}\left(\frac{\partial \psi''}{\partial \xi_k}\right)_{\boldsymbol{\xi}=\mathbf{0}},$$

the coefficients in (3.1.18) are determined by (3.1.20), and we thus obtain the expressions of the differential operators

$$\begin{aligned}
\mathsf{A}_1 &= \frac{\sin\psi}{\sin\theta}\frac{\partial}{\partial \varphi} + \cos\psi\,\frac{\partial}{\partial \theta} - \cot\theta\,\sin\psi\,\frac{\partial}{\partial \psi}, \\
\mathsf{A}_2 &= \frac{\cos\psi}{\sin\theta}\frac{\partial}{\partial \varphi} - \sin\psi\,\frac{\partial}{\partial \theta} - \cot\theta\,\cos\psi\,\frac{\partial}{\partial \psi}, \\
\mathsf{A}_3 &= \frac{\partial}{\partial \psi}.
\end{aligned} \tag{3.1.21}$$

Now, the Hermitian operators defined by (3.1.8) take the form

$$\begin{aligned}
\mathsf{H}_+ &\equiv \mathsf{H}_1 + i\,\mathsf{H}_2 = e^{-i\psi}\left(-\frac{1}{\sin\theta}\frac{\partial}{\partial\varphi} + \cot\theta\,\frac{\partial}{\partial\psi} + i\,\frac{\partial}{\partial\theta}\right), \\
\mathsf{H}_- &\equiv \mathsf{H}_1 - i\,\mathsf{H}_2 = e^{i\psi}\left(\frac{1}{\sin\theta}\frac{\partial}{\partial\varphi} - \cot\theta\,\frac{\partial}{\partial\psi} + i\,\frac{\partial}{\partial\theta}\right), \\
\mathsf{H}_3 &= i\,\frac{\partial}{\partial\psi}\,.
\end{aligned} \tag{3.1.22}$$

To obtain a form of the infinitesimal generators of $\mathsf{SU}(2)$ analogous to (3.1.13), we consider the two-dimensional space $\mathbb{C}^2$ of complex vectors

$$z = \begin{bmatrix} z_1 \\ z_2 \end{bmatrix}, \qquad z_1, z_2 \in \mathbb{C}\,.$$

The transformations of $\mathsf{SU}(2)$ leave invariant the quadratic form $z_1\overline{z_1} + z_2\overline{z_2} \equiv z_1 z^1 + z_2 z^2$. The vectors with the vertex on the sphere

$$z_1\overline{z_1} + z_2\overline{z_2} = r^2 \tag{3.1.23}$$

allow a parametrization of the form

$$\begin{aligned}
z_1 &= r\,e^{i\varphi_1}\,\cos\theta\,, \\
z_2 &= r\,e^{i\varphi_2}\,\sin\theta\,,
\end{aligned} \tag{3.1.24}$$

with

$$0 \le r < \infty\,, \quad 0 \le \theta \le \pi/2\,, \quad 0 \le \varphi_1, \varphi_2 < 2\pi\,.$$

Let $f(z_1, z_2, \overline{z_1}, \overline{z_2})$ be the square integrable functions defined on the manifold (3.1.23) with $r = 1$. Thus, in the same manner as in Chap. 3 Sec. 1.1.2, we obtain the differential operators

$$\mathsf{A}_{ij} = z_i\,\frac{\partial}{\partial z_j} - z^j\,\frac{\partial}{\partial z^i}\,, \quad z^i \equiv \overline{z_i}\,, \quad i = 1, 2, \tag{3.1.25}$$

that satisfy the commutation relations

$$[\mathsf{A}_{ij}, \mathsf{A}_{kl}] = \delta_{jk}\mathsf{A}_{il} - \delta_{il}\mathsf{A}_{jk}\,. \tag{3.1.26}$$

By denoting $\partial_i \equiv \partial/\partial z_i$ and $\overline{\partial_j} \equiv \partial/\partial z^j$, for $r = 1$ in (3.1.24), we have

$$\begin{aligned}
\partial/\partial r &= z_1\,\partial_1 + z_2\,\partial_2 + \overline{z_1}\,\overline{\partial_1} + \overline{z_2}\,\overline{\partial_2} \equiv 0\,, \\
\partial/\partial\theta &= -e^{i\varphi_1}\sin\theta\,\partial_1 + e^{i\varphi_2}\cos\theta\,\partial_2 - e^{-i\varphi_1}\sin\theta\,\overline{\partial_1} + e^{-i\varphi_2}\cos\theta\,\overline{\partial_2} \\
&= -\tan\theta\,z_1\,\partial_1 + \cot\theta\,z_2\,\partial_2 - \tan\theta\,\overline{z_1}\,\overline{\partial_1} + \cot\theta\,\overline{z_2}\,\overline{\partial_2}\,, \\
\partial/\partial\varphi_1 &= i\,z_1\,\partial_1 - i\,\overline{z_1}\,\overline{\partial_1}\,, \\
\partial/\partial\varphi_2 &= i\,z_2\,\partial_2 - i\,\overline{z_2}\,\overline{\partial_2}\,.
\end{aligned}$$

The first two relations lead to

$$z_1\,\partial_1 + \overline{z_1}\,\overline{\partial_1} = -(z_2\,\partial_2 + \overline{z_2}\,\overline{\partial_2}) = -\sin\theta\,\cos\theta\,\frac{\partial}{\partial\theta},$$

such that

$$\begin{aligned}\partial_1 &= \frac{1}{2z_1}\left(-\sin\theta\,\cos\theta\frac{\partial}{\partial\theta} - i\,\frac{\partial}{\partial\varphi_1}\right),\\ \partial_2 &= \frac{1}{2z_2}\left(\sin\theta\,\cos\theta\frac{\partial}{\partial\theta} - i\,\frac{\partial}{\partial\varphi_2}\right),\end{aligned} \tag{3.1.27}$$

and by substituting into (3.1.25) we finally obtain

$$\begin{aligned}\mathsf{A}_{11} &= -i\,\frac{\partial}{\partial\varphi_1},\\ \mathsf{A}_{12} &= \frac{1}{2}\,e^{i(\varphi_1-\varphi_2)}\left(\frac{\partial}{\partial\theta} - i\,\tan\theta\,\frac{\partial}{\partial\varphi_1} - i\,\cot\theta\,\frac{\partial}{\partial\varphi_2}\right),\\ \mathsf{A}_{21} &= \frac{1}{2}\,e^{-i(\varphi_1-\varphi_2)}\left(-\frac{\partial}{\partial\theta} - i\,\tan\theta\,\frac{\partial}{\partial\varphi_1} - i\,\cot\theta\,\frac{\partial}{\partial\varphi_2}\right),\\ \mathsf{A}_{22} &= -i\,\frac{\partial}{\partial\varphi_2}.\end{aligned} \tag{3.1.28}$$

We mention that the Cartan basis for the Lie algebra of the $\mathsf{SU}(2)$ group is formed by the operators

$$\mathsf{E}_+ = \frac{1}{\sqrt{2}}\,\mathsf{A}_{12}\,,\quad \mathsf{E}_- = \frac{1}{\sqrt{2}}\,\mathsf{A}_{21}\,,\quad \mathsf{H}_0 = \frac{1}{2}\,(\mathsf{A}_{11} - \mathsf{A}_{22})\,. \tag{3.1.29}$$

1.2.2 The $\mathsf{SU}(3)$ group

According to Definition 2.3.1, $\mathsf{SU}(3)$ is the group of transformations in $\mathbb{C}^3$ that leaves invariant the quadratic form

$$z_1 z^1 + z_2 z^2 + z_3 z^3 = 1\,,\quad z^k = \overline{z_k}\,,\quad k = 1,2,3. \tag{3.1.30}$$

The differential operators associated to the infinitesimal generators of the group are of the form (3.1.25), with $i, j = 1, 2, 3$, and satisfy the commutation relations (3.1.26). Because only eight of these differential operators are independent we introduce the new differential operators

$$\mathsf{B}_{ik} = \mathsf{A}_{ik} - \frac{1}{3}\,\delta_{ik}\,(\mathsf{A}_{11} + \mathsf{A}_{22} + \mathsf{A}_{33})\,, \tag{3.1.31}$$

which correspond to the infinitesimal generators in the Okubo basis (2.3.10), reordered, and satisfy the relation

$$\mathsf{B}_{11} + \mathsf{B}_{22} + \mathsf{B}_{33} = 0\,.$$

The relation between the operators (3.1.31) and the Cartan basis of the Lie algebra associated to the $\mathsf{SU}(3)$ group is given by the formulae

$$\begin{aligned}\mathsf{H}_1 &= \frac{1}{2\sqrt{3}}\left(\mathsf{A}_{22} - \mathsf{A}_{33}\right) = \frac{1}{2\sqrt{3}}\left(\mathsf{B}_{22} - \mathsf{B}_{33}\right), \\ \mathsf{H}_2 &= \frac{1}{6}\left(\mathsf{A}_{22} + \mathsf{A}_{33} - 2\mathsf{A}_{11}\right) = -\frac{1}{2}\,\mathsf{B}_{11}\,,\end{aligned} \tag{3.1.32}$$

and

$$\begin{aligned}\mathsf{E}_1 &= \frac{1}{\sqrt{6}}\,\mathsf{A}_{23} = \frac{1}{\sqrt{6}}\,\mathsf{B}_{23}\,, \quad \mathsf{E}_{-1} = \frac{1}{\sqrt{6}}\,\mathsf{A}_{32} = \frac{1}{\sqrt{6}}\,\mathsf{B}_{32}\,, \\ \mathsf{E}_2 &= \frac{1}{\sqrt{6}}\,\mathsf{A}_{21} = \frac{1}{\sqrt{6}}\,\mathsf{B}_{21}\,, \quad \mathsf{E}_{-2} = \frac{1}{\sqrt{6}}\,\mathsf{A}_{12} = \frac{1}{\sqrt{6}}\,\mathsf{B}_{12}\,, \\ \mathsf{E}_3 &= \frac{1}{\sqrt{6}}\,\mathsf{A}_{31} = \frac{1}{\sqrt{6}}\,\mathsf{B}_{31}\,, \quad \mathsf{E}_{-3} = \frac{1}{\sqrt{6}}\,\mathsf{A}_{13} = \frac{1}{\sqrt{6}}\,\mathsf{B}_{13}\,.\end{aligned} \tag{3.1.33}$$

Compare the equations (3.1.32) and (3.1.33) with (2.3.49) and (2.3.50) to find the exact relation between B_{ik} and $\mathbf{X}^l_j$.

The equation (3.1.30) represents the unit sphere in $\mathbb{C}^3$, and the position of a point on this sphere is determined by five independent parameters. Indeed, the components z_1, z_2, and z_3 of the unit vector $z \in \mathbb{C}^3$ are complex numbers, which can be represented in the form

$$z_k = \xi_k\, e^{i\varphi_k}\,, \quad 0 \le \varphi_k < 2\pi\,, \quad k = 1, 2, 3. \tag{3.1.34}$$

The equation (3.1.30) imposes to the variables ξ_1, ξ_2, and ξ_3 the condition $\xi_1^2 + \xi_2^2 + \xi_3^2 = 1$, that represents the equation of the unit sphere in the space $\mathcal{E}_3$. Hence, the variables ξ_1, ξ_2, and ξ_3 can be expressed in terms of two independent variables which determine the position of a point on the unit sphere in $\mathcal{E}_3$. We will thus consider two systems of orthogonal curvilinear coordinates, i.e. the system of *polar coordinates*

$$\xi_1 = \cos\theta\,, \quad \xi_2 = \sin\theta\,\cos\psi\,, \quad \xi_3 = \sin\theta\,\sin\psi\,, \tag{3.1.35}$$

where $\theta \in [0, \pi]$, $\psi \in [0, 2\pi)$, and the system of *elliptical coordinates*

$$\begin{aligned}\xi_1 &= \left(1 - k'^2 \cos^2\psi'\right)^{1/2} \cos\psi\,, \\ \xi_2 &= \sin\psi'\,\sin\psi\,, \\ \xi_3 &= \left(1 - k^2\cos^2\psi\right)^{1/2} \cos\psi'\,,\end{aligned} \tag{3.1.36}$$

with $\psi \in [0, 2\pi)$, $\psi' \in [0, \pi]$, $k = \sin f$, and $k' = \cos f$. Here, $2f$ is the distance between the two foci of the basic ellipse. It is easy to see that the system of polar coordinates can be regarded as a particular case of the system of elliptical coordinates for $k^2 = 0$, $k'^2 = 1$, and $\psi' = \theta$, with the variables ξ_1, ξ_2, and ξ_3 reordered. The elliptical parametrization (3.1.36) can also be expressed in the algebraic form

$$\xi_i^2 = \frac{(\rho - e_i)(\rho' - e_i)}{(e_i - e_j)(e_i - e_k)}, \quad i \neq j \neq k \neq i, \quad i, j, k = 1, 2, 3, \tag{3.1.37}$$

where

$$\begin{aligned} \rho &= e_1 + (e_2 - e_1)\cos^2\psi, \\ \rho' &= e_3 - (e_3 - e_2)\cos^2\psi', \\ e_1 &= -1 - k^2, \quad e_2 = 2k^2 - 1, \quad e_3 = 2 - k^2. \end{aligned} \tag{3.1.38}$$

Note that $e_1 + e_2 + e_3 = 0$, $e_1 \leq \rho \leq e_2 \leq \rho' \leq e_3$, the algebraic form of the system of elliptical coordinates being obtained by a cyclic permutation.

In terms of the variables of the parametrization (3.1.34) the differential operators of the type (3.1.25) become

$$\mathsf{A}_{jk} = \frac{1}{2}\, e^{i(\varphi_j - \varphi_k)} \left[\xi_j \frac{\partial}{\partial \xi_k} - \xi_k \frac{\partial}{\partial \xi_j} - i \left(\frac{\xi_j}{\xi_k} \frac{\partial}{\partial \varphi_k} + \frac{\xi_k}{\xi_j} \frac{\partial}{\partial \varphi_j} \right) \right], \tag{3.1.39}$$

with $j, k = 1, 2, 3$. This relation gives us the possibility to determine the expressions of the differential operators A_{jk} in the two systems of coordinates (3.1.35) and (3.1.36). Thus, in the parametrization (3.1.35), restricted to the domain $0 \leq \theta, \psi \leq \pi/2$, the operators of the Cartan basis from (3.1.32) and (3.1.33), have the expressions

$$\begin{aligned} \mathsf{H}_1 &= -i\,\frac{1}{2\sqrt{3}} \left(\frac{\partial}{\partial \varphi_2} - \frac{\partial}{\partial \varphi_3} \right), \\ \mathsf{H}_2 &= i\,\frac{1}{6} \left(2\,\frac{\partial}{\partial \varphi_1} - \frac{\partial}{\partial \varphi_2} - \frac{\partial}{\partial \varphi_3} \right), \\ \mathsf{E}_1 &= \frac{1}{2\sqrt{6}}\, e^{i(\varphi_2 - \varphi_3)} \left[\frac{\partial}{\partial \psi} - i \tan\psi\, \frac{\partial}{\partial \varphi_2} - i \cot\psi\, \frac{\partial}{\partial \varphi_3} \right], \\ \mathsf{E}_2 &= \frac{1}{2\sqrt{6}}\, e^{i(\varphi_2 - \varphi_1)} \left[\cos\psi\, \frac{\partial}{\partial \theta} + \cot\theta\, \sin\psi\, \frac{\partial}{\partial \psi} \right. \\ &\qquad \left. - i\,\frac{\cot\theta}{\cos\psi} \frac{\partial}{\partial \varphi_2} - i \tan\theta\, \cos\psi\, \frac{\partial}{\partial \varphi_1} \right], \\ \mathsf{E}_3 &= \frac{1}{2\sqrt{6}}\, e^{i(\varphi_3 - \varphi_1)} \left[-\sin\psi\, \frac{\partial}{\partial \theta} - \cot\theta\, \cos\psi\, \frac{\partial}{\partial \psi} \right. \\ &\qquad \left. - i\,\frac{\cot\theta}{\sin\psi} \frac{\partial}{\partial \varphi_3} - i \tan\theta\, \sin\psi\, \frac{\partial}{\partial \varphi_1} \right], \end{aligned} \tag{3.1.40}$$

and $\mathsf{E}_{-\alpha} = \mathsf{E}_\alpha^\dagger$, $\alpha = 1, 2, 3$.

2. Invariants and differential equations

2.1 Preliminary considerations

2.1.1 Transformation groups

Let us consider the transformations of a set M onto itself, defined in Chap. 1 Sec. 1.2.3. We will denote by $x\mathsf{T}$ or $\mathsf{T}x$ the result of the application of a transformation T to an element $x \in M$, and the mapping itself by $x \to x\mathsf{T}$, respectively $x \to \mathsf{T}x$. The transformation T is called *right transformation*, in the first case, and *left transformation* in the latter. The set of right transformations of M form the group $\mathsf{S}_r(M)$, while the set of left transformations of M form the group $\mathsf{S}_l(M)$ that is identical to the *symmetric* group of M, defined in Chap. 1 Sec. 1.2.8, and denoted by $\mathsf{S}(M)$. Other examples of transformation groups are: $\mathsf{GL}(n, \mathbb{R})$, $\mathsf{O}(n)$, the group of translations in the Euclidean space etc.

DEFINITION 3.2.1 *The group of transformations of a set M is called* **effective** *if* $\forall\, \mathsf{T} \neq \mathsf{E}$, $\exists\, x \in M$ *such that* $\mathsf{T}x \neq x$.

In general, we can represent any group G as a transformation group of the set $M \equiv G$, if we associate to each element $g_0 \in G$ a right transformation $\mathsf{R}(g_0)$ acting in the set G according to the relation

$$\mathsf{R}(g_0)\, g = g\, g_0^{-1}\,, \quad \forall g \in G\,.$$

This transformation is called *right translation* on G, and the set of all right translations $\{\mathsf{R}(g) | g \in G\}$ forms a group.

THEOREM 3.2.1 *Any group G is isomorphic to the group of right translation on G.*

DEFINITION 3.2.2 *The isomorphism $g \leftrightarrow \mathsf{R}(g)$ is called the* **right regular representation** *of G.*

Analogously, one can construct the *left regular representation* of G, if to each element $g_0 \in G$ we associate a left transformation $\mathsf{S}(g_0)$, which acts in G according to the relation

$$\mathsf{S}(g_0)\, g = g_0\, g\,, \quad \forall g \in G\,,$$

and is called *left translation* on G. If the group G is finite and $n = |G|$, then a left translation of G is nothing else but a permutation of n elements. Hence, the group of left translations on G is a subgroup of S_n, and from the isomorphism $g \leftrightarrow \mathsf{S}(g)$ it follows Cayley's Theorem 1.1.21 for finite groups.

Note that, if we consider the inverse of all the elements in a group G we obtain a group G^{-1}, that has precisely the same elements as G, but in a different order, such that we have $G \simeq G^{-1}$. Consequently, it makes no difference if we define the right and left translations, respectively, by

$$\mathsf{R}(g_0)\, g = g\, g_0\,, \quad \mathsf{S}(g_0)\, g = g_0^{-1}\, g\,, \quad \forall g \in G\,.$$

Thus, one can easily see that the construction of regular representations shown in Chap. 1 Sec. 1.3.5 corresponds to Definition 3.2.2.

DEFINITION 3.2.3 *Let G be an effective transformation group of a set M. If $\forall\, x, y \in M$, $\exists\, g \in G$ such that $gx = y$, then G is called* **transitive transformation group** *of M, while the set M for which such a group does exist is called* **homogeneous space**.

The set $O_x \equiv \{gx | g \in G\}$, where $x \in M$ is fixed, is called the *trajectory*, or *orbit*, of x determined by G. If the set M contains only one orbit, then G is a transitive group and M is a homogeneous space. If the group G is not transitive, then there exists a partition of M in terms of transitivity (orbit) classes, so that two elements of the same orbit are related by a transformation of G.

Let $a \in M$ be a fixed element, and G a transitive transformation group of M. The set of elements $h \in G$ that leave invariant the element a (i.e., $ha = a$) forms a subgroup $H \subset G$, which is called the *stationary subgroup* of a. To prove that H is a subgroup we use the relations

$$h_1 a = a\,, \quad h_2 a = a\,, \quad \forall\, h_1, h_2 \in H\,,$$

from which it follows that

$$h_1^{-1} a = a\,, \quad (h_1 h_2) a = h_1 (h_2 a) = a\,,$$

such that $h_1^{-1} \in H$ and $h_1 h_2 \in H$. Note that, by definition, H has to contain the neutral element of G. Now, if the effect of a transformation $g \in G$ is $a \to ga = x$, where $x \in M$, then any other transformation $g_1 \in G$, for which $g_1 a = x$, is of the form $g_1 = gh$, $h \in H$. We thus obtain a left coset gH with respect to the subgroup H. The group G is transitive so that there exists a one-to-one mapping between the elements $x \in M$ and the corresponding left cosets. At the same time, the transformations $x \to gx$ are equivalent to the left multiplication by $g \in G$ (if x is associated to the coset $g_1 H$, then gx is associated to the coset $gg_1 H$).

Let $H \subset G$ be the stationary subgroup of $a \in M$, and $g \in G$ a transformation $ga = b$, $b \in M$. Then, the element b is invariant with respect to all the transformations of the form ghg^{-1}, where $h \in H$. The stationary subgroup of b is the subgroup gHg^{-1}, that is conjugate to the stationary subgroup of a. Thus, for any homogeneous space M, the class of the conjugate subgroups of G corresponds to the class of stationary subgroups of different elements of M

and, reciprocally, to each class of conjugate subgroups does correspond a certain homogeneous space. This space is the space of left cosets with respect to H. To each element $g \in G$ we may associate a transformation in M, which changes a coset $g_1 H$ into the coset $gg_1 H$, so that the group G becomes a transitive transformation group of M, while the subgroup $H \subset G$ becomes the stationary subgroup of the coset $eH = H$, where $e \in G$ is the neutral element. We thus obtain $M = G/H$. Note that, the same procedure can be applied in the case of right translations.

DEFINITION 3.2.4 *Let $H \subset G$ be the stationary subgroup of $a \in M$. Then, the set $\{hy|h \in H, y \in M\}$ is called* **sphere** *centred about a, and passing through y.*

If M is realized as a space of left cosets with respect to H, then a sphere is the set of elements of the form HgH, where $g \in G$ is fixed.

Let G be a transformation group of a set M, and let $\mathcal{L}$ be the linear space of functions $f(x)$, $x \in M$. Then, the space $\mathcal{L}$ is called *invariant* with respect to translations of G if, together with each function $f(x)$, it also contains all the functions $f(gx)$, $g \in G$. A representation of G by a translation operator in $\mathcal{L}$ is defined by the relation

$$\mathsf{T}(g)\, f(x) = f(g^{-1}x)\,, \quad g \in G\,, \quad f \in \mathcal{L}\,, \tag{3.2.1}$$

and it is a unitary representation. If G is regarded as a group of translations, then the representation operators are of the form

$$\mathsf{S}(g_0)\, f(g) = f(g_0^{-1} g)\,, \tag{3.2.2}$$

for the left translations $g \to g_0 g$, and

$$\mathsf{R}(g_0)\, f(g) = f(g g_0)\,, \tag{3.2.3}$$

for the right translations $g \to g g_0^{-1}$. If the group G admits the Hurwitz integral, then $\mathcal{L}$ becomes the space of square integrable functions; S is the *left regular representation*, while R is the *right regular representation*.

THEOREM 3.2.2 *Any irreducible representation T of a group G is equivalent to a representation associated to the translation operators in a certain space of numerical functions $f(g)$ defined on G.*

For many groups, in particular for all compact groups, all the irreducible representations are obtained be decomposing a regular representation into a direct sum of irreducible representations (see also Chap. 1 Sec. 1.3.5 for the case of finite groups).

2.1.2 Representations of class 1

Let us consider a representation T of a group G in the linear space $\mathcal{L}$, and a subgroup $H \subset G$. Then, a vector $\mathbf{a} \in \mathcal{L}$ is called *invariant vector* with respect to the subgroup H if $\mathsf{T}(h)\,\mathbf{a} = \mathbf{a}$, $\forall\, h \in H$.

DEFINITION 3.2.5 *The representation* T *is of* **class 1** *with respect to the subgroup* H *if there exist in* $\mathcal{L}$ *nonzero vectors, that are invariant with respect to* H*, and the restriction of the representation* T *to the subgroup* H *is unitary.*

DEFINITION 3.2.6 *If in the space* $\mathcal{L}$ *of a representation of class 1 with respect to* H *does exist only one normalized invariant vector, then the subgroup* H *is called* **massive**.

Let T be a representation of class 1 of a group G in a representation space $\mathcal{L}$ and H a massive subgroup of G corresponding to the invariant vector $\mathbf{a} \in \mathcal{L}$. If we associate to each vector $\mathbf{x} \in \mathcal{L}$ a function on G, defined by the scalar product

$$f(g) = < \mathsf{T}(g)\,\mathbf{x}, \mathbf{a} > , \tag{3.2.4}$$

where $\mathsf{T}(g)$, $g \in G$, are the operators of the representation T, then:

DEFINITION 3.2.7 *The function* $f(g)$ *is called* **spherical function** *of the representation* T *with respect to the subgroup* H.

The representation T is equivalent to the right regular representation R, defined in (3.2.3), acting in the space of spherical functions. Because the restriction of the representation T to the subgroup H is unitary, we have

$$\begin{aligned} f(hg) &= < \mathsf{T}(hg)\,\mathbf{x}, \mathbf{a} > = < \mathsf{T}(h)\mathsf{T}(g)\,\mathbf{x}, \mathbf{a} > = \\ &= < \mathsf{T}(g)\,\mathbf{x}, \mathsf{T}(h^{-1})\,\mathbf{a} > = < \mathsf{T}(g)\,\mathbf{x}, \mathbf{a} > = f(g) \,, \end{aligned} \tag{3.2.5}$$

where $h \in H$. Consequently, the spherical functions $f(g)$ of the representation T are constant on the right cosets Hg, $g \in G$. The set of all these cosets forms a homogeneous space for the translation group of G, while H is a stationary subgroup. It follows that $f(g)$ is a function defined on $M = G/H$.

DEFINITION 3.2.8 *The spherical function*

$$\alpha(g) = < \mathsf{T}(g)\,\mathbf{a}, \mathbf{a} > , \tag{3.2.6}$$

associated to the invariant vector $\mathbf{a} \in \mathcal{L}$*, is called* **zonal spherical function** *(or* **zonal harmonic***).*

From (3.2.5) and (3.2.6) we obtain

$$\alpha(h_1 g h_2) = \alpha(g h_2) \,, \quad h_1, h_2 \in H \,;$$

but $\mathsf{T}(h_2)\,\mathbf{a} = \mathbf{a}$, such that $\alpha(h_1 g h_2) = \alpha(g)$, the zonal harmonics being thus constant on the two-sided cosets HgH, or on the corresponding spheres (see Definition 3.2.4).

2.1.3 Invariant operators

Let $\mathcal{L}$ be the linear space of a representation T of a group G, and let A be an operator that commutes with the operators in the representation T

$$\mathsf{AT}(g) = \mathsf{T}(g)\mathsf{A}\,, \quad \forall\, g \in G.$$

If $\mathbf{x} \in \mathcal{L}$ is an eigenvector of the operator A, corresponding to an eigenvalue λ, then all the vectors $\mathsf{T}(g)\mathbf{x}$ are eigenvectors of A, for the same eigenvalue λ, that is

$$\mathsf{A}\mathbf{x} = \lambda\mathbf{x} \implies \mathsf{AT}(g)\mathbf{x} = \mathsf{T}(g)\mathsf{A}\mathbf{x} = \lambda\mathsf{T}(g)\mathbf{x}\,, \quad \forall\, g \in G.$$

Hence, the eigenvectors of A, which correspond to a given eigenvalue, form an invariant subspace in the representation space. If the representation T is irreducible, then the operator A cannot have but only one eigenvalue, and $\mathsf{A} = \lambda\mathsf{E}$, according to Schur's Lemma 1.1.2.

Let T be a representation of a group G in a Hilbert space $\mathcal{H}$, which can be decomposed into a direct sum of nonequivalent irreducible representations T_i, defined on the subspaces $\mathcal{H}_i \subset \mathcal{H}$.

THEOREM 3.2.3 *If $\mathcal{F} \subset \mathcal{H}$ is an invariant subspace, such that the restriction of the representation T to $\mathcal{F}$ is fully reducible, then $\mathcal{F}$ is an orthogonal direct sum of certain subspaces $\mathcal{H}_i$.*

It results that any minimal invariant subspace $\mathcal{F} \subset \mathcal{H}$ coincides with one of the subspaces $\mathcal{H}_i \subset \mathcal{H}$.

Let G be a transformation group of a set M, and let $\mathcal{L}$ be a linear space of functions defined on M, invariant with respect to the translation group of G.

DEFINITION 3.2.9 *An operator A, acting in $\mathcal{L}$, is called* **invariant with respect to the group** *G, if it commutes with all the translation operators $\mathsf{T}(g)$, defined in* (3.2.1).

To each eigenvalue λ of an invariant operator A does correspond a representation T_λ of G, which is constructed in the space $\mathcal{L}_\lambda$ comprising the eigenfunctions associated to the eigenvalue λ, and we have

$$\mathsf{T}_\lambda(g)\, f(x) = f(g^{-1}x)\,, \quad g \in G\,, \quad f \in \mathcal{L}\,, \quad x \in M\,. \tag{3.2.7}$$

If we choose a basis $\{\mathbf{e}_k(x)\} \subset \mathcal{L}_\lambda$, and a fixed point $a \in M$ then, from (3.2.7), we obtain

$$\mathsf{T}_\lambda(g)\, \mathbf{e}_k(a) \equiv \mathbf{e}_k(g^{-1}a) = \sum_j t_{jk}(g)\, \mathbf{e}_j(a)\,.$$

If the set M is homogeneous, then $x = g_x^{-1}a$ and

$$\mathbf{e}_k(x) = \mathbf{e}_k(g_x^{-1}a) = \sum_j t_{jk}(g_x)\, \mathbf{e}_j(a)\,;$$

hence, the elements of $\mathcal{L}_\lambda$ are linear combinations of the matrix elements of the representation T_λ.

2.2 Invariant differential operators

2.2.1 The SO(3) group

The group of rotations in the Euclidean space $\mathcal{E}_3$ is not transitive because the points situated at various distances to the origin of coordinates cannot be obtained one from the other by rotations about the origin. However, this group is transitive if we consider the transformations of the unit sphere into itself. Hence, a sphere in $\mathcal{E}_3$ is a homogeneous space relative to the rotation group $\mathsf{O}(3)$, while the spheres centred about the origin of coordinates represent transitivity classes in $\mathcal{E}_3$. Let us choose the point of coordinates $(0,0,1)$ on the unit sphere (denoted by $\mathcal{S}_3$). The stationary subgroup of this point is the group $\mathsf{SO}(2)$ of rotations about the x_3 axis, and $\mathcal{S}_3 = \mathsf{O}(3)/\mathsf{SO}(2)$. Also, the spherical functions are the functions $f(\theta,\varphi)$ defined in (3.1.1). The differential operators (3.1.8) represent the infinitesimal generators of the left regular representation of $\mathsf{SO}(3)$, and we may construct the differential operator

$$\mathsf{H}^2 \equiv \mathsf{H}_+\mathsf{H}_- - \mathsf{H}_3 + \mathsf{H}_3^2 = -\frac{1}{\sin\theta}\frac{\partial}{\partial\theta}\left(\sin\theta\frac{\partial}{\partial\theta}\right) - \frac{1}{\sin^2\theta}\frac{\partial^2}{\partial\varphi^2}, \quad (3.2.8)$$

which is invariant with respect to $\mathsf{SO}(3)$. Such second degree invariant operators constructed from the infinitesimal generators of the Lie algebra associated to a group are called *Casimir operators*. The operator H^2, defined in (3.2.8), coincides with the angular part of Laplace's operator in three dimensions, in spherical coordinates,

$$\Delta \equiv \frac{\partial^2}{\partial x_1^2} + \frac{\partial^2}{\partial x_2^2} + \frac{\partial^2}{\partial x_3^2} = \frac{1}{r^2}\frac{\partial}{\partial r}\left(r^2\frac{\partial}{\partial r}\right) + \frac{1}{r^2}\mathsf{H}^2. \quad (3.2.9)$$

Therefore, the eigenfunctions of H^2, which satisfy the eigenvalue equation

$$\mathsf{H}^2 f(\theta,\varphi) = \lambda\, f(\theta,\varphi), \quad (3.2.10)$$

are the *spherical harmonics*

$$Y_{lm}(\theta,\varphi) = (-1)^m\left[\frac{2l+1}{4\pi}\frac{(l-m)!}{(l+m)!}\right]^{1/2} P_l^m(\cos\theta)\, e^{im\varphi}, \quad 0 \le m \le l, \quad (3.2.11)$$

where P_l^m are the *associated Legendre polynomials*, and the eigenvalues have the form $\lambda = l(l+1)$, $l = 0,1,2,\ldots$. Also, the spherical harmonics satisfy the relation

$$Y_{l,-m}(\theta,\varphi) = (-1)^m\,\overline{Y_{lm}(\theta,\varphi)}.$$

The action of the differential operators (3.1.8) on the spherical harmonics is defined by the relations

$$\begin{aligned} \mathsf{H}_3\, Y_{lm} &= m\, Y_{lm}\,, \\ \mathsf{H}_+ Y_{lm} &= \sqrt{(l-m)(l+m+1)}\; Y_{l,m+1}\,, \\ \mathsf{H}_- Y_{lm} &= \sqrt{(l+m)(l-m+1)}\; Y_{l,m-1}\,. \end{aligned} \tag{3.2.12}$$

Because the operator H^2 is an invariant operator, to each eigenvalue l does correspond a representation T_l of $\mathsf{SO}(3)$. The basis of the representation space $\mathcal{L}_l$ is formed by the functions $\{Y_{lm}\}_{|m|\leq l}$, for a fixed l, while the dimension of this space is $2l+1$. The space $\mathcal{L}$ of the regular representation is thus decomposed into an orthogonal direct sum of invariant subspaces $\mathcal{L}_l$, while the regular representation is decomposed into a direct sum of nonequivalent irreducible representations T_l, defined in the subspaces $\mathcal{L}_l$. Consequently, because the group $\mathsf{SO}(3)$ is compact, we have obtained all its irreducible representations. Note that the representation space $\mathcal{L}$ is a Hilbert space, the scalar product in this space being defined by the relation

$$< f_1, f_2 >= \int_0^{\pi/2} \sin\theta\, d\theta \int_0^{2\pi} \overline{f_1(\theta,\varphi)}\, f_2(\theta,\varphi)\, d\varphi\,. \tag{3.2.13}$$

Hence, we may construct a basis of $\mathcal{L}$, of the form

$$\{Y_{lm}\}\,, \quad |m| \leq l\,, \quad l = 0,1,2,\dots\,, \tag{3.2.14}$$

which is orthonormal with respect to the scalar product (3.2.13), that is

$$\int_0^{\pi/2} \sin\theta\, d\theta \int_0^{2\pi} \overline{Y_{lm}(\theta,\varphi)}\; Y_{l'm'}(\theta,\varphi)\, d\varphi = \delta_{ll'}\delta_{mm'}\,.$$

The product of two spherical harmonics of equal arguments may be written in the form

$$Y_{l_1m_1}(\theta,\varphi)\, Y_{l_2m_2}(\theta,\varphi) = \sum_{l=|l_1-l_2|}^{l_1+l_2} \langle l, m_1+m_2|l_2m_2|l_1m_1\rangle\, Y_{l,m_1+m_2}(\theta,\varphi), \tag{3.2.15}$$

where

$$\langle lm|l_2m_2|l_1m_1\rangle = \int_0^{\pi/2} \sin\theta\, d\theta \int_0^{2\pi} \overline{Y_{lm}(\theta,\varphi)}\, Y_{l_2m_2}(\theta,\varphi)\, Y_{l_1m_1}(\theta,\varphi)\, d\varphi\,, \tag{3.2.16}$$

are called *Gaunt's coefficients.*

2.2.2 The $\mathsf{SU}(2)$ group

Let us consider the infinitesimal generators of the Lie algebra associated to the $\mathsf{SU}(2)$ group defined in (3.1.22), that satisfy the commutation relations

$$[\mathsf{H}_3, \mathsf{H}_+] = \mathsf{H}_+ , \quad [\mathsf{H}_3, \mathsf{H}_-] = \mathsf{H}_- , \quad [\mathsf{H}_+, \mathsf{H}_-] = 2\,\mathsf{H}_3 . \tag{3.2.17}$$

The Cartan subalgebra of $\mathsf{SU}(2)$ is formed by only one linear operator H_3. In a finite dimensional linear space $\mathcal{L}$, the operator H_3 admits an eigenvector $\mathbf{f} \in \mathcal{L}$, of weight λ (i.e., $\mathsf{H}_3\mathbf{f} = \lambda\mathbf{f}$). The action of the operators H_+ and H_- on this eigenvector is the same as that of the operators E_1, respectively E_{-1}, introduced in Chap. 2 Sec. 3.1.3, and is expressed by adding $+1$, respectively -1, to the weight λ, such that, if $\mathbf{f}' = \mathsf{H}_+\mathbf{f}$ and $\mathbf{f}'' = \mathsf{H}_-\mathbf{f}$, then we have

$$\begin{aligned} \mathsf{H}_3\mathbf{f}' &= (\lambda + 1)\,\mathbf{f}' , \\ \mathsf{H}_3\mathbf{f}'' &= (\lambda - 1)\,\mathbf{f}'' . \end{aligned}$$

Let us denote by j the maximum weight ($\mathsf{H}_3\mathbf{f}_j = j\,\mathbf{f}_j$, $\mathsf{H}_+\mathbf{f}_j = \mathbf{0}$) and let us consider the vector $\mathbf{f}_{j-1} \equiv (1/\alpha_j)\mathsf{H}_-\mathbf{f}_j$, where $\alpha_j = \text{const.}$; if $\mathsf{H}_-\mathbf{f}_j \neq \mathbf{0}$, then the vector $\mathbf{f}_{j-1}$ is an eigenvector of the operator H_3,

$$\mathsf{H}_3\mathbf{f}_{j-1} = (j-1)\,\mathbf{f}_{j-1} .$$

Then, we introduce the vector $\mathbf{f}_{j-2} \equiv (1/\alpha_{j-1})\mathsf{H}_-\mathbf{f}_{j-1}$ and repeat the procedure until we obtain $\mathsf{H}_-\mathbf{f}_{j-k} = \mathbf{0}$, respectively until we reach the minimum weight $j-k$. We thus construct the set of $k+1$ vectors $\{\mathbf{f}_i | i = j-k, j-k+1, \ldots, j\}$, which are linearly independent because they correspond to different eigenvalues. The set of these vectors is also finite, because $k+1 \leq n = \dim \mathcal{L}$. By using the notations introduced here, we may write $\mathsf{H}_-\mathbf{f}_m = \alpha_m \mathbf{f}_{m-1}$.

The application of the operator H_+ upon the vector $\mathbf{f}_{j-1}$ can be expressed in the form

$$\mathsf{H}_+\mathbf{f}_{j-1} = \mathsf{H}_+\left(\frac{1}{\alpha_j}\mathsf{H}_-\mathbf{f}_j\right) = \frac{1}{\alpha_j}\left(2\mathsf{H}_3 + \mathsf{H}_-\mathsf{H}_+\right)\mathbf{f}_j = \frac{2j}{\alpha_j}\,\mathbf{f}_j ,$$

and we will denote $\beta_j = 2j/\alpha_j$. If $\mathsf{H}_+\mathbf{f}_p = \beta_{p+1}\mathbf{f}_{p+1}$, then

$$\begin{aligned} \mathsf{H}_+\mathbf{f}_{p-1} &= \frac{1}{\alpha_p}\mathsf{H}_+\mathsf{H}_-\mathbf{f}_p = \frac{1}{\alpha_p}\left(2\mathsf{H}_3 + \mathsf{H}_-\mathsf{H}_+\right)\mathbf{f}_p = \\ &= \frac{2p}{\alpha_p}\mathbf{f}_p + \mathsf{H}_-\beta_{p+1}\mathbf{f}_{p+1} = \left(\frac{2p}{\alpha_p} + \alpha_{p+1}\beta_{p+1}\right)\mathbf{f}_p \equiv \beta_p\mathbf{f}_p . \end{aligned}$$

Therefore, the constants α_p and β_p satisfy the relations

$$\alpha_j\beta_j = 2j ,$$

$$\dots\dots\dots\dots\dots\dots\dots\dots$$

$$\alpha_p\beta_p - \alpha_{p+1}\beta_{p+1} = 2p\,,$$

$$\dots\dots\dots\dots\dots\dots\dots\dots$$

$$-\alpha_{l+1}\beta_{l+1} = 2l\,,$$

where $l = j - k$. By summing these relations we obtain

$$0 = 2[j + (j-1) + \ldots + (j-k)] = \frac{1}{2}\,(k+1)(2j-k)\,.$$

Because $k \geq 0$ by definition, it follows that $j = k/2$, $k = 0, 1, 2, \ldots$, so that the eigenvalues of H_3 are positive integers and half-integers. The number of these eigenvalues is equal to $k + 1 = 2j + 1$; the minimum eigenvalue is $l = j - k = -j$, while the maximum eigenvalue is j. If the eigenvectors of H_3, for a given maximum eigenvalue j, form an orthonormal system, then one can determine the expressions of the constants α and we obtain the relations

$$\begin{aligned} \mathsf{H}_3\,\mathbf{f}_{jm} &= m\,\mathbf{f}_{jm}\,, \\ \mathsf{H}_+\mathbf{f}_{jm} &= \alpha_{j,m+1}\,\mathbf{f}_{j,m+1}\,, \\ \mathsf{H}_-\mathbf{f}_{jm} &= \alpha_{jm}\,\mathbf{f}_{j,m-1}\,, \end{aligned} \tag{3.2.18}$$

where $|m| \leq j$, and $\alpha_{jm} = \sqrt{(j+m)(j-m+1)}$. Here, the eigenvectors of H_3 have been labelled by two indices in order to emphasize the maximum eigenvalue.

THEOREM 3.2.4 *The unitary irreducible representations of* $\mathsf{SU}(2)$ *are characterized by a positive number* j*, integer or half-integer. The unitary space of the representation characterized by the weight* j *has the dimension* $2j+1$*, and the infinitesimal generators are acting upon the orthonormal basis* $\{\mathbf{f}_{jm}\}$*,* $|m| \leq j$*, according to (3.2.18).*

The Casimir operator $\mathsf{H}^2 \equiv \mathsf{H}_1^2 + \mathsf{H}_2^2 + \mathsf{H}_3^2 = \mathsf{H}_+\mathsf{H}_- - \mathsf{H}_3 + \mathsf{H}_3^2$ has the eigenvalues $j(j+1)$ and is an invariant operator in the linear space of dimension $2j + 1$. By solving the equation

$$\mathsf{H}^2\mathbf{f}_{jm} = j(j+1)\,\mathbf{f}_{jm}\,, \tag{3.2.19}$$

where the infinitesimal generators have the form (3.1.22), we obtain the basis of the subspace $\mathcal{L}_{jm}$ (the *generalized harmonic functions*). The space $\mathcal{F}$ of the regular representation, defined in Chap. 3 Sec.1.2.1, can be decomposed into an orthogonal direct sum of invariant subspaces $\mathcal{F}_{jm}$ with bases $\{\mathbf{f}_{mn}^j\}$, $|n| \leq j$, for fixed j and m.

We can also use other methods to obtain the basis of a representation space for the regular representation of a compact group. Thus, in the homogeneous set M we define the element of arc

$$ds^2 = g_{ij}dx^i dx^j\,, \tag{3.2.20}$$

where g_{ij} is the metric tensor, and the invariant *Laplace-Beltrami operator*

$$\Delta_M = \frac{1}{\sqrt{g}} \frac{\partial}{\partial x^i} \left(\sqrt{g}\, g^{ij} \frac{\partial}{\partial x^j} \right). \tag{3.2.21}$$

Here, $g \equiv \det [g_{ij}]$ and $g^{ij} = (g^{-1})_{ij}$. The eigenvalues of the operator Δ_M are $-n(n + \nu - 1)$, where n is a positive integer, while ν is the dimension of the space. The eigenfunctions of Δ_M constitute the searched basis. Thus, in the case of the $\mathsf{SU}(2)$ group the homogeneous set is the surface of the unit sphere $|z_1|^2 + |z_2|^2 = 1$ in $\mathbb{C}^2$ and, in terms of the parametrization (3.1.24), the element of the arc becomes

$$ds^2 = |dz_1|^2 + |dz_2|^2 = d\theta^2 + \cos^2\theta\, d\varphi_1^2 + \sin^2\theta\, d\varphi_2^2 . \tag{3.2.22}$$

Also, the Laplace-Beltrami operator will be

$$\Delta_3 = \frac{1}{\sin\theta\, \cos\theta} \frac{\partial}{\partial\theta} \left(\sin\theta\, \cos\theta \frac{\partial}{\partial\theta} \right) + \frac{1}{\cos^2\theta} \frac{\partial^2}{\partial\varphi_1^2} + \frac{1}{\sin^2\theta} \frac{\partial^2}{\partial\varphi_2^2} ,$$

with the eigenvalues $-n(n+2)$, where n is a positive integer. Hence, the eigenfunctions are the solutions of the equation

$$[\Delta_3 + n(n+2)]Y = 0 . \tag{3.2.23}$$

In this equation the variables can be completely separated and the general solution is obtained in terms of elementary solutions. Thus, the general solution of (3.2.23) has the form

$$Y^n_{m_1 m_2} = e^{i(m_1\varphi_1 + m_2\varphi_2)}\, d^{n/2}_{(m_1+m_2)/2,(m_1-m_2)/2}(2\theta) , \tag{3.2.24}$$

where

$$\begin{aligned} d^j_{m'm}(\alpha) &= \left[\frac{(j+m')!(j-m')!}{(j+m)!(j-m)!} \right]^{1/2} \\ &\quad \times \left(\cos\frac{\alpha}{2} \right)^{m'+m} \left(\sin\frac{\alpha}{2} \right)^{m'-m} P^{m'-m,m'+m}_{j-m'}(\cos\alpha) , \end{aligned}$$

$P^{lm}_k(\cos\alpha)$ being *Jacobi's polynomials*. With the weight $\sqrt{g} = \sin\theta\, \cos\theta$, the functions $Y^n_{m_1 m_2}$ form a complete orthogonal system of functions. The infinitesimal generators of the Lie algebra of the $\mathsf{SU}(2)$ group have the form (3.1.28), while the Casimir operator is

$$\begin{aligned} \mathsf{H}^2 &= \mathsf{H}_0^2 + \mathsf{E}_+\mathsf{E}_- + \mathsf{E}_-\mathsf{E}_+ \\ &= -\frac{1}{4} \left[\frac{\partial^2}{\partial\theta^2} + 2\cot 2\theta \frac{\partial}{\partial\theta} + \frac{1}{\cos^2\theta} \frac{\partial^2}{\partial\varphi_1^2} + \frac{1}{\sin^2\theta} \frac{\partial^2}{\partial\varphi_2^2} \right] . \end{aligned} \tag{3.2.25}$$

From (3.2.23) and (3.2.24), by comparing the expressions of the operators Δ_3 and H^2, we obtain the relations

$$\begin{aligned} \mathsf{H}^2 Y^n_{m_1 m_2} &= \lambda(\lambda+1)\, Y^n_{m_1 m_2}\,, \quad \lambda = n/2\,, \\ \mathsf{H}_3 Y^n_{m_1 m_2} &= [(m_1 - m_2)/2]\, Y^n_{m_1 m_2}\,. \end{aligned} \tag{3.2.26}$$

The first equation in (3.2.26) is identical to (3.2.19) for $j = n/2$, while the second one, compared to (3.2.18), gives $m = (m_1 - m_2)/2$. We thus obtain the basis of the subspace $\mathcal{F}_{jm}$ in the form

$$\left\{ Y^{2j}_{n+m,n-m}(\varphi_1, \theta, \varphi_2) \right\}. \tag{3.2.27}$$

2.2.3 The $\mathsf{SU}(3)$ group

In the case of the $\mathsf{SU}(3)$ group the homogeneous set M is the five-dimensional hypersphere, the infinitesimal generators having the expressions (3.1.40). We can construct two independent invariant operators (a quadratic one and a cubic one)

$$\begin{aligned} \mathsf{C}_2 &= \mathsf{H}_1^2 + \mathsf{H}_2^2 + \sum_{\alpha=1}^{3} (\mathsf{E}_\alpha \mathsf{E}_{-\alpha} + \mathsf{E}_{-\alpha}\mathsf{E}_\alpha) \\ &= -\frac{1}{12}\left[\frac{\partial^2}{\partial\theta^2} + (3\cot\theta - \tan\theta)\frac{\partial}{\partial\theta} + \frac{1}{\sin^2\theta}\frac{\partial^2}{\partial\xi^2} \right. \\ &\quad + \frac{\cot\xi - \tan\xi}{\sin^2\theta}\frac{\partial}{\partial\xi} + \frac{1}{\cos^2\theta}\frac{\partial^2}{\partial\varphi_1^2} \\ &\quad \left. + \frac{1}{\sin^2\theta\cos^2\xi}\frac{\partial^2}{\partial\varphi_2^2} + \frac{1}{\sin^2\theta\sin^2\xi}\frac{\partial^2}{\partial\varphi_3^2} - \frac{1}{3}\mathsf{C}_1^2 \right], \\ \mathsf{C}_3 &= \frac{1}{162}\,\mathsf{C}_1\left(27\,\mathsf{C}_2 - \mathsf{C}_1^2 + 9\right), \end{aligned} \tag{3.2.28}$$

where

$$\mathsf{C}_1 = -i\left(\frac{\partial}{\partial\varphi_1} + \frac{\partial}{\partial\varphi_2} + \frac{\partial}{\partial\varphi_3} \right).$$

We choose $\{\mathsf{H}_1, \mathsf{E}_1, \mathsf{E}_{-1}\}$ to be the infinitesimal generators of the subgroup $\mathsf{SU}(2)$ in the canonical decomposition $\mathsf{SU}(3) \supset \mathsf{U}(1) \times \mathsf{SU}(2)$, while H_2 is the infinitesimal generator of $\mathsf{U}(1)$. Then, the Casimir operator of this $\mathsf{SU}(2)$ subgroup is

$$\mathsf{H}^2 = -\frac{1}{4}\left[\frac{\partial^2}{\partial\xi^2} + 2\cot 2\xi\,\frac{\partial}{\partial\xi} + \frac{1}{\cos^2\xi}\frac{\partial^2}{\partial\varphi_2^2} + \frac{1}{\sin^2\xi}\frac{\partial^2}{\partial\varphi_3^2} \right]. \tag{3.2.29}$$

Also, in the case of parametrization (3.1.35) the Laplace-Beltrami operator has the form

$$\Delta_5 = \frac{1}{3}\,\mathsf{C}_1^2 - 12\,\mathsf{C}_2\,,$$

and its eigenvalues are $-n(n+4)$, where n is a positive integer, such that the spherical harmonics are obtained from the equation

$$\Delta_5 Y + n(n+4)Y = 0\,. \tag{3.2.30}$$

By separating the variables and taking into account (3.2.23) and (3.2.24), we obtain the solutions

$$Y^n_{m_1m_2m_3j} = \Theta^n_{m_1j}(\theta)\, e^{i(m_1\varphi_1+m_2\varphi_2+m_3\varphi_3)}\, d^j_{(m_2+m_3)/2,(m_2-m_3)/2}(2\xi)\,, \tag{3.2.31}$$

where the functions Θ satisfy the differential equation

$$\frac{1}{\sin^3\theta\cos\theta}\frac{d}{d\theta}\left(\sin^3\theta\cos\theta\frac{d\Theta}{d\theta}\right)+\left[n(n+4)-\frac{m_1^2}{\cos^2\theta}-\frac{4j(j+1)}{\sin^2\theta}\right]\Theta = 0. \tag{3.2.32}$$

The change of variable $\theta \longrightarrow \cos 2\theta$ reduces the equation (3.2.32) to the *Papperitz equation*, which can be integrated by means of *hypergeometric functions*. Then, a change of function of the form $\Theta \rightarrow \Theta/\sin\theta$ leads us to an equation for the functions d. Finally, the solution (3.2.31) takes the form

$$\begin{aligned} Y^n_{m_1m_2m_3j} &= \frac{1}{\sin\theta}\, e^{i(m_1\varphi_1+m_2\varphi_2+m_3\varphi_3)}\, d^{(n+1)/2}_{(m_1+2j+1)/2,(m_1-2j-1)/2}(2\theta) \\ &\quad \times d^j_{(m_2+m_3)/2,(m_2-m_3)/2}(2\xi)\,, \end{aligned} \tag{3.2.33}$$

that also satisfies the relations

$$\begin{aligned} \mathsf{C}_1 Y^n_{m_1m_2m_3j} &= (p-q)\, Y^n_{m_1m_2m_3j}\,, \\ \mathsf{C}_2 Y^n_{m_1m_2m_3j} &= \frac{1}{9}(p^2+q^2+pq+3p+3q)\, Y^n_{m_1m_2m_3j}\,, \\ \mathsf{C}_3 Y^n_{m_1m_2m_3j} &= \frac{1}{162}[(p-q)(2p^2+2q^2+5pq+9p+9q+9)]\, Y^n_{m_1m_2m_3j}\,, \end{aligned}$$

where $p+q = n$ and $p-q = m_1+m_2+m_3$. It can be shown that any irreducible representation of $\mathsf{SU}(3)$ is determined by the two nonnegative integers p and q. The dimension of such a representation is

$$N(p,q) = \frac{1}{2}(p+1)(q+1)(p+q+2)\,, \tag{3.2.34}$$

corresponding to the formula (2.3.45). The functions (3.2.33) form a complete orthogonal system, with the weight $\sqrt{g} = \cos\theta\,\sin^3\theta\,\cos\xi\,\sin\xi$, and constitute the basis of the representation space for the irreducible representations of $\mathsf{SU}(3)$.

In the case of elliptical parametrization (3.1.37) we use the invariant operators

$$
\begin{aligned}
\mathsf{C}^{(2)} &= \frac{1}{2}\,\mathsf{B}_{ij}\mathsf{B}_{ji}\,, \quad \left[\frac{1}{3}(p^2+pq+q^2+3p+3q)\right], \\
\mathsf{C}^{(3)} &= \mathsf{B}_{ij}\mathsf{B}_{jk}\mathsf{B}_{ki}\,, \left[\frac{1}{9}(p-q)(2p^2+3pq+2q^2)-p(q+2)-2q(q+2)\right], \\
& \qquad\qquad\qquad\qquad\qquad\qquad\qquad\qquad\qquad\qquad\qquad\qquad (3.2.35) \\
\mathsf{C}^{(2)}_{\mathsf{T}} &= \frac{1}{2}[\mathsf{B}_{12}\mathsf{B}_{21}+\mathsf{B}_{21}\mathsf{B}_{12}+(\mathsf{B}_{11})^2+(\mathsf{B}_{22})^2-\frac{1}{2}(\mathsf{B}_{11}+\mathsf{B}_{22})^2]\,, \\
& \qquad\qquad\qquad\qquad\qquad\qquad\qquad\qquad \left[T(T+1)\right],
\end{aligned}
$$

and the infinitesimal generators of the Cartan subalgebra

$$
\begin{aligned}
&\mathsf{B}_{11}\,, \quad [Q]\,, \\
&\mathsf{B}_{33}\,, \quad [-Y]\,.
\end{aligned}
$$

Here, the operators B_{ij} have been defined in (3.1.31), and we have enclosed within square brackets the eigenvalues associated to each operator. The operator $\mathsf{C}^{(2)}_{\mathsf{T}}$ is the Casimir operator of the subgroup $\mathsf{SU}(2)_{\mathsf{T}}$ in the canonical decomposition of the form (2.3.58). Note that the $\mathsf{SU}(3)$ group admits three different canonical decompositions of the form (2.3.58), so that the operator $\mathsf{C}^{(2)}_{\mathsf{T}}$ can be replaced by one of the operators

$$
\begin{aligned}
\mathsf{C}^{(2)}_{\mathsf{V}} &= \frac{1}{2}[\mathsf{B}_{31}\mathsf{B}_{13}+\mathsf{B}_{13}\mathsf{B}_{31}+(\mathsf{B}_{11})^2+(\mathsf{B}_{33})^2-\frac{1}{2}(\mathsf{B}_{11}+\mathsf{B}_{33})^2]\,, \\
& \qquad\qquad\qquad\qquad\qquad\qquad\qquad\qquad \left[V(V+1)\right], \\
\mathsf{C}^{(2)}_{\mathsf{U}} &= \frac{1}{2}[\mathsf{B}_{23}\mathsf{B}_{32}+\mathsf{B}_{32}\mathsf{B}_{23}+(\mathsf{B}_{33})^2+(\mathsf{B}_{22})^2-\frac{1}{2}(\mathsf{B}_{33}+\mathsf{B}_{22})^2]\,, \\
& \qquad\qquad\qquad\qquad\qquad\qquad\qquad\qquad \left[U(U+1)\right].
\end{aligned}
$$

As well, we may consider the operator formed by the linear combination with constant coefficients

$$
\mathsf{C}_{\mathsf{UVT}} = a_1\,\mathsf{C}^{(2)}_{\mathsf{U}} + a_2\,\mathsf{C}^{(2)}_{\mathsf{V}} + a_3\,\mathsf{C}^{(2)}_{\mathsf{T}}\,, \qquad (3.2.36)
$$

which is equivalent to the operators $\mathsf{C}^{(2)}_{\mathsf{T}}$, $\mathsf{C}^{(2)}_{\mathsf{V}}$, and $\mathsf{C}^{(2)}_{\mathsf{U}}$. The operator $\mathsf{C}_{\mathsf{UVT}}$ commutes with the operators $\mathsf{C}^{(2)}$, $\mathsf{C}^{(3)}$, B_{11}, and B_{33}. At the same time, we may choose the coefficients in (3.2.36) such that $a_1+a_2+a_3=0$, because any operator of the form (3.2.36) with $a_1+a_2+a_3\neq 0$ can be reduced to an operator with $a_1 + a_2 + a_3 = 0$, by a linear combination with the operator $\mathsf{C}^{(2)} = \mathsf{C}_{\mathsf{U}}^{(2)} + \mathsf{C}_{\mathsf{V}}^{(2)} + \mathsf{C}_{\mathsf{T}}^{(2)}$. One can thus obtain the system of operators

$$
\begin{aligned}
&\frac{1}{2}\,\mathsf{A}_{ij}\mathsf{A}_{ji}\,, \quad \left[u(u+2) + m^2 = \frac{1}{2}\,[p(p+2) + q(q+2)]\right], \\
&\frac{1}{2}\,\mathsf{A}_{11}\,, \quad [m_1]\,, \quad \frac{1}{2}\mathsf{A}_{22}\,, \quad [m_2]\,, \quad \frac{1}{2}\mathsf{A}_{33}\,, \quad [m_3]\,, \\
&\frac{1}{2}\,\{e_1\left[\mathsf{A}_{23}\mathsf{A}_{32} + \mathsf{A}_{32}\mathsf{A}_{23} - (\mathsf{A}_{11})^2 - \mathsf{A}_{22}\mathsf{A}_{33}\right] \\
&\quad + e_2\left[\mathsf{A}_{31}\mathsf{A}_{13} + \mathsf{A}_{13}\mathsf{A}_{31} - (\mathsf{A}_{22})^2 - \mathsf{A}_{33}\mathsf{A}_{11}\right] \\
&\quad + e_3\left[\mathsf{A}_{12}\mathsf{A}_{21} + \mathsf{A}_{21}\mathsf{A}_{12} - (\mathsf{A}_{33})^2 - \mathsf{A}_{11}\mathsf{A}_{22}\right]\}\,, \quad \left[\varepsilon^{u(\tau)}_{m_1 m_2 m_3}(k^2)\right],
\end{aligned}
\tag{3.2.37}
$$

that is equivalent to the system comprising the operators $\mathsf{C}^{(2)}$, $\mathsf{C}^{(3)}$, $\mathsf{C}_{\mathsf{UVT}}$, B_{11}, and B_{33}. In (3.2.37) we used the notations $u = (p+q)/2 = 0, 1, \ldots$ and $m = m_1 + m_2 + m_3 = (p-q)/2$, $|m_i| = 0, 1/2, 1, \ldots, u$, while the coefficients e_j correspond to the elliptical parametrization (3.1.38). The eigenvalues of the last operator in (3.2.37) depend on the parameter $0 \leq k^2 \leq 1$, defined in (3.1.36), and satisfy the relations

$$
\begin{aligned}
\varepsilon^{u(\tau)}_{m_1 m_2 m_3}(0) &= 3T(T+1) - 3m_3^2 - u(u+2)\,, \\
\varepsilon^{u(\tau)}_{m_1 m_2 m_3}(1) &= -3U(U+1) + 3m_1^2 + u(u+2)\,.
\end{aligned}
$$

Generally, the operator $\mathsf{C}_{\mathsf{UVT}}$ has the eigenvalues

$$
\varepsilon^{u(\tau)}_{m_1 m_2 m_3}(k^2) + a_1 m_1^2 + a_2 m_2^2 + a_3 m_3^2\,,
$$

where $k^2 = (a_2 - a_1)/(a_3 - a_1)$, and the symbol $\tau = 1, 2, \ldots, u+1-|m_1|-|m_2|-|m_3|$ is used to distinguish between the eigenvalues which have identical indices m_1, m_2, m_3. The invariant cubic operator is

$$
\mathsf{A}_{ij}\mathsf{A}_{jk}\mathsf{A}_{ki} = \mathsf{B}_{ij}\mathsf{B}_{jk}\mathsf{B}_{ki} + \mathsf{A}_{ii}\mathsf{A}_{jk}\mathsf{A}_{kj} - \frac{2}{9}(\mathsf{A}_{ii})^3\,,
$$

and has the eigenvalues

$$
(2m-1)[3u(u+2) + m^2] + 2m = p^3 - q^3 - pq + 2p(p-1) - 4q(q+1)\,.
$$

The operators (3.2.37) can also be expressed as differential operators

$$
\begin{aligned}
\mathsf{A}_{11} &= -i\frac{\partial}{\partial \alpha_1}\,, \quad [2m_1]\,, \\
\mathsf{A}_{22} &= -i\frac{\partial}{\partial \alpha_2}\,, \quad [2m_2]\,,
\end{aligned}
$$

$$\mathsf{A}_{33} = -i\frac{\partial}{\partial \alpha_3}\,, \qquad [2m_3]\,,$$

$$\frac{1}{2}\mathsf{A}_{ij}\mathsf{A}_{ji} - \frac{1}{4}(\mathsf{A}_{ii})^2 = \frac{1}{\rho-\rho'}\left\{\frac{\partial}{\partial\rho}\left[P(\rho)\frac{\partial}{\partial\rho}\right] + \frac{\partial}{\partial\rho'}\left[P(\rho')\frac{\partial}{\partial\rho'}\right]\right\} + \frac{m_1^2}{\xi_1^2} + \frac{m_2^2}{\xi_2^2} + \frac{m_3^2}{\xi_3^2}\,, \qquad [u(u+2)],$$

$$\frac{1}{\rho-\rho'}\left\{\frac{\partial}{\partial\rho}\left[P(\rho)\frac{\partial}{\partial\rho}\right] - \frac{\partial}{\partial\rho'}\left[P(\rho')\frac{\partial}{\partial\rho'}\right]\right\} + \frac{\rho+\rho'-e_1}{\xi_1^2}\,m_1^2 + \frac{\rho+\rho'-e_2}{\xi_2^2}\,m_2^2 + \frac{\rho+\rho'-e_3}{\xi_3^2}\,m_3^2\,, \qquad \left[\varepsilon^{u(\tau)}_{m_1m_2m_3}(k^2)\right],$$

where $P(\rho) = (\rho-e_1)(\rho-e_2)(\rho-e_3)$. Here, we used the notations from (3.1.37) and omitted the expression of the last operator in terms of the A_{ij}. Now, we seek for a system of eigenfunctions which is common to all these operators, by applying the method of separation of variables to functions of the form

$$\Lambda^{u(\tau)}_{m_1m_2m_3}(\rho)\,\Lambda'^{u(\tau)}_{m_1m_2m_3}(\rho')\,e^{2i(m_1\alpha_1+m_2\alpha_2+m_3\alpha_3)}. \tag{3.2.38}$$

Thus, the first function in (3.2.38) satisfies the differential equation

$$\left\{\frac{d}{d\rho}\left[P(\rho)\frac{d}{d\rho}\right] - u(u+2)\rho + \varepsilon^{u(\tau)}_{m_1m_2m_3}(k^2) - 9\left(\frac{k^2m_1^2}{\rho-e_1} - \frac{kk'm_2^2}{\rho-e_2} + \frac{k'^2m_3^2}{\rho-e_3}\right)\right\}\Lambda^{u(\tau)}_{m_1m_2m_3}(\rho) = 0.$$

By the change of function

$$\Lambda^{u(\tau)}_{m_1m_2m_3}(\rho) = (\rho-e_1)^{|m_1|}\,(\rho-e_2)^{|m_2|}\,(\rho-e_3)^{|m_3|}\,R^{u(\tau)}_{m_1m_2m_3}(\rho)\,,$$

we obtain *Heun's differential equation* with four special regular points

$$\left\{P(\rho)\frac{d^2}{d\rho^2} + \left[(2m_0+3)\rho^2 + 2G_1\rho + 2G_2 + g_2\right]\frac{d}{d\rho} + \varepsilon^{u(\tau)}_{m_1m_2m_3}(k^2) - (u-m_0)(u+m_0+2)\rho + (2m_0+1)G_1\right\}R^{u(\tau)}_{m_1m_2m_3}(\rho) = 0\,, \tag{3.2.39}$$

where

$$m_0 = |m_1| + |m_2| + |m_3|\,,$$

$$
\begin{aligned}
G_1 &= e_1|m_1| + e_2|m_2| + e_3|m_3|\,,\\
G_2 &= e_2e_3|m_1| + e_3e_1|m_2| + e_1e_2|m_3|\,,\\
g_2 &= e_2e_3 + e_3e_1 + e_1e_2 = -(e_1^2+e_2^2+e_3^2)/2 = -3(1-k^2k'^2)\,.
\end{aligned}
$$

The solution of this equation is searched under the form

$$
\sum_{j=0}^{\infty} A_j \rho^{u-m_0-j}\,, \tag{3.2.40}
$$

being thus led to the recurrence relation

$$
\begin{aligned}
&(j-2u+2)A_j + [\varepsilon(k^2) + (2u+3-2j)G_1]A_{j-1}\\
&+(u+2-m_0-j)[2G_2 + (u+2-m_0-j)g_2]A_{j-2}\\
&\quad -(u+2-m_0-j)(u+3-m_0-j)e_1e_2e_3A_{j-3} = 0\,,
\end{aligned}
$$

where $A_l \equiv 0$ for $l < 0$. The eigenvalues $\varepsilon(k^2)$ are determined by imposing the condition that the series in (3.2.40) be finite, that is a condition of the form $A_{u-m_0+1} = 0$. Consequently, we obtain an algebraic equation of degree $u - m_0 + 1$, whose distinct roots are indexed by means of the symbol $\tau = 1, 2, \ldots, u - m_0 + 1$.

The explicit form of the eigenfunctions (3.2.38) will be given by

$$
\xi_1^{2|m_1|}\xi_2^{2|m_2|}\xi_3^{2|m_3|}e^{2i(m_1\alpha_1+m_2\alpha_2+m_3\alpha_3)}\sum_{j=0}^{u-m_0}A_j\rho^{u-m_0-j}\sum_{j'=0}^{u-m_0}A_{j'}\rho'^{u-m_0-j'}\,. \tag{3.2.41}
$$

Also, by the aid of the identity

$$
\frac{z_1z^1}{\gamma-e_1} + \frac{z_2z^2}{\gamma-e_2} + \frac{z_3z^3}{\gamma-e_3} = \frac{(\rho-\gamma)(\rho'-\gamma)}{P(\gamma)}\,,
$$

these eigenfunctions can be rewritten in the form

$$
\xi_1^{2|m_1|}\xi_2^{2|m_2|}\xi_3^{2|m_3|}e^{2i(m_1\alpha_1+m_2\alpha_2+m_3\alpha_3)}\prod_{j=1}^{u-m_0}\left(\frac{z_1z^1}{\gamma_j-e_1} + \frac{z_2z^2}{\gamma_j-e_2} + \frac{z_3z^3}{\gamma_j-e_3}\right),
$$

where γ_l are the roots of the polynomials in (3.2.41).

We thus obtain the system of eigenfunctions which is common to the operators (3.2.37), defined on the unit sphere in $\mathbb{C}^3$ in elliptical coordinates. These eigenfunctions form an orthogonal basis in the representation space of the $\mathsf{SU}(3)$ group. The eigenfunctions are analytic functions of the parameter k^2, allowing to pass from the decomposition $\mathsf{SU}(3) \supset \mathsf{SU}(2)_\mathsf{T} \times \mathsf{U}(1)_\mathsf{T}$, when $k^2 = 0$, to the decomposition $\mathsf{SU}(3) \supset \mathsf{SU}(2)_\mathsf{U} \times \mathsf{U}(1)_\mathsf{U}$, when $k^2 = 1$. The dimension of the $\mathsf{SU}(3)$ irreducible representation defined by the numbers p and q, is

$$
N(p,q) = \frac{1}{2}(p+1)(q+1)(p+q+2) = (u+1)[(u+1)^2 - m^2]\,.
$$

3. Symmetry groups of certain differential equations

3.1 Central functions. Characters

3.1.1 Central functions

In general, the matrices of the operators in a representation T of a group G are defined by the relation (1.1.2). If a group G is compact and we choose the basis in the space $\mathcal{L}$, introduced in Chap. 3 Sec. 2.1.2, such that the vector $\mathbf{e}_1$ is identical to a vector $\mathbf{a} \in \mathcal{L}$ that is invariant with respect to a subgroup $H \subset G$, then the matrix element $t_{11}(g)$ is a zonal harmonic

$$t_{11}(g) = < \mathsf{T}(g)\,\mathbf{e}_1, \mathbf{e}_1 > \equiv < \mathsf{T}(g)\,\mathbf{a}, \mathbf{a} > , \tag{3.3.1}$$

while the other matrix elements in the first row are called associated spherical harmonics

$$t_{1j}(g) = < \mathsf{T}(g)\,\mathbf{e}_j, \mathbf{e}_1 > \equiv < \mathsf{T}(g)\,\mathbf{e}_j, \mathbf{a} > , \quad j > 1 .$$

These functions are constant on the right cosets Hg, $g \in G$. The elements of the first column

$$t_{k1}(g) = < \mathsf{T}(g)\,\mathbf{e}_1, \mathbf{e}_k > \equiv < \mathsf{T}(g)\,\mathbf{a}, \mathbf{e}_k > , \quad k > 1 ,$$

are constant on the left cosets gH, $g \in G$, that is

$$t_{k1}(gh) = t_{k1}(g) , \quad \forall h \in H , \quad \forall g \in G ,$$

and they are also called associated spherical harmonics. If the representation T is unitary, then we have

$$t_{1k}(g) = \overline{t_{k1}(g)} , \quad k = 1, 2, \ldots , n. \tag{3.3.2}$$

We partition the set of all irreducible unitary representations of a compact group G into equivalence classes and we select one representation from each class. The set of representations $\{\mathsf{T}^\alpha\}$ thus obtained is called a *complete system of nonequivalent irreducible unitary representations* of G. All the representations contained in the system $\{\mathsf{T}^\alpha\}$ can be found in the decomposition of the regular representation of G into a direct sum of irreducible representations. We denote by A the set of indices α.

THEOREM 3.3.1 *Let $\{\mathsf{T}^\alpha | \alpha \in A\}$ be a complete system of nonequivalent irreducible unitary representations of a compact group G, d_α the dimension of the representation T^α, and $t^\alpha_{ij}(g)$, $1 \le i, j \le d_\alpha$, the matrix elements of the matrices of operators in T^α. Then, the functions*

$$\left\{ \sqrt{d_\alpha}\, t^\alpha_{ij}(g) \right\}^{d_\alpha}_{i,j=1} , \quad \alpha \in A , \tag{3.3.3}$$

form a complete orthonormal set on G, relative to the normalized invariant measure dg of this group.

From the completeness of the system (3.3.3) it follows that any square integrable function $f(g)$, defined on G, can be expanded into a series that converges on the average

$$f(g) = \sum_{\alpha \in A} \sum_{i,j=1}^{d_\alpha} c_{ij}^{\alpha}\, t_{ij}^{\alpha}(g)\,, \tag{3.3.4}$$

the coefficients being given by the relations

$$c_{ij}^{\alpha} = d_\alpha \int f(g)\, \overline{t_{ij}^{\alpha}(g)}\, dg\,.$$

Note that *Parseval's relation*

$$\int |f(g)|^2 dg = \sum_{\alpha \in A} \frac{1}{d_\alpha} \sum_{i,j=1}^{d_\alpha} |c_{ij}^{\alpha}|^2$$

is also satisfied.

DEFINITION 3.3.1 *A function $f(g)$, defined on a group G, is called a* **central function** *if the relation*

$$f(gg_1) = f(g_1 g)\,, \tag{3.3.5}$$

or the equivalent relation

$$f(g_1^{-1} g g_1) = f(g)\,, \tag{3.3.6}$$

take place $\forall g, g_1 \in G$.

From (3.3.6) it follows that the central functions are constant on the conjugacy classes of G.

3.1.2 Characters

Let T be a representation of a group G, comprising the matrices $\mathbf{t}(g)$, $g \in G$. The *character* of the representation T, defined in (1.1.34), is a central function (property 2) in Chap. 1 Sec. 1.3.7), because

$$\chi(g_1^{-1} g g_1) = \mathrm{Tr}\,\mathbf{t}(g_1^{-1} g g_1) = \mathrm{Tr}\,[\mathbf{t}^{-1}(g_1)\mathbf{t}(g)\mathbf{t}(g_1)] = \mathrm{Tr}\,\mathbf{t}(g) = \chi(g)\,.$$

THEOREM 3.3.2 *Any square integrable central function $f(g)$, defined on a compact group G, admits a series expansion in terms of the characters $\chi_\alpha(g)$ of the nonequivalent unitary irreducible representations of G, of the form*

$$f(g) = \sum_{\alpha \in A} c_\alpha\, \chi_\alpha(g)\,, \quad c_\alpha = \int f(g)\, \overline{\chi_\alpha(g)}\, dg\,. \tag{3.3.7}$$

The system of characters $\{\chi_\alpha(g)\}_{\alpha\in A}$ forms a complete orthonormal system of functions with respect to the scalar product

$$< \chi_\alpha(g), \chi_\beta(g) > \equiv \int \chi_\alpha(g)\, \overline{\chi_\beta(g)}\, dg = \delta_{\alpha\beta} \tag{3.3.8}$$

and constitute an orthonormal basis in the Hilbert space of square integrable central functions. The Theorem 3.3.2, applied to the character $\chi(g)$ of an arbitrary representation T of a compact group G, represents the property 4) from Chap. 1 Sec. 1.3.7. Thus, in the case of the decomposition of T into a direct sum of nonequivalent irreducible representations, the characters satisfy the relation

$$\chi(g) = \sum_{\alpha\in A} c_\alpha\, \chi_\alpha(g)\,, \tag{3.3.9}$$

where

$$c_\alpha = \int \chi(g)\, \overline{\chi_\alpha(g)}\, dg \tag{3.3.10}$$

is the multiplicity of the irreducible representation T^α in the decomposition

$$\mathsf{T} = \sum_{\alpha\in A}^{\oplus} c_\alpha\, \mathsf{T}^\alpha\,.$$

3.2 The $\mathsf{SO}(3)$, $\mathsf{SU}(2)$, and $\mathsf{SU}(3)$ groups

3.2.1 The $\mathsf{SO}(3)$ group

For a fixed l, the system of spherical harmonics $\{Y_{lm}\}_{|m|\leq l}$ defined in (3.2.11), forms the basis of the space of an irreducible representation of $\mathsf{SO}(3)$, that is these functions are transformed one into another by a rotation of the system of coordinates about the origin. To prove this property of spherical harmonics we consider the function

$$f_{lm}(x_1, x_2, x_3) = r^l\, Y_{lm}(\theta, \varphi)\,, \quad r^2 = x_1^2 + x_2^2 + x_3^2\,,$$

which is the solution of the Laplace equation $\Delta f_{lm} = 0$, in the spherical coordinates $\{r, \theta, \varphi\}$. In Cartesian coordinates, $f_{lm}(x_1, x_2, x_3)$ is a homogeneous polynomial of degree 1 in the variables $\{x_1, x_2, x_3\}$. After a rotation, the system of coordinates $Ox_1x_2x_3$ becomes $Ox_1'x_2'x_3'$ and, because the formulae for the transformation of coordinates are linear and homogeneous, by substituting into $f_{lm}(x_1, x_2, x_3)$ the variables $\{x_1, x_2, x_3\}$ by their expressions in terms of $\{x_1', x_2', x_3'\}$ we also obtain a homogeneous polynomial of degree 1. This polynomial is the solution of Laplace's equation $\Delta' f_{lm} = 0$, where Δ' is Laplace's operator in the new spherical coordinates $\{r, \theta', \varphi'\}$. Therefore, $f_{lm} = r^l\, F_{lm}(\theta', \varphi')$, are eigenfunctions of the eigenvalue problem

$$\mathsf{H}'^2 F_{lm} = l(l+1)\, F_{lm}\,, \tag{3.3.11}$$

where H'^2 is the operator (3.2.8) in the variables θ' and φ'. Note that, a rotation about the origin of coordinates preserves the distances so that f_{lm} and F_{lm} exhibit the same dependence on r.

The eigenfunctions of the angular part of Δ' (respectively H'^2) are the spherical harmonics $\{Y_{lm}(\theta', \varphi')\}_{|m|\leq l}$, which form a complete orthonormal set of functions. Hence, we may write

$$F_{lm}(\theta', \varphi') = \sum_{|n|\leq l} D^l_{nm}\, Y_{ln}(\theta', \varphi').$$

But $F_{lm}(\theta', \varphi') = Y_{lm}(\theta, \varphi)$ (the angular part in f_{lm}) and we obtain

$$Y_{lm}(\theta, \varphi) = \sum_{|n|\leq l} D^l_{nm}\, Y_{ln}(\theta', \varphi').$$

If T^l is a representation of $\mathsf{SO}(3)$, then the action the representation operator $\mathsf{T}^l(g)$, $g \in \mathsf{SO}(3)$, is given by (3.1.2) and, in the case of spherical harmonics we have

$$\mathsf{T}^l(g)\, Y_{lm}(\theta', \varphi') = Y_{lm}(\theta, \varphi) = \sum_{|n|\leq l} D^l_{nm}(g)\, Y_{ln}(\theta', \varphi'), \tag{3.3.12}$$

where we used the general relation (3.1.12) in the form $\mathsf{T}^l(g)\, f(g\omega) = f(\omega)$, corresponding to the transformation $\theta, \varphi \xrightarrow{g} \theta', \varphi'$. The coefficients $D^l_{nm}(g)$ represent the matrix elements of the operator $\mathsf{T}^l(g)$ in the space $\mathcal{L}^2$ with the basis $\{Y_{lm}\}$. The system of coordinates $Ox_1x_2x_3$ is transformed into $Ox'_1x'_2x'_3$ by the rotation g^{-1} (the rotation of the geometric space), and by inverting the relation (3.3.12) it results

$$Y_{lm}(\theta', \varphi') = \sum_{|n|\leq l} D^l_{nm}(g^{-1})\, Y_{ln}(\theta, \varphi). \tag{3.3.13}$$

The matrices $[D^l_{nm}(g^{-1})]$ are square matrices of order $2l+1$, and the set of these matrices, for all the possible rotations, form, for any value of $l = 0, 1, 2, \ldots$, a unitary irreducible representation of $\mathsf{SO}(3)$ in the space $\mathcal{E}_3$.

For the Eulerian parametrization of $\mathsf{SO}(3)$ we consider the Euler angles defined by the following decomposition of a rotation of the system of coordinates:

- a rotation of angle α about the axis Ox_3 ($Ox_1x_2x_3 \to O\overline{x}_1\overline{x}_2\overline{x}_3$, $\overline{x}_3 = x_3$),
- a rotation of angle β about the axis $O\overline{x}_2$ ($O\overline{x}_1\overline{x}_2\overline{x}_3 \to O\widetilde{x}_1\widetilde{x}_2\widetilde{x}_3$, $\widetilde{x}_2 = \overline{x}_2$),
- a rotation of angle γ about the axis $O\widetilde{x}_3$ ($O\widetilde{x}_1\widetilde{x}_2\widetilde{x}_3 \to Ox'_1x'_2x'_3$, $x'_3 = \widetilde{x}_3$).

For a rotation of angle α about the axis Ox_3 ($\beta = \gamma = 0$) we obtain

$$\begin{aligned} Y_{lm}(\theta, \varphi') &= (-1)^m \left[\frac{2l+1}{4\pi}\frac{(l-m)!}{(l+m)!}\right]^{1/2} e^{im\varphi'} P_l^m(\cos\theta) \\ &= (-1)^m \left[\frac{2l+1}{4\pi}\frac{(l-m)!}{(l+m)!}\right]^{1/2} e^{im(\varphi-\alpha)} P_l^m(\cos\theta) \\ &= e^{-im\alpha}\, Y_{lm}(\theta, \varphi)\,. \end{aligned} \tag{3.3.14}$$

The comparison of (3.3.13) and (3.3.14) leads us to

$$D^l_{nm}(\alpha, 0, 0) = e^{-im\alpha}\delta_{nm}\,,$$

and analogously,

$$D^l_{nm}(0, 0, \gamma) = e^{-im\gamma}\delta_{nm}\,.$$

Thus, we obtain the relations

$$\begin{aligned} Y_{lm}(\theta', \varphi') &= e^{-im\gamma}\, Y_{lm}(\widetilde{\theta}, \widetilde{\varphi})\,, \\ Y_{lm}(\widetilde{\theta}, \widetilde{\varphi}) &= \sum_{|n|\le l} d^l_{nm}(\beta)\, Y_{ln}(\overline{\theta}, \overline{\varphi})\,, \\ Y_{lm}(\overline{\theta}, \overline{\varphi}) &= e^{-im\alpha}\, Y_{lm}(\theta, \varphi)\,, \end{aligned}$$

which lead to the matrix elements

$$D^l_{nm}(\alpha, \beta, \gamma) = e^{-im\alpha}\, d^l_{nm}(\beta)\, e^{-im\gamma}\,, \tag{3.3.15}$$

where

$$\begin{aligned} d^l_{nm}(\beta) &\equiv D^l_{nm}(0, \beta, 0) \\ &= \sum_p (-1)^p \frac{\sqrt{(l+m)!(l-m)!(l+n)!(l-n)!}}{(l+n-p)!(l-m-p)!(m-n+p)!p!} \\ &\times \left(\cos\frac{\beta}{2}\right)^{2l+n-m-2p} \left(\sin\frac{\beta}{2}\right)^{m-n+2p}\,. \end{aligned} \tag{3.3.16}$$

The matrices $\mathbf{D}^l$ are unitary, hence satisfy the relations

$$\begin{aligned} [(\mathbf{D}^l)^\dagger(\alpha, \beta, \gamma)]_{nm} &= [(\mathbf{D}^l)^{-1}(\alpha, \beta, \gamma)]_{nm} = D^l_{nm}(-\gamma, -\beta, -\alpha) \\ &= (-1)^{n-m}\, D^l_{-m,-n}\,. \end{aligned} \tag{3.3.17}$$

To determine the character of an irreducible representation we take into account the fact that the traces of the matrices belonging to the same conjugacy class are equal. These classes are determined by the rotation angle about a given axis (see Chap. 2 Sec 1.1.4), hence the character of a representation depends only on this angle. Thus, we choose a particular matrix, e.g., the matrix associated to a rotation of angle α about the axis Ox_3, to obtain the expression of the character in the form

$$\chi^l(\alpha) = \mathsf{Tr}\,\mathbf{D}^l(\alpha,0,0) = \sum_{m=-l}^{l} e^{-im\alpha} = \frac{\sin\left(l+\frac{1}{2}\right)\alpha}{\sin\frac{\alpha}{2}}, \tag{3.3.18}$$

for $l = 0, 1, 2, \ldots$, while the orthogonality relation (3.3.8) becomes

$$< \chi^l, \chi^{l'} > = \frac{1}{\pi}\int_0^{\pi} \chi^l(\alpha)\,\overline{\chi^{l'}(\alpha)}\,(1-\cos\alpha)\,d\alpha = \delta_{ll'}\,.$$

3.2.2 The $\mathsf{SU}(2)$ group

Let us now consider a unitary irreducible representation T^j of $\mathsf{SU}(2)$ in the representation space $\mathcal{F}$, with the canonical base $\{\mathbf{f}^j_{mn}\}$, defined in Chap. 3 Sec. 2.2.2. All the representations T^j, $j = 0, 1/2, 1, 3/2, \ldots$, are contained in the decomposition of the regular representation of $\mathsf{SU}(2)$ into a direct sum of irreducible representations. At the same time, there exists a close connection between the matrix elements of the operators in the regular representation and the functions in the canonical basis (see (3.3.1)). We denote by $t^j_{mn}(\mathsf{U})$, $|m|, |n| \leq j$, $\mathsf{U} \in \mathsf{SU}(2)$, the matrix elements of the representation operator $\mathsf{T}^j(\mathsf{U})$. These matrix elements are numerical functions belonging to the space $\mathcal{F}$, so that they can be expressed as linear combinations of the elements in the canonical basis.

Thus, if we apply the operators of the regular representation T to the functions $t^j_{mn}(\mathsf{U})$ then, according to (3.1.14), we obtain

$$\mathsf{T}(\mathsf{U}')\,t^j_{mk}(\mathsf{U}) = t^j_{mk}(\mathsf{U}\mathsf{U}') = [\mathbf{t}^j(\mathsf{U})\mathbf{t}^j(\mathsf{U}')]_{mk} = \sum_{n=-j}^{j} t^j_{mn}(\mathsf{U})\,t^j_{nk}(\mathsf{U}')\,. \tag{3.3.19}$$

We consider the exponential parametrization of $\mathsf{SU}(2)$, differentiate the relation (3.3.19) with respect to a parameter ξ'_r, and set $\boldsymbol{\xi} = \mathbf{0}$, where $\boldsymbol{\xi} = (\xi_1, \xi_2, \xi_3)$. Because $t^j_{mk}(\mathsf{U})$ does not depend on ξ'_r, we may write

$$\mathsf{A}_r\,t^j_{mk}(\mathsf{U}) = \sum_{n=-j}^{j} \left[\frac{\partial t^j_{nk}(\mathsf{U}')}{\partial \xi'_r}\right]_{\boldsymbol{\xi}'=\mathbf{0}} t^j_{mn}(\mathsf{U}) = \sum_{n=-j}^{j} (\mathbf{a}^j_r)_{nk}\,t^j_{mn}(\mathsf{U})\,, \tag{3.3.20}$$

revealing the matrices of the infinitesimal generators of weight j. We can also use the Hermitian infinitesimal generators $\mathsf{H}_r = i\mathsf{A}_r$. and write

$$\mathsf{H}_r\, t^j_{mk}(\mathsf{U}) = \sum_{n=-j}^{j} (\mathbf{h}^j_r)_{nk}\, t^j_{mn}(\mathsf{U})\,. \tag{3.3.21}$$

The advantage of using the Hermitian infinitesimal generators consists in the fact that their matrices $\mathbf{h}^j_r$, corresponding to a unitary irreducible representation of weight j in a canonical base, have the standard form

$$\mathbf{h}^j_1 = \frac{1}{2}\begin{bmatrix} 0 & \alpha_{jj} & 0 & \ldots & 0 & 0 \\ \alpha_{jj} & 0 & \alpha_{j,j-1} & \ldots & 0 & 0 \\ . & . & . & \ldots & . & . \\ 0 & 0 & 0 & \ldots & 0 & \alpha_{j,-j+1} \\ 0 & 0 & 0 & \ldots & \alpha_{j,-j+1} & 0 \end{bmatrix},$$

$$\mathbf{h}^j_2 = \frac{1}{2}\begin{bmatrix} 0 & -\alpha_{jj} & 0 & \ldots & 0 & 0 \\ \alpha_{jj} & 0 & -\alpha_{j,j-1} & \ldots & 0 & 0 \\ . & . & . & \ldots & . & . \\ 0 & 0 & 0 & \ldots & 0 & -\alpha_{j,-j+1} \\ 0 & 0 & 0 & \ldots & \alpha_{j,-j+1} & 0 \end{bmatrix}, \tag{3.3.22}$$

$$\mathbf{h}^j_3 = \begin{bmatrix} j & 0 & 0 & \ldots & 0 & 0 \\ 0 & j-1 & 0 & \ldots & 0 & 0 \\ . & . & . & \ldots & . & . \\ 0 & 0 & 0 & \ldots & -j+1 & 0 \\ 0 & 0 & 0 & \ldots & 0 & -j \end{bmatrix},$$

where $\alpha_{jm} = \sqrt{(j+m)(j-m+1)}$, $|m| \leq j$. Note that the form of these matrices results from (3.2.18) with $\mathsf{H}_\pm = \mathsf{H}_1 \pm i\mathsf{H}_2$, taking into account the orthogonality of the system of eigenfunctions $\mathbf{f}^j_{mn}$. By giving particular values to the index r in (3.3.21), we obtain the relations

$$\begin{aligned} \mathsf{H}_3\, t^j_{mn}(\mathsf{U}) &= n\, t^j_{mn}(\mathsf{U})\,, \\ \mathsf{H}_+\, t^j_{mn}(\mathsf{U}) &= \alpha_{j,n+1}\, t^j_{m,n+1}(\mathsf{U})\,, \\ \mathsf{H}_-\, t^j_{mn}(\mathsf{U}) &= \alpha_{j,n}\, t^j_{m,n-1}(\mathsf{U})\,, \\ \mathsf{H}^2\, t^j_{mn}(\mathsf{U}) &= j(j+1)\, t^j_{mn}(\mathsf{U})\,, \end{aligned} \tag{3.3.23}$$

of the same form as the relations defining the action of the operators H_3, H_+, H_- on the canonical basis. Consequently, the matrix elements differ only by a factor from the elements of the canonical basis in the subspace $\mathcal{F}_{jm} \subset \mathcal{F}$, and it can be shown that

$$f^j_{mn}(\mathsf{U}) = \sqrt{2j+1}\, t^j_{mn}(\mathsf{U})\,. \tag{3.3.24}$$

Now, from (3.3.23) one obtains the analytical form of the matrix elements of the operators $\mathsf{T}^j(\mathsf{U})$ in the canonical basis

$$t^j_{mn}(\mathsf{U}) = e^{-i(m\varphi + n\psi)}\, d^j_{mn}(\theta)\,, \tag{3.3.25}$$

where the functions $d^j_{mn}(\theta)$ are given by (3.3.16), such that

$$t^j_{mn}(\mathsf{U}) = D^j_{mn}(\varphi, \theta, \psi)\,. \tag{3.3.26}$$

These matrix elements satisfy the orthogonality relations

$$< t^j_{mn}(\mathsf{U}), t^{j'}_{m'n'}(\mathsf{U}) > \equiv \int_{\mathsf{SU}(2)} t^j_{mn}(\mathsf{U})\, \overline{t^{j'}_{m'n'}(\mathsf{U})}\, d\mathsf{U} = \frac{1}{2j+1}\, \delta_{jj'}\, \delta_{mm'}\, \delta_{nn'}, \tag{3.3.27}$$

and are related to the spherical functions by the formulae

$$\begin{aligned} t^j_{0n}\left(\varphi, \theta, \frac{\pi}{2} - \psi\right) &= \sqrt{\frac{4\pi}{2j+1}}\, Y_{jn}(\theta, \psi)\,, \\ t^j_{n0}\left(\frac{\pi}{2} - \varphi, \theta, \psi\right) &= \sqrt{\frac{4\pi}{2j+1}}\, Y_{jn}(\theta, \varphi)\,. \end{aligned} \tag{3.3.28}$$

From (3.3.26) it follows that, for integer j, the irreducible representations of $\mathsf{SU}(2)$ coincide with the irreducible representations of $\mathsf{SO}(3)$, that is a consequence of Theorem 2.2.1. For half-integer j, the rotation defined by the angles α, β, γ corresponds to the matrix $\pm\mathbf{D}^j(\alpha, \beta, \gamma)$. Note that, this does not imply that the signs of the matrices $\mathbf{D}^j$, associated to different rotations, can be changed individually; one has to change simultaneously the sign of all the matrices in the representation.

The character of the representation T^j has the form

$$\chi^j(\varphi) = \mathsf{Tr}\, \mathbf{D}^j(\varphi, 0, 0) = \sum_{m=-j}^{j} e^{-im\varphi} = \frac{\sin\left(j + \frac{1}{2}\right)\varphi}{\sin\frac{\varphi}{2}}\,, \tag{3.3.29}$$

for $j = 0, 1/2, 1, \ldots$.

3.2.3 The $\mathsf{SU}(3)$ group

For the $\mathsf{SO}(2)$ and $\mathsf{SU}(2)$ groups the basis vectors in the representation space are labelled by two numbers (l, m), respectively (j, n). In the case of the $\mathsf{SU}(3)$ group we need five numbers, that are:

- the labels of the irreducible representation (p, q),
- the labels (j, m) associated to the $\mathsf{SU}(2)$ group in the canonical decomposition $\mathsf{SU}(3) \subset \mathsf{SU}(2) \times \mathsf{U}(1)$,

- one label, which we will denote by Y, associated to the group $\mathsf{U}(1)$ in the canonical decomposition.

All these labels are obtained from the eigenvalues of five operators which commute between them, the eigenfunctions of these operators constituting the basis of the representation space.

We choose as infinitesimal generators of the subgroup $\mathsf{SU}(2) \subset \mathsf{SU}(3)$ the operators

$$\mathsf{j}_1 = \frac{1}{2}\,\lambda_1\,, \quad \mathsf{j}_2 = \frac{1}{2}\,\lambda_2\,, \quad \mathsf{j}_3 = \frac{1}{2}\,\lambda_3\,,$$

where $\lambda_k = 2\,\mathsf{F}_k$ ($k = 1,2,3$) have been defined in Chap. 2 Sec.3.2.2. For the subgroup $\mathsf{U}(1) \subset \mathsf{SU}(3)$ we choose the infinitesimal generator

$$\mathsf{Y} = \frac{2}{\sqrt{3}}\,\mathsf{F}_8 = \frac{1}{\sqrt{3}}\,\lambda_8\,.$$

First, we consider that the basis in the representation space is formed by the eigenvectors

$$\begin{aligned} \mathsf{Y}\,\Psi(m,Y) &= Y\,\Psi(m,Y)\,, \\ \mathsf{j}_3\,\Psi(m,Y) &= m\,\Psi(m,Y)\,. \end{aligned} \tag{3.3.30}$$

Then, the Casimir operator of the $\mathsf{SU}(2)$ subgroup, $\mathsf{j}^2 = \mathsf{j}_1^2 + \mathsf{j}_2^2 + \mathsf{j}_3^2$, provides another label which is the eigenvalue in the equation

$$\mathsf{j}^2\,\Psi(j,m,Y) = j(j+1)\,\Psi(j,m,Y)\,. \tag{3.3.31}$$

At the same time, the functions $\Psi(j,m,Y)$ are eigenfunctions of the operators $\mathsf{C}^{(2)}$ and $\mathsf{C}^{(3)}$, defined in (3.2.35). Hence, the basis of the representation space is formed by the set of functions $\{\Psi^{pq}(j,m,Y)\}$, where the labels p and q determine the irreducible representation of $\mathsf{SU}(3)$ having the dimension $N(p,q)$ given by (3.2.34).

In the parametrization (2.3.62) the elements of the matrices associated to the operators in a representation have the form

$$\mathcal{D}^{pq}_{j'm'Y',jmY}(\boldsymbol{\alpha}) = \big\langle \Psi^{pq}(j,m,Y), \mathsf{T}_2\mathsf{T}_3\mathsf{T}'_2\,\Psi^{pq}(j',m',Y')\big\rangle\,, \tag{3.3.32}$$

and can be calculated by using the decomposition

$$\begin{aligned} \mathcal{D}^{pq}_{j'm'Y',jmY}(\boldsymbol{\alpha}) = \sum_{m'',m'''} &\big\langle \Psi^{pq}(j,m,Y), \mathsf{T}_2\,\Psi^{pq}(j,m'',Y)\big\rangle \\ &\times \big\langle \Psi^{pq}(j,m'',Y), \mathsf{T}_3\,\Psi^{pq}(j',m''',Y')\big\rangle \\ &\times \big\langle \Psi^{pq}(j',m''',Y'), \mathsf{T}'_2\,\Psi^{pq}(j',m',Y')\big\rangle\,. \end{aligned} \tag{3.3.33}$$

The matrix elements for the operators T_2 and T'_2 are given by the relations (3.3.25), because these operators correspond to the subgroup $\mathsf{SU}(2) \subset \mathsf{SU}(3)$.

Also, the operator

$$\exp\left(-i\frac{2}{\sqrt{3}}\,\alpha_8\mathsf{F}_8\right)$$

has the eigenvalues $\exp(-i\alpha_8 Y)$, such that (3.3.33) becomes

$$\begin{aligned}\mathcal{D}^{pq}_{j'm'Y',jmY}(\boldsymbol{\alpha}) = \sum_{m''} e^{-i\alpha_8 Y}\, D^{j}_{mm''}(\alpha_1,\alpha_2,\alpha_3)\\ \times\left\langle \Psi^{pq}(j,m'',Y), e^{-2i\alpha_7\mathsf{F}_7}\,\Psi^{pq}(j',m''',Y')\right\rangle\\ \times D^{j}_{m'''m''}(\alpha_4,\alpha_5,\alpha_6)\,, \qquad (3.3.34)\end{aligned}$$

where

$$m''' = m'' + (Y'-Y)/2\,.$$

Note that the operators in the Gell-Mann basis satisfy relations of the form

$$e^{-i\pi\mathsf{F}_3}\,\mathsf{F}_7\,e^{i\pi\mathsf{F}_3} = \mathsf{F}_6\,,$$

so that we may redefine α_3 and α_4 in (2.3.62) to absorb the exponentials $\exp(\pm i\pi\mathsf{F}_3)$, and replace F_7 by F_6. Also, note that the interchange of the first two components of the vector $\mathbf{z}$ in (2.3.60) allows the replacement of F_6 by F_4. Hence, we may rewrite (3.3.34) as

$$\begin{aligned}\mathcal{D}^{pq}_{j'm'Y',jmY}(\boldsymbol{\alpha}) = \sum_{m''} e^{-i\alpha_8 Y}\, D^{j}_{mm''}(\alpha_1,\alpha_2,\alpha_3)\\ \times\left\langle \Psi^{pq}(j,m'',Y), e^{-2i\alpha_4\mathsf{F}_4}\,\Psi^{pq}(j',m''',Y')\right\rangle\\ \times D^{j}_{m'''m''}(\alpha_4,\alpha_5,\alpha_6)\,, \qquad (3.3.35)\end{aligned}$$

where we have used the same symbols for the redefined parameters. It can be shown that the matrix element in (3.3.35) has the expression

$$\begin{aligned}&\left\langle \Psi^{pq}(j,m,\delta), e^{-2i\nu\mathsf{F}_4}\,\Psi^{pq}(j',m',\delta')\right\rangle\\ &= N\sum_{n,r,s}(-i\,\sin\nu)^{2(s+\delta'+j-r-n)}(\cos\nu)^{p+q-m'-\delta'+2(n-j-s)}\\ &\times\left[\frac{(s+m'+\delta')!}{s!r!(s+m'+r-j')!(s+m'+r+j'+1)!(m-\delta+r)!}\right]^{1/2}\\ &\times\frac{C^{(s+m'+\delta'+r)/2\;\;(r+m'+s-\delta')/2\;\;j'}_{(s+m'+\delta'-r)/2\;\;(r+m'-s-\delta')/2\;\;m}}{n!(\delta+q-j-s)!(j-m-r)!(s+m'+\delta'-n)!}\\ &\times{}_4F_3\left[\begin{matrix}-s, & -j+m+r, & j+\delta+m'+\delta'+s-p-n, & -n; & 1\\ 1+\delta+q-j-s, & 1+s+m'+\delta'-n, & 1+j+\delta-r-n & & \end{matrix}\right]\end{aligned}$$

$$\times \frac{1}{(j+\delta-r-n)!(p-\delta-j-s-m'-\delta'+n)!}, \tag{3.3.36}$$

with

$$N = \left[\frac{(2j+1)(j-m)!(j+m)!(j-\delta)!(j+\delta)!(p-\delta-j)!(q+\delta-j)!}{(p+j-\delta+1)!(q+j+\delta+1)!} \right]^{1/2}$$
$$\times [(p-\delta'-j')!(q+\delta'-j')!(p+j'-\delta'+1)!(q+j'+\delta'+1)!]^{1/2}.$$

Here, ${}_4F_3$ is a *hypergeometric function*, and $C^{a\ b\ c}_{a'\ b'\ c'}$ are *Clebsch-Gordan coefficients* for the $\mathsf{SU}(2)$ group (the definition of the Clebsch-Gordan coefficients will be given in the next Section).

3.3 Direct products of irreducible representations

3.3.1 Overview

Generally, the direct product of two irreducible representations of a group G (see Definition 1.1.32) can be decomposed into a direct sum of irreducible representations. Thus, let T^1 and T^2 be two irreducible representations of a group G in the linear spaces $\mathcal{L}_1$ and $\mathcal{L}_2$, respectively. The direct product of representations, denoted by $\mathsf{T}^1 \otimes \mathsf{T}^2$, is a reducible representation in the linear space $\mathcal{L} = \mathcal{L}_1 \otimes \mathcal{L}_2$. This representation can be decomposed into the direct sum

$$\mathsf{T}^1 \otimes \mathsf{T}^2 = \sum_{\alpha \in A} c_\alpha \, \mathsf{T}^\alpha, \tag{3.3.37}$$

where $\{\mathsf{T}^\alpha | \alpha \in A\}$ is the complete set of nonequivalent irreducible representations of the group G. If the group G is compact, then according to Theorem 3.3.2, applied to the functions $\mathsf{T}(g) = \mathsf{T}^1(g) \otimes \mathsf{T}^2(g)$, $g \in G$, respectively from the relations (3.3.9) and (3.3.10) satisfied by the characters, and the property 5) specified in Chap. 1 Sec. 1.3.7, it follows that

$$\chi_1(g)\,\chi_2(g) = \sum_{\alpha \in A} c_\alpha \, \chi_\alpha(g), \tag{3.3.38}$$

where

$$c_\alpha = \int \overline{\chi_\alpha(g)}\,\chi_1(g)\,\chi_2(g)\,dg \tag{3.3.39}$$

is the multiplicity of the representation T^α in the decomposition of the direct product of representations. The representation space $\mathcal{L}$ can also be decomposed into a direct sum of subspaces $\mathcal{L}_\alpha$, in which the representations T^α are realized. Let us now assume that we have defined the bases $\{\mathbf{e}_i | i \in [1, n_1]\} \subset \mathcal{L}_1$, $\{\mathbf{f}_j | j \in [1, n_2]\} \subset \mathcal{L}_2$, and $\{\mathbf{e}_i \otimes \mathbf{f}_j\} \subset \mathcal{L} = \mathcal{L}_1 \otimes \mathcal{L}_2$. Then, the matrices of the operators in the direct product T are

$$[\mathbf{t}(g)]_{\alpha\beta,\gamma\delta} = [\mathbf{t}^1(g)]_{\alpha\beta}\,[\mathbf{t}^2(g)]_{\gamma\delta}\,. \tag{3.3.40}$$

The matrix $\mathbf{t}(g)$ can be brought to the block diagonal form (1.1.20) by a similarity transformation

$$\mathbf{t}'(g) = \mathbf{S}^{-1}\,\mathbf{t}(g)\,\mathbf{S}\,,$$

where $\mathbf{S}$ is a matrix of constant elements. It is said that the matrix $\mathbf{t}'(g)$ is the transform of the matrix $\mathbf{t}(g)$ by the matrix $\mathbf{S}$, and the matrices $\mathbf{t}'(g)$ and $\mathbf{t}(g)$ are called *similar matrices*. The matrices $\mathbf{t}'(g)$ and $\mathbf{t}(g)$ correspond to the same operator $\mathsf{T}(g)$, but in two different bases of the representation space $\mathcal{L}$. The relation between these two bases is also determined by the matrix $\mathbf{S}$, and has the form

$$\{\mathbf{e}_i \otimes \mathbf{f}_j\} \xrightarrow{\mathbf{S}} \{\mathbf{h}_k\} \subset \mathcal{L} \tag{3.3.41}$$

or

$$\mathbf{h}_k = \sum_{i=1}^{n_1}\sum_{j=1}^{n_2} S_{ij,k}\,\mathbf{e}_i \otimes \mathbf{f}_j\,, \quad k = 1, 2, \ldots, n, \tag{3.3.42}$$

where $n = n_1 n_2$. The matrix elements $S_{ij,k}$ are called *Clebsch-Gordan coefficients*, and the decomposition (3.3.37) is called *Clebsch-Gordan series*.

Finally, we mention that there is no method to derive the expression of the Clebsch-Gordan coefficients, in the general case, and such formulae have to be found for each group, separately.

3.3.2 The $\mathsf{SU}(2)$ and $\mathsf{SO}(3)$ groups

In the case of the $\mathsf{SU}(2)$ group, it can be shown that the product of two unitary irreducible representations of weights j_1 and j_2, decomposes into a direct sum of irreducible representations of weights $j = j_1+j_2, j_1+j_2-1, \ldots, |j_1-j_2|$, each representation of a given weight appearing only once. Hence, the decomposition (3.3.37) becomes

$$\mathsf{T}^{j_1}(\mathsf{U}) \otimes \mathsf{T}^{j_2}(\mathsf{U}) = \sum_{j=|j_1-j_2|}^{j_1+j_2} \mathsf{T}^{j}(\mathsf{U})\,, \quad \mathsf{U} \in \mathsf{SU}(2)\,. \tag{3.3.43}$$

If $\{\mathbf{f}_{j_1 m_1} \| m_1| \leq j_1\}$ is the canonical base in the space $\mathcal{L}_1$, which is the representation space for the unitary irreducible representation T^{j_1}, and $\{\mathbf{f}_{j_2 m_2} \| m_2| \leq j_2\}$ is the canonical base in the space $\mathcal{L}_2$ associated to T^{j_2}, then (3.3.42) takes the form

$$\mathbf{f}_{jm} = \sum_{m_2=-j_2}^{j_2}\sum_{m_1=-j_1}^{j_1} C^{\,j_1\,j_2\,j}_{m_1 m_2 m}\,\mathbf{f}_{j_1 m_1} \otimes \mathbf{f}_{j_2 m_2}\,, \tag{3.3.44}$$

where $|m| \leq j$, $|j_1 - j_2| \leq j \leq j_1 + j_2$, and $\{\mathbf{f}_{jm}\}$ is the union of the canonical bases corresponding to the linear spaces associated to the irreducible representations in which the direct product $\mathsf{T}^{j_1} \otimes \mathsf{T}^{j_2}$ is decomposed. This canonical base is determined by the action of the infinitesimal generators

$$\begin{aligned} \mathsf{H}_3\, \mathbf{f}_{jm} &= m\, \mathbf{f}_{jm}\,, \\ \mathsf{H}_+ \mathbf{f}_{jm} &= \alpha_{j,m+1}\, \mathbf{f}_{j,m+1}\,, \\ \mathsf{H}_- \mathbf{f}_{jm} &= \alpha_{jm}\, \mathbf{f}_{j,m-1}\,, \end{aligned}$$

which has been defined in (3.2.18). The infinitesimal generators of the representation T^j are constructed from the corresponding infinitesimal generators of the representations T^{j_1} and T^{j_2}, according to the relations

$$\mathsf{H}_\pm = \mathsf{H}^1_\pm \otimes \mathsf{T}^{j_2}(\mathsf{E}) + \mathsf{T}^{j_1}(\mathsf{E}) \otimes \mathsf{H}^2_\pm\,,$$

$$\mathsf{H}_3 = \mathsf{H}^1_3 \otimes \mathsf{T}^{j_2}(\mathsf{E}) + \mathsf{T}^{j_1}(\mathsf{E}) \otimes \mathsf{H}^2_3\,,$$

where E is the identity element of $\mathsf{SU}(2)$. Thus, applying the operator H_3 to the relation (3.3.44) we obtain

$$m\, \mathbf{f}_{jm} = \sum_{m_2=-j_2}^{j_2} \sum_{m_1=-j_1}^{j_1} C^{j_1\, j_2\, j}_{m_1 m_2 m}\, (m_1 + m_2)\, \mathbf{f}_{j_1 m_1} \otimes \mathbf{f}_{j_2 m_2}\,,$$

as well as

$$\sum_{m_2=-j_2}^{j_2} \sum_{m_1=-j_1}^{j_1} C^{j_1\, j_2\, j}_{m_1 m_2 m}\, (m_1 + m_2 - m)\, \mathbf{f}_{j_1 m_1} \otimes \mathbf{f}_{j_2 m_2} = \mathbf{0}\,.$$

Because the basis $\{\mathbf{f}_{j_1 m_1} \otimes \mathbf{f}_{j_2 m_2}\}$ comprises a system of linearly independent vectors, it follows that $C^{j_1\, j_2\, j}_{m_1 m_2 m} = 0$ if $m \neq m_1 + m_2$. Consequently, the Clebsch-Gordan coefficients vanish, unless $|j_1 - j_2| \leq j \leq j_1 + j_2$ and $m = m_1 + m_2$. The matrices associated to the representation operators are unitary, therefore the matrix of Clebsch-Gordan coefficients is also unitary. The canonical bases $\{\mathbf{f}_{jm}\}$ are determined up to a phase factor $\exp(i\theta)$, such that the Clebsch-Gordan coefficients are also defined up to a phase factor. By convention, this factor is determined by the condition

$$C^{j_1\, j_2\, j}_{j_1, j-j_1, j} \geq 0\,, \tag{3.3.45}$$

so that all the Clebsch-Gordan coefficients are real. The condition that the matrix of Clebsch-Gordan coefficients is unitary (for j_1 and j_2 fixed), can be written in the form

$$\sum_{m_1=-j_1}^{j_1} \sum_{m_2=-j_2}^{j_2} C^{j_1\, j_2\, j}_{m_1 m_2 m}\, C^{j_1\, j_2\, j'}_{m_1 m_2 m'} = \delta_{jj'}\, \delta_{mm'} \,,$$

$$\sum_{j=|j_1-j_2|}^{j_1+j_2} \sum_{m=-j}^{j} C^{j_1\, j_2\, j}_{m_1 m_2 m}\, C^{j_1\, j_2\, j}_{m_1' m_2' m} = \delta_{m_1 m_1'}\, \delta_{m_2 m_2'} \,, \tag{3.3.46}$$

where $m = m_1 + m_2$ and $m' = m_1' + m_2'$. Hence, we may invert the relation (3.3.44) to obtain

$$\mathbf{f}_{j_1 m_1} \otimes \mathbf{f}_{j_2 m_2} = \sum_{j=|j_1-j_2|}^{j_1+j_2} \sum_{m=-j}^{j} C^{j_1\, j_2\, j}_{m_1 m_2 m}\, \mathbf{f}_{jm} \,. \tag{3.3.47}$$

Applying the operators $\mathsf{T}^j(\mathsf{U}) = \mathsf{T}^{j_1}(\mathsf{U}) \otimes \mathsf{T}^{j_2}(\mathsf{U})$ to (3.3.44), we obtain a relation in terms of the matrix elements of the representation operators

$$\sum_{n=-j}^{j} t^j_{nm}(\mathsf{U})\, \mathbf{f}_{jn}$$
$$= \sum_{m_2=-j_2}^{j_2} \sum_{m_1=-j_1}^{j_1} \sum_{n_2=-j_2}^{j_2} \sum_{n_1=-j_1}^{j_1} C^{j_1\, j_2\, j}_{m_1 m_2 m}\, t^{j_1}_{n_1 m_1}(\mathsf{U})\, t^{j_2}_{n_2 m_2}(\mathsf{U})\, \mathbf{f}_{j_1 n_1} \otimes \mathbf{f}_{j_2 n_2},$$

and by substituting (3.3.44) we are lead to a relation between the matrix elements of the representation operators

$$\sum_{n=-j}^{j} t^j_{nm}(\mathsf{U})\, C^{j_1\, j_2\, j}_{n_1 n_2 n} = \sum_{m_1=-j_1}^{j_1} \sum_{m_2=-j_2}^{j_2} C^{j_1\, j_2\, j}_{m_1 m_2 m}\, t^{j_1}_{n_1 m_1}(\mathsf{U})\, t^{j_2}_{n_2 m_2}(\mathsf{U}) \,.$$

Multiplying by $C^{j_1\, j_2\, j}_{n_1' n_2' m}$, summing over j and m, and employing the orthogonality properties (3.3.46), we may write

$$t^{j_1}_{n_1 n_1'}(\mathsf{U})\, t^{j_2}_{n_2 n_2'}(\mathsf{U}) = \sum_{j=|j_1-j_2|}^{j_1+j_2} \sum_{m=-j}^{j} \sum_{n=-j}^{j} C^{j_1\, j_2\, j}_{n_1' n_2' m}\, C^{j_1\, j_2\, j}_{n_1 n_2 n}\, t^j_{nm}(\mathsf{U}) \,. \tag{3.3.48}$$

From (3.3.27) we have

$$\int \left| t^j_{nm}(\mathsf{U}) \right|^2 \, d\mathsf{U} = \frac{1}{2j+1} \,,$$

such that, if we multiply (3.3.48) by $\overline{t^j_{nm}(\mathsf{U})}$ and integrate, it follows

$$\int \overline{t^j_{nm}(\mathsf{U})}\, t^{j_1}_{n_1n_1'}(\mathsf{U})\, t^{j_2}_{n_2n_2'}(\mathsf{U})\, d\mathsf{U} = \frac{1}{2j+1} C^{j_1\, j_2\, j}_{n_1 n_2 n}\, C^{j_1\, j_2\, j}_{n_1' n_2' m}\,. \tag{3.3.49}$$

By means of this relation we can compute the values of the Clebsch-Gordan coefficients, replacing the matrix elements by the corresponding expressions (3.3.25). We thus obtain

$$C^{j_1\, j_2\, j}_{j_1, j-j_1, j} = \left[\frac{(2j+1)!(2j_1)!}{(j+j_1+j_2+1)!(j+j_1-j_2)!}\right]^{1/2}. \tag{3.3.50}$$

The other coefficients, computed from the relation

$$\int \overline{t^j_{mj}(\mathsf{U})}\, t^{j_1}_{m_1 j_1}(\mathsf{U})\, t^{j_2}_{m_2, j-j_1}(\mathsf{U})\, d\mathsf{U} = \frac{1}{2j+1} C^{j_1\, j_2\, j}_{m_1 m_2 m}\, C^{j_1\, j_2\, j}_{j_1, j-j_1, j}\,,$$

are expressed in the form

$$\begin{aligned} &C^{j_1\, j_2\, j}_{m_1 m_2 m} \\ &= (-1)^{j_1+j_2-j}\left[\frac{(2j+1)(j_1+j_2-j)!(j_1-j_2+j)!(j-j_1+j_2)!(j_2+m_2)!(j_2-m_2)!}{(j+j_1+j_2+1)!(j+m_1+m_2)!(j-m_1-m_2)!(j_1+m_1)!(j_1-m_1)!}\right]^{1/2} \\ &\times \sum_{p=\min(j_2-m_2, j_1+j_2-j)}^{\max(0, j_1-j-m_2)} (-1)^p \frac{(j_1+j_2-m_1-m_2-p)!(j+m_1+m_2+p)!}{(j_1+j_2-j-p)!p!(j_2-m_2-p)!(j-j_1+m_2+p)!}\,. \end{aligned} \tag{3.3.51}$$

The Clebsch-Gordan coefficients are also related to Wigner's 3-j symbol

$$C^{j_1\, j_2\, j}_{m_1 m_2 m} = (-1)^{m+j_1-j_2}\sqrt{2j+1}\begin{pmatrix} j_1 & j_2 & j \\ m_1 & m_2 & -m \end{pmatrix}. \tag{3.3.52}$$

We mention the following properties of the Clebsch-Gordan coefficients

$$\begin{aligned} C^{j_1\, j_2\, j}_{-m_1, -m_2, -m} &= C^{j_2\, j_1\, j}_{m_2 m_1 m} = (-1)^{j_1+j_2-j} C^{j_1\, j_2\, j}_{m_1 m_2 m}\,, \\ C^{j\, j_2\, j_1}_{m, -m_2, m_1} &= (-1)^{j-j_1-m_2}\sqrt{\frac{2j_1+1}{2j+1}}\, C^{j_1\, j_2\, j}_{m_1 m_2 m}\,, \\ C^{j_1\, j\, j_2}_{-m_1, m, m_2} &= (-1)^{j-j_2+m_1}\sqrt{\frac{2j_2+1}{2j+1}}\, C^{j_1\, j_2\, j}_{m_1 m_2 m}\,. \end{aligned} \tag{3.3.53}$$

For integer j the formula (3.3.51) gives the Clebsch-Gordan coefficients of the $\mathsf{SO}(3)$ group. Also, the matrix elements (3.3.15) satisfy a relation analogous to (3.3.49).

Finally, we mention that the Gaunt coefficients (3.2.16) can be expressed in terms of the $\mathsf{SO}(3)$ Clebsch-Gordan coefficients in the form

$$\langle lm|l_2m_2|l_1m_1\rangle = \left[\frac{(2l_1+1)(2l_2+1)}{4\pi(2l+1)}\right]^{1/2} C^{l_1\; l_2\; l}_{m_1\, m_2\, m}\, C^{l_1\, l_2\, l}_{0\;\, 0\;\, 0}\,. \tag{3.3.54}$$

3.3.3 The $\mathsf{SU}(3)$ group

Similarly to the $\mathsf{SU}(2)$ group, in the case of the $\mathsf{SU}(3)$ group, we use the eigenvalues of a set of independent operators to label the elements of the basis in the representation space of an irreducible representation. A set of such operators is formed by $\mathsf{G}^3 \equiv \mathsf{C}^{(3)}$, $\mathsf{F}^2 \equiv \mathsf{C}^{(2)} = \sum_{i=1}^{8} \mathsf{F}_i^2$, $\mathsf{T}^2 \equiv \mathsf{j}^2$, $\mathsf{T}_3 \equiv \mathsf{j}_3$, and Y (see Chap. 3 Sec.3.2.3). We will denote by $\mathsf{D}(p,q)$ the representation of dimension $N(p,q)$, such that the decomposition of the direct product of two irreducible representations into a direct sum of irreducible representations can be written in the form

$$\mathsf{D}(p_1,q_1) \otimes \mathsf{D}(p_2,q_2) = \sum_{P,Q}^{\oplus} \sigma(P,Q)\, \mathsf{D}(P,Q)\,, \tag{3.3.55}$$

where the multiplicities $\sigma(P,Q)$ are non-negative integers. The vectors in the basis of the representation space for the product of representations are completely determined by the eigenvalues of ten linearly independent operators, that commute one another,

$$\begin{aligned}&\mathsf{G}^3(1)\,, \quad \mathsf{G}^3(2)\,, \quad \mathsf{F}^2(1)\,, \quad \mathsf{F}^2(2)\,,\\ &\mathsf{T}^2(1)\,, \quad \mathsf{T}^2(2)\,, \quad \mathsf{T}_3(1)\,, \quad \mathsf{T}_3(2)\,, \quad \mathsf{Y}(1)\,, \quad \mathsf{Y}(2)\,.\end{aligned} \tag{3.3.56}$$

If, in the Gell-Mann basis, we associate to the direct product of irreducible representations the operators $\mathsf{F}_i = \mathsf{F}_i(1) + \mathsf{F}_i(2)$, $i = 1, 2, \ldots, 8$, we obtain nine linearly independent operators, which commute one another,

$$\mathsf{G}^3\,, \quad \mathsf{G}^3(1)\,, \quad \mathsf{G}^3(2)\,, \quad \mathsf{F}^2\,, \quad \mathsf{F}^2(1)\,, \quad \mathsf{F}^2(2)\,, \quad \mathsf{T}^2\,, \quad \mathsf{T}_3\,, \quad \mathsf{Y}\,, \tag{3.3.57}$$

and we need one more operator. This operator, denoted by Γ, which cannot be obtained from the above mentioned infinitesimal generators, is necessary to distinguish between them the various representations $\mathsf{D}(P,Q)$, with the same P and the same Q, when $\sigma(P,Q) > 1$. We will denote by μ a pair of eigenvalues of the operators G^3 and F^2, by ν the eigenvalues corresponding to the operators T^2, T_3, and Y, and by γ an eigenvalue of the operator Γ. With these notations, the basis of the representation space of $\mathsf{D}(p_1,q_1)$ is written as $\{\Psi_{\nu_1}^{\mu_1}\}$, for the representation $\mathsf{D}(p_2,q_2)$ we have the basis $\{\Psi_{\nu_2}^{\mu_2}\}$, while for the direct product of these two representations the common eigenvectors of the set of operators (3.3.57) and the operator Γ, will be denoted by $\Phi_{\nu}^{\mu_1\mu_2\mu_\gamma}$. The dimension of the representation $\mathsf{D}(p_i,q_i)$ is N_i. In the product space, of dimension N_1N_2, we may choose as basis the vectors $\{\Psi_{\nu_1}^{\mu_1}\,\Psi_{\nu_2}^{\mu_2}\}$, corresponding to the direct product of two linear spaces, or the direct sum of the bases $\{\Phi_{\nu}^{\mu_1\mu_2\mu_\gamma}\}$, corresponding to the decomposition (3.3.55). We may normalize these bases to obtain orthonormal sets of vectors, which are related by a unitary transformation of the form

$$\Phi_{\nu}^{\mu_1\mu_2\mu_\gamma} = \sum_{\nu_1}\sum_{\nu_2} \begin{pmatrix} \mu_1 & \mu_2 & \mu_\gamma \\ \nu_1 & \nu_2 & \nu \end{pmatrix} \Psi_{\nu_1}^{\mu_1}\, \Psi_{\nu_2}^{\mu_2}\,, \tag{3.3.58}$$

where

$$\begin{pmatrix} \mu_1 & \mu_2 & \mu_\gamma \\ \nu_1 & \nu_2 & \nu \end{pmatrix}$$

are the Clebsch-Gordan coefficients of the $\mathsf{SU}(3)$ group. Note that, the vectors $\Psi_{\nu_1}^{\mu_1}\, \Psi_{\nu_2}^{\mu_2}$ are common eigenvectors of the operators (3.3.56), and also of the operators $\mathsf{T}_3 = \mathsf{T}_3(1) + \mathsf{T}_3(2)$ and $\mathsf{Y} = \mathsf{Y}(1) + \mathsf{Y}(2)$, but they are not eigenvectors of the operators G^3, F^2, T^2, and Γ, from (3.3.57). We may use the Clebsch-Gordan coefficients for the $\mathsf{SU}(2)$ group, to construct eigenvectors of the form

$$\Omega^{\mu_1\mu_2}_{T(1)\,Y(1),T(2)\,Y(2),T\,T_3\,Y} = \sum_{T_3(1)}\sum_{T_3(2)} C^{T(1)\,T(2)\,T}_{T_3(1)\,T_3(2)\,T_3}\, \Psi_{\nu_1}^{\mu_1}\, \Psi_{\nu_2}^{\mu_2}\,, \tag{3.3.59}$$

which are common to the operators $\mathsf{G}^3(1)$, $\mathsf{G}^3(2)$, $\mathsf{F}^2(1)$, $\mathsf{F}^2(2)$, T^2, T_3, and Y. In (3.3.59), the indices represent the eigenvalues of the corresponding operators. Then, the vectors Φ are obtained as linear combinations of the vectors Ω, of the form

$$\Phi_{\nu}^{\mu_1\mu_2\mu_\gamma} = \sum_{T(1)}\sum_{Y(1)}\sum_{T(2)}\sum_{Y(2)} \left(\begin{matrix} \mu_1 & \mu_2 \\ T(1)Y(1) & T(2)Y(2) \end{matrix} \middle| \begin{matrix} \mu_\gamma \\ TY \end{matrix} \right) \times \Omega^{\mu_1\mu_2}_{T(1)Y(1),T(2)Y(2),TT_3Y}\,, \tag{3.3.60}$$

where the coefficients of the vectors Ω are called *isoscalar factors*. By comparing the relations (3.3.59) and (3.3.60) with the relation (3.3.58) we are led to a formula for the $\mathsf{SU}(3)$ Clebsch-Gordan coefficients expressed in terms of the $\mathsf{SU}(2)$ Clebsch-Gordan coefficients, defined in (3.3.51), and the isoscalar factors

$$\begin{pmatrix} \mu_1 & \mu_2 & \mu_\gamma \\ \nu_1 & \nu_2 & \nu \end{pmatrix} = C^{T(1)\,T(2)\,T}_{T_3(1)\,T_3(2)\,T_3} \left(\begin{matrix} \mu_1 & \mu_2 \\ T(1)Y(1) & T(2)Y(2) \end{matrix} \middle| \begin{matrix} \mu_\gamma \\ TY \end{matrix} \right). \tag{3.3.61}$$

Thus, the computation of the $\mathsf{SU}(3)$ Clebsch-Gordan coefficients is reduced to the computation of the isoscalar factors. These factors are uniquely determined if we define the phase factors of the basis vectors for the irreducible representations in the series (3.3.55), with respect to the basis vectors for the direct product $\mathsf{D}(p_1, q_1) \otimes \mathsf{D}(p_2, q_2)$. If we impose the condition that these phase factors be real, then the isoscalar factors are also real.

The $\mathsf{SU}(3)$ Clebsch-Gordan coefficients satisfy the orthogonality relations

$$\sum_{\nu_1}\sum_{\nu_2} \begin{pmatrix} \mu_1 & \mu_2 & \mu_\gamma \\ \nu_1 & \nu_2 & \nu \end{pmatrix} \begin{pmatrix} \mu_1 & \mu_2 & \mu'_{\gamma'} \\ \nu_1 & \nu_2 & \nu' \end{pmatrix} = \delta_{\mu\mu'}\,\delta_{\gamma\gamma'}\,\delta_{\nu\nu'}\,,$$

$$\sum_{\mu}\sum_{\gamma}\sum_{\nu}\begin{pmatrix}\mu_1 & \mu_2 & \mu_\gamma \\ \nu_1 & \nu_2 & \nu\end{pmatrix}\begin{pmatrix}\mu_1 & \mu_2 & \mu_\gamma \\ \nu_1' & \nu_2' & \nu\end{pmatrix} = \delta_{\nu_1\nu_1'}\,\delta_{\nu_2\nu_2'}\,. \tag{3.3.62}$$

From these orthogonality relations and the unitarity conditions (3.3.46) we derive the orthogonality relations satisfied by the isoscalar factors

$$\sum_{T(1)}\sum_{Y(1)}\sum_{T(2)}\sum_{Y(2)}\left(\begin{matrix}\mu_1 & \mu_2 \\ T(1)Y(1) & T(2)Y(2)\end{matrix}\middle|\begin{matrix}\mu_\gamma \\ TY\end{matrix}\right)$$

$$\times\left(\begin{matrix}\mu_1 & \mu_2 \\ T(1)Y(1) & T(2)Y(2)\end{matrix}\middle|\begin{matrix}\mu'_{\gamma'} \\ TY'\end{matrix}\right) = \delta_{\mu\mu'}\,\delta_{\gamma\gamma'}\,\delta_{YY'}\,,$$

$$\sum_{\mu}\sum_{\gamma}\sum_{Y}\left(\begin{matrix}\mu_1 & \mu_2 \\ T(1)Y(1) & T(2)Y(2)\end{matrix}\middle|\begin{matrix}\mu_\gamma \\ TY\end{matrix}\right)$$

$$\times\left(\begin{matrix}\mu_1 & \mu_2 \\ T'(1)Y'(1) & T'(2)Y'(2)\end{matrix}\middle|\begin{matrix}\mu_\gamma \\ TY\end{matrix}\right) = \delta_{T(1)T'(1)}\delta_{T(2)T'(2)}\delta_{Y(1)Y'(1)}\delta_{Y(2)Y'(2)}.$$

Also, in the parametrization (2.3.62), the matrix elements (3.3.35) satisfy the relations

$$\int \overline{\mathcal{D}^{\mu_1}_{\nu_1\lambda_1}(\boldsymbol{\alpha})}\,\mathcal{D}^{\mu_2}_{\nu_2\lambda_2}(\boldsymbol{\alpha})\,d\boldsymbol{\alpha} = \frac{1}{N(\mu_1)}\,\delta_{\mu_1\mu_2}\,\delta_{\nu_1\nu_2}\,\delta_{\lambda_1\lambda_2}\,,$$

$$\int \overline{\mathcal{D}^{\mu_3}_{\nu_3\lambda_3}(\boldsymbol{\alpha})}\,\mathcal{D}^{\mu_2}_{\nu_2\lambda_2}(\boldsymbol{\alpha})\,\mathcal{D}^{\mu_1}_{\nu_1\lambda_1}(\boldsymbol{\alpha})\,d\boldsymbol{\alpha} \tag{3.3.63}$$

$$= \frac{1}{N(\mu_3)}\sum_{\gamma}\begin{pmatrix}\mu_1 & \mu_2 & \mu_{3\gamma} \\ \nu_1 & \nu_2 & \nu_3\end{pmatrix}\begin{pmatrix}\mu_1 & \mu_2 & \mu_{3\gamma} \\ \lambda_1 & \lambda_2 & \lambda_3\end{pmatrix},$$

and

$$\mathcal{D}^{\mu_1}_{\nu_1\lambda_1}(\boldsymbol{\alpha})\,\mathcal{D}^{\mu_2}_{\nu_2\lambda_2}(\boldsymbol{\alpha}) = \sum_{\mu,\nu,\lambda,\gamma}\begin{pmatrix}\mu_1 & \mu_2 & \mu_\gamma \\ \nu_1 & \nu_2 & \nu\end{pmatrix}\begin{pmatrix}\mu_1 & \mu_2 & \mu_\gamma \\ \lambda_1 & \lambda_2 & \lambda\end{pmatrix}\mathcal{D}^{\mu}_{\nu\lambda}(\boldsymbol{\alpha})\,,$$

$$\sum_{\nu_1,\nu_2,\lambda_1,\lambda_2}\begin{pmatrix}\mu_1 & \mu_2 & \mu_\gamma \\ \nu_1 & \nu_2 & \nu\end{pmatrix}\begin{pmatrix}\mu_1 & \mu_2 & \mu'_{\gamma'} \\ \lambda_1 & \lambda_2 & \lambda\end{pmatrix}\mathcal{D}^{\mu_1}_{\nu_1\lambda_1}(\boldsymbol{\alpha})\,\mathcal{D}^{\mu_2}_{\nu_2\lambda_2}(\boldsymbol{\alpha}) \tag{3.3.64}$$

$$= \delta_{\mu\mu'}\,\delta_{\gamma\gamma'}\,\mathcal{D}^{\mu}_{\nu\lambda}(\boldsymbol{\alpha})\,.$$

Here, we have used the notation

$$\mathcal{D}^{\mu}_{\nu\lambda}(\boldsymbol{\alpha}) \equiv \mathcal{D}^{pq}_{jmY,\,j'm'Y'}(\boldsymbol{\alpha})\,.$$

By the aid of (3.3.63) and (3.3.64) one can establish the following symmetry properties of the $\mathsf{SU}(3)$ Clebsch-Gordan coefficients:

$$\begin{aligned}
\begin{pmatrix} \mu_1 & \mu_2 & \mu_{3\gamma} \\ \nu_1 & \nu_2 & \nu_3 \end{pmatrix} &= \xi_1 \begin{pmatrix} \mu_2 & \mu_1 & \mu_{3\gamma} \\ \nu_2 & \nu_1 & \nu_3 \end{pmatrix}, \\
\begin{pmatrix} \mu_1 & \mu_2 & \mu_{3\gamma} \\ \nu_1 & \nu_2 & \nu_3 \end{pmatrix} &= \xi_2 (-1)^{T_3(1)+Y(1)/2} \sqrt{\frac{N(\mu_3)}{N(\mu_2)}} \begin{pmatrix} \mu_1 & \overline{\mu}_3 & \overline{\mu}_{2\gamma'} \\ \nu_1 & -\nu_3 & -\nu_2 \end{pmatrix}, \quad (3.3.65) \\
\begin{pmatrix} \mu_1 & \mu_2 & \mu_{3\gamma} \\ \nu_1 & \nu_2 & \nu_3 \end{pmatrix} &= \xi_3 \begin{pmatrix} \overline{\mu}_1 & \overline{\mu}_2 & \overline{\mu}_{3\gamma} \\ -\nu_1 & -\nu_2 & -\nu_3 \end{pmatrix},
\end{aligned}$$

where $\xi_1, \xi_2, \xi_3 = \pm 1$ depend only on μ_1, μ_2, and μ_3, and are chosen such that the Clebsch-Gordan coefficient for the representation of maximum weight in the series (3.3.55) is positive. We have denoted by $\overline{\mu}$ the pair of eigenvalues corresponding to the representation which is the contragredient of the representation associated to the pair of eigenvalues μ.

The irreducible representations contained in the decomposition (3.3.55) as well as the multiplicities $\sigma(P, Q)$ are determined by the *Speiser method*, based on the weight diagrams introduced in Chap. 2 Sec. 3.1.4. Thus, we choose an oblique system of coordinates, the angle between the two axes labelled by p and q being of $\pi/3$ (see Fig. 3.1). In this system of coordinates, each irreducible representation of $\mathsf{SU}(3)$ $\mathsf{D}(p, q)$ is associated to a point of coordinates (p, q). Note that $p, q \in \mathbb{N}$. By successive reflections of Fig. 3.1 with respect to the axes p, q, and Y, respectively, we obtain Fig 3.2. Note that, the axes T_3 and Y in Figs. 3.1 and 3.2 correspond to a system of coordinates determined by the eigenvalues of the operators T_3 and Y. If the unit of length on the axis T_3 is l, then the unit of length on the axes p and q is $l/\sqrt{3}$, while the unit of length along the Y axis is $2l/\sqrt{3}$. To obtain the terms in the decomposition of the direct product $\mathsf{D}(p_1, q_1) \otimes \mathsf{D}(p_2, q_2)$, we must know the weight diagram of one of the two irreducible representations. Note that, the relations between the eigenvalues T_3, Y and the weights m_1, m_2, defined in Chap. 2 Sec. 3.1.4, are

$$T_3 = \sqrt{3}\, m_1 \,, \quad Y = 2\, m_2 \,.$$

Suppose that we know the weight diagram of $\mathsf{D}(p_1, q_1)$. We superpose this weight diagram over Fig 3.2 such that the weight $T_3 = Y = 0$ is placed at the point (p_2, q_2), that corresponds to the irreducible representation $\mathsf{D}(p_2, q_2)$, in the first sextant. Thus, the points in the weight diagram of $\mathsf{D}(p_1, q_1)$ will coincide with the lattice points in Fig 3.2. The non-hatched sextans are considered as positive, while the hatched ones are considered as negative. Finally, we apply the following theorem:

THEOREM 3.3.3 (SPEISER) *Each irreducible representation associated to a weight in the weight diagram of* $\mathsf{D}(p_1, q_1)$*, by means of a lattice point in*

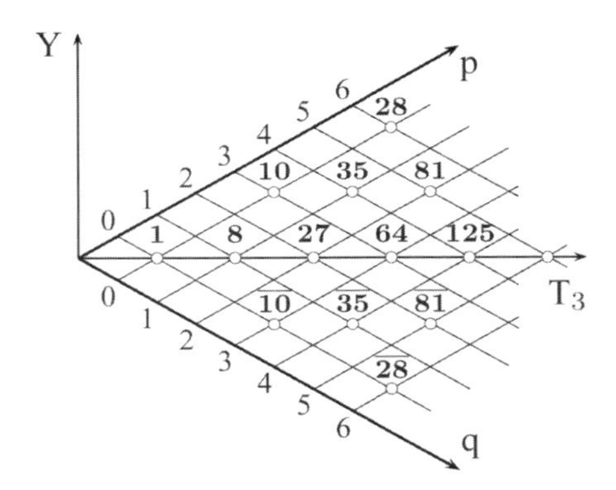

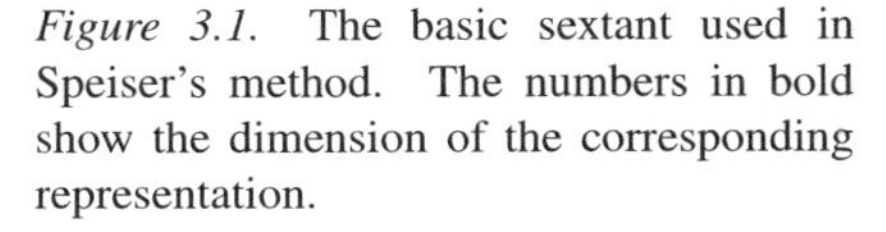
Figure 3.1. The basic sextant used in Speiser's method. The numbers in bold show the dimension of the corresponding representation.

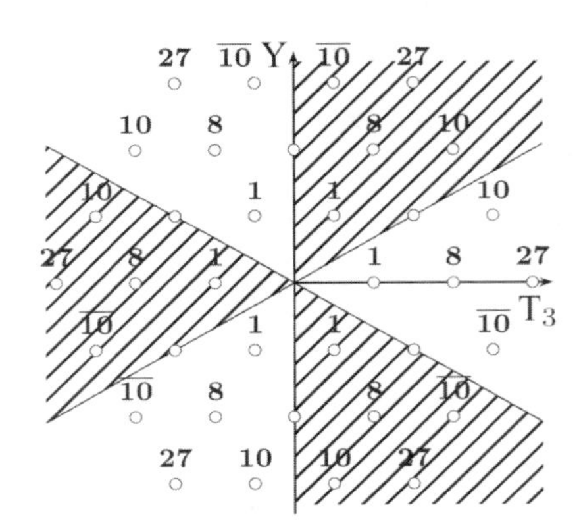

Figure 3.2. The complete diagram for Speiser's method is obtained by successive reflections of the basic sextant with respect to the axes p, q, and Y.

the above described manner, is contained in the decomposition of the direct product $\mathsf{D}(p_1, q_1) \otimes \mathsf{D}(p_2, q_2)$ *into irreducible representations, so many times as the multiplicity of the respective weight.*

The contributions of the weights in the negative sextants are subtracted from the decomposition, while the contributions of the weights on the p and q axes are neglected.

To clarify the meaning of Speiser's theorem let us consider, for example, the decomposition of the direct product $\mathsf{D}(1,1) \otimes \mathsf{D}(3,0)$. Suppose that we know the weight diagram of $\mathsf{D}(1,1)$, given in Fig. 2.7A. We place this weight diagram on the lattice Fig 3.2, with the origin of coordinates at the point $(3,0)$ (see Fig. 3.3) and, according to Speiser's theorem we have

$$8 \otimes 10 = 35 \ominus 10 \oplus 8 \oplus 27 \oplus 2 \times 10 = 35 \oplus 27 \oplus 10 \oplus 8\,,$$

where we have denoted by $N = N(p,q)$ the representation $\mathsf{D}(p,q)$ of dimension N. Note that the contragredient of a representation is denoted by $\overline{N}$. Equally well, we may use the weight diagram of $\mathsf{D}(3,0)$, given in Fig. 2.7C, to obtain the decomposition of the direct product $10 \otimes 8$ (see Fig. 3.4)

$$\begin{aligned} 10 \otimes 8 &= 35 \oplus 10 \ominus 1 \oplus 1 \oplus 8 \oplus 27 \oplus \overline{10} \ominus \overline{10} \\ &= 35 \oplus 27 \oplus 10 \oplus 8\,. \end{aligned}$$

Finally, we may write the decomposition (3.3.55) in the form

$$\mathsf{D}(1,1) \otimes \mathsf{D}(3,0) = \mathsf{D}(3,0) \otimes \mathsf{D}(1,1) = \mathsf{D}(4,1) \oplus \mathsf{D}(2,2) \oplus \mathsf{D}(3,0) \oplus \mathsf{D}(1,1).$$

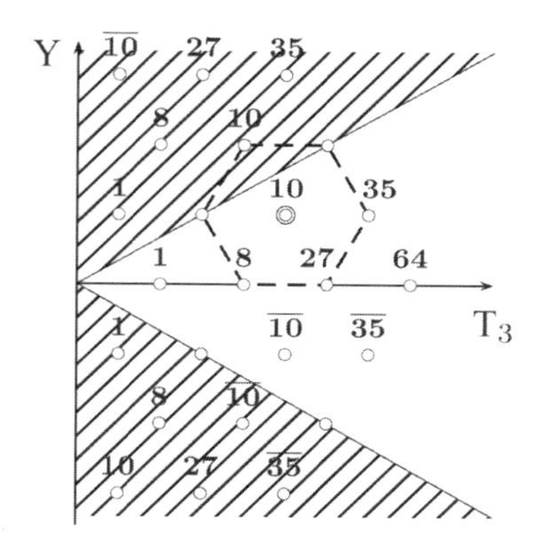

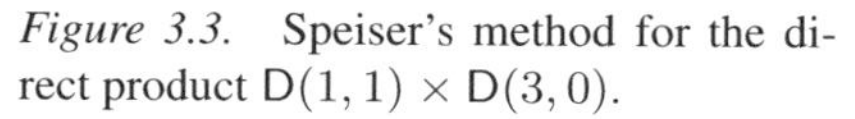
Figure 3.3. Speiser's method for the direct product $\mathsf{D}(1,1) \times \mathsf{D}(3,0)$.

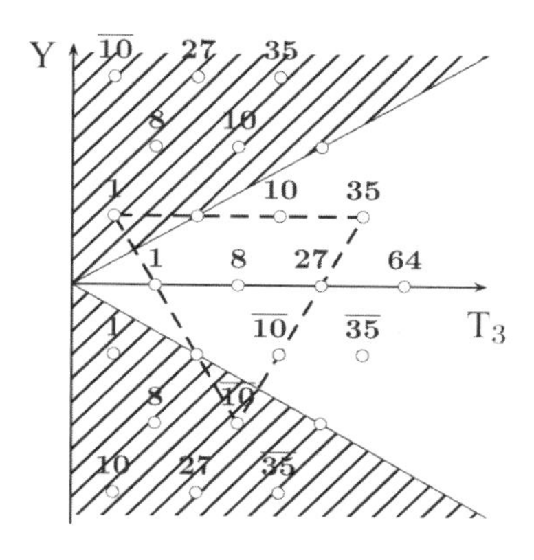

Figure 3.4. Speiser's method for the direct product $\mathsf{D}(3,0) \times \mathsf{D}(1,1)$.

Note that, the direct product of representations is commutative, that is

$$\mathsf{D}(p_1, q_1) \otimes \mathsf{D}(p_2, q_2) = \mathsf{D}(p_2, q_2) \otimes \mathsf{D}(p_1, q_1),$$

the use of Speiser's theorem in practical applications being thus simplified.

3.3.4 Class parameters of $\mathsf{U}(n)$ and $\mathsf{SU}(n)$

We choose the parameters

$$a = \cos\varphi\, e^{i\alpha}, \quad b = -\sin\varphi\, e^{-i\beta},$$

in the Cayley-Klein parametrization of $\mathsf{SU}(2)$, detailed in Chap. 2 Sec. 2.1.2, and write the matrix (2.2.3) in the form

$$\mathbf{u}_1 = \begin{bmatrix} c\,e^{i\alpha} & -s\,e^{-i\beta} \\ s\,e^{i\beta} & c\,e^{-i\alpha} \end{bmatrix}, \tag{3.3.66}$$

where $c = \cos\varphi$ and $s = \sin\varphi$. The angle φ is called *longitude* angle, while α and β are called *latitude* angles. We now introduce the diagonal matrix

$$\mathbf{D}(\alpha, -\alpha) = \begin{bmatrix} e^{i\alpha} & 0 \\ 0 & e^{-i\alpha} \end{bmatrix},$$

and the unimodular unitary matrix

$$\mathbf{u}(\varphi, \sigma) = \begin{bmatrix} c & -s\,e^{-i\sigma} \\ s\,e^{i\sigma} & c \end{bmatrix}, \quad \sigma = \alpha + \beta, \tag{3.3.67}$$

such that

$$\mathbf{u}_1 = \mathbf{D}(\alpha, -\alpha)\, \mathbf{u}(\varphi, \sigma)\,.$$

The angle σ can be regarded as a latitude angle, a change in the sign of φ being equivalent to the replacement of σ by $\sigma + \pi$ in $\mathbf{u}(\varphi, \sigma)$. The matrix associated to an element of the $\mathsf{U}(2)$ group is obtained by multiplying the matrix elements of $\mathbf{u}_1$ by $\exp(i\delta/2)$, such that $\det \mathbf{u}_1 = \det \mathbf{u} = \exp i\delta$. Thus, it becomes evident that the parametrization of the $\mathsf{U}(2)$ group can be considered of the form $\mathbf{D}(\delta_1, \delta_2)\, \mathbf{u}(\varphi, \sigma)$, where we may choose one of the two angles δ_1, δ_2 as a latitude angle. Consequently, we will assume that the matrix associated to an element of the $\mathsf{U}(2)$ group is $\mathbf{D}(\delta_1, \delta_2)\, \mathbf{u}(\varphi, \sigma)$, where φ and δ_2 are longitude angles ($\varphi \in [-\pi, \pi]$, $\delta_2 \in [-\pi, \pi]$), while σ and δ_1 are latitude angles ($\sigma \in [-\pi/2, \pi/2]$, $\delta_1 \in [-\pi/2, \pi/2]$). Note that the parameter domain is closed and bounded, the closure of this domain being obtained by identifying the ends of the interval $[-\pi, \pi]$.

For the parametrization of the $\mathsf{U}(n)$ group ($n > 2$), we denote by p and q, $q > p$, two natural numbers, and by $\mathbf{u}_{pq}(\varphi, \sigma)$ a unimodular unitary matrix of order n, which satisfies the conditions:

1 all the diagonal elements have the value 1, excepting the elements of indices pp and qq, which are equal to $\cos\varphi$;

2 all the non-diagonal elements are zero, with the exception of the element in column p and row q, which is equal to $s\,\exp(i\sigma)$, and the element in column q and row p, which is equal to $-s\,\exp(-i\sigma)$, where $s = \sin\varphi$.

In terms of these matrices, the matrix associated to an element of the $\mathsf{U}(n)$ group can be written in the form

$$\mathbf{u} = \mathbf{u}^{(n-1)}\mathbf{u}_{12}(\theta_{n-2}, \sigma_{n-1})\mathbf{u}_{13}(\theta_{n-3}, \sigma_{n-2}) \ldots \mathbf{u}_{1,n-1}(\theta_1, \sigma_2)\mathbf{u}_{1n}(\varphi_1, \sigma_1), \tag{3.3.68}$$

where

$$\mathbf{u}^{(n-1)} = \begin{bmatrix} e^{i\delta_1} & \mathbf{0} \\ \mathbf{0} & \mathbf{v} \end{bmatrix},$$

and $\mathbf{v}$ is a unitary matrix of order $n-1$. The matrices associated to the elements of $\mathsf{SU}(n)$ are obtained from (3.3.68) by imposing the condition $\det \mathbf{u} = 1$.

For instance, in the case of the $\mathsf{U}(3)$ group the matrix $\mathbf{v}$ has the form

$$\mathbf{v} = \mathbf{D}(\delta_2, \varphi_3) \begin{bmatrix} c_2 & -s_2\, e^{-i\sigma_3} \\ s_2\, e^{i\sigma_3} & c_2 \end{bmatrix}, \qquad c_2 = \cos\varphi_2\,, \qquad s_2 = \sin\varphi_2\,,$$

such that $\mathbf{u}^{(2)} = \mathbf{D}(\delta_1, \delta_2, \varphi_3)\mathbf{u}_{23}(\varphi_2, \sigma_3)$, and the matrix associated to an element $\mathsf{U} \in \mathsf{U}(3)$ is

$$\mathbf{u} = \mathbf{D}(\delta_1, \delta_2, \varphi_3)\, \mathbf{u}_{23}(\varphi_2, \sigma_3)\, \mathbf{u}_{12}(\theta_1, \sigma_2)\, \mathbf{u}_{13}(\varphi_1, \sigma_1)\,,$$

where

$$\mathbf{u}_{23}(\varphi,\sigma)=\begin{bmatrix}1 & 0 & 0\\ 0 & c & -s\,e^{-i\sigma}\\ 0 & s\,e^{i\sigma} & c\end{bmatrix},$$

$$\mathbf{u}_{13}(\varphi,\sigma)=\begin{bmatrix}c & 0 & -s\,e^{-i\sigma}\\ 0 & 1 & 0\\ s\,e^{i\sigma} & 0 & c\end{bmatrix},\quad \mathbf{u}_{12}(\varphi,\sigma)=\begin{bmatrix}c & -s\,e^{-i\sigma} & 0\\ s\,e^{i\sigma} & c & 0\\ 0 & 0 & 1\end{bmatrix}.$$

Thus, the elements of the $\mathsf{U}(3)$ group are determined by nine parameters, the parameter domain being bounded and closed. From these nine parameters, the angles φ_1, φ_2, and φ_3 are longitude angles varying in the interval $[-\pi,\pi]$, while the other six angles (σ_1, σ_2, σ_3, δ_1, δ_2, and θ_1) are latitude angles varying in the interval $[-\pi/2,\pi/2]$.

From the factorization (3.3.68) of an arbitrary element of $\mathsf{U}(n)$, one can deduce the matrix corresponding to an element of $\mathsf{O}(n)$ or $\mathsf{SO}(n)$. To do this, we have to choose all the angles σ and δ equal to zero, and $\varphi_n=0$, in the case of a proper rotation, or $\varphi_n=-\pi$, in the case of an inversion. For instance, the matrices of $\mathsf{SO}(3)$ are factorized in the form $\mathbf{R}_{23}(\varphi_2)\mathbf{R}_{12}(\theta)\mathbf{R}_{13}(\varphi_1)$, which is different from the Eulerian parametrization (2.1.26) expressed by the factorization $\mathbf{R}_{12}(\psi)\mathbf{R}_{23}^{\dagger}(\theta)\mathbf{R}_{12}(\varphi)$. In fact, the repetition of the plane x_1x_2 in the Eulerian parametrization precludes the generalization of this parametrization to n-dimensional rotation matrices for $n>3$.

Let us consider a matrix representation of $\mathsf{U}(n)$ in the n-dimensional complex space $\mathcal{C}^n$. From Theorem 2.1.1 it results that any matrix $\mathbf{u}$ in the representation can be reduced to the diagonal form changing the basis of the representation space $\mathcal{C}^n$ by means of a unitary matrix $\mathbf{v}$. Being unitary, the matrix $\mathbf{v}$ necessarily belongs to this matrix representation of $\mathsf{U}(n)$, and thus corresponds to an element $\mathsf{V}\in\mathsf{U}(n)$. The set of all matrices of the form $\mathbf{v}^{\dagger}\mathbf{u}\mathbf{v}$, for V going over $\mathsf{U}(n)$, constitutes a class of $\mathsf{U}(n)$, and one can state

THEOREM 3.3.4 *Any class of the* $\mathsf{U}(n)$ *group contains diagonal matrices.*

These diagonal matrices have the form $\mathbf{D}(\alpha_1,\alpha_2,\ldots,\alpha_n)$, where

$$e^{i\alpha_1}=z_1\,,\ldots,e^{i\alpha_n}=z_n$$

are characteristic numbers (eigenvalues). The order of characteristic numbers is not relevant because all the elements in a class have the same characteristic numbers.

Any unitary n-dimensional matrix can be written in the form

$$\mathbf{u}=\mathbf{D}(\delta_1,\delta_2,\ldots,\delta_n)\mathbf{u}_1\,,\quad \mathbf{u}_1\to\mathsf{U}_1\in\mathsf{SU}(n)\,,$$

and can be factorized into a product of unimodular unitary matrices of dimension n, $\mathbf{u}_{pq}(\theta,\sigma)$ or $\mathbf{u}_{pq}(\varphi,\sigma)$, with $p<q$. Let us apply this result to the matrix associated to the element $\mathsf{V}^\dagger \in \mathsf{U}(n)$, in the relation $\mathbf{v}^\dagger\mathbf{u}\mathbf{v} = \mathbf{D}(\boldsymbol{\alpha})$, where $\boldsymbol{\alpha} = (\alpha_1,\ldots,\alpha_n)$ and $\mathsf{U} \in \mathsf{U}(n)$. It follows that

$$\mathbf{D}(\boldsymbol{\delta})\,\mathbf{v}_1\,\mathbf{u}_1\,\mathbf{v}_1^\dagger\,\mathbf{D}(\boldsymbol{\delta})^\dagger = \mathbf{D}(\boldsymbol{\alpha})$$

or, equivalently,

$$\mathbf{v}_1\,\mathbf{u}_1\,\mathbf{v}_1^\dagger = \mathbf{D}(\boldsymbol{\delta})^\dagger\,\mathbf{D}(\boldsymbol{\alpha})\,\mathbf{D}(\boldsymbol{\delta}) = \mathbf{D}(\boldsymbol{\alpha})\,,$$

because all the diagonal matrices commute. Hence,

$$\mathbf{u} = \mathbf{v}_1^\dagger\,\mathbf{D}(\boldsymbol{\alpha})\,\mathbf{v}_1\,,$$

and we may consider as parameters of the $\mathsf{U}(n)$ group:

1 n longitude angles α; when these angles are known, independently of the order in which they are arranged, we know the class of $\mathsf{U}(n)$, to which the corresponding element belongs. Because the order of the angles α is immaterial, any class function defined on $\mathsf{U}(n)$ is a symmetric function of these angles, or, equivalently of the characteristic numbers $z_1,\ldots,z_n$. The parameters $\alpha_1,\ldots,\alpha_n$ are called class parameters of $\mathsf{U}(n)$.

2 $n(n-1)$ parameters in $\mathbf{v}_1$, $n-1$ being longitude angles, while the remaining $(n-1)^2$ parameters are latitude angles. Once the values of the parameters α are known, these $n(n-1)$ parameters localise the element of the group inside the class to which it belongs.

Note that, in the case of the $\mathsf{SU}(n)$ group, the condition $\det\mathbf{D}(\boldsymbol{\alpha}) = 1$ ($\det\mathbf{u} = \det\mathbf{v}_1^\dagger\det\mathbf{D}(\boldsymbol{\alpha})\det\mathbf{v} = \det\mathbf{D}(\boldsymbol{\alpha})$) leads us to the relation $\alpha_1 + \alpha_2 + \ldots + \alpha_n = 0$, or $z_1z_2\ldots z_n = 1$, and thus reducing the number of class parameters to $n-1$.

In terms of these parameters, the volume element for the invariant integral on the $\mathsf{U}(2)$ group has the expression

$$\frac{1}{8\pi^3}\,\sin 2\varphi\,d\varphi\,d\sigma\,d\delta_1\,d\delta_2 \tag{3.3.69}$$

and, by integrating over the angle $\delta_2 \in [-\pi,\pi]$ we obtain the volume element for the invariant integral over the $\mathsf{SU}(2)$ group

$$\frac{1}{4\pi^2}\,\sin 2\varphi\,d\varphi\,d\sigma\,d\delta_1\,. \tag{3.3.70}$$

It can be shown that, in this parametrization, the volume element for the $\mathsf{SO}(3)$ group, defined in (2.1.41) for the Eulerian parametrization, becomes

$$\cos\theta\,d\varphi_1\,d\theta\,d\varphi_2\,,\quad \theta\in[-\pi/2,\pi/2]\,, \tag{3.3.71}$$

or, separating the class parameter

$$\frac{1}{2\pi^2}\sin^2\frac{\alpha}{2}\cos\theta\,d\alpha\,d\theta\,d\varphi\,,\quad \alpha\in[0,\pi]\,,\quad \theta\in\left[-\frac{\pi}{2},\frac{\pi}{2}\right]\,,\quad \varphi\in[-\pi,\pi]\,.$$

Generally, the class dependent part in the volume element for the invariant integral over the $\mathsf{U}(n)$ group has the form

$$dV=\frac{\overline{\Delta(z)}\,\Delta(z)}{(2\pi)^n\,n!}\,d\boldsymbol{\alpha}\,, \tag{3.3.72}$$

where

$$\Delta(z)=\prod_{k<j}(z_k-z_j)\,,\quad z_k=e^{i\alpha_k}\,,\quad \overline{z}_k=1/z_k\,,\quad \alpha_k\in[-\pi,\pi]\,,$$

for $k=1,2,\ldots,n$.

Because the relation

$$\mathsf{Tr}\,\mathbf{u}=\mathsf{Tr}\left[\mathbf{v}_1^\dagger\,\mathbf{D}(\boldsymbol{\alpha})\,\mathbf{v}_1\right]=\mathsf{Tr}\,\mathbf{D}(\boldsymbol{\alpha})$$

takes place, the character of the fundamental representation of the $\mathsf{U}(n)$ group can be expressed as a function of the class parameters

$$\chi(\boldsymbol{\alpha})=\sum_{k=1}^{n}e^{i\alpha_k}=\sum_{k=1}^{n}z_k\,. \tag{3.3.73}$$

Thus, the relations (3.3.72) and (3.3.73) enable us to calculate the multiplicities (3.3.39). Note that, in the case of the $\mathsf{SU}(n)$ group we have to use the relation $\alpha_1+\alpha_2+\ldots+\alpha_n=0$.

3.3.5 Projection operators

Let T be a unitary representation of a compact group G, in a Hilbert space $\mathcal{H}$, and $\chi(g)$ the character of the representation T. We denote by $\{\mathbf{e}_i\}_{i=1}^n$, $n=\dim\mathsf{T}$, the basis of the space $\mathcal{H}$, and by $t_{ij}(g)$ the matrix elements of the representation operators $\mathsf{T}(g)$ in this basis.

Let us consider the linear operators, continuous in $\mathcal{H}$,

$$\mathsf{P}_{ij}^{\mathsf{T}}=\frac{n}{V}\int\overline{t_{ij}(g)}\,\mathsf{T}(g)\,dg\,,\quad \mathsf{P}^{\mathsf{T}}=\frac{n}{V}\int\overline{\chi(g)}\,\mathsf{T}(g)\,dg\,, \tag{3.3.74}$$

where $V=\int dg$. These operators are called *projection operators*, and they verify the relations

$$\mathsf{P}^{\mathsf{T}}=\sum_{i=1}^{n}\mathsf{P}_{ii}^{\mathsf{T}}\,,$$

$$\mathsf{T}(g)\,\mathsf{P}^{\mathsf{T}}_{jl} = \sum_{i=1}^{n} t_{ij}(g)\,\mathsf{P}^{\mathsf{T}}_{il}\,, \quad \forall g \in G\,, \tag{3.3.75}$$

$$\mathsf{P}^{\mathsf{T}}_{jl}\,\mathsf{T}(g) = \sum_{i=1}^{n} t_{li}(g)\,\mathsf{P}^{\mathsf{T}}_{ji}\,, \quad \forall g \in G\,.$$

If T' is a unitary irreducible representation of G, nonequivalent to T, then

$$\overline{\mathsf{P}^{\mathsf{T}}_{ij}}\,\mathsf{P}^{\mathsf{T}'}_{lk} = 0\,, \quad \mathsf{P}^{\mathsf{T}}\,\mathsf{P}^{\mathsf{T}'} = 0\,. \tag{3.3.76}$$

We also have the relations

$$\mathsf{P}^{\mathsf{T}}_{ij}\,\mathsf{P}^{\mathsf{T}}_{lk} = \begin{cases} 0\,, & j \neq l\,, \\ \mathsf{P}^{\mathsf{T}}_{ik}\,, & j = l\,, \end{cases} \qquad \mathsf{P}^{\mathsf{T}}_{ii}\,\mathsf{P}^{\mathsf{T}}_{jj} = \begin{cases} 0\,, & i \neq j\,, \\ \mathsf{P}^{\mathsf{T}}_{ii}\,, & i = j\,, \end{cases} \tag{3.3.77}$$

$$\overline{\mathsf{P}^{\mathsf{T}}_{ij}} = \mathsf{P}^{\mathsf{T}}_{ji}\,, \quad \mathsf{P}^{\mathsf{T}}\,\mathsf{T}(g) = \mathsf{T}(g)\,\mathsf{P}^{\mathsf{T}}\,, \quad \forall g \in G\,. \tag{3.3.78}$$

One can easily show that the relations (3.3.75) and (3.3.77) imply $\mathsf{P}^{\mathsf{T}}\mathsf{P}^{\mathsf{T}} = \mathsf{P}^{\mathsf{T}}$, such that the projection operators P^{T} are idempotent

$$\left(\mathsf{P}^{\mathsf{T}}\right)^{n} = \mathsf{P}^{\mathsf{T}}\,.$$

Let $\mathcal{M}^{\mathsf{T}}_{i}$ and $\mathcal{M}^{\mathsf{T}}$ be two sets defined in the form

$$\begin{aligned} \mathcal{M}^{\mathsf{T}}_{i} &= \{\mathbf{x} |\, \mathbf{x} \in \mathcal{H}\,, \quad \mathsf{P}^{\mathsf{T}}_{ii}\mathbf{x} = \mathbf{x}\}\,, \\ \mathcal{M}^{\mathsf{T}} &= \{\mathbf{x} |\, \mathbf{x} \in \mathcal{H}\,, \quad \mathsf{P}^{\mathsf{T}}\mathbf{x} = \mathbf{x}\}\,. \end{aligned} \tag{3.3.79}$$

These two sets are closed linear subspaces of the Hilbert space $\mathcal{H}$. If T and T' are two nonequivalent representations of G, then the subspaces $\mathcal{M}^{\mathsf{T}'}_{k}$ and $\mathcal{M}^{\mathsf{T}}_{i}$ are orthogonal, and the subspaces $\mathcal{M}^{\mathsf{T}'}$ and $\mathcal{M}^{\mathsf{T}}$ are also orthogonal. Now, for $i \neq k$, the subspaces $\mathcal{M}^{\mathsf{T}}_{i}$ and $\mathcal{M}^{\mathsf{T}}_{k}$ are orthogonal. The space $\mathcal{M}^{\mathsf{T}}$ is the orthogonal direct sum of the subspaces $\mathcal{M}^{\mathsf{T}}_{i}$, $i = 1, 2, \ldots, n$, because for $\mathbf{x}_j \in \mathcal{M}^{\mathsf{T}}_{j}$,

$$\mathsf{P}^{\mathsf{T}}\mathbf{x}_j = \mathsf{P}^{\mathsf{T}}\mathsf{P}^{\mathsf{T}}_{jj}\mathbf{x}_j = \left(\sum_{i=1}^{n} \mathsf{P}^{\mathsf{T}}_{ii}\right)\mathsf{P}^{\mathsf{T}}_{jj}\mathbf{x}_j = \sum_{i=1}^{n}\left(\mathsf{P}^{\mathsf{T}}_{ii}\mathsf{P}^{\mathsf{T}}_{jj}\right)\mathbf{x}_j = \mathsf{P}^{\mathsf{T}}_{jj}\mathbf{x}_j = \mathbf{x}_j\,,$$

that is $\mathbf{x}_j \in \mathcal{M}^{\mathsf{T}}$, and $\mathcal{M}^{\mathsf{T}}$ contains all the sums of the form $\mathbf{x} = \mathbf{x}_1 + \mathbf{x}_2 + \ldots + \mathbf{x}_n$. Reciprocally, if $\mathbf{x} \in \mathcal{M}^{\mathsf{T}}$ then, from $\mathbf{x}_i = \mathsf{P}^{\mathsf{T}}_{ii}\mathbf{x}$, it follows that

$$\mathbf{x} = \mathsf{P}^{\mathsf{T}}\mathbf{x} = \left(\sum_{i=1}^{n} \mathsf{P}^{\mathsf{T}}_{ii}\right)\mathbf{x} = \sum_{i=1}^{n} \mathbf{x}_i\,,$$

because

$$\mathsf{P}^{\mathsf{T}}_{ii}\mathbf{x}_i = \left(\mathsf{P}^{\mathsf{T}}_{ii}\right)^2 \mathbf{x} = \mathsf{P}^{\mathsf{T}}_{ii}\mathbf{x} = \mathbf{x}_i \,.$$

The Hilbert space $\mathcal{H}$ is the orthogonal direct sum of the subspaces $\mathcal{M}^{\mathsf{T}}$,

$$\mathcal{H} = \sum_{\mathsf{T}}^{\oplus} \mathcal{M}^{\mathsf{T}} \,,$$

where T goes over the set of all irreducible representations of G, defined in Chap. 3 Sec. 3.1.1.

A representation S of G in the Hilbert space $\mathcal{H}$ is called a *multiple of an irreducible representation* T of G, defined in the same Hilbert space $\mathcal{H}$, if the space $\mathcal{H}$ is the orthogonal direct sum of closed linear subspaces $\mathcal{H}_i$, $i = 1, 2, \ldots, n = \dim \mathsf{T}$, having the property that for each $\mathcal{H}_i$ and for each $\mathbf{x}_i \in \mathcal{H}_i$ there exist the vectors $\mathbf{x}_j \in \mathcal{H}_j$, $j = 1, 2, \ldots, n$, such that

$$\mathsf{S}(g)\,\mathbf{x}_j = \sum_{k=1}^{n} t_{kj}(g)\,\mathbf{x}_k \,, \quad \forall g \in G \,. \tag{3.3.80}$$

THEOREM 3.3.5 *Let* $\{\mathbf{x}_1, \mathbf{x}_2, \ldots, \mathbf{x}_n\}$ *be a set of vectors in* $\mathcal{H}$*. Then, the relation* (3.3.80) *is satisfied if and only if*

$$\mathsf{P}^{\mathsf{T}}_{kj}\,\mathbf{x}_j = \mathbf{x}_k \,, \quad j, k = 1, 2, \ldots, n = \dim \mathsf{T} \,. \tag{3.3.81}$$

In particular, if S is a multiple of the representation T, then $\mathcal{H}_i = \mathcal{M}^{\mathsf{T}}_i$. If a representation S of G in the space $\mathcal{H}$ is a multiple of the irreducible representation T of G, then the space $\mathcal{H}$ is the orthogonal direct sum of a family of subspaces $\mathcal{M}_k$, such that each subspace $\mathcal{M}_k$ is invariant relative to the representation S, while the restriction of S to each of the subspaces $\mathcal{M}_k$ is equivalent to the representation T.

THEOREM 3.3.6 *Let* S *be a continuous unitary representation of a group* G *in a Hilbert space* $\mathcal{H}$*. Each of the subspaces* $\mathcal{M}^{\mathsf{T}}$ *is invariant with respect to all the operators* $\mathsf{S}(g) \in \mathsf{S}$, $g \in G$*, and if* $\mathcal{M}^{\mathsf{T}} \neq \emptyset$*, then the restriction of* S *to each subspace* $\mathcal{M}^{\mathsf{T}}$ *is a unitary representation of* G *in* $\mathcal{M}^{\mathsf{T}}$*, multiple of an irreducible representation* T.

Hence, each continuous unitary irreducible representation of a compact group G in a Hilbert space is the direct sum of the unitary representations of G, multiple of the irreducible unitary representations, of finite dimension, of G. Consequently, the operator P^{T} projects the whole space $\mathcal{H}$ onto the subspace $\mathcal{M}^{\mathsf{T}}$ in which a restriction of the representation S, multiple of the irreducible representation T, is realized.

4. Methods of study of certain differential equations

4.1 Ordinary differential equations

4.1.1 Reduction of systems of ordinary differential equations

The earliest studies of continuous transformation groups are due to Sophus Lie, who introduced these groups (known as *Lie groups*) to develop methods for solving ordinary differential equations. One of Lie's results is that the order of an ordinary differential equation can be reduced by one if the equation is invariant with respect to a one-parameter Lie group. Actually, an ordinary differential equation of order n is equivalent to a system of n ordinary differential equations of the first order, such that Lie's result also holds for systems of ordinary differential equations of the first order, in the sense that the number of equations can be reduced by one, if the system of equations is invariant with respect to a one-parameter Lie group.

Let us consider the system of ordinary differential equations of the first order

$$\frac{dx_i}{dt} = f_i(x_1, x_2, \ldots, x_n)\,, \quad i = 1, 2, \ldots, n, \tag{3.4.1}$$

where t is a real variable, and the functions $f_1, f_2, \ldots, f_n$ are of class C^1 in a vicinity U of a point $\mathbf{x}^0 = (x_1^0, x_2^0, \ldots, x_n^0) \in \mathcal{E}_n$. Let G be a Lie group of transformations

$$\varphi : V \to U\,, \quad V \subseteq U \subseteq \mathcal{E}_n\,,$$

which depend on only one real parameter s, and are defined by the relation

$$\mathbf{x} = \varphi(\widetilde{\mathbf{x}}, s)\,, \quad \widetilde{\mathbf{x}} \in V\,, \quad \mathbf{x} \in U\,, \quad s \in \mathbb{R}\,, \tag{3.4.2}$$

where $\widetilde{\mathbf{x}} = (\widetilde{x}_1, \widetilde{x}_2, \ldots, \widetilde{x}_n)$, $\mathbf{x} = (x_1, x_2, \ldots, x_n)$, and the functions $\varphi = \{\varphi_1, \varphi_2, \ldots, \varphi_n\}$ are of class C^2 with respect to all the variables $\widetilde{x}_1, \widetilde{x}_2, \ldots, \widetilde{x}_n$, and s. By a transformation of the form (3.4.2), the system of differential equations (3.4.1) becomes

$$\frac{d\widetilde{x}_i}{dt} = \widetilde{f}_i(\widetilde{\mathbf{x}}, s)\,, \quad \widetilde{\mathbf{x}} \in V\,, \quad i = 1, 2, \ldots, n, \tag{3.4.3}$$

where

$$\widetilde{f}_i(\widetilde{\mathbf{x}}, s) = \sum_{k=1}^{n} \left(\mathbf{A}^{-1}\right)_{ik} f_k[\varphi(\widetilde{\mathbf{x}}, s)]\,, \quad \mathbf{A} = [A_{ik}] = \left[\frac{\partial \varphi_i(\widetilde{\mathbf{x}}, s)}{\partial \widetilde{x}_k}\right]. \tag{3.4.4}$$

DEFINITION 3.4.1 *If $\widetilde{f}_k(\widetilde{\mathbf{x}}, s) = f_k(\widetilde{\mathbf{x}})$, $k = 1, 2, \ldots, n$, $\forall \varphi \in G$, then $d\widetilde{x}_k/dt = f_k(\widetilde{\mathbf{x}})$, and the system of differential equations is called* **invariant with respect to the group** G.

The system of differential equations (3.4.1), where the vectorial function $\mathbf{f} : U \to \mathcal{E}_n$ is invariant with respect to a one-parameter Lie group of transformations, can be reduced to a system of at the most $n-1$ differential equations. Indeed, in a vicinity of the point $\mathbf{x}^0 \in U$, we may introduce a new system of coordinates $\mathbf{y} \equiv (y_1, y_2, \ldots, y_n)$, such that the system (3.4.1) becomes

$$\frac{dy_k}{dt} = g_k(y_1, y_2, \ldots, y_{n-1}), \quad k = 1, 2, \ldots, n-1, \tag{3.4.5}$$

where the functions g_k are independent of y_n. This last unknown is expressed in the form

$$y_n(t) = y_n(0) + \int_0^t g_n[y_1(\tau), y_2(\tau), \ldots, y_{n-1}(\tau)]\, d\tau\,. \tag{3.4.6}$$

Hence, the integration of the initial system of n differential equations is reduced to the integration of the system (3.4.5), which has at most $n-1$ differential equations.

4.2 The linear equivalence method

The *linear equivalence method*, introduced by I. Toma, enables to study – both numerically and qualitatively – the solutions of a system of nonlinear ordinary differential equations. This method associates to a given system of nonlinear differential equations two linear equivalents: a linear partial differential equation and an infinite system of linear differential equations, thus allowing the application of the methods specific to the linear case. While the method can be applied to larger classes of nonlinear differential operators, we shall present it here only for systems of ordinary differential equations of the type (3.4.1).

4.2.1 Outline of the linear equivalence method

We assume that $\mathbf{x} = (x_1, x_2, \ldots, x_n)$ and $\mathbf{f}(\mathbf{x}) = (f_1(\mathbf{x}), f_2(\mathbf{x}), \ldots, f_n(\mathbf{x}))$ are expressed as column vectors, such that the autonomous system of ordinary differential equations (3.4.1) can be written in matrix form as

$$\frac{d\mathbf{x}}{dt} = \mathbf{f}(\mathbf{x}), \quad \mathbf{x} \in \mathbb{R}^n\,. \tag{3.4.7}$$

We also assume that the functions $f_1, f_2, \ldots, f_n$ allow series expansions of the form

$$f_j(\mathbf{x}) = \sum_{|\boldsymbol{\mu}| \geq 1} f_{j\boldsymbol{\mu}}\, \mathbf{x}^{\boldsymbol{\mu}}, \quad \mathbf{x}^{\boldsymbol{\mu}} \equiv x_1^{\mu_1} x_2^{\mu_2} \ldots x_n^{\mu_n}\,, \tag{3.4.8}$$

where $\boldsymbol{\mu} = \{\mu_1, \mu_2, \ldots, \mu_n\}$ is an n-dimensional index, and $|\boldsymbol{\mu}| = \sum \mu_k$.

The main idea of the linear equivalence method is to make use of the exponential mapping

$$v(t, \boldsymbol{\xi}) = e^{\langle \mathbf{x}, \boldsymbol{\xi} \rangle}, \quad \langle \mathbf{x}, \boldsymbol{\xi} \rangle = \sum_{j=1}^{n} x_j \xi_j, \quad \boldsymbol{\xi} \in \mathbb{R}^n, \tag{3.4.9}$$

depending on n parameters $\xi_1, \xi_2, \ldots, \xi_n$, to transform a nonlinear system of ordinary differential equations into a linear partial differential equation, that is of first order with respect to the independent variable t. In our case, the system of ordinary differential equations (3.4.7) is transformed into

$$\frac{\partial v}{\partial t} - \langle \boldsymbol{\xi}, \mathbf{f}(\mathbf{D}_{\boldsymbol{\xi}}) \rangle \, v = 0, \tag{3.4.10}$$

where the differential operator

$$\langle \boldsymbol{\xi}, \mathbf{f}(\mathbf{D}_{\boldsymbol{\xi}}) \rangle = \sum_{j=1}^{n} \xi_j f_j(\mathbf{D}_{\boldsymbol{\xi}}) \tag{3.4.11}$$

is formally obtained by substituting

$$\mathbf{x}^{\boldsymbol{\mu}} \to \frac{\partial^{|\boldsymbol{\mu}|}}{\partial \boldsymbol{\xi}^{\boldsymbol{\mu}}} \equiv \frac{\partial^{\mu_1}}{\partial \xi_1^{\mu_1}} \frac{\partial^{\mu_2}}{\partial \xi_2^{\mu_2}} \cdots \frac{\partial^{\mu_n}}{\partial \xi_n^{\mu_n}},$$

into (3.4.8). Indeed, from (3.4.9) we have

$$\frac{\partial v}{\partial t} = \left(\sum_{j=1}^{n} \xi_j \frac{dx_j}{dt} \right) v = \sum_{j=1}^{n} \xi_j \left(v \frac{dx_j}{dt} \right)$$

and, by means of (3.4.7) and (3.4.8) we calculate

$$v \frac{dx_j}{dt} = v\, f_j(\mathbf{x}) = v \sum_{|\boldsymbol{\mu}| \geq 1} f_{j\boldsymbol{\mu}} \mathbf{x}^{\boldsymbol{\mu}} = \sum_{|\boldsymbol{\mu}| \geq 1} f_{j\boldsymbol{\mu}} \frac{\partial^{|\boldsymbol{\mu}|} v}{\partial \boldsymbol{\xi}^{\boldsymbol{\mu}}} = f_j(\mathbf{D}_{\boldsymbol{\xi}})\, v,$$

which demonstrates (3.4.10).

It can be shown that, the linear partial differential equation (3.4.10) is consistent on *exponential type functional spaces*. If one associates to (3.4.7) some Cauchy conditions

$$\mathbf{x}(t_0) = \mathbf{x}_0, \tag{3.4.12}$$

then the corresponding conditions for v are

$$v(t_0, \boldsymbol{\xi}) = e^{\langle \mathbf{x}_0, \boldsymbol{\xi} \rangle}. \tag{3.4.13}$$

It has been proved by I. Toma that the exponential mapping (3.4.9) establishes a one-to-one correspondence between the solutions of the nonlinear Cauchy problem (3.4.7), (3.4.12), and the solutions of the linear problem (3.4.10), (3.4.13), that are the analytic with respect to $\boldsymbol{\xi}$. Thus, the linear equivalence method reduces the study of a nonlinear problem to a linear one, which is always more convenient.

If we assume that v has the form

$$v(t, \boldsymbol{\xi}) = \sum_{|\gamma| \geq 1} v_\gamma(t) \frac{\boldsymbol{\xi}^\gamma}{\gamma!}, \quad \gamma = \{\gamma_1, \gamma_2, \ldots, \gamma_n\}, \quad \gamma! \equiv \gamma_1! \gamma_2! \ldots \gamma_n!, \tag{3.4.14}$$

then $v_\gamma(t)$ are the solutions of the infinite linear system of ordinary differential equations, with constant coefficients,

$$\frac{d}{dt} v_\gamma = \sum_{j=1}^{n} j \sum_{|\mu| \geq 1} f_{j\mu} v_{\gamma+\mu-\mathbf{e}_j}, \quad \mathbf{e}_j = \{\delta_{jk}\}_{k=1}^{n}, \tag{3.4.15}$$

satisfying the Cauchy conditions

$$v_\gamma(t_0) = \mathbf{x}_0^\gamma, \quad |\gamma| \in \mathbb{N}^*. \tag{3.4.16}$$

It has also been proved that the first n components of the solution of the linear Cauchy problem (3.4.15), (3.4.16) coincide with the solution of the nonlinear Cauchy problem (3.4.7), (3.4.12). Therefore, the linear equivalence method associates to the nonlinear system of differential equations (3.4.7), the two linear equivalents (3.4.10) and (3.4.15).

4.2.2 The linear equivalent partial differential equation

Let $\mathbf{g}(\mathbf{x}) = (g_1(\mathbf{x}), g_2(\mathbf{x}), \ldots, g_n(\mathbf{x}))$ be a vector function whose components are analytic with respect to the components of the vector variable $\mathbf{x} = (x_1, x_2, \ldots, x_n)$. Then, the *Lie derivative* of an analytic scalar function $\lambda(\mathbf{x})$ is defined as

$$\mathsf{L}_\mathbf{g} \lambda \equiv \langle \mathbf{g}, \operatorname{grad} \lambda \rangle . \tag{3.4.17}$$

This definition immediately yields

$$\langle \boldsymbol{\xi}, \mathbf{f}(\mathbf{D}_\boldsymbol{\xi}) \rangle v = \mathsf{L}_\mathbf{f} v, \tag{3.4.18}$$

and it is now obvious that

THEOREM 3.4.1 (I. TOMA) *The linear partial differential equation* (3.4.10) *can be expressed in terms of the Lie derivative as*

$$\frac{\partial v}{\partial t} - \mathsf{L}_\mathbf{f} v = 0. \tag{3.4.19}$$

Therefore, the analytic solution of (3.4.10), that also satisfies (3.4.13), allows both expansions

$$v(t,\boldsymbol{\xi}) = 1 + \sum_{k=1}^{\infty} \mathsf{L}_{\mathbf{f}}{}^{k}\left(e^{\langle \mathbf{x}_0,\boldsymbol{\xi}\rangle}\right)\frac{(t-t_0)^k}{k!}, \tag{3.4.20}$$

and

$$v(t,\boldsymbol{\xi}) = 1 + \sum_{k=1}^{\infty} (\langle \boldsymbol{\xi}, \mathbf{f}(\mathbf{D}_{\boldsymbol{\xi}})\rangle)^k \left(e^{\langle \mathbf{x}_0,\boldsymbol{\xi}\rangle}\right)\frac{(t-t_0)^k}{k!}. \tag{3.4.21}$$

Thus, according to (3.4.9), the solution of the nonlinear initial problem (3.4.7), (3.4.12) can be written in the form

$$\mathbf{x}(t) = \operatorname{grad}_{\boldsymbol{\xi}} v(t,\boldsymbol{\xi})\Big|_{\boldsymbol{\xi}=\mathbf{0}}, \tag{3.4.22}$$

where v has to be replaced either by (3.4.20) or, equally well, by (3.4.21).

4.2.3 The linear equivalent system of differential equations

In matrix notation, the infinite linear system of differential equations (3.4.15) reads

$$\frac{d\mathbf{V}}{dt} = \mathbf{A}\mathbf{V}, \quad \mathbf{V} = [\mathbf{V}_j]_{j\in\mathbb{N}^*}, \quad \mathbf{V}_j = [v_\gamma(t)]_{|\gamma|=j}, \tag{3.4.23}$$

where $\mathbf{A}$ is a triangular block matrix of the form

$$\mathbf{A} = \begin{bmatrix} \mathbf{A}_{11} & \mathbf{A}_{12} & \mathbf{A}_{13} & \cdots & \cdots & \cdots & \cdots \\ \mathbf{0} & \mathbf{A}_{22} & \mathbf{A}_{23} & \cdots & \cdots & \cdots & \cdots \\ \mathbf{0} & \mathbf{0} & \mathbf{A}_{33} & \cdots & \cdots & \cdots & \cdots \\ \cdots & \cdots & \cdots & \cdots & \cdots & \cdots & \cdots \\ \mathbf{0} & \mathbf{0} & \mathbf{0} & \cdots & \mathbf{A}_{jj} & \mathbf{A}_{j,j+1} & \cdots \\ \cdots & \cdots & \cdots & \cdots & \cdots & \cdots & \cdots \end{bmatrix}. \tag{3.4.24}$$

The rectangular matrices $\mathbf{A}_{j,j+k-1}$, $j, k \in \mathbb{N}^*$, depend only on those coefficients $f_{j\boldsymbol{\mu}}$ for which $|\boldsymbol{\mu}| = k$; in the case $n = 2$, these matrices have the dimensions $(j+1)\times(j+k)$. The square matrices $\mathbf{A}_{jj}$, along the diagonal, depend only on the linear part of the differential operator; more precisely, $\mathbf{A}_{11}$ is the associated Jacobian matrix. Now, the Cauchy conditions (3.4.16) take the form

$$\mathbf{V}(t_0) = [\mathbf{x}_0^{\gamma}]_{|\gamma|\in\mathbb{N}^*}. \tag{3.4.25}$$

THEOREM 3.4.2 (I. TOMA) *The solution of the nonlinear Cauchy problem* (3.4.7), (3.4.12) *coincides with the first n components of the vector* $\mathbf{V}$, *which is the solution of the linear Cauchy problem* (3.4.23), (3.4.25).

The proof of this theorem is constructive, as it is based on the construction of the inverse matrix for the considered differential operator. Thus, it can be shown that the Theorem 3.4.2 also holds for systems with variable coefficients. In the case of systems with constant coefficients, the inverse matrix allows a simplified exponential type form and thus the solution of the problem (3.4.23), (3.4.25) may be written as

$$\mathbf{V}(t) = e^{\mathbf{A}(t-t_0)}\mathbf{V}(t_0)\,. \tag{3.4.26}$$

The local consistency of the first n components of $\mathbf{V}$, that is the solution $\mathbf{x}$, has also been demonstrated.

It should be mentioned that the block structure of the matrix $\mathbf{A}$ enables a step-by-step evaluation of the powers $\mathbf{A}^k$. A secondary result is that the first n rows of $\mathbf{A}^k$ can be partitioned in blocks, each block being computable by a finite number of steps.

Let P_n be the operator that associates to a matrix its first n rows. Then, on the basis of Theorem 3.4.2 and equation (3.4.26), the solution of the nonlinear problem (3.4.7), (3.4.12) can be expressed as a Taylor series expansion

$$\mathbf{x}(t) = \mathbf{x}_0 + \sum_{k=1}^{\infty} \frac{(t-t_0)^k}{k!}\,\mathsf{P}_n\mathbf{A}^k\,[\mathbf{x}_0^{\gamma}]_{|\gamma|\in\mathbb{N}^*}\,. \tag{3.4.27}$$

This result enables a direct comparison with the Fliess expansions, derived in a different functional frame, to solve optimal control problems. The Fliess expansion corresponding to the solution of the nonlinear problem (3.4.7), (3.4.12) is the Taylor series

$$x_j(t) = x_{j0} + \sum_{k=1}^{\infty} \mathsf{L}_{\mathbf{f}}{}^k(x_{j0})\,\frac{(t-t_0)^k}{k!}\,, \quad j = 1, 2, \ldots, n, \tag{3.4.28}$$

where $\mathsf{L}_{\mathbf{f}}{}^k(x_{j0})$ is the k-th order iterative of the Lie derivative $\mathsf{L}_{\mathbf{f}}(x_{j0})$, and by comparing (3.4.28) and (3.4.27) we get

THEOREM 3.4.3 (I. TOMA) *The k-th order iterative of the Lie derivative with respect to* $\mathbf{f}$ *can be evaluated in terms of the matrix of linear equivalence method according to the formula*

$$\begin{bmatrix} \mathsf{L}_{\mathbf{f}}^k(x_{10}) \\ \mathsf{L}_{\mathbf{f}}^k(x_{20}) \\ \vdots \\ \mathsf{L}_{\mathbf{f}}^k(x_{n0}) \end{bmatrix} = \mathsf{P}_n\mathbf{A}^k\,[\mathbf{x}_0^{\gamma}]_{|\gamma|\in\mathbb{N}^*}\,. \tag{3.4.29}$$

It is well-known that the evaluation of higher order Lie derivatives involves a large amount of calculations. The advantage of Theorem 3.4.3 is that all these calculations can be carried out in a simple matrix formulation, with a finite number of matrix multiplications in each step. Indeed, on the grounds of the structure of matrix $\mathbf{A}$ and its specific properties one has

$$\mathsf{P}_n \mathbf{A}^m = [\mathbf{S}_{km}]_{k\in\mathbb{N}^*} , \tag{3.4.30}$$

where

$$\begin{aligned} \mathbf{S}_{1m} &= \mathbf{A}_{11}^m , \\ \mathbf{S}_{km} &= \sum_{|\gamma|=m-k-1} \mathbf{A}_{11}^{\gamma_1}\mathbf{A}_{12}\mathbf{A}_{22}^{\gamma_2}\mathbf{A}_{23}\dots\mathbf{A}_{k-1,k}\mathbf{A}_{kk}^{\gamma_k} , \quad k \geq 2 , \end{aligned} \tag{3.4.31}$$

$\gamma = \{\gamma 1, \gamma_2, \dots, \gamma_k\}$ being a k-dimensional index that satisfies the relation $|\gamma| = \sum \gamma_j = m - k - 1$. The evaluation of $\mathbf{S}_{km}$ becomes easier if we know the eigenvalues of $\mathbf{A}_{kk}$. From the general linear equivalence method it is known that these eigenvalues are

$$\langle \boldsymbol{\tau}, \boldsymbol{\lambda} \rangle \equiv \sum_{j=1}^{n} \tau_j \lambda_j , \tag{3.4.32}$$

where $\boldsymbol{\lambda} = \{\lambda_1, \lambda_2, \dots, \lambda_n\}$ are the eigenvalues of the matrix $\mathbf{A}_{11}$, and $\boldsymbol{\tau} = \{\tau_1, \tau_2, \dots, \tau_n\}$ is an n-dimensional index, with $\tau_j \in \mathbb{N}$ and $|\boldsymbol{\tau}| = \sum \tau_j = k$.

Another possibility is to get the *normal representations* of the linear equivalence method, which are useful for a qualitative study of solutions of nonlinear differential equations. Thus, it has been shown that

THEOREM 3.4.4 (I. TOMA) *The solution of the nonlinear Cauchy problem* (3.4.7), (3.4.12) *may be expressed as a series with respect to the Cauchy data* $\mathbf{x}_0$

$$x_j(t) = x_{j0} + \sum_{|\gamma|\geq 1} u_\gamma^j(t)\mathbf{x}_0^\gamma , \quad j = 1, 2, \dots, n, \tag{3.4.33}$$

where for every given $|\gamma| = k$ *the vectors*

$$\mathbf{U}_k^j \equiv \left[u_\gamma^j(t)\right]_{|\gamma|=k}$$

satisfy the finite linear system

$$\begin{aligned}
\frac{d\mathbf{U}_1^j}{dt} &= \mathbf{A}_{11}^{\mathrm{T}}\mathbf{U}_1^j\,,\\
\frac{d\mathbf{U}_2^j}{dt} &= \mathbf{A}_{22}^{\mathrm{T}}\mathbf{U}_2^j + \mathbf{A}_{12}^{\mathrm{T}}\mathbf{U}_1^j\,,\\
&\cdots\cdots\cdots\cdots\cdots\cdots\cdots\cdots\\
\frac{d\mathbf{U}_k^j}{dt} &= \mathbf{A}_{kk}^{\mathrm{T}}\mathbf{U}_k^j + \mathbf{A}_{k,k-1}^{\mathrm{T}}\mathbf{U}_{k-1}^j + \ldots + \mathbf{A}_{1k}^{\mathrm{T}}\mathbf{U}_1^j\,,
\end{aligned} \tag{3.4.34}$$

and the Cauchy conditions

$$\mathbf{U}_1^j(t_0) = [\delta_{jm}]_{m=\overline{1,n}}\,, \quad \mathbf{U}_s^j(t_0) = \mathbf{0}\,, \quad s = 2, 3, \ldots, k. \tag{3.4.35}$$

It is easy to see that the finite linear system (3.4.34) can be solved by block partitioning. The procedure is greatly simplified by knowing the eigenvalues of the square blocks, that can be obtained from (3.4.32). Furthermore, the coefficients $u_\gamma^j(t)$ in (3.4.33) remain unchanged if we consider higher orders in the normal representation of the linear equivalence method.

Note that, according to the classical theories for systems of ordinary differential equations, the nonlinear Cauchy problem (3.4.7), (3.4.12) locally allows a unique solution $\mathbf{x}(t, \mathbf{x}_0)$, $\mathbf{x} \in (C^0(I))^n$, where $I \equiv [t_0 - h, t_0 + h]$, with h being specific to the problem. The normal representation of the linear equivalence method shows that, $\forall t \in I$ the solution $\mathbf{x}(t, \mathbf{x}_0)$ is an element of the Lie algebra generated by the Cauchy data $\mathbf{x}_0$.

4.2.4 The nonlinear damped pendulum

The linear equivalence method has been efficiently applied to the study of different mechanical systems like the Bernoulli-Euler bar, and the nonlinear pendulum. Here, we will illustrate the usefulness of the method in the study of long-term behaviour of a nonlinear damped pendulum.

This mechanical system is governed by the nonlinear differential equation of the second order

$$\ddot{y} + 2\lambda\dot{y} + \omega^2 \sin y = 0\,, \tag{3.4.36}$$

where $\lambda = k'/2m > 0$, k' is the damping coefficient, m the mass of pendulum, and ω the circular frequency ($\omega = 2\pi\nu$, ν being the frequency of oscillation). In this case, $n = 2$ such that the exponential mapping (3.4.9) takes the form

$$v(t, \sigma, \xi) = e^{\sigma y(t) + \xi \dot{y}(t)}\,, \tag{3.4.37}$$

leading to the linear equivalent partial differential equation

$$\frac{\partial v}{\partial t} - \sigma\frac{\partial v}{\partial \xi} + \xi\left[2\lambda\frac{\partial v}{\partial \xi} + \omega^2(\sin \mathrm{D}_\sigma)v\right] = 0\,. \tag{3.4.38}$$

Also, by using a series expansion of the form (3.4.14)

$$v(t,\sigma,\xi) = 1 + \sum_{j+k=1}^{\infty} v_{jk}(t)\,\frac{\sigma^j}{j!}\,\frac{\xi^k}{k!}\,, \tag{3.4.39}$$

we obtain the linear equivalent system of differential equations

$$\frac{dv_{jk}}{dt} = j\,v_{j-1,k+1} - k\Bigg[2\lambda\,v_{jk} + \omega^2\,v_{j+1,k-1} + \omega^2\sum_{m=1}^{\infty}\frac{(-1)^m}{(2m+1)!}\,v_{j+2m+1,k-1}\Bigg]\,, \tag{3.4.40}$$

Note that, here we have denoted the parameters of the exponential mapping by σ and ξ, instead of ξ_1 and ξ_2. Also, note that the "sin" function has to be understood in terms of its series expansion.

The nonlinear differential operator in (3.4.36) is odd with respect to y, and thus the matrix $\mathbf{A}$, defined in (3.4.24), has a simplified form with the blocks $\mathbf{A}_{jk}$, for which $j+k$ is odd, being null matrices. The remaining nonzero blocks are of the form

$$\left[(\mathbf{A}_{2j-1,2j-1})_q^m\right]_{2j\times 2j} = \left[(2j-m)\,\delta_q^{m+1} - 2\lambda(q-1)\delta_q^m - \omega^2\delta_{q+1}^m\right]\,, \tag{3.4.41}$$

$$\left[(\mathbf{A}_{2j-1,2j+2k-1})_q^m\right]_{2j\times(2j+2k)} = \left[q\,\frac{(-1)^{k+1}\omega^2}{(2k+1)!}\,\delta_{q+1}^m\right]\,,$$

for $j,k\in\mathbb{N}^*$, where we have marked the dimensions of the matrices (note that $m=1,2,\ldots,2j$ in both expressions).

If we associate to (3.4.36) some arbitrary Cauchy conditions

$$y(0) = \alpha\,,\quad \dot{y}(0) = \beta\,, \tag{3.4.42}$$

then the solution of the Cauchy problem (3.4.36), (3.4.42) can be represented in the form

$$y(t) = \sum_{j=1}^{\infty}\sum_{k=0}^{2j-1} u_{2j-1-k,k}(t)\,\alpha^{2j-1-k}\,\beta^k\,, \tag{3.4.43}$$

where $u_{2j-1-k,k}(t)$ are the components of the vectors $\mathbf{U}_{2j-1}(t)$ which satisfy the finite linear problems

$$\frac{d\mathbf{U}_1}{dt} = \mathbf{A}_{11}^{\mathrm{T}}\,\mathbf{U}_1 \quad \mathbf{U}_1(0) = \begin{bmatrix}1\\0\end{bmatrix}\,, \tag{3.4.44}$$

$$\frac{d\mathbf{U}_{2j-1}}{dt} = \mathbf{A}_{1,2j-1}^{\mathrm{T}}\,\mathbf{U}_1 + \ldots\mathbf{A}_{2j-3,2j-1}^{\mathrm{T}}\,\mathbf{U}_{2j-3} + \mathbf{A}_{2j-1,2j-1}^{\mathrm{T}}\,\mathbf{U}_{2j-1}\,,$$

with $\mathbf{U}_{2j-1}(0) = \mathbf{0}$, $j \geq 2$, the matrices $\mathbf{A}_{2j-1,2j-1-k}$ being defined in (3.4.41). To study the third order effects we should go as far as $j = 2$, and thus we have to solve two linear problems:

$$\frac{d\mathbf{U}_1}{dt} = \begin{bmatrix} 0 & -\omega^2 \\ 1 & -2\lambda \end{bmatrix} \mathbf{U}_1 , \quad \mathbf{U}_1(0) = \begin{bmatrix} 1 \\ 0 \end{bmatrix}, \tag{3.4.45}$$

and

$$\frac{d\mathbf{U}_3}{dt} = \begin{bmatrix} 0 & -\omega^2 & 0 & 0 \\ 3 & -2\lambda & -2\omega^2 & 0 \\ 0 & 2 & -4\lambda & -3\omega^2 \\ 0 & 0 & 1 & -6\lambda \end{bmatrix} \mathbf{U}_3 + \frac{\omega^2}{6} \begin{bmatrix} u_{01} \\ 0 \\ 0 \\ 0 \end{bmatrix}, \quad \mathbf{U}_3(0) = \mathbf{0} . \tag{3.4.46}$$

In the case of subcritical damping ($\chi = \lambda/\omega < 1$), the eigenvalues of the matrix $\mathbf{A}_{11}$ are complex conjugated ($-\lambda \pm i\mu$, with $\mu = \omega\sqrt{1-\chi^2}$). We solve the equation (3.4.45) to get u_{01}, which we substitute into (3.4.46), to obtain the solution

$$\begin{aligned} y(t) \cong \alpha \left(\frac{\omega\chi}{\mu} \sin \mu t + \cos \mu t \right) e^{-\lambda t} + \beta \frac{\sin \mu t}{\mu} e^{-\lambda t} \\ + \alpha^3 u_{30}(t) + \alpha^2 \beta\, u_{21}(t) + \alpha\beta^2 u_{12}(t) + \beta^3 u_{03}(t) , \end{aligned} \tag{3.4.47}$$

where

$$\begin{aligned} u_{30}(t) &= \frac{e^{-\lambda t}}{3p} \left[-(11 + 12\chi^2) \frac{\chi\mu}{\omega} \cos \mu t + (6 + 25\chi^2 - 12\chi^4) \sin \mu t \right] \\ &\qquad + \varphi_1(\mu t) , \\ u_{21}(t) &= \frac{e^{-\lambda t}}{\omega p} \left[-(2 + 12\chi^2) \frac{\mu}{\omega} \cos \mu t + 9\chi \sin \mu t \right] + \varphi_2(\mu t) , \\ u_{12}(t) &= \frac{e^{-\lambda t}}{\omega^2 p} \left[-9\chi \frac{\mu}{\omega} \cos \mu t + (2 + 3\chi^2) \sin \mu t \right] + \varphi_3(\mu t) , \\ u_{03}(t) &= \frac{e^{-\lambda t}}{\omega^3 p} \left[-2 \frac{\mu}{\omega} \cos \mu t + \chi \sin \mu t \right] + \varphi_4(\mu t) , \end{aligned} \tag{3.4.48}$$

and $p = 16\omega^2 \chi (4 - 3\chi^2) \sqrt{1 - \chi^2}$,

$$\varphi(\zeta) = e^{-3\zeta} \left(A_j \cos 3\zeta + B_j \sin 3\zeta + C_j \cos \zeta + D_j \sin \zeta \right) , \quad j = \overline{1,4} .$$

The constant coefficients A_j, B_j, C_j, and D_j follow from the initial conditions for $u_{jk}(t)$. The solution (3.4.47) and (3.4.48) shows that in the case of subcritical damping, the pendulum motion is qualitatively the same as in the linear case, which is quasi-periodical of period $2\pi/\mu$.

In the case of supercritical damping ($\chi = \lambda/\omega > 1$) the solution is obtained by replacing $\sqrt{1-\chi^2}$ by $\sqrt{\chi^2-1}$, and the trigonometric functions sin and cos by the hyperbolic functions sinh and cosh, respectively. The case of critical damping ($\chi = \lambda/\omega = 1$), has to be treated separately, due to the degeneracy of eigenvalues.

4.3 Partial differential equations

4.3.1 The symmetry group of a system of partial differential equations

At first, we will define the invariance of a system of partial differential equations with respect to a group of transformations G. We assume that the system of equations contains p independent variables $\mathbf{x} \equiv (x_1, x_2, \ldots, x_p) \in X = \mathbb{R}^p$, and q dependent variables $\mathbf{u} \equiv (u_1, u_2, \ldots, u_q) \in U = \mathbb{R}^q$. A solution of this system of partial differential equations is a mapping $\mathbf{f} : X \to U$, denoted by $\mathbf{u} = \mathbf{f}(\mathbf{x})$, which determines a p-dimensional manifold (the *graph of the solution*) in the space $X \times U$. We denote this graph by

$$\Gamma = \{(\mathbf{x}, \mathbf{u}) | \mathbf{u} = \mathbf{f}(\mathbf{x}), \mathbf{x} \in X\}.$$

Let G be a group of transformations acting on the space $X \times U$, and let $g\Gamma = \{g(\mathbf{x}, \mathbf{u}) | (\mathbf{x}, \mathbf{u}) \in \Gamma\}$ be the image of the graph Γ by the transformation $g \in G$. The image $g\Gamma$ represents the graph of a function $\widetilde{\mathbf{u}} = (g \cdot \mathbf{f})(\widetilde{\mathbf{x}})$, called the *transform of* $\mathbf{f}$ *by* g. Supposing that the transformation $g \in G$ is defined by the relations

$$g(\mathbf{x}, \mathbf{u}) = (\varphi_g(\mathbf{x}, \mathbf{u}), \psi_g(\mathbf{x}, \mathbf{u})) \equiv (\widetilde{\mathbf{x}}, \widetilde{\mathbf{u}}), \tag{3.4.49}$$

it results that the graph $g\Gamma$ is determined by

$$\begin{aligned} \widetilde{\mathbf{x}} &= \varphi_g(\mathbf{x}, \mathbf{f}(\mathbf{x})) = \varphi_g \circ (\mathbf{I} \times \mathbf{f})(\mathbf{x}), \\ \widetilde{\mathbf{u}} &= \psi_g(\mathbf{x}, \mathbf{f}(\mathbf{x})) = \psi_g \circ (\mathbf{I} \times \mathbf{f})(\mathbf{x}), \end{aligned} \tag{3.4.50}$$

where $\mathbf{I}$ is the identity mapping on X. For g sufficiently close to the neutral element of G, using the theorem of implicit functions, we may locally solve for $\mathbf{x}$ the first equation in (3.4.50) to obtain

$$\mathbf{x} = [\varphi_g \circ (\mathbf{I} \times \mathbf{f})]^{-1}(\widetilde{\mathbf{x}}).$$

Then, by substituting the expression of $\mathbf{x}$ into the second equation in (3.4.50), we find the transform of $\mathbf{f}$ by g

$$\widetilde{\mathbf{u}} = (g \cdot \mathbf{f})(\widetilde{\mathbf{x}}) = \psi_g \circ (\mathbf{I} \times \mathbf{f})([\varphi_g \circ (\mathbf{I} \times \mathbf{f})]^{-1}(\widetilde{\mathbf{x}})). \tag{3.4.51}$$

In terms of these notations, we say that the system of partial differential equations is invariant with respect to the group of transformations G if, for any local solution $\mathbf{u} = \mathbf{f}(\mathbf{x})$, the transformed function $g \cdot \mathbf{f}$ is defined $\forall g \in G$, and $\widetilde{\mathbf{u}} = (g \cdot \mathbf{f})(\widetilde{\mathbf{x}})$ is also a solution of the considered system. In such cases the group G is called the *symmetry group* of the system of partial differential equations.

Further, we will establish the relation between the derivatives of the function $\mathbf{f}$ and the derivatives of the transform $g \cdot \mathbf{f}$. Each component of the function $\mathbf{f} : X \to U$, $u^l = f^l(\mathbf{x})$, $l = 1, 2, \ldots, q$, generates $p_k = C^k_{p+k-1}$ different partial derivatives of order k, of the form

$$u^l_{\mathbf{J}} = \partial_{\mathbf{J}} f^l(\mathbf{x}) \equiv \frac{\partial^k f^l}{\partial x_1^{j_1} \partial x_2^{j_2} \ldots \partial x_p^{j_p}}, \tag{3.4.52}$$

where $\mathbf{J} \equiv \{j_1, j_2, \ldots, j_p\}$ is a p-dimensional index, and j_i are non-negative integers satisfying $|\mathbf{J}| = j_1 + j_2 + \ldots + j_p = k$. Hence, the function $\mathbf{f}$ generates a set of $q\, p_k$ different derivatives of order k. Let U_k be the space of dimension $q\, p_k$, which represents all these derivatives, and let $U^{(k)} = U \times U_1 \times \ldots \times U_k$ be the space of all derivatives of the function $\mathbf{f}$ up to the k-th order, inclusively. We may thus consider that the function $\mathbf{f} : X \to U$ generates a function $\mathrm{pr}^{(k)}\, \mathbf{f} : X \to U^{(k)}$, called the *k-th prolongation* of $\mathbf{f}$, which has the graph

$$\Gamma^{(k)} = \left\{ \left(\mathbf{x}, \{u^l_{\mathbf{J}}\}_{|\mathbf{J}|=\overline{0,k}\,,\, l=\overline{1,q}} \right) \middle| u^l_{\mathbf{J}} = \partial_{\mathbf{J}} f^l(\mathbf{x}), \mathbf{x} \in X \right\} .$$

So, the k-th prolongation of $\mathbf{f}$ is a vector $\mathrm{pr}^{(k)}\, \mathbf{f}(\mathbf{x}) \in U^{(k)}$ with $n = q(1 + p_1 + p_2 + \ldots + p_k) = q\, p_{k+1}$ components, that represents the value of $\mathbf{f}$ and its derivatives up to the order k, inclusively, at the point $\mathbf{x} \in X$.

Also, the symmetry group G which acts on the space $X \times U$ induces a local action on the space $X \times U^{(k)}$, called the *k-th prolongation* of G, and denoted by $\mathrm{pr}^{(k)}G$. By definition, $\mathrm{pr}^{(k)}g(\mathbf{x}, \mathbf{u}^{(k)}) = (\widetilde{\mathbf{x}}, \widetilde{\mathbf{u}}^{(k)})$, where $\mathbf{u}^{(k)} \in U^{(k)}$, and $\widetilde{u}^l_{\mathbf{J}} = \partial_{\mathbf{J}}(g \cdot f)^l(\widetilde{\mathbf{x}})$. Hence, the transform of a derivative of $\mathbf{f}$ is equal to the corresponding derivative of the transform of $\mathbf{f}$ by g, defined in (3.4.51). If X is the infinitesimal generator of a one-parameter subgroup of G, that is $\{\exp(t\mathsf{X})\} \subset G$, then the k-th prolongation of the infinitesimal generator is

$$\mathrm{pr}^{(k)}\mathsf{X} = \frac{d}{dt} \left(\mathrm{pr}^{(k)}\, e^{t\mathsf{X}} \right) \bigg|_{t=0} . \tag{3.4.53}$$

We suppose that the system of partial differential equations is of order k, and it is represented by m partial differential equations

$$\Omega^i \left(\mathbf{x}, \mathbf{u}^{(k)} \right) = 0\,, \quad i = 1, 2, \ldots, m\,, \quad \mathbf{u}^{(k)} \in U^{(k)} . \tag{3.4.54}$$

This system of equations can be regarded as defining a manifold $S_\Omega \subset X \times U^{(k)}$, given by the vanishing of a smooth function $\mathbf{\Omega} : X \times U^{(k)} \to \mathbb{R}^m$, that is

$$S_\Omega = \left\{ (\mathbf{x}, \mathbf{u}^{(k)}) \,\middle|\, \mathbf{\Omega}(\mathbf{x}, \mathbf{u}^{(k)}) = \mathbf{0} \right\}.$$

The solution of the system (3.4.54) is a smooth function $\mathbf{f} : X \to U$, such that $\mathbf{\Omega}(\mathbf{x}, \mathsf{pr}^{(k)}\, \mathbf{f}(\mathbf{x})) = \mathbf{0}$. It can be shown that G is a symmetry group of the system (3.4.54) if the manifold S_Ω is invariant with respect to the group $\mathsf{pr}^{(k)} G$. Thus, the condition that G is a symmetry group of (3.4.54) is established by admitting that:

1 $\mathbf{\Omega}(\mathbf{x}, \mathbf{u}^{(k)}) = \mathbf{0}$ is a system of partial differential equations of order k, in p independent variables and q dependent variables, such that the Jacobian of $\mathbf{\Omega}$ has everywhere the maximum rank;

2 at any point $(\mathbf{x}_0, \mathbf{u}_0^{(k)}) \in S_\Omega$ there exists a solution $\mathbf{u} = \mathbf{f}(\mathbf{x})$ defined in a vicinity of $\mathbf{x}_0$, such that $\mathbf{u}_0^{(k)} = \mathsf{pr}^{(k)}\, \mathbf{f}(\mathbf{x}_0)$;

3 G is a connected local group of transformations acting on $X \times U$.

In the frame of these hypotheses it can be proved that G is a symmetry group of the system (3.4.54) if and only if each infinitesimal generator X of G satisfies the relation

$$\mathsf{pr}^{(k)}\mathsf{X} \left[\mathbf{\Omega}(\mathbf{x}, \mathbf{u}^{(k)}) \right] = \mathbf{0}, \tag{3.4.55}$$

when $\mathbf{\Omega}(\mathbf{x}, \mathbf{u}^{(k)}) = \mathbf{0}$.

If we suppose that the function $\mathbf{\Omega} : X \times U^{(k)} \to \mathbb{R}^m$ is differentiable, then its total derivative with respect to the variable x_i is expressed by the function $\mathsf{D}_i \mathbf{\Omega} : X \times U^{(k+1)} \to \mathbb{R}^m$, $i = 1, 2, \ldots, p$, the operator D_i being of the form

$$\mathsf{D}_i = \frac{\partial}{\partial x_i} + \sum_{l=1}^{q} \sum_{\mathbf{J}} u^l_{\mathbf{J}_i} \frac{\partial}{\partial u^l_{\mathbf{J}}}, \tag{3.4.56}$$

where $\mathbf{J}_i = \{j_1, j_2, \ldots, j_{i-1}, j_i + 1, j_{i+1}, \ldots, j_p\}$, and the sum is over all $\mathbf{J}$ with $|\mathbf{J}| = j_1 + j_2 + \ldots + j_p \leq k$. For instance, the total derivative with respect to the variable $x \in \mathbb{R}$ of a function $f(x, u, u_x, \ldots)$, where $u = u(x)$, $u_x = du/dx, \ldots$, can be expressed in terms of the operator D_x by the formula

$$\mathsf{D}_x f = \left(\frac{\partial}{\partial x} + u_x \frac{\partial}{\partial u} + u_{xx} \frac{\partial}{\partial u_x} + \ldots \right) f. \tag{3.4.57}$$

If the p-dimensional index $\mathbf{J} = \{j_1, j_2, \ldots, j_p\}$ is given, then we denote $\mathsf{D}^{\mathbf{J}} = \mathsf{D}_1^{j_1} \mathsf{D}_2^{j_2} \ldots \mathsf{D}_p^{j_p}$, where $\mathsf{D}_i^{j_i}$ represents the total derivative of order j_i with respect to the variable x_i.

THEOREM 3.4.5 (P. J. OLVER) *Let* X *be the infinitesimal generator of a one-parameter subgroup of G, expressed as a smooth vector field on* $X \times U$

$$\mathsf{X} = \sum_{i=1}^{p} \xi_i(\mathbf{x}, \mathbf{u}) \frac{\partial}{\partial x_i} + \sum_{l=1}^{q} \varphi_l(\mathbf{x}, \mathbf{u}) \frac{\partial}{\partial u^l} . \qquad (3.4.58)$$

Then, the k*-th order prolongation of the infinitesimal generator* X*, defined in (3.4.53), is the vector field*

$$\mathsf{pr}^{(k)}\mathsf{X} = \mathsf{X} + \sum_{l=1}^{q} \sum_{\mathbf{J}} \varphi_l^{\mathbf{J}}(\mathbf{x}, \mathbf{u}) \frac{\partial}{\partial u_{\mathbf{J}}^l} , \qquad (3.4.59)$$

on $X \times U^{(k)}$*, where the second sum is over all* p*-dimensional indexes* $\mathbf{J} = \{j_1, j_2, \ldots, j_p\}$ *with* $0 < j_1 + j_2 + \ldots + j_p \leq k$*. Also, the coefficients* $\varphi_l^{\mathbf{J}}(\mathbf{x}, \mathbf{u})$ *are given by the following formula*

$$\varphi_l^{\mathbf{J}} = D^{\mathbf{J}} \left(\varphi_l - \sum_{i=1}^{p} u_i^l \, \xi_i \right) + \sum_{i=1}^{p} u_{\mathbf{J},i}^l \xi_i , \qquad (3.4.60)$$

where $u_i^l = \partial u^l / \partial x_i$*,* $l = 1, 2, \ldots, q$*.*

Knowing the expression of the k-th order prolongation (3.4.58) of the infinitesimal generators, we may use the condition (3.4.55) to find the coefficients ξ_i and φ_l, which in turn determine the symmetry group of the system of partial differential equations (3.4.54).

4.3.2 The method of separation of variables

The application of the group theory to the study of partial differential equations has led to the substantiation of the *method of separation of variables*, a method frequently used to solve such equations. One can thus establish a close connection between the symmetry group of a partial differential equation, the systems of coordinates in which the equation admits solutions by separation of variables, and the properties of the separated solutions. We will present the results obtained in this direction by means of one of the simplest example of partial differential equations, that is the *Helmholtz equation* (or the equation of stationary waves) in two dimensions

$$\Omega \, u \equiv \left(\Delta + k^2\right) u = 0 , \quad \Delta = \partial_{xx} + \partial_{yy} , \qquad (3.4.61)$$

where $u(x, y)$ is an analytic function, defined in the domain $\mathcal{D} \subseteq \mathbb{R}^2$, and $k^2 \in \mathbb{R}$ is a constant.

Let us apply the method of separation of variables to (3.4.61), assuming that, in Cartesian coordinates, the solution can be written in the form

$$u(x, y) = X(x) \, Y(y) . \qquad (3.4.62)$$

By substituting into (3.4.61) we obtain the ordinary differential equations

$$\frac{X''}{X} = -\frac{Y''}{Y} - k^2 = -\lambda^2\,, \tag{3.4.63}$$

where λ^2 is the separation constant. A basis for the solutions of the first equation in (3.4.63) is formed by the functions

$$X_{1,2}(x) = e^{\pm i\lambda x}\,, \quad \lambda \neq 0\,,$$

while for the solutions of the second equation we obtain the basis

$$Y_{1,2}(y) = e^{\pm i\sqrt{k^2-\lambda^2}\,y}\,, \quad k^2 - \lambda^2 \neq 0\,.$$

Consequently, the solutions of Helmholtz's equation (3.4.61) in Cartesian coordinates read

$$u_\lambda^{jl}(x,y) = X_j(x)\,Y_l(y)\,, \quad j,l = 1,2, \quad \lambda \in \mathbb{R} \setminus \{0, \pm k\}\,. \tag{3.4.64}$$

In polar coordinates ($x = r\cos\theta$, $y = r\sin\theta$), Helmholtz's equation becomes

$$\left(\partial_{rr} + \frac{1}{r}\,\partial_r + \frac{1}{r^2}\,\partial_{\theta\theta} + k^2\right) u(r,\theta) = 0\,. \tag{3.4.65}$$

In this case, we search a separated solution of the form

$$u(r,\theta) = R(r)\,\Theta(\theta)\,, \tag{3.4.66}$$

which we substitute into (3.4.65) to obtain the ordinary differential equations

$$\frac{1}{R}\left(r^2R'' + r\,R' + r^2k^2R\right) = -\frac{\Theta''}{\Theta} = \mu^2\,. \tag{3.4.67}$$

The solutions of the first equation in (3.4.67), which is called *Bessel's equation*, are determined by a basis comprising the Bessel functions of the first kind and order $\pm\mu$, that is $R_{1,2}(r) = J_{\pm\mu}(kr)$, while the basis for the second equation is formed by the functions

$$\Theta_{1,2}(\theta) = e^{\pm i\mu\theta}\,, \quad \mu \in \mathbb{R}\,.$$

Hence, in polar coordinates, the separated solutions of (3.4.65) are

$$u_\mu^{jl}(r,\theta) = R_j(r)\,\Theta_l(\theta)\,, \quad j,l = 1,2, \quad \mu \in \mathbb{R}\,. \tag{3.4.68}$$

Now, let us see the connection between the symmetry group of Helmholtz's equation and the form of separated solutions in Cartesian and polar coordinates.

The symmetry group of Helmholtz's equation can be obtained by using the general method described in the previous section. Thus, in Cartesian coordinates, the infinitesimal generator (3.4.58) of the symmetry group is

$$\mathsf{L} = \xi(x,y,u)\,\partial_x + \eta(x,y,u)\,\partial_y + \varphi(x,y,u)\,\partial_u\,, \tag{3.4.69}$$

and the second order prolongation of L is given by (3.4.59)

$$\mathsf{pr}^{(2)}\mathsf{L} = \mathsf{L} + \varphi^x \frac{\partial}{\partial u_x} + \varphi^y \frac{\partial}{\partial u_y} + \varphi^{xx} \frac{\partial}{\partial u_{xx}} + \varphi^{xy} \frac{\partial}{\partial u_{xy}} + \varphi^{yy} \frac{\partial}{\partial u_{yy}}\,. \tag{3.4.70}$$

Also, the condition (3.4.55) becomes

$$\varphi^{xx} + \varphi^{yy} + k^2\varphi = 0\,,$$

and by means of the formulae (3.4.60) we calculate

$$\begin{aligned}\varphi^{xx} &= \mathsf{D}_x^2(\varphi - u_x\xi - u_y\eta) + u_{xxx}\xi + u_{xxy}\eta\,,\\ \varphi^{yy} &= \mathsf{D}_y^2(\varphi - u_x\xi - u_y\eta) + u_{xyy}\xi + u_{yyy}\eta\,.\end{aligned}$$

Finally, we obtain the equation

$$\begin{aligned}\varphi_{xx} + u_x(2\varphi_{xu} - \xi_{xx}) - u_y\eta_{xx} + u_x^2(\varphi_{uu} - 2\xi_{xu}) - 2u_xu_y\eta_{xu} - u_x^3\xi_{uu}\\ - u_x^2u_y\eta_{uu} + u_{xx}(\varphi_u - 2\xi_x) - 2u_{xy}\eta_x - 3u_xu_{xx}\xi_u - u_{xx}u_y\eta_u - 2u_{xy}u_x\eta_u\\ + \varphi_{yy} + u_y(2\varphi_{yu} - \eta_{yy}) - u_x\xi_{yy} + u_y^2(\varphi_{uu} - 2\eta_{yu}) - 2u_xu_y\xi_{yu}\\ - u_y^3\eta_{uu} - u_y^2u_x\xi_{uu} + u_{yy}(\varphi_u - 2\eta_y) - 2u_{xy}\xi_y - 3u_yu_{yy}\eta_u\\ - u_xu_{yy}\xi_u - 2u_{xy}u_y\xi_u + k^2\varphi = 0\,,\end{aligned}$$

in which we have to replace $u_{xx} = -u_{yy} - k^2u$. Then, by vanishing the coefficients of different derivatives of the function u (considered as independent variables) we are led to the system of independent equations

$$\begin{aligned}&\xi_u = \eta_u = 0\,, \quad \xi_x = \eta_y = \frac{1}{2}\varphi_u\,, \quad \eta_x + \xi_y = 0\,,\\ &2\varphi_{xu} - \Delta\xi = 0\,, \quad 2\varphi_{yu} - \Delta\eta = 0\,, \quad \Delta\varphi + k^2\varphi = 0\,.\end{aligned} \tag{3.4.71}$$

From the first sequence of equations in (3.4.71) we obtain $\varphi_{uu} = 0$ and, because x, y, and u are considered independent variables, we may choose $\varphi(x,y,u) = f(x,y)\,u + g(x,y)$, where the functions f and g satisfy Helmholtz's equation (3.4.61). Thus, the system (3.4.71) becomes

$$\begin{aligned}&\xi_u = \eta_u = 0\,, \quad \xi_x = \eta_y = \frac{1}{2}f\,, \quad \eta_x + \xi_y = 0\,,\\ &\Delta\xi = 2f_x\,, \quad \Delta\eta = 2f_y\,, \quad \Delta f + k^2 f = 0\,, \quad \Delta g + k^2 g = 0\,,\end{aligned}$$

having the general solution

$$f = 0\,, \quad \xi = \gamma y + \alpha\,, \quad \eta = -\gamma x + \beta\,,$$

where α, β, and γ are arbitrary constants (the parameters of the symmetry group). Also, the infinitesimal generator (3.4.69) takes the form

$$\mathsf{L} = \alpha\partial_x + \beta\partial_y + \gamma(y\partial_x - x\partial_y) + g\partial_u\,. \tag{3.4.72}$$

Therefore, the Lie algebra associated to the symmetry group of Helmholtz's equation (3.4.61) is formed by the three-dimensional real subalgebra

$$\mathcal{E}(2) = \{\alpha\mathsf{P}_1 + \beta\mathsf{P}_2 + \gamma\mathsf{M}|\alpha, \beta, \gamma \in \mathbb{R}\}\,, \tag{3.4.73}$$

where

$$\mathsf{P}_1 = \partial_x\,, \quad \mathsf{P}_2 = \partial_y\,, \quad \mathsf{M} = y\partial_x - x\partial_y\,, \tag{3.4.74}$$

and the infinite-dimensional subalgebra of operators $g(x, y)\partial_u$. The operators (3.4.74) constitute the basis of the Lie algebra $\mathcal{E}(2)$ and satisfy the commutation relations

$$[\mathsf{P}_1, \mathsf{P}_2] = 0\,, \quad [\mathsf{M}, \mathsf{P}_1] = \mathsf{P}_2\,, \quad [\mathsf{M}, \mathsf{P}_2] = -\mathsf{P}_1\,. \tag{3.4.75}$$

It follows that the Lie algebra $\mathcal{E}(2)$ is isomorphic to the Lie algebra associated to the Euclidean group $\mathsf{E}(2)$, which is the group of rigid motions of the plane. The elements of $\mathsf{E}(2)$ are plane rotations about an axis and plane translations, defined by the relations

$$\begin{gathered} \mathbf{x}' = g\mathbf{x} = (x\,\cos\theta - y\,\sin\theta + a,\; x\,\sin\theta + y\,\cos\theta + b)\,, \\ \mathbf{x}', \mathbf{x} \in \mathbb{R}^2\,, \quad g \in \mathsf{E}(2)\,, \quad a, b \in \mathbb{R}\,, \quad \theta \in [0, 2\pi) \end{gathered} \tag{3.4.76}$$

(a detailed treatment of $\mathsf{E}(2)$ will be given in Chap. 4 Sec. 2.2.2).

In terms of the operators (3.4.74), Helmholtz's equation (3.4.61) can be written in the form

$$\Omega u \equiv \left(\mathsf{P}_1^2 + \mathsf{P}_2^2 + k^2\right) u = 0\,, \tag{3.4.77}$$

making thus evident that

$$[\Omega, \mathsf{P}_1] = [\Omega, \mathsf{P}_2] = [\Omega, \mathsf{M}] = 0\,.$$

Besides the symmetry operators of the first order (3.4.72), one can determine symmetry operators of the second order (S), which satisfy the relation

$$[\mathsf{S}, \Omega] = \mathsf{U}\,\Omega\,,$$

where U is a differential operator of the first order. It can be shown that the operators

$$\mathsf{P}_1^2\,, \quad \mathsf{P}_1\mathsf{P}_2\,, \quad \mathsf{M}^2\,, \quad [\mathsf{M}, \mathsf{P}_1]_+\,, \quad [\mathsf{M}, \mathsf{P}_2]_+\,, \tag{3.4.78}$$

where $[\mathsf{M}, \mathsf{P}_j]_+ = \mathsf{M}\mathsf{P}_j + \mathsf{P}_j\mathsf{M}$, $j = 1, 2$, are symmetry operators of the second order for Helmholtz's equation (3.4.61). Because a symmetry operator of the first order is a particular case of a symmetry operator of the second order, we associate to the operator Ω the linear space

$$\mathcal{S} = \{\mathsf{S} \,|\, [\mathsf{S}, \Omega] = \mathsf{U}\,\Omega\},$$

whose basis is formed by the operators (3.4.74) and (3.4.78).

The connection between the symmetry group of the partial differential equation and the method of separation of variables results from the fact that the solutions (3.4.64) satisfy relations of the form

$$\mathsf{P}_1 u_\lambda^{11} = i\lambda\, u_\lambda^{11}\,, \quad \mathsf{P}_2 u_\lambda^{11} = i\sqrt{k^2 - \lambda^2}\, u_\lambda^{11}\,.$$

Thus, in Cartesian coordinates, we can characterize the separated solutions of (3.4.61), observing that they are eigenfunctions common to both symmetry operators $\mathsf{P}_1, \mathsf{P}_2 \in \mathcal{E}(2)$. In polar coordinates, the symmetry operator M in (3.4.74) has the form $\mathsf{M} = -\partial_\theta$, and the separated solutions (3.4.68) satisfy the relation

$$\mathsf{M} u_\mu^{11} = -i\mu\, u_\mu^{11}\,,$$

hence they are eigenfunctions of the symmetry operator $\mathsf{M} \in \mathcal{E}(2)$, the separation constant playing the role of an eigenvalue.

In both cases, the solution of (3.4.61) is factorized such that one of the factors is an eigenfunction of one of the operators in $\mathcal{E}(2)$, the corresponding operator being of diagonal form ($\mathsf{P}_1 = \partial_x$ or $\mathsf{M} = -\partial_\theta$) in the space of the solutions of the considered equation. Hence, if we choose an operator $\mathsf{L} \in \mathcal{E}(2)$ and if we can determine a system of coordinates (x', y') such that $\mathsf{L} = \partial_{x'}$, then in this system of coordinates we can separate the variables of (3.4.61), and the separated solution is an eigenfunction of L. It can be shown that, for any operator $\mathsf{L} = A(x,y)\partial_x + B(x,y)\partial_y$, where the functions $A(x,y)$ and $B(x,y)$ are analytic in $\mathcal{D}$, it is always possible to find, in a vicinity of a given point $(x_0, y_0) \in \mathcal{D}$, a system of coordinates (x', y') such that $\mathsf{L} = \partial_{x'}$. In this system of coordinates, the operator Ω in (3.4.61) becomes

$$\Omega \equiv \Delta + k^2 = B_{11}\partial_{x'x'} + B_{12}\partial_{x'y'} + B_{22}\partial_{y'y'} + C_1\partial_{x'} + C_2\partial_{y'} + k^2\,, \quad (3.4.79)$$

where B_{ij} and C_j are analytic functions defined in a vicinity of the point (x_0, y_0). If $\mathsf{L} \in \mathcal{E}(2)$ is a symmetry operator of (3.4.61), then $[\mathsf{L}, \Omega] = 0$, and by evaluating this commutator in the system of coordinates (x', y') it follows that the functions B_{ij} and C_j are independent of x'. Thus, the Helmholtz equation (3.4.61) will admit separated solutions of the form

$$u_\lambda = e^{i\lambda x'}\,\varphi(y')\,,$$

in which the function $\varphi(y')$ satisfies the ordinary differential equation

$$B_{22}\varphi'' + (i\lambda B_{12} + C_2)\,\varphi' + \left(-\lambda^2 B_{11} + i\lambda C_1 + k^2\right)\varphi = 0\,. \qquad (3.4.80)$$

The solutions u_λ will be characterized by the eigenvalue equation

$$\mathsf{L}\, u_\lambda = i\lambda\, u_\lambda\,. \qquad (3.4.81)$$

Let $\mathcal{F}$ be the space of analytic functions defined on $\mathcal{D}$. The action of the elements of the Euclidean group $\mathsf{E}(2)$ in $\mathcal{F}$ is defined by means of the representation T whose operators satisfy the relation

$$\mathsf{T}(g)\, f(\mathbf{x}) = e^{\theta\mathsf{M}}\, e^{a\mathsf{P}_1 + b\mathsf{P}_2}\, f(\mathbf{x}) = f(g^{-1}\mathbf{x})\,, \quad \mathbf{x}\in\mathcal{D}\,, \quad f\in\mathcal{F}\,.$$

The solutions of (3.4.61) form a subspace $\mathcal{F}_0 \subset \mathcal{F}$, which is the kernel of the operator Ω. Because $\mathsf{E}(2)$ is the symmetry group of (3.4.61), the subspace $\mathcal{F}_0$ is an invariant subspace for the representation T, that is $\forall\, u_\lambda \in \mathcal{F}_0$ we have $\mathsf{T}(g)u_\lambda \in \mathcal{F}_0$. Also, the solutions u_λ satisfy the eigenvalue equation (3.4.81), so that we obtain

$$\mathsf{T}(g)\,\mathsf{L}\, u_\lambda = \mathsf{T}(g)\mathsf{L}\mathsf{T}(g^{-1})\mathsf{T}(g)u_\lambda = i\lambda\,\mathsf{T}(g)u_\lambda\,, \qquad (3.4.82)$$

implying that the functions $\mathsf{T}(g)u_\lambda$ are eigenfunctions of the operator $\mathsf{L}^g = \mathsf{T}(g)\mathsf{L}\mathsf{T}(g^{-1})$. This operator is a symmetry operator of Helmholtz's equation (3.4.61) and it can be shown that it is an operator of the first order, hence $\mathsf{L}^g \in \mathcal{E}(2)$. If u_λ is a separated solution in the variables $\mathbf{x} = (x, y)$, then $\mathsf{T}(g)u_\lambda$ is a separated solution in the variables $\mathbf{x}' = g\mathbf{x}$, $g \in \mathsf{E}(2)$. Because the systems of coordinates (x, y) and (x', y') are obtained one from the other by a transformation corresponding to an element in the symmetry group $\mathsf{E}(2)$, we consider that these systems of coordinates are equivalent. Also, the coordinates and the eigenfunctions associated to the operator $\mathsf{L} \in \mathcal{E}(2)$ are the same as the coordinates and the eigenfunctions associated to the operator $c\mathsf{L}$, $c \in \mathbb{R}\setminus\{0\}$. Thus, all the separable systems of coordinates associated to the operators $c\,\mathsf{L}^g$, $g \in \mathsf{E}(2)$, are equivalent to the separable system of coordinates associated to the operator $\mathsf{L} \in \mathcal{E}(2)$.

The action of $\mathsf{E}(2)$ on the Lie algebra $\mathcal{E}(2)$, defined by $\mathsf{L} \xrightarrow{g} \mathsf{L}^g$, constitutes the *adjoint representation* of $\mathsf{E}(2)$, and generates a partitioning of the Lie algebra $\mathcal{E}(2)$ into *orbits* (one-dimensional subspaces). We say that an operator $\mathsf{K} \in \mathcal{E}(2)$ lies on the same orbit as the operator L if $\exists\, g \in \mathsf{E}(2)$ and $c \in \mathbb{R}\setminus\{0\}$, such that $\mathsf{K} = c\,\mathsf{L}^g$.

It can be proved that the Lie algebra $\mathcal{E}(2)$ contains only two orbits represented by the operators P_1 and M. Therefore, any operator in $\mathcal{E}(2)$ lies on one of these two orbits. Consequently, there exist only two non-equivalent systems of coordinates, associated to symmetry operators of the first order, in which the solutions of Helmholtz's equation (3.4.61) can be separated: the Cartesian system of coordinates and the polar system of coordinates. These systems of coordinates are called subgroup coordinates because they are associated respectively, with the diagonalization of the infinitesimal generators for the subgroup of translations ($\mathsf{P}_1 = \partial_x$), and the subgroup of rotations ($\mathsf{M} = -\partial_\theta$) from $\mathsf{E}(2)$.

The algebra of the symmetry operators of the second order (3.4.78) contains four orbits, represented by the operators P_2^2, M^2, $[\mathsf{M}, \mathsf{P}_2]_+$, and $\mathsf{M}^2 - a^2\mathsf{P}_2^2$, $a \neq 0$. To these orbits there correspond the non-equivalent systems of coordinates: Cartesian (x, y), polar (r, θ), parabolic ($x = (\xi^2 - \eta^2)/2$, $y = \xi\eta$), and elliptic ($x = k \cosh\alpha \cos\beta$, $y = k \sinh\alpha \sin\beta$). One can prove that these are the only systems of coordinates in which we can apply the method of separation of variables to solve Helmholtz's equation (3.4.61). Note that Helmholtz's equation is a partial differential equation of the second order, hence the maximum number of non-equivalent systems of coordinates in which the solutions can be separated is determined by the number of orbits in the algebra of the symmetry operators of the second order.

The results obtained for the particular case of Helmholtz's equation have a general character. Thus, let

$$\Omega u \equiv \sum_{i,j=1}^{n} A_{ij}(\mathbf{x}) \frac{\partial^2 u}{\partial x_i \, \partial x_j} + \sum_{i=1}^{n} B_i(\mathbf{x}) \frac{\partial u}{\partial x_i} + C(\mathbf{x}) u = 0 \qquad (3.4.83)$$

be a partial differential equation of the second order, linear and homogeneous, in n variables. Here, $\mathbf{x} = (x_1, x_2, \ldots, x_n)$, and $u(\mathbf{x})$, $A_{ij}(\mathbf{x})$, $B_i(\mathbf{x})$, $C(\mathbf{x})$ are analytic functions defined in a domain $\mathcal{D} \subseteq \mathbb{R}^n$. All the systems of coordinates $(y_1, y_2, \ldots, y_n)$, in which we can apply the method of separation of variables to solve the equation (3.4.83), are characterized by the system of n eigenvalue equations

$$\mathsf{S}_j u = \lambda_j \, u \,, \quad j = 1, 2, \ldots, n-1, \quad \Omega u = 0 \,, \qquad (3.4.84)$$

where $\mathsf{S}_1, \mathsf{S}_2, \ldots, \mathsf{S}_{n-1} \in \mathcal{S}$ are symmetry operators of the second order (for the equation (3.4.83)), which commute between them ($[\mathsf{S}_j, \mathsf{S}_k] = 0$), and λ_1, λ_2, ..., λ_{n-1} are separation constants. For such a systems of coordinates the solution of (3.4.83) has the form

$$u = Y_1(y_1) \, Y_2(y_2) \, \ldots \, Y_n(y_n) \,.$$

4.3.3 Similarity solutions of partial differential equations

Likewise the case of systems of ordinary differential equations (see Chap. 3 Sec. 4.1.1), the invariance of a system of partial differential equations with respect to a Lie group of transformations enables us to find some particular solutions of the corresponding system of equations. These solutions, called *similarity* or *invariant* solutions, are invariant with respect to a subgroup of the full symmetry group admitted by the system of partial differential equations, and are obtained by solving a reduced system of equations comprising a smaller number of independent variables. Here, we will illustrate this technique by detailing the solution of Navier's equation (also known as the Navier-Lamé equation) in classical elasticity, as developed by T. Özer.

For a linear, homogeneous, and isotropic elastic medium the Navier equation has the form

$$(\lambda + \mu)\nabla\, \nabla \cdot \mathbf{u} + \mu\, \nabla^2 \mathbf{u} + \rho\, \mathbf{p} = \rho\, \frac{\partial^2 \mathbf{u}}{\partial t^2}\,, \tag{3.4.85}$$

or equivalently

$$(\lambda + 2\mu)\nabla\, \nabla \cdot \mathbf{u} - \mu\, \nabla \times \nabla \times \mathbf{u} + \rho\, \mathbf{p} = \rho\, \frac{\partial^2 \mathbf{u}}{\partial t^2}\,, \tag{3.4.86}$$

where λ and μ are the Lamé constants, $\mathbf{u}$ is the displacement vector, ρ is the density of the medium, and $\mathbf{p}$ is the body force per unit mass. We can distinguish two cases: $\partial^2\mathbf{u}/\partial t^2 = \mathbf{0}$ (elastostatics problems), when (3.4.85) provides the distribution of a static force field, and $\partial^2\mathbf{u}/\partial t^2 \neq \mathbf{0}$ (elastodynamics problems), when (3.4.85) describes the propagation of elastic waves in the medium. In the case of an axially-symmetric problem, the system of equations (3.4.85) reduces to two coupled equations for the parallel (w) and perpendicular (u) components of $\mathbf{u}$. By assuming that the symmetry axis is Oz, and in cylindrical coordinates (r, θ, z), we obtain

$$\begin{aligned}
&(\lambda + 2\mu)\left(u_{rr} - \frac{1}{r^2}\,u + \frac{1}{r}\,u_r + w_{rz}\right) + \mu\,\frac{\partial}{\partial z}\,(u_z - w_r) + \rho p^{(r)} = \rho u_{tt},\\
&(\lambda + 2\mu)\left(u_{zr} + \frac{1}{r}\,u_z + w_{zz}\right) - \mu\left[\frac{\partial}{\partial r}\,(u_z - w_r) + \frac{1}{r}\,(u_z - w_r)\right]\\
&\qquad\qquad + \rho p^{(z)} = \rho w_{tt},
\end{aligned} \tag{3.4.87}$$

where the parallel coordinate is z, the perpendicular component is r, and $p^{(z)}$ and $p^{(r)}$ are the parallel and perpendicular components of $\mathbf{p}$, respectively. Note that the axially-symmetry means invariance with respect to rotations about the z axis, which form a Lie group, and due to this invariance the number of independent variables has been reduced to three (the displacement $\mathbf{u}$ does not depend on θ). We will now continue by considering separately the two cases of elastostatics and elastodynamics problems.

In the case of elastostatics problems $w_{tt} = u_{tt} = 0$, and we assume that the elastic medium is subjected to a single force (F) acting along Oz at the origin of coordinates, such that the components of the body force are

$$p^{(r)} = 0\,, \quad \rho\, p^{(z)} = F\,\frac{1}{2\pi r}\,\delta(r)\,\delta(z)\,,$$

and the system (3.4.87) becomes

$$\begin{aligned} &u_{rr} + \frac{1}{r}\,u_r - \frac{1}{r^2}\,u + w_{rz} + \kappa\,(u_{zz} - w_{rz}) = 0, \\ &u_{zr} + \frac{1}{r}\,u_z + w_{zz} - \kappa\left[u_{rz} - w_{rr} + \frac{1}{r}\,(u_z - w_r)\right] = -\frac{F}{\rho c_0^2}\,\frac{\delta(r)}{2\pi r}\,\delta(z). \end{aligned} \qquad (3.4.88)$$

Here, $\kappa = \mu/(\lambda + 2\mu)$, $c_0^2 = (\lambda + 2\mu)/\rho$, and δ is Dirac's delta function. Also, c_0 is the speed of a longitudinal (pressure) wave in an elastic medium of density ρ, characterized by the Lamé constants λ and μ.

Now, the infinitesimal generator (3.4.58) is

$$\mathsf{L} = \xi\,\partial_r + \eta\,\partial_z + \zeta\,\partial_u + \tau\,\partial_w\,, \qquad (3.4.89)$$

and we assume that the coefficients are functions of the form

$$\begin{aligned} \xi &= \xi(r,z)\,, \qquad \eta = \eta(r,z)\,, \\ \zeta &= f(r,z)\,u + g(r,z)\,w\,, \\ \tau &= h(r,z)\,u + k(r,z)\,w\,. \end{aligned} \qquad (3.4.90)$$

The system (3.4.88) contains partial differential equations of the second order, so that we need the second order prolongation of L to determine the full symmetry group of (3.4.88). Thus, from (3.4.59) we have

$$\begin{aligned} \mathsf{pr}^{(1)}\mathsf{L} &= \mathsf{L} + \zeta_r\,\frac{\partial}{\partial u_r} + \zeta_z\,\frac{\partial}{\partial u_z} + \tau_r\,\frac{\partial}{\partial w_r} + \tau_z\,\frac{\partial}{\partial w_z}\,, \\ \mathsf{pr}^{(2)}\mathsf{L} &= \mathsf{pr}^{(1)}\mathsf{L} + \zeta_{rr}\,\frac{\partial}{\partial u_{rr}} + \zeta_{rz}\,\frac{\partial}{\partial u_{rz}} + \zeta_{zz}\,\frac{\partial}{\partial u_{zz}} \\ &\qquad + \tau_{rr}\,\frac{\partial}{\partial w_{rr}} + \tau_{rz}\,\frac{\partial}{\partial w_{rz}} + \tau_{zz}\,\frac{\partial}{\partial w_{zz}}\,. \end{aligned} \qquad (3.4.91)$$

By applying the operator $\mathsf{pr}^{(2)}\mathsf{L}$ to the system of equations (3.4.88), we obtain a new system of partial differential equations in which we have to vanish the coefficients of different derivatives of the functions u and w, considered as independent variables. We are thus led to a system of partial differential equations satisfied by the functions ξ, η, ζ, and τ, whose solution determines two infinitesimal generators

$$\mathsf{X}_1 = r\,\partial_r + z\,\partial_z\,, \quad \mathsf{X}_2 = u\,\partial_u + w\,\partial_w\,, \qquad (3.4.92)$$

such that the infinitesimal generator L takes the form

$$\mathsf{L} = \alpha\,\mathsf{X}_1 + \beta\,\mathsf{X}_2\,, \tag{3.4.93}$$

where α and β are arbitrary constants.

Let us assume that $u = \varphi(r,z)$ is a solution of (3.4.88). We use the infinitesimal generator (3.4.93) to form the expression

$$\mathsf{L}\,(u - \varphi(r,z))|_{u=\varphi} = 0\,, \tag{3.4.94}$$

that is

$$\alpha r \varphi_r + \alpha z \varphi_z - \beta\varphi = 0\,. \tag{3.4.95}$$

The characteristic system of ordinary differential equations associated to (3.4.95) has the form

$$\frac{dr}{\alpha r} = \frac{dz}{\alpha z} = \frac{d\varphi}{\beta\varphi}\,, \tag{3.4.96}$$

giving the general solution of (3.4.95) as

$$\varphi(r,z) = r^{\gamma}\,f(z/r)\,, \tag{3.4.97}$$

with $\gamma = \alpha/\beta$. In a similar manner, by taking $w = \psi(r,z)$, we obtain

$$\psi(r,z) = r^{\gamma}\,g(z/r)\,. \tag{3.4.98}$$

Note that f and g are arbitrary functions of class C^2. The system (3.4.88) comprises one homogeneous and one inhomogeneous equation. If we introduce a new variable $s = z/r$, and substitute $u = \varphi(r,z)$ and $w = \psi(r,z)$ into (3.4.88) we obtain a system of ordinary differential equations which has to be satisfied by the functions f and g:

$$\begin{aligned} r^{-2+\gamma}\,[(-1+\gamma^2)\,f(s) + (s - 2\,s\,\gamma)\,f'(s) - (1 - \gamma - \kappa + \gamma\,\kappa)g'(s) \\ +(s^2+\kappa)f''(s) - (s - s\,\kappa)g''(s)] = 0\,, \\ r^{-2+\gamma}\,[\gamma^2\,\kappa\,g(s) + \gamma(1-\kappa)f'(s) + s\,\kappa(1-2\,\gamma)g'(s) - s(1-\kappa)f''(s) \\ +(1+s^2\,\kappa)g''(s)] = -\frac{F}{\rho c_0^2}\,\frac{1}{2\pi r^3 s}\,\delta\left(\frac{1}{s}\right)\delta(s)\,, \end{aligned} \tag{3.4.99}$$

where we have used the relations

$$\delta(r) = \delta(z/s) = \frac{1}{z}\,\delta(1/s)\,, \quad \delta(z) = \delta(rs) = \frac{1}{r}\,\delta(s)\,.$$

It is evident that the compatibility of (3.4.99) imposes the restriction $\gamma = -1$. Finally, by solving the reduced system (3.4.99) for f and g, and substituting into

$$u = f(s)/r\,, \quad w = g(s)/r\,,$$

we obtain the general form of the displacement components

$$
\begin{aligned}
u(r,z) &= \frac{F(1-\kappa)}{8\pi\kappa c_0^2\rho}\,\frac{rz}{(r^2+z^2)^{3/2}}\,,\\
w(r,z) &= \frac{F}{4\pi\kappa c_0^2\rho}\,\frac{1}{(r^2+z^2)^{1/2}} - \frac{F(1-\kappa)}{8\pi\kappa c_0^2\rho}\,\frac{r^2}{(r^2+z^2)^{3/2}}\,.
\end{aligned}
$$

In the case of axially-symmetric elastodynamics problems, neglecting the body forces, the propagation of elastic waves in a linear, homogeneous, and isotropic medium is described by the equation

$$
\begin{aligned}
&(\lambda+2\mu)\left(u_{rr}-\frac{1}{r^2}\,u+\frac{1}{r}\,u_r+w_{rz}\right)+\mu\,\frac{\partial}{\partial z}\,(u_z-w_r)-\rho u_{tt}=0\,,\\
&\qquad\qquad\qquad\qquad\qquad\qquad\qquad\qquad\qquad\qquad\qquad\qquad\qquad\qquad (3.4.100)\\
&(\lambda+2\mu)\left(u_{zr}+\frac{1}{r}\,u_z+w_{zz}\right)-\mu\left[\frac{\partial}{\partial r}\,(u_z-w_r)+\frac{1}{r}\,(u_z-w_r)\right]\\
&\qquad\qquad\qquad\qquad\qquad\qquad\qquad\qquad\qquad\qquad\qquad -\rho w_{tt}=0\,.
\end{aligned}
$$

Also, the form of these equations has been obtained by considering a quasi-static problem. We assume a time dependence of the form $\exp{(-i\omega t)}$ for the displacement components, where ω is the angular frequency, and rearrange the equations (3.4.100) to obtain the *spectral Navier equations* of classical elasticity

$$
\begin{aligned}
&U_{rr}+\frac{1}{r}\,U_r-\frac{1}{r^2}\,U+W_{rz}+\kappa\,(U_{zz}-W_{rz})+\chi\,U=0\,,\\
&\qquad\qquad\qquad\qquad\qquad\qquad\qquad\qquad\qquad\qquad\qquad\qquad\qquad\qquad (3.4.101)\\
&U_{rz}+\frac{1}{r}\,U_z+W_{zz}-\kappa\left[U_{rz}-W_{rr}+\frac{1}{r}\,(U_z-W_r)\right]+\chi\,W=0\,,
\end{aligned}
$$

where

$$
u=U\,e^{-i\omega t}\,,\quad w=W\,e^{-i\omega t}\,,\quad \kappa=\frac{\mu}{\lambda+2\mu}\,,\quad \chi=\frac{\rho\omega^2}{\lambda+2\mu}\,.
$$

Then, we follow the same procedure as in the case of elastostatics problem, that is applying the second order prolongation (3.4.91) of the infinitesimal generator (3.4.89) to the equations in (3.4.101) to find the coefficients of the infinitesimal generator (3.4.89) in terms of two arbitrary constants α and β

$$
\xi=0\,,\quad \eta=\alpha\,,\quad \zeta=\beta U\,,\quad \tau=\beta W\,, \qquad (3.4.102)
$$

such that we have

$$
\mathsf{L}=\alpha\,\partial_z+\beta\,(U\,\partial_U+W\,\partial_W)\,. \qquad (3.4.103)
$$

Again, we consider two arbitrary solutions of (3.4.101), $\varphi(r, z)$ and $\psi(r, z)$, and solve the partial differential equations derived from expressions of the form (3.4.94). These led us to the similarity solutions

$$U(r, z) = \varphi(r)\, e^{\gamma z}\,, \quad W(r, z) = \psi(r)\, e^{\gamma z}\,, \tag{3.4.104}$$

with $\gamma = \alpha/\beta$. Note that we have used the same notations as in the case of elastostatics problem but the meaning of the symbols is different. The two problems are related in the quasi-static limit ($\omega \to 0$) when the solutions of the elastodynamics problem merge into the corresponding solutions of the elastostatics problem.

By substituting (3.4.104) into (3.4.101) we obtain the reduced system of ordinary differential equations

$$\begin{gathered}\varphi'' + \frac{1}{r}\,\varphi' + \left(\chi + \gamma^2\,\kappa - \frac{1}{r^2}\right)\varphi + \gamma(1-\kappa)\,\psi' = 0\,,\\ \gamma(1-\kappa)\,\varphi' + (1-\kappa)\,\frac{\gamma}{r}\,\varphi + \kappa\,\psi'' + \frac{\kappa}{r}\,\psi' + (\gamma^2 + \chi)\,\psi = 0\,.\end{gathered} \tag{3.4.105}$$

The general solution of this system has been given by T. Özer in the form

$$\begin{aligned}\varphi(r) &= C_1\, I_1(r\,M) + C_2\, K_1(r\,M)\,,\\ \psi(r) &= C_1 + \frac{\gamma(\lambda+2\mu)M(1-I_0(r\,M))C_1}{\rho\omega^2+\gamma\lambda+2\gamma^2\mu} - \frac{\gamma C_2}{M}\,K_0(r\,M)\,,\\ M &= \left(-\frac{\rho\omega^2+\gamma^2(\lambda+\mu)}{\lambda+2\mu}\right)^{1/2}\,,\end{aligned} \tag{3.4.106}$$

where I and K are the modified Bessel functions of the first and the second kind, respectively. Also, the condition that the components of the stress tensor vanish in the limit $r, z \to \infty$ gives $C_1 = 0$, such that the similarity solutions of the spectral Navier equations are

$$U(r, z) = K_1(r\,M)\, e^{\gamma z}\,, \quad W(r, z) = -\frac{\gamma}{M}\, K_0(r\,M)\, e^{\gamma z}\,. \tag{3.4.107}$$

Chapter 4

APPLICATIONS IN MECHANICS

1. Classical models of mechanics

1.1 Lagrangian formulation of classical mechanics

1.1.1 Configuration space. Hamilton's principle

To set up the basic definitions and notations used in classical mechanics, we begin this chapter with an outline of the Lagrangian and Hamiltonian formulations. Then, we analyse in detail the symmetry properties of the basic equations in both formulations, and show a demonstration of Noether's theorem and its reciprocal. Following these preliminaries, the main part of the chapter is dedicated to the study of the one-to-one correspondence between symmetry properties of mechanical systems, arising from the invariance of the Lagrangian with respect to certain Lie groups, and conservation laws.

The spatial position of a *discrete mechanical system* $\mathcal{S}$, consisting of n free particles, is specified by $3n$ Cartesian coordinates, or n radius vectors. When the mechanical system is subjected to some constraints the $3n$ coordinates have to satisfy certain relations, such that the number of independent coordinates becomes less than $3n$. Thus, we assume that there are $m < 3n$ independent *geometrical constraints* (*holonomic*) of the form

$$f_k(\mathbf{r}_1, \mathbf{r}_2, \ldots, \mathbf{r}_n; t) = 0, \quad k = 1, 2, \ldots, m, \tag{4.1.1}$$

where $\mathbf{r}_j = (x_j, y_j, z_j)$ denotes the radius vector of the particle j, and the functions f_k are of class C^2. Since the constraints (4.1.1) are independent, the matrix

$$\left[\frac{\partial(f_1, f_2, \ldots, f_m)}{\partial(x_1, y_1, z_1, x_2, \ldots, z_n)}\right]$$

contains at least one non-vanishing determinant of order m (i.e., the rank of the matrix is m) and, from the theorem of implicit functions, it follows that

m coordinates can be expressed as functions of the other $s = 3n - m$ independent coordinates, denoted by q_k $(k = 1, 2, \ldots, s)$. The coordinates q_k are called *generalized coordinates* (*Lagrange's coordinates*) of the considered mechanical system. The set of all values which can be assigned to the generalized coordinates defines an s-dimensional space (*configuration space*, or *Lagrange's space*) denoted by Λ_s.

To each representative point $P(q_1, q_2, \ldots, q_s) \in \Lambda_s$ it corresponds a position of the mechanical system $\mathcal{S}$ in the real space $\mathcal{E}_3$ (in accordance with the considered holonomic constraints) and reciprocally, by means of the one-to-one mapping

$$\mathbf{r}_i = \mathbf{r}_i(q_1, q_2, \ldots, q_s; t), \quad i = 1, 2, \ldots, n, \tag{4.1.2}$$

where $\mathbf{r}_i$ are functions of class C^2, and the time (t) is regarded as a parameter. For an interval of time $t \in [t_0, t_1]$ the mechanical system $\mathcal{S}$ takes a sequence of positions in the space $\mathcal{E}_3$, which corresponds to a sequence of points in the space Λ_s. Consequently, the evolution of $\mathcal{S}$ is represented by a system of equations of the form

$$q_k = q_k(t), \quad k = 1, 2, \ldots, s, \quad t \in [t_0, t_1], \tag{4.1.3}$$

which defines a trajectory of the *representative point* P in the space Λ_s, and, by means of (4.1.2), the trajectories of all the constituent particles in the space $\mathcal{E}_3$. The derivatives of the generalized coordinates with respect to the time $\dot{q}_k$ $(k = 1, 2, \ldots, s)$ are the *generalized velocities*. Hence, the state of the discrete mechanical system $\mathcal{S}$ at a given instant is determined by the generalized coordinates $\{q_k\}$ and the generalized velocities $\{\dot{q}_k\}$.

In the Newtonian mechanics the equations of motion of a representative point are obtained from *Hamilton's principle*. According to this principle, a discrete mechanical system $\mathcal{S}$ is characterized by a scalar function $\mathcal{L} = \mathcal{L}(\mathbf{q}, \dot{\mathbf{q}}; t)$, called *Lagrangian* (or *Lagrange's kinetic potential*), where we have used the notations $\mathbf{q}$ and $\dot{\mathbf{q}}$ for the sets of generalized coordinates $\{q_1, q_2, \ldots, q_s\}$ and generalized velocities $\{\dot{q}_1, \dot{q}_2, \ldots, \dot{q}_s\}$, respectively. The motion of the representative point $P(\mathbf{q})$ between two representative points, determined by the generalized coordinates corresponding to the instants t_0 and t_1 (therefore, the evolution of the mechanical system $\mathcal{S}$), takes place in such a way that the functional

$$\mathcal{A} = \int_{t_0}^{t_1} \mathcal{L}(\mathbf{q}, \dot{\mathbf{q}}; t)\, dt \tag{4.1.4}$$

is stationary (i.e., the synchronous variation $\delta\mathcal{A}$ on different arbitrary paths with fixed ends is equal to zero). The functional $\mathcal{A}$ is called *Lagrangian action*.

1.1.2 Lagrange's equations

We note that

$$\delta\mathcal{L} = \frac{\partial\mathcal{L}}{\partial q_k}\delta q_k + \frac{\partial\mathcal{L}}{\partial\dot{q}_k}\delta\dot{q}_k = \frac{\partial\mathcal{L}}{\partial q_k}\delta q_k + \frac{d}{dt}\left(\frac{\partial\mathcal{L}}{\partial\dot{q}_k}\delta q_k\right) - \frac{d}{dt}\left(\frac{\partial\mathcal{L}}{\partial\dot{q}_k}\right)\delta q_k,$$

where we have used the summation convention for dummy Latin indices from 1 to s, and the commutability of operators $\delta(d/dt) = (d/dt)\delta$. The arbitrary paths for the variation $\delta\mathcal{A}$ have fixed ends, that is $(\delta q_k)_{t=t_0} = (\delta q_k)_{t=t_1} = 0$, and we may write

$$\delta\mathcal{A} = \int_{t_0}^{t_1}\delta\mathcal{L}\,dt = \int_{t_0}^{t_1}\left[\frac{\partial\mathcal{L}}{\partial q_k} - \frac{d}{dt}\left(\frac{\partial\mathcal{L}}{\partial\dot{q}_k}\right)\right]\delta q_k\,dt\,.$$

Finally, by cancelling the first synchronous variation of the Lagrangian action (4.1.4), and taking into account the fact that the variations δq_k are arbitrary, we obtain *Lagrange's equations of the second kind*

$$\frac{d}{dt}\left(\frac{\partial\mathcal{L}}{\partial\dot{q}_k}\right) - \frac{\partial\mathcal{L}}{\partial q_k} = 0\,,\quad k = 1, 2, \ldots, s\,, \tag{4.1.5}$$

which form a system of s ordinary differential equations of second order. The solutions of this system of equations describe the time dependence of the generalized coordinates $q_k = q_k(t)$, $k = 1, 2, \ldots, s$, and define the integral trajectories along which the Lagrangian action is stationary. Therefore, Lagrange's equations (4.1.5) represent the equations of motion of the representative point P in the space Λ_s. In the calculus of variations, the equations (4.1.5) are also called *Euler's equations* corresponding to the problem of finding the extrema of the functional $\mathcal{A}$.

All these results have been obtained for the case of *natural mechanical systems* of particles, when the Lagrangian $\mathcal{L}$ has the form

$$\mathcal{L} = T + U\,, \tag{4.1.6}$$

where T represents the total kinetic energy of the mechanical system, and U is a scalar function which can be a *simple potential*, if $U = U(\mathbf{q})$, or a *simple quasi-potential*, if $U = U(\mathbf{q}; t)$. In such cases, the generalized force is respectively a *conservative* or a *quasi-conservative force*, and is given by

$$Q_k = \frac{\partial U}{\partial q_k}\,,\quad k = 1, 2, \ldots, s\,. \tag{4.1.7}$$

If $U = U(\mathbf{q}, \dot{\mathbf{q}})$, or $U = U(\mathbf{q}, \dot{\mathbf{q}}; t)$, such that

$$Q_k = \frac{\partial U}{\partial q_k} - \frac{d}{dt}\left(\frac{\partial U}{\partial\dot{q}_k}\right)\,,\quad k = 1, 2, \ldots, s\,, \tag{4.1.8}$$

we have a *generalized*, or a *quasi-generalized potential*, respectively. For mechanical systems, which are not natural, Lagrange's equations take the more general form

$$\frac{d}{dt}\left(\frac{\partial T}{\partial \dot{q}_k}\right) - \frac{\partial T}{\partial q_k} = Q_k\,, \quad k = 1, 2, \ldots, s\,, \tag{4.1.9}$$

and come up as a consequence of the principle of virtual work

$$\sum_{i=1}^{n} (m_i \ddot{\mathbf{r}}_i - \mathbf{F}_i) \cdot \delta \mathbf{r}_i = 0\,, \tag{4.1.10}$$

where $\mathbf{F}_i$ denotes the force acting upon the particle i in the space $\mathcal{E}_3$, and m_i is the mass of particle i. In this case, the generalized forces are defined as

$$Q_k = \sum_{i=1}^{n} \mathbf{F}_i \cdot \frac{\partial \mathbf{r}_i}{\partial q_k}\,, \quad k = 1, 2, \ldots, s\,. \tag{4.1.11}$$

The total kinetic energy T is given by a sum of homogeneous functions of second, first, and zeroth degree

$$T = T_2 + T_1 + T_0\,, \tag{4.1.12}$$

where

$$\begin{aligned} T_2 &= \frac{1}{2} g_{jk} \dot{q}_j \dot{q}_k\,, \quad g_{jk} = \sum_{i=1}^{n} m_i \frac{\partial \mathbf{r}_i}{\partial q_j} \cdot \frac{\partial \mathbf{r}_i}{\partial q_k}\,, \\ T_1 &= g_j \dot{q}_j\,, \quad g_j = \sum_{i=1}^{n} m_i \frac{\partial \mathbf{r}_i}{\partial q_j} \cdot \frac{\partial \mathbf{r}_i}{\partial t}\,, \\ T_0 &= g_0\,, \quad g_0 = \sum_{i=1}^{n} m_i \left(\frac{\partial \mathbf{r}_i}{\partial t}\right)^2 . \end{aligned} \tag{4.1.13}$$

When the mechanical system is subjected to scleronomic constraints (which do not explicitly depend on time) $\partial \mathbf{r}_i / \partial t = \mathbf{0}$, $U(\mathbf{q})$ is a potential, and

$$T = T_2 = \frac{1}{2} \sum_{i=1}^{n} m_i \left(\frac{\partial \mathbf{r}_i}{\partial q_j} \dot{q}_j\right)^2 . \tag{4.1.14}$$

The mapping between the positions of the discrete mechanical system in the space $\mathcal{E}_3$ and the representative points in Λ_s is one-to-one. Therefore, the matrix

$$\left[\frac{\partial(x_1, y_1, z_1, x_2, \ldots, z_n)}{\partial(q_1, q_2, \ldots, q_s)}\right]$$

is of rank s, and the linear system of equations

$$\frac{\partial \mathbf{r}_i}{\partial q_j}\,\dot{q}_j = \mathbf{0}\,, \quad i = 1, 2, \ldots, n\,,$$

in generalized velocities, has only trivial solutions. It follows that the kinetic energy $T = T_2$ is a *positive definite quadratic form*, which vanishes only when all the generalized velocities are zero.

In the case of scleronomic constraints we also have

$$\begin{aligned}\frac{d}{dt}\left(\frac{\partial T}{\partial \dot{q}_k}\right) &= \frac{d}{dt}\left(g_{jk}\,\dot{q}_j\right) = g_{jk}\,\ddot{q}_j + \frac{\partial g_{jk}}{\partial q_l}\,\dot{q}_j\dot{q}_l \\ &= g_{jk}\,\ddot{q}_j + \frac{1}{2}\left(\frac{\partial g_{jk}}{\partial q_l} + \frac{\partial g_{kl}}{\partial q_j}\right)\dot{q}_j\dot{q}_l\,, \\ \frac{\partial T}{\partial q_k} &= \frac{\partial g_{jl}}{\partial q_k}\,\dot{q}_j\dot{q}_l\,,\end{aligned}$$

and, by introducing the *Christoffel symbols of the first kind*

$$[jl, k] = \frac{1}{2}\left(\frac{\partial g_{jk}}{\partial q_l} + \frac{\partial g_{kl}}{\partial q_j} - \frac{\partial g_{jl}}{\partial q_k}\right), \quad j, k, l = 1, 2, \ldots, s\,, \tag{4.1.15}$$

we may rewrite (4.1.9) in the form

$$g_{jk}\,\ddot{q}_j + [jl, k]\,\dot{q}_j\dot{q}_l = Q_k\,, \quad k = 1, 2, \ldots, s\,. \tag{4.1.16}$$

Now, we introduce the *Christoffel symbols of the second kind*

$$\left\{ \begin{matrix} m \\ j \ \ l \end{matrix} \right\} = g^{km}\,[jl, k]\,, \tag{4.1.17}$$

where g^{km} are the components of the inverse of g_{jk} ($g_{jk}g^{km} = \delta_j^m$), and the *generalized forces*

$$\widetilde{Q}_m = Q_k\,g^{km}\,, \tag{4.1.18}$$

to obtain the normal form of Lagrange's equations

$$\ddot{q}_m + \left\{ \begin{matrix} m \\ j \ \ l \end{matrix} \right\}\dot{q}_j\dot{q}_l = \widetilde{Q}_m\,, \quad m = 1, 2, \ldots, s\,. \tag{4.1.19}$$

Note that the normal form of Lagrange's equations reveals the *generalized accelerations*. Hence, the unknown functions (4.1.3) are obtained as solutions of a system of s ordinary differential equations of the second order. Generally, the solutions of this system depend on $2s$ integration constants, which can be determined by means of the Cauchy initial conditions

$$q_k(t_0) = q_k^0\,, \quad \dot{q}_k(t_0) = \dot{q}_k^0\,, \quad k = 1, 2, \ldots, s\,, \tag{4.1.20}$$

i.e., by knowing the initial position ($\mathbf{q}^0$) and velocity ($\dot{\mathbf{q}}^0$) of the representative point in Λ_s.

1.1.3 Gauge transformations

Let be a transformation of the Lagrangian

$$\mathcal{L}'(\mathbf{q},\dot{\mathbf{q}};t)=\mathcal{L}(\mathbf{q},\dot{\mathbf{q}};t)+\widetilde{\mathcal{L}}(\mathbf{q},\dot{\mathbf{q}};t)\,. \tag{4.1.21}$$

The Lagrangian $\mathcal{L}'\in C^2$ is also a kinetic potential if it satisfies (4.1.5), and, by substituting (4.1.21) into (4.1.5), we obtain a similar condition for the function $\widetilde{\mathcal{L}}$

$$\frac{d}{dt}\left(\frac{\partial\widetilde{\mathcal{L}}}{\partial\dot{q}_k}\right)-\frac{\partial\widetilde{\mathcal{L}}}{\partial q_k}\equiv 0\,,\quad k=1,2,\ldots,s\,. \tag{4.1.22}$$

Note that the function $\widetilde{\mathcal{L}}$ must have the form $\widetilde{\mathcal{L}}=a_k(\mathbf{q};t)\,\dot{q}_k+a_0(\mathbf{q};t)$, otherwise (4.1.22) will contain generalized accelerations with non-zero coefficients. Hence, from the condition (4.1.22) we obtain

$$\left(\frac{\partial a_k}{\partial q_j}-\frac{\partial a_j}{\partial q_k}\right)\dot{q}_j+\frac{\partial a_k}{\partial t}-\frac{\partial a_0}{\partial q_k}\equiv 0\,,\quad \forall\dot{q}_k\,.$$

It follows that

$$\frac{\partial a_k}{\partial q_j}=\frac{\partial a_j}{\partial q_k}\,,\quad \frac{\partial a_k}{\partial t}=\frac{\partial a_0}{\partial q_k}\,,$$

and, consequently

$$a_k=\frac{\partial\varphi}{\partial q_k}\,,\quad a_0=\frac{\partial\varphi}{\partial t}\,,\quad \varphi(\mathbf{q};t)\in C^2\,.$$

Therefore, the function $\widetilde{\mathcal{L}}$ must have the form

$$\widetilde{\mathcal{L}}(\mathbf{q},\dot{\mathbf{q}};t)=\frac{\partial\varphi}{\partial q_k}\,\dot{q}_k+\frac{\partial\varphi}{\partial t}=\frac{d\varphi}{dt}\,. \tag{4.1.23}$$

In conclusion, by applying a transformation of the form

$$\mathcal{L}'(\mathbf{q},\dot{\mathbf{q}};t)=\mathcal{L}(\mathbf{q},\dot{\mathbf{q}};t)+\frac{d\varphi}{dt}\,,\quad \varphi(\mathbf{q};t)\in C^2\,, \tag{4.1.24}$$

called *gauge transformation*, to the Lagrangian $\mathcal{L}$ we obtain a new Lagrangian $\mathcal{L}'$ which satisfies the same Lagrange's equations as $\mathcal{L}$.

1.2 Hamiltonian formulation of classical mechanics

1.2.1 Phase space. Hamilton's equations

We consider a natural mechanical system of particles, subject to scleronomic constraints. In this case, $U(\mathbf{q})$ is a potential, and by applying Euler's theorem to the total kinetic energy T, which is a homogeneous function of the second degree of the generalized velocities, we obtain

$$T = \frac{1}{2} \frac{\partial T}{\partial \dot{q}_k} \dot{q}_k = \frac{1}{2} \frac{\partial \mathcal{L}}{\partial \dot{q}_k} \dot{q}_k = \frac{1}{2} p_k \dot{q}_k \,. \tag{4.1.25}$$

Here, we have used *Legendre's transform*

$$\frac{\partial \mathcal{L}}{\partial \dot{q}_k} = p_k \,, \quad k = 1, 2, \ldots, s \,, \tag{4.1.26}$$

to define the *generalized momenta* p_k. We note that $\partial \mathcal{L} / \partial \dot{q}_k$ is a linear function of the generalized velocities, hence (4.1.26) represents a linear system of algebraic equations, the unknowns being the generalized velocities. This linear system has nontrivial solutions because $\mathcal{L} = T + U$, and the total kinetic energy is a positive definite quadratic form with a non-vanishing discriminant. By solving the system (4.1.26) we obtain

$$\dot{q}_k = \dot{q}_k(\mathbf{q}, \mathbf{p}; t) \,, \quad k = 1, 2, \ldots, s \,, \tag{4.1.27}$$

where $\mathbf{p} = \{p_1, p_2, \ldots, p_s\}$. In this case, *Hamilton's function* (also called *Hamiltonian*)

$$H = p_k \dot{q}_k - \mathcal{L} \tag{4.1.28}$$

takes the form

$$H = H(\mathbf{q}, \mathbf{p}; t) \,. \tag{4.1.29}$$

For scleronomic constraints, from (4.1.25) we obtain $H = T - U$. Taking into account the fact that the functions (4.1.27) are linear in the generalized momenta $\mathbf{p}$, and using (4.1.12) for the total kinetic energy T, we may conclude that the Hamiltonian is also of the form $H = H_2 + H_1 + H_0$, that is a sum of homogeneous functions of second, first, and zeroth degree.

From Lagrange's equations (4.1.5) and Legendre's transform (4.1.26) we get

$$\frac{\partial \mathcal{L}}{\partial q_k} = \dot{p}_k \,, \quad k = 1, 2, \ldots, s, \tag{4.1.30}$$

and consequently, the system of $2s$ ordinary differential equations of the first order (4.1.26) and (4.1.30) is equivalent to Lagrange's equations (4.1.5). Now, a state of the mechanical system $\mathcal{S}$ at a given instant t, is determined by $2s$ *canonical coordinates* (the generalized coordinates $\mathbf{q}$ and the generalized momenta $\mathbf{p}$), and to each state we may associate a representative point $P(\mathbf{q}, \mathbf{p})$ in a $2s$-dimensional space Γ_{2s}, called *phase space* (*Gibbs' space*). Then, the trajectory of the representative point P in Γ_{2s} will be determined by the equations (4.1.26) and (4.1.30).

From the definition of the Hamiltonian (4.1.28) we have

$$\frac{\partial H}{\partial q_k} + \dot{p}_k = \left(p_j - \frac{\partial \mathcal{L}}{\partial \dot{q}_j}\right)\frac{\partial \dot{q}_j}{\partial q_k} - \left(\frac{\partial \mathcal{L}}{\partial q_k} - \dot{p}_k\right),$$

$$\frac{\partial H}{\partial p_k} - \dot{q}_k = \left(p_j - \frac{\partial \mathcal{L}}{\partial \dot{q}_j}\right)\frac{\partial \dot{q}_j}{\partial p_k},$$

and, as a consequence of (4.1.26) and (4.1.30), we obtain the system of differential equations

$$\frac{\partial H}{\partial q_k} = -\dot{p}_k\,, \quad \frac{\partial H}{\partial p_k} = \dot{q}_k\,, \quad k = 1, 2, \ldots, s\,. \tag{4.1.31}$$

Note that (4.1.27) is a linear system of algebraic equations for the unknowns p_k ($k = 1, 2, \ldots, s$), which must have a non-zero determinant ($\det\left[\partial \dot{q}_j / \partial p_k\right] \neq 0$) to allow the return to the linear system (4.1.26). In this case, if the second subsystem in (4.1.31) is satisfied, we are led to a linear system of algebraic equations for the unknowns $(p_j - \partial \mathcal{L}/\partial \dot{q}_j)$, which admits only trivial solutions. If the first subsystem in (4.1.31) is also satisfied it follows that the system of differential equations (4.1.26) and (4.1.30), the system of Lagrange's equations (4.1.5), and the system of differential equations (4.1.31), are mutually equivalent. The equations (4.1.31) represent the *canonical equations* of the analytical mechanics (*Hamilton's equations*), and determine the trajectory of the representative point $P(\mathbf{q}, \mathbf{p})$ in the phase space Γ_{2s}.

We may also obtain Hamilton's equations from a variational principle. By substituting the definition of Hamilton's function (4.1.28) into the action integral (4.1.4) we get the functional

$$\mathcal{A} = \int_{t_0}^{t_1} (p_k\,\dot{q}_k - H)\,dt\,, \tag{4.1.32}$$

where H has the form (4.1.29). The equation (4.1.32) represents the *canonical form* of Hamilton's principle, and Euler's equations for this functional yield the canonical system (4.1.31).

Hamilton's equations represent a system of $2s$ linear differential equations of the first order, in the canonical form, for the unknown functions $q_k = q_k(t)$ and $p_k = p_k(t)$, $k = 1, 2, \ldots, s$. The solutions of this system depend on $2s$ integration constants, which can be determined from the initial conditions

$$q_k(t_0) = q_k^0\,, \quad p_k(t_0) = p_k^0\,, \quad k = 1, 2, \ldots, s\,, \tag{4.1.33}$$

that is, by knowing the initial position of the representative point in the phase space Γ_{2s}.

Note that

$$\frac{dH}{dt} = \frac{\partial H}{\partial q_k}\dot{q}_k + \frac{\partial H}{\partial p_k}\dot{p}_k + \frac{\partial H}{\partial t}\,, \tag{4.1.34}$$

and along the system trajectories, when Hamilton's equations (4.1.31) are satisfied, we have

$$\frac{dH}{dt}=\frac{\partial H}{\partial t}\,. \tag{4.1.35}$$

In particular, in the case of scleronomic constraints, when $H=T-U$ and $\partial H/\partial t=0$, the Hamiltonian remains constant during the motion of the mechanical system $\mathcal{S}$; i.e., H is a first integral (also named *integral of motion*, in mechanics) of Hamilton's equations.

1.2.2 Lagrange brackets. Poisson brackets. First integrals

We assume that two integration constants of Hamilton's equations, denoted by a and b, can be regarded as parameters, such that $q_k = q_k(t;a,b)$ and $p_k=p_k(t;a,b)$. Now, we may introduce the *Lagrange bracket* defined as

$$\{a,b\}=\det\left[\frac{\partial(q_k,p_k)}{\partial(a,b)}\right]=\frac{\partial q_k}{\partial a}\frac{\partial p_k}{\partial b}-\frac{\partial q_k}{\partial b}\frac{\partial p_k}{\partial a}\,. \tag{4.1.36}$$

Taking into account that we may reverse the order of differential operators, and by means of Hamilton's equations (4.1.31), it is easy to calculate

$$\begin{aligned}
&\frac{d}{dt}\{a,b\}=\\
&=\frac{d}{dt}\left(\frac{\partial q_k}{\partial a}\right)\frac{\partial p_k}{\partial b}+\frac{\partial q_k}{\partial a}\frac{d}{dt}\left(\frac{\partial p_k}{\partial b}\right)-\frac{d}{dt}\left(\frac{\partial q_k}{\partial b}\right)\frac{\partial p_k}{\partial a}-\frac{\partial q_k}{\partial b}\frac{d}{dt}\left(\frac{\partial p_k}{\partial a}\right)\\
&=\frac{\partial}{\partial a}\left(\frac{\partial H}{\partial p_k}\right)\frac{\partial p_k}{\partial b}-\frac{\partial q_k}{\partial a}\frac{\partial}{\partial b}\left(\frac{\partial H}{\partial q_k}\right)-\frac{\partial}{\partial b}\left(\frac{\partial H}{\partial p_k}\right)\frac{\partial p_k}{\partial a}+\frac{\partial q_k}{\partial b}\frac{\partial}{\partial a}\left(\frac{\partial H}{\partial q_k}\right)\\
&=\left(\frac{\partial^2 H}{\partial q_j\partial p_k}\frac{\partial q_j}{\partial a}+\frac{\partial^2 H}{\partial p_j\partial p_k}\frac{\partial p_j}{\partial a}\right)\frac{\partial p_k}{\partial b}-\ldots=0\,.
\end{aligned}$$

THEOREM 4.1.1 (LAGRANGE) *Along the trajectory of the representative point $P(\mathbf{q},\mathbf{p})$, the Lagrange bracket is constant ($\{a,b\}=$ const.).*

Let $\varphi(\mathbf{q},\mathbf{p};t)$ and $\psi(\mathbf{q},\mathbf{p};t)$ be two functions of class C^1. The *Poisson bracket* of these two functions is defined as the determinant

$$[\varphi,\psi]=\det\left[\frac{\partial(\varphi,\psi)}{\partial(q_k,p_k)}\right]=\frac{\partial\varphi}{\partial q_k}\frac{\partial\psi}{\partial p_k}-\frac{\partial\varphi}{\partial p_k}\frac{\partial\psi}{\partial q_k}\,. \tag{4.1.37}$$

By means of this definition, it is easy to prove the following properties of Poisson's bracket:

$$\begin{aligned}
&[\varphi,C]=0\,,\quad C=\text{ const.}\,,\\
&[\varphi,\psi]=-[\psi,\varphi]\,,\quad [-\varphi,\psi]=-[\varphi,\psi]\,,
\end{aligned}$$

$$\frac{\partial}{\partial t}[\varphi,\psi]=\left[\frac{\partial\varphi}{\partial t},\psi\right]+\left[\varphi,\frac{\partial\psi}{\partial t}\right],$$
$$[\varphi+\psi,\chi]=[\varphi,\chi]+[\psi,\chi],\quad [\varphi\psi,\chi]=\varphi[\psi,\chi]+\psi[\varphi,\chi],$$
$$[\varphi,\psi]=[P_k,\psi]\,\frac{\partial\varphi}{\partial P_k},$$

where $\chi=\chi(\mathbf{q},\mathbf{p};t)$ is a function of class C^1 and, in the last property, $\varphi=\varphi(\mathbf{P})$, $\mathbf{P}\equiv\{P_k(\mathbf{q},\mathbf{p})|k=1,2,\ldots,s\}$. In terms of Poisson's brackets, we may rewrite Hamilton's equations (4.1.31) in the symmetrical form

$$\dot q_k=[q_k,H],\quad \dot p_k=[p_k,H],\quad k=1,2,\ldots,s. \tag{4.1.38}$$

The state of a discrete mechanical system at a given instant t is determined by the canonical coordinates

$$q_k=q_k(t;\mathbf{q}^0,\mathbf{p}^0),\quad p_k=p_k(t;\mathbf{q}^0,\mathbf{p}^0),\quad k=1,2,\ldots,s, \tag{4.1.39}$$

where $\mathbf{q}^0=\{q_1^0,q_2^0,\ldots,q_s^0\}$ and $\mathbf{p}^0=\{p_1^0,p_2^0,\ldots,p_s^0\}$ are the initial canonical coordinates at the instant $t=t_0$. A physical quantity (momentum, angular momentum, energy) $A=A(\mathbf{q},\mathbf{p};t)$ will be determined at any instant t by the corresponding canonical coordinates. We describe the evolution of this quantity by means of its total derivative with respect to the time

$$\frac{dA}{dt}=\frac{\partial A}{\partial q_k}\dot q_k+\frac{\partial A}{\partial p_k}\dot p_k+\frac{\partial A}{\partial t},$$

and, by employing the canonical equations (4.1.31) and the definition (4.1.37), we obtain

$$\frac{dA}{dt}=[A,H]+\frac{\partial A}{\partial t}, \tag{4.1.40}$$

along the trajectory of the representative point in the phase space Γ_{2s}.

We may write the canonical system (4.1.31) in the form

$$\frac{dt}{1}=\frac{dq_1}{\dfrac{\partial H}{\partial p_1}}=\frac{dq_2}{\dfrac{\partial H}{\partial p_2}}=\ldots=\frac{dq_s}{\dfrac{\partial H}{\partial p_s}}=\frac{dp_1}{-\dfrac{\partial H}{\partial q_1}}=\frac{dp_2}{-\dfrac{\partial H}{\partial q_2}}=\ldots=\frac{dp_s}{-\dfrac{\partial H}{\partial q_s}},$$

which shows that to integrate this system is the same as to find $2s$ first integrals

$$f_k(\mathbf{q},\mathbf{p};t)=f_k(\mathbf{q}^0,\mathbf{p}^0;t_0),\ g_k(\mathbf{q},\mathbf{p};t)=g_k(\mathbf{q}^0,\mathbf{p}^0;t_0),\ k=1,2,\ldots,s, \tag{4.1.41}$$

with $\det\left[\partial(f_k,g_l)/\partial(q_i,p_j)\right]\neq 0$, which led us to the solutions (4.1.39). Note that, for scleronomic constrains, the Hamiltonian H is a first integral, so we have to find only $2s-1$ first integrals.

The formula (4.1.40) provides the necessary and sufficient condition for a function $f(\mathbf{q}, \mathbf{p}; t)$ to be a first integral of the canonical system

$$[f, H] + \frac{\partial f}{\partial t} = 0. \tag{4.1.42}$$

Indeed, from (4.1.41) we obtain (4.1.42). Also, to the partial differential equation of the first order (4.1.42) we may associate the characteristic system of differential equations (the Lagrange-Charpit method), which is exactly the canonical system. The partial derivative of (4.1.42) with respect to the time has the form

$$\left[\frac{\partial f}{\partial t}, H\right] + \left[f, \frac{\partial H}{\partial t}\right] + \frac{\partial}{\partial t}\left(\frac{\partial f}{\partial t}\right) = 0,$$

and, if $\partial H/\partial t = 0$ (i.e., H is a first integral) and f is a first integral, then all the derivatives $\partial f/\partial t$, $\partial^2 f/\partial t^2, \ldots$ are also first integrals (not necessarily independent).

By means of the definition of Poisson's bracket (4.1.37) it is easy to verify the *Poisson-Jacobi identity*

$$[[\varphi, \psi], \chi] + [[\psi, \chi], \varphi] + [[\chi, \varphi], \psi] = 0, \tag{4.1.43}$$

where φ, ψ, and χ are functions of class C^2. If φ and ψ are first integrals, that is

$$[\varphi, H] + \frac{\partial \varphi}{\partial t} = 0, \quad [\psi, H] + \frac{\partial \psi}{\partial t} = 0,$$

then from the Poisson-Jacobi identity (4.1.43) we obtain

$$[[\varphi, \psi], H] + \frac{\partial}{\partial t}[\varphi, \psi] = 0.$$

THEOREM 4.1.2 (JACOBI-POISSON) *If φ and ψ are first integrals of the canonical system, then their Poisson bracket $[\varphi, \psi]$ is also a first integral (independent or not of φ and ψ) of the canonical system.*

1.2.3 The Hamilton-Jacobi method

If $S = S(t)$ is a primitive of the Lagrangian $\mathcal{L}$ we may write

$$\frac{dS(t)}{dt} = \mathcal{L}, \quad S = \int \mathcal{L}\, dt, \tag{4.1.44}$$

and, by comparing (4.1.44) and the definition (4.1.4), it follows that

$$\mathcal{A} = S(t)\Big|_{t_0}^{t_1}. \tag{4.1.45}$$

By assuming that $S = S(\mathbf{q};t)$, we may write the total derivative of S with respect to the time in the form

$$\frac{dS}{dt} = \frac{\partial S}{\partial q_k}\,\dot{q}_k + \frac{\partial S}{\partial t}\,.$$

Taking into account the definition (4.1.44) and Lagrange's equations (4.1.5) we also have

$$\frac{\partial S}{\partial q_k} = \int \frac{\partial \mathcal{L}}{\partial q_k}\,dt = \int \frac{d}{dt}\left(\frac{\partial \mathcal{L}}{\partial \dot{q}_k}\right) dt = \frac{\partial \mathcal{L}}{\partial \dot{q}_k} = p_k\,, \tag{4.1.46}$$

so that finally, we obtain

$$\frac{\partial S}{\partial t} = \mathcal{L} - p_k\,\dot{q}_k = -H(\mathbf{q},\mathbf{p};t)\,. \tag{4.1.47}$$

Here, we have also employed Legendre's transform (4.1.26) and the definition of the Hamiltonian (4.1.28). Actually, Eq.(4.1.47) represents a partial differential equation of the first order (the *Hamilton-Jacobi equation*), which can be written in the more explicit form

$$\frac{\partial S}{\partial t} + H\left(q_1, q_2, \ldots, q_s, \frac{\partial S}{\partial q_1}, \frac{\partial S}{\partial q_2}, \ldots, \frac{\partial S}{\partial q_s}; t\right) = 0\,, \tag{4.1.48}$$

where we have replaced the generalized momenta p_k in H by $\partial S/\partial q_k$. We call *complete integral* of the Hamilton-Jacobi equation a function

$$S = S(q_1, q_2, \ldots, q_s;\ t;\ a_1, a_2, \ldots, a_s) \equiv S(\mathbf{q};\ t;\ \mathbf{a})$$

depending on s *essential integration constants* $\{a_1, a_2, \ldots, a_s\}$, and with a non-vanishing Hessian, that is

$$\det\left[\frac{\partial^2 S}{\partial q_k\,\partial a_j}\right] \neq 0\,. \tag{4.1.49}$$

A complete integral of (4.1.48) can be given, for instance, by a combination of s particular solutions.

THEOREM 4.1.3 (HAMILTON-JACOBI) *If $S(\mathbf{q};\ t;\ \mathbf{a})$ is a complete integral of the Hamilton-Jacobi equation, then the functions $q_k = q_k(t)$, $p_k = p_k(t)$, $k = 1, 2, \ldots, s$, obtained from the equations*

$$\frac{\partial S}{\partial a_k} = b_k\,, \quad \frac{\partial S}{\partial q_k} = p_k\,, \quad k = 1, 2, \ldots, s\,, \tag{4.1.50}$$

are solutions of the canonical system.

Here, b_k represents the *conjugate constant* of the constant a_k. When (4.1.49) is satisfied, the first set of equations in (4.1.50) yields the generalized coordinates $q_k = q_k(t; \mathbf{a}, \mathbf{b})$. Then from the second set we obtain the generalized momenta $p_k = p_k(t; \mathbf{a}, \mathbf{b})$. Consequently, the Hamilton-Jacobi theorem establishes the method by which the problem of finding the solution of a system of differential equations of the first order (in our case Hamilton's equations) is replaced by the problem of solving a single partial differential equation of the first order (in our case the Hamilton-Jacobi equation). According to Lagrange's theorem (Theorem 4.1.1), the Lagrange brackets for the integration constants a_k and b_k are

$$\{a_j, a_k\} = 0\,, \ \{b_j, b_k\} = 0\,, \ \{b_j, a_k\} = \delta_{jk}\,, \quad j, k, = 1, 2, \ldots, s\,, \tag{4.1.51}$$

where δ_{jk} is the Kronecker symbol.

1.3 Invariance of the Lagrange and Hamilton equations

1.3.1 Point transformations. Invariance of Lagrange's equations

Let

$$\widetilde{q}_j = \widetilde{q}_j(\mathbf{q}; t)\,, \quad j = 1, 2, \ldots, s\,, \tag{4.1.52}$$

be a point transformation of the generalized coordinates. By assuming that this transformation can be inverted, that is the condition

$$\det\left[\frac{\partial q_k}{\partial \widetilde{q}_j}\right] \neq 0 \tag{4.1.53}$$

is satisfied, we may write

$$q_k = q_k(\widetilde{\mathbf{q}}; t)\,, \quad k = 1, 2, \ldots, s\,, \tag{4.1.54}$$

where $\widetilde{\mathbf{q}} = \{\widetilde{q}_1, \widetilde{q}_2, \ldots, \widetilde{q}_s\}$. In this case, we also have

$$\dot{q}_k = \frac{\partial q_k}{\partial \widetilde{q}_l}\dot{\widetilde{q}}_l + \frac{\partial q_k}{\partial t} = \dot{q}_k(\widetilde{\mathbf{q}}, \dot{\widetilde{\mathbf{q}}}; t)\,, \tag{4.1.55}$$

with $\dot{\widetilde{\mathbf{q}}} = \{\dot{\widetilde{q}}_1, \dot{\widetilde{q}}_2, \ldots, \dot{\widetilde{q}}_s\}$, and consequently

$$\frac{\partial \dot{q}_k}{\partial \dot{\widetilde{q}}_l} = \frac{\partial q_k}{\partial \widetilde{q}_l}\,. \tag{4.1.56}$$

Now, by taking into account (4.1.55) and assuming that q_k are functions of class C^2, we derive

$$\frac{\partial \dot{q}_k}{\partial \widetilde{q}_j} = \frac{\partial^2 q_k}{\partial \widetilde{q}_j \partial \widetilde{q}_l}\dot{\widetilde{q}}_l + \frac{\partial^2 q_k}{\partial \widetilde{q}_j \partial t} = \frac{\partial}{\partial \widetilde{q}_l}\left(\frac{\partial q_k}{\partial \widetilde{q}_j}\right)\dot{\widetilde{q}}_l + \frac{\partial}{\partial t}\left(\frac{\partial q_k}{\partial \widetilde{q}_j}\right) = \frac{d}{dt}\left(\frac{\partial q_k}{\partial \widetilde{q}_j}\right)\,,$$

and, consequently, we obtain the commutation relation of differential operators

$$\frac{\partial}{\partial \widetilde{q}_j} \frac{d}{dt} = \frac{d}{dt} \frac{\partial}{\partial \widetilde{q}_j} . \tag{4.1.57}$$

A transformation of the form (4.1.54) will also change the kinetic energy $T = T(\mathbf{q}, \dot{\mathbf{q}}; t) = \widetilde{T}(\widetilde{\mathbf{q}}, \dot{\widetilde{\mathbf{q}}}; t)$, such that

$$\begin{aligned}
\frac{\partial \widetilde{T}}{\partial \widetilde{q}_j} &= \frac{\partial T}{\partial q_k} \frac{\partial q_k}{\partial \widetilde{q}_j} + \frac{\partial T}{\partial \dot{q}_k} \frac{\partial \dot{q}_k}{\partial \widetilde{q}_j} = \frac{\partial T}{\partial q_k} \frac{\partial q_k}{\partial \widetilde{q}_j} + \frac{\partial T}{\partial \dot{q}_k} \frac{d}{dt} \left(\frac{\partial q_k}{\partial \widetilde{q}_j} \right) , \\
\frac{\partial \widetilde{T}}{\partial \dot{\widetilde{q}}_j} &= \frac{\partial T}{\partial \dot{q}_k} \frac{\partial \dot{q}_k}{\partial \dot{\widetilde{q}}_j} = \frac{\partial T}{\partial \dot{q}_k} \frac{\partial q_k}{\partial \widetilde{q}_j} ,
\end{aligned}$$

where the last forms follow from the relations (4.1.56) and (4.1.57). We may also introduce the generalized forces, which, in the transformed system of coordinates, have the form

$$\widetilde{Q}_j = \sum_{i=1}^{n} \mathbf{F}_i \cdot \frac{\partial \mathbf{r}_i}{\partial \widetilde{q}_j} = \sum_{i=1}^{n} \mathbf{F}_i \cdot \frac{\partial \mathbf{r}_i}{\partial q_k} \frac{\partial q_k}{\partial \widetilde{q}_j} = Q_k \frac{\partial q_k}{\partial \widetilde{q}_j} .$$

Hence, we obtain the relation

$$\frac{d}{dt} \left(\frac{\partial \widetilde{T}}{\partial \dot{\widetilde{q}}_j} \right) - \frac{\partial \widetilde{T}}{\partial \widetilde{q}_j} - \widetilde{Q}_j = \left[\frac{d}{dt} \left(\frac{\partial T}{\partial \dot{q}_k} \right) - \frac{\partial T}{\partial q_k} - Q_k \right] \frac{\partial q_k}{\partial \widetilde{q}_j} . \tag{4.1.58}$$

It follows that, if Lagrange's equations (4.1.9) are satisfied in a system of generalized coordinates $(\mathbf{q})$, then they are also satisfied in the new system of generalized coordinates $(\widetilde{\mathbf{q}})$. The reciprocal implication is also true, and it follows by analyzing the linear homogeneous system of algebraic equations (4.1.58), which, according to (4.1.53), admits only trivial solutions. Consequently, we may infer that the form of Lagrange's equations (4.1.9) is *invariant* with respect to a point transformation of the form (4.1.52-4.1.53).

Note that, when $\partial q_k / \partial t = 0$, by taking into account (4.1.56), we obtain

$$\frac{\partial \widetilde{T}}{\partial \dot{\widetilde{q}}_j} \dot{\widetilde{q}}_j = \frac{\partial T}{\partial \dot{q}_k} \frac{\partial q_k}{\partial \widetilde{q}_j} \dot{\widetilde{q}}_j = \frac{\partial T}{\partial \dot{q}_k} \dot{q}_k . \tag{4.1.59}$$

Therefore, in the case of scleronomic constraints, the form (4.1.25) of the kinetic energy is also invariant with respect to the point transformation (4.1.52-4.1.53).

1.3.2 Canonical transformations. Invariance of Hamilton's equations

Let

$$Q_j = Q_j(\mathbf{q}, \mathbf{p}; t) , \quad P_j = P_j(\mathbf{q}, \mathbf{p}; t) , \quad j = 1, 2, \ldots, s , \tag{4.1.60}$$

be a transformation of the canonical variables, where Q_j and P_j are functions of class C^2, having the Jacobian

$$\det\left[\frac{\partial(\mathbf{Q},\mathbf{P})}{\partial(\mathbf{q},\mathbf{p})}\right] \neq 0\,, \tag{4.1.61}$$

where $\mathbf{Q} = \{Q_1, Q_2, \ldots, Q_s\}$ and $\mathbf{P} = \{P_1, P_2, \ldots, P_s\}$. As a result of this transformation the Hamiltonian becomes $H = H(\mathbf{q},\mathbf{p};t) = \widetilde{H}(\mathbf{Q},\mathbf{P};t)$. The transformation defined by (4.1.60-4.1.61) is called *canonical transformation*, if Hamilton's equations (4.1.31) remain unaltered by the passage from one set of canonical variables to another, that is, in the new variables Hamilton's equations are

$$\frac{\partial\widetilde{H}}{\partial Q_j} = -\dot{P}_j\,,\quad \frac{\partial\widetilde{H}}{\partial P_j} = \dot{Q}_j\,,\quad j = 1,2,\ldots,s\,. \tag{4.1.62}$$

The theorem of implicit functions and the condition (4.1.61) ensure the existence of the inverse transformation

$$q_k = q_k(\mathbf{Q},\mathbf{P};t)\,,\quad p_k = p_k(\mathbf{Q},\mathbf{P};t)\,,\quad k = 1,2,\ldots,s\,, \tag{4.1.63}$$

the functions q_k and p_k also being of class C^2. The canonical variables $(\mathbf{q},\mathbf{p})$ and $(\mathbf{Q},\mathbf{P})$ correspond to the same representative point in the phase space Γ_{2s}. When they do not explicitly depend on time, the canonical transformations (4.1.60-4.1.63) are called *complete canonical transformations*.

The canonical transformations allow us to simplify Hamilton's equations of complicated forms. In particular, if we can find a canonical transformation such that, along the trajectory of the representative point, the new Hamiltonian $\widetilde{H}$ is constant (i.e., $\widetilde{H}$ does not depend on $\mathbf{Q}$ and $\mathbf{P}$), then the solution of the canonical system (4.1.62) has the form

$$Q_j = A_j\,,\quad P_j = B_j\,,\quad j = 1,2,\ldots,s\,, \tag{4.1.64}$$

where A_j and B_j are constants. In this case, the functions (4.1.60) are first integrals (integrals of motion) of Hamilton's equations (4.1.31).

Let

$$\omega = p_k\,dq_k - P_j\,dQ_j \tag{4.1.65}$$

be a differential one-form. By examining the definition (4.1.32), that leads us to the canonical form of Hamilton's principle, and subsequently to Hamilton's equations, we deduce that a necessary and sufficient condition for the transformation (4.1.60) to be a complete canonical transformation, is $d\omega = 0$. This means that the one-form (4.1.65) is a total differential, and on the basis of the *Poincaré lemma* we have $d\,(d\omega) = 0$, where the operator d denotes *Cartan's operator*. According to the reciprocal of this lemma, if $d\omega = 0$, then there is a scalar function $\psi(q,p)$ such that

$$\omega = d\psi\,. \tag{4.1.66}$$

In the particular case when $\psi = 0$, the canonical transformation is called *homogeneous*.

A *point transformation* is a transformation of the form

$$q_k = q_k(\mathbf{Q}) , \quad p_k = p_k(\mathbf{Q}, \mathbf{P}) , \quad k = 1, 2, \dots , s . \tag{4.1.67}$$

A sufficient condition for (4.1.67) to be a canonical transformation is $d\omega = 0$, that is

$$p_k \frac{\partial q_k}{\partial Q_j} dQ_j = P_j \, dQ_j ,$$

and therefore

$$p_k \frac{\partial q_k}{\partial Q_j} = P_j , \quad j = 1, 2, \dots , s . \tag{4.1.68}$$

Since $\psi = 0$, such a point canonical transformation is also homogeneous.

Now, by taking into account (4.1.65) and (4.1.66) we may write the relation

$$p_k \, dq_k - P_j \left(\frac{\partial Q_j}{\partial q_k} dq_k + \frac{\partial Q_j}{\partial p_k} dp_k \right) = \frac{\partial \psi}{\partial q_k} dq_k + \frac{\partial \psi}{\partial p_k} dp_k ,$$

from which we obtain the equations

$$p_k - P_j \frac{\partial Q_j}{\partial q_k} = \frac{\partial \psi}{\partial q_k} , \quad -P_j \frac{\partial Q_j}{\partial p_k} = \frac{\partial \psi}{\partial p_k} , \quad k = 1, 2, \dots , s .$$

By eliminating the function $\psi(\mathbf{q}, \mathbf{p})$, or similarly by using the independence of the second order mixed derivatives with respect to the order of differentiation (for instance, the equality of the mixed derivatives

$$\begin{aligned} \frac{\partial^2 \psi}{\partial q_j \partial q_k} &= - \frac{\partial P_l}{\partial q_j} \frac{\partial Q_l}{\partial q_k} - P_l \frac{\partial^2 Q_l}{\partial q_j \partial q_k} , \\ \frac{\partial^2 \psi}{\partial q_k \partial q_j} &= - \frac{\partial P_l}{\partial q_k} \frac{\partial Q_l}{\partial q_j} - P_l \frac{\partial^2 Q_l}{\partial q_k \partial q_j} , \end{aligned}$$

assuming that P_l, Q_l, and ψ are functions of class C^2), we obtain *sufficient conditions* for the considered transformation to be a *complete canonical transformation*. In total, there are $2[s(s-1)/2] + s^2 = s(2s-1)$ such conditions, which, in terms of Lagrange's brackets (4.1.36), have the form

$$\{q_j, q_k\} = 0 , \ \{p_j, p_k\} = 0 , \ \{q_j, p_k\} = \delta_{jk} , \quad j, k = 1, 2, \dots , s , \tag{4.1.69}$$

where δ_{jk} represents the Kronecker symbol. Here, q_j and p_j ($j = 1, 2, \dots , s$) stand for integration constants in the solutions (4.1.60) of the canonical system (4.1.62).

We note that along the integral curves of Hamilton's equations (4.1.31), in terms of Poisson's brackets, we may write the relations

$$\begin{aligned}\dot{Q}_j &= \frac{\partial Q_j}{\partial q_k}\dot{q}_k + \frac{\partial Q_j}{\partial p_k}\dot{p}_k = \frac{\partial Q_j}{\partial q_k}\frac{\partial H}{\partial p_k} - \frac{\partial Q_j}{\partial p_k}\frac{\partial H}{\partial q_k} = [Q_j, H]\,,\\ \dot{P}_j &= [P_j, H]\,.\end{aligned}$$

Also, we have

$$dQ_j = \frac{\partial Q_j}{\partial q_k}dq_k + \frac{\partial Q_j}{\partial p_k}dp_k\,, \quad dP_j = \frac{\partial P_j}{\partial q_k}dq_k + \frac{\partial P_j}{\partial p_k}dp_k\,.$$

By means of these relations and employing Lagrange's brackets (4.1.69), we obtain the chain of relations

$$\begin{aligned}&\dot{Q}_j\,dP_j - \dot{P}_j\,dQ_j\\ &= \left[\{q_k, q_l\}\frac{\partial H}{\partial p_k} + \{q_l, p_k\}\frac{\partial H}{\partial q_k}\right]dq_l + \left[\{p_l, p_k\}\frac{\partial H}{\partial q_k} + \{q_l, p_k\}\frac{\partial H}{\partial p_k}\right]dp_l\\ &= \frac{\partial H}{\partial q_k}\,dq_k + \frac{\partial H}{\partial p_k}\,dp_k = dH - \frac{\partial H}{\partial t}\,dt\,,\end{aligned}$$

which leads us back to the canonical system (4.1.62), with $H = \widetilde{H}$. Once more, this result proves that (4.1.66) is a sufficient condition for the considered transformations to be complete canonical transformations, the Hamiltonian preserving its form in the new coordinates.

When the time appears explicitly, a sufficient condition for the transformations (4.1.60-4.1.63) to be canonical transformations, is that the differential one-form

$$\Omega = p_k\,dq_k - P_j\,dQ_j - (H - \widetilde{H})dt \tag{4.1.70}$$

be a total differential, that is

$$\Omega = dW\,, \tag{4.1.71}$$

where W is a function of class C^2. This affirmation can be proved easily by examining the canonical form of Hamilton's principle. If the transformation (4.1.60) satisfies the condition

$$\det\left[\frac{\partial(Q_1, Q_2, \ldots, Q_s)}{\partial(p_1, p_2, \ldots, p_s)}\right] \neq 0\,, \tag{4.1.72}$$

then from the first subsystem of transformations we obtain $p_k = p_k(\mathbf{q}, \mathbf{Q}; t)$, $k = 1, 2, \ldots, s$, and by substituting into the second subsystem of transformations, we get the transformations $P_j = P_j(\mathbf{q}, \mathbf{Q}; t)$, $j = 1, 2, \ldots, s$. In this case, $W = W(\mathbf{q}, \mathbf{Q}; t)$ such that the condition (4.1.71) leads us to the relation

$$\left(p_k - \frac{\partial W}{\partial q_k}\right)dq_k - \left(P_j + \frac{\partial W}{\partial Q_j}\right)dQ_j + \left(\widetilde{H} - H - \frac{\partial W}{\partial t}\right)dt = 0\,,$$

from which it follows that

$$p_k = \frac{\partial W}{\partial q_k}\,, \quad P_j = -\frac{\partial W}{\partial Q_j}\,, \quad k, j = 1, 2, \ldots, s\,, \quad \widetilde{H} = H + \frac{\partial W}{\partial t}\,. \tag{4.1.73}$$

If

$$\det\left[\frac{\partial^2 W}{\partial Q_j \partial q_k}\right] \neq 0\,, \tag{4.1.74}$$

then, from the first subsystem of equations in (4.1.73) we obtain the generalized coordinates $\mathbf{Q}$. The substitution of $\mathbf{Q}$ into the second subsystem in (4.1.73) completely determines the new canonical variables $(\mathbf{Q}, \mathbf{P})$, as functions of the initial canonical variables $(\mathbf{q}, \mathbf{p})$, which in turn determine the transformation (4.1.60). Also, the condition (4.1.74) allows us to determine the generalized coordinates $\mathbf{q}$ from the second subsystem in (4.1.73), and by substituting them into the first subsystem in (4.1.73) we get the inverse canonical transformation (4.1.63). The last equation in (4.1.73) points out the relation between the Hamilton's functions corresponding to the two systems of canonical variables. Consequently, every function $W(\mathbf{q}, \mathbf{Q}; t)$, which satisfies the condition (4.1.74), defines a canonical transformation determined by (4.1.73). The function W is called the *generating function of canonical transformations.*

When a canonical transformation of the form (4.1.60) leads to $\widetilde{H} = 0$, it follows that

$$\frac{\partial W}{\partial t} + H = 0\,,$$

where

$$H = H(\mathbf{q}, \mathbf{p}; t) = H\left(q_1, q_2, \ldots, q_s; \frac{\partial W}{\partial q_1}, \frac{\partial W}{\partial q_2}, \ldots, \frac{\partial W}{\partial q_s}; t\right).$$

In this case, the function W is equivalent to the function S in the Hamilton-Jacobi equation (4.1.48). Hence, the problem of determining the canonical transformations which render $\widetilde{H} = 0$, is equivalent to that of solving the canonical system.

It can be shown that the Lagrange and Poisson brackets are invariant with respect to canonical transformations, that is

$$\{a, b\}_{\mathbf{q},\mathbf{p}} = \{a, b\}_{\mathbf{Q},\mathbf{P}}\,, \quad [\varphi, \psi]_{\mathbf{q},\mathbf{p}} = [\varphi, \psi]_{\mathbf{Q},\mathbf{P}}\,, \tag{4.1.75}$$

where we have marked the corresponding canonical variables.

Now, we introduce the $2s \times 2s$ matrix

$$\mathbf{J} = \left[\frac{\partial(\mathbf{Q}, \mathbf{P})}{\partial(\mathbf{q}, \mathbf{p})}\right], \tag{4.1.76}$$

corresponding to the Jacobian (4.1.61), and the block matrix

$$\mathbf{I} = \begin{bmatrix} \mathbf{0} & -\mathbf{e}_s \\ \mathbf{e}_s & \mathbf{0} \end{bmatrix}, \tag{4.1.77}$$

where $\mathbf{e}_s$ denotes the $s \times s$ identity matrix. With these notations, the following theorem holds

THEOREM 4.1.4 *The necessary and sufficient condition for the transformation (4.1.60) to be a canonical transformation consists in the existence of a scalar constant $c \neq 0$, such that*

$$\mathbf{J}^{\mathrm{T}} \mathbf{I} \mathbf{J} = c \mathbf{I}. \tag{4.1.78}$$

The matrix $\mathbf{I}$ is called *canonical matrix.* From (4.1.77) we have $\mathbf{I}^2 = -\mathbf{e}_{2s}$, where $\mathbf{e}_{2s}$ denotes the $2s \times 2s$ identity matrix, such that $\mathbf{I}^{-1} = -\mathbf{I}$. Hence, the canonical matrix is nonsingular.

The matrix $\mathbf{J}$, which satisfies the condition (4.1.78), is called *Symplectic matrix.* It is also nonsingular because $(\det \mathbf{J})^2 = c$. It can be shown that the set of $2s \times 2s$ Symplectic matrices form a group (the *Symplectic group*, see Chap. 4 Sec. 2.2), and consequently, the set of canonical transformations also form a group.

When $c = 1$ the corresponding canonical transformation is called *univalent.* In fact, the invariance property (4.1.75) holds for univalent transformations.

If there is a scalar constant $c \neq 0$, such that the differential one-form $\omega_c = c\, p_k\, dq_k - P_j\, dQ_j$ satisfies the condition (4.1.66), then this condition is not only sufficient, but it is also necessary for the considered transformation to be complete canonical. Also, if we replace the Kronecker symbol δ_{jk} by $c\, \delta_{jk}$ in (4.1.69), these conditions become sufficient and necessary. In this case, we obtain $\widetilde{H} = c\, H$. All the conditions (4.1.65), (4.1.66), and (4.1.69) are sufficient and necessary for univalent transformations. We mention that analogous affirmations hold for the differential one-form (4.1.70).

1.3.3 Infinitesimal canonical transformations

Let

$$\begin{aligned} Q_j &= q_j + \delta q_j = q_j + f_j(\mathbf{q}, \mathbf{p})\delta t\,, \\ P_j &= p_j + \delta p_j = p_j + g_j(\mathbf{q}, \mathbf{p})\delta t\,, \\ j &= 1, 2, \ldots, s\,, \end{aligned} \tag{4.1.79}$$

be an *infinitesimal transformation* of canonical variables. This transformation is canonical if the differential one-form

$$P_j dQ_j - p_j dq_j = (p_j + g_j \delta t)(dq_j + df_j \delta t) - p_j dq_j \cong (p_j df_j + g_j dq_j)\delta t$$

is a total differential. Because

$$p_j df_j + g_j dq_j = d(p_j f_j) - (f_j dp_j - g_j dq_j)\,,$$

it is sufficient to prove that in plus there is function $K = K(\mathbf{q}, \mathbf{p})$, of class C^2, such that

$$f_j dp_j - g_j dq_j = dK(\mathbf{q}, \mathbf{p}) = \frac{\partial K}{\partial q_j}\, dq_j + \frac{\partial K}{\partial p_j}\, dp_j\,,$$

from which it follows that $f_j = \partial K / \partial p_j, g_j = -\partial K / \partial q_j$. Hence, the *complete canonical infinitesimal transformations* have the form

$$Q_j = q_j + \frac{\partial K}{\partial p_j}\,\delta t\,,\quad P_j = p_j - \frac{\partial K}{\partial q_j}\,\delta t\,,\quad j = 1, 2, \ldots, s\,. \tag{4.1.80}$$

Let us consider a function $F = F(\mathbf{q}, \mathbf{p})$, of class C^1. An infinitesimal canonical transformation (4.1.80) will induce a variation of the function F of the form

$$\delta F = \frac{\partial F}{\partial q_j}\delta q_j + \frac{\partial F}{\partial p_j}\delta p_j = \left(\frac{\partial F}{\partial q_j}\frac{\partial K}{\partial p_j} - \frac{\partial F}{\partial p_j}\frac{\partial K}{\partial q_j} \right) \delta t = [F, K]\delta t\,. \tag{4.1.81}$$

In particular, for $F = q_k$ and $F = p_k$, we obtain respectively

$$\delta q_k = [q_k, K]\,\delta t\,,\quad \delta p_k = [p_k, K]\,\delta t\,,\quad k = 1, 2, \ldots, s\,. \tag{4.1.82}$$

Consequently, we may express the variations of the canonical coordinates (that is, the infinitesimal transformation determined by the generating function $K(\mathbf{q}, \mathbf{p})$, of class C^1) in terms of Poisson's brackets (4.1.82).

If we set $K = H$ in (4.1.82), we obtain the *canonical equations in terms of finite differences*. Thus, we are led to an iterative method to solve the canonical equations, which allows us to find a finite number of points along the trajectory of the representative point in the phase space.

1.4 Noether's theorem and its reciprocal

1.4.1 Symmetry transformations

In 1918 Emmy Noether proved her famous theorem relating the symmetry properties of a physical system and conservation laws, when the equations of motion emerge from a variational principle. Thus, she established a general method to derive conservation laws in classical mechanics, but this method can be applied equally well in the case of relativistic mechanics, quantum mechanics, electromagnetism etc.

Let be a physical system whose states are determined by means of the *independent variables* $\mathbf{x} = \{x_i\}$ $(i = 1, 2, \ldots, n)$ and the *dependent variables* (functions of state) $\{u^\alpha\}$ $(\alpha = 1, 2, \ldots, m)$. The derivatives of the functions of state are denoted by $u_i^\alpha = \partial u^\alpha / \partial x_i$. Note that there is only one independent variable in classical mechanics, namely the time t, while the functions of state and their derivatives correspond to the generalized coordinates $\mathbf{q}$ and generalized velocities $\dot{\mathbf{q}}$, respectively. We assume that the equations of motion derive from the functional

$$\mathcal{F} = \int_\Omega \mathcal{L}(x_i, u^\alpha, u_i^\alpha) dx \,, \tag{4.1.83}$$

where $dx = dx_1 dx_2 \ldots dx_n$ is the volume element in the domain Ω and the density function $\mathcal{L}$, corresponding to the unit volume, is of class C^2. To obtain the equations of motion we introduce the variations of the independent variables

$$x_i' = x_i + \delta x_i \,, \quad i = 1, 2, \ldots, n \,, \tag{4.1.84}$$

where δx_i are arbitrary infinitesimal functions of class C^2; these functions may have the form $\varepsilon \eta_i(\mathbf{x})$, ε being a small parameter, and η_i arbitrary functions of class C^2. The equations (4.1.84) define an *infinitesimal transformation* of the domain Ω into Ω'. Consequently, the state functions and their derivatives are transformed according to the relations

$$u'^\alpha(\mathbf{x}') = u^\alpha(\mathbf{x}) + \delta u^\alpha(\mathbf{x}) \,, \quad u'^\alpha_i(\mathbf{x}') = u_i^\alpha(\mathbf{x}) + \delta u_i^\alpha(\mathbf{x}) \,, \tag{4.1.85}$$

and the variation of the functional (4.1.83) is given by

$$\delta \mathcal{F} = \int_{\Omega'} \mathcal{L}(x_i', u'^\alpha, u'^\alpha_i) dx' - \int_\Omega \mathcal{L}(x_i, u^\alpha, u_i^\alpha) dx \,. \tag{4.1.86}$$

The Jacobian of the transformation (4.1.84) is

$$\begin{aligned} J = \det \left[\frac{\partial(\mathbf{x}')}{\partial(\mathbf{x})} \right] &= \det \left[\frac{\partial(x_j + \delta x_j)}{\partial(x_k)} \right] = \det \left[\delta_{jk} + \frac{\partial}{\partial x_k} \delta x_j \right] \\ &= 1 + \frac{\partial}{\partial x_j} \delta x_j + \mathcal{O}(\varepsilon^2) \,, \end{aligned} \tag{4.1.87}$$

where we have used *Einstein's summation convention* for dummy Latin indices from 1 to n. The volume elements associated to the domains Ω and Ω' are related by the equation $dx' = J dx$, and for the inverse transformation we will have

$$J^{-1} = 1 - \frac{\partial}{\partial x_j} \delta x_j + \mathcal{O}(\varepsilon^2) \,. \tag{4.1.88}$$

With these definitions, and using *Einstein's summation convention* for dummy Greek indices from 1 to m, we may write the Taylor series expansion

$$\begin{aligned} &\mathcal{L}(x_i', u'^\alpha, u'^\alpha_i) = \mathcal{L}(x_i + \delta x_i, u^\alpha + \delta u^\alpha, u_i^\alpha + \delta u_i^\alpha) \\ &= \mathcal{L}(x_i, u^\alpha, u_i^\alpha) + \frac{\partial \mathcal{L}}{\partial x_i} \delta x_i + \frac{\partial \mathcal{L}}{\partial u^\alpha} \delta u^\alpha + \frac{\partial \mathcal{L}}{\partial u_i^\alpha} \delta u_i^\alpha + \mathcal{O}(\varepsilon^2) \,. \end{aligned}$$

Hence, the variation of the functional $\mathcal{F}$ can be written in the form

$$\delta\mathcal{F} = \int_\Omega \left(\mathcal{L} \frac{\partial}{\partial x_i} \delta x_i + \frac{\partial \mathcal{L}}{\partial x_i} \delta x_i + \frac{\partial \mathcal{L}}{\partial u^\alpha} \delta u^\alpha + \frac{\partial \mathcal{L}}{\partial u_i^\alpha} \delta u_i^\alpha \right) dx\,. \tag{4.1.89}$$

As $\delta u_i^\alpha \neq \partial(\delta u^\alpha)/\partial x_i$, we introduce the variations

$$\delta_* u^\alpha(\mathbf{x}') = u'^\alpha(\mathbf{x}') - u^\alpha(\mathbf{x}')\,, \quad \delta_* u_i^\alpha(\mathbf{x}') = u_i'^\alpha(\mathbf{x}') - u_i^\alpha(\mathbf{x}') \tag{4.1.90}$$

to integrate by parts the last term in (4.1.89). Thus, we have the Taylor series

$$\delta_* u^\alpha(\mathbf{x}') = \delta_* u^\alpha(\mathbf{x} + \delta\mathbf{x}) = \delta_* u^\alpha(\mathbf{x}) + \frac{\partial}{\partial x_i} (\delta_* u^\alpha)\, \delta x_i + \mathcal{O}(\varepsilon^2)\,,$$

where $\delta_* u^\alpha$ and δx_i are variations of order ε. By neglecting terms of order ε^2 we have $\delta_* u^\alpha(\mathbf{x}') = \delta_* u^\alpha(\mathbf{x})$. Therefore, from (4.1.85) and (4.1.89) we obtain

$$\begin{aligned} \delta u^\alpha(\mathbf{x}) &= [u'^\alpha(\mathbf{x}') - u^\alpha(\mathbf{x}')] + [u^\alpha(\mathbf{x}') - u^\alpha(\mathbf{x})] = \delta_* u^\alpha(\mathbf{x}) + u_i^\alpha \delta x_i\,, \\ \delta u_i^\alpha(\mathbf{x}) &= [u_i'^\alpha(\mathbf{x}') - u_i^\alpha(\mathbf{x}')] + [u_i^\alpha(\mathbf{x}') - u_i^\alpha(\mathbf{x})] = \delta_* u_i^\alpha(\mathbf{x}) + u_{ij}^\alpha \delta x_j\,, \end{aligned} \tag{4.1.91}$$

where u_{ij}^α are mixed derivatives of second order. From the second equation in (4.1.90) we have

$$\delta_* u_i'^\alpha(\mathbf{x}') = \frac{\partial}{\partial x_i'} [u'^\alpha(\mathbf{x}') - u^\alpha(\mathbf{x}')] = \frac{\partial}{\partial x_i'} \delta_* u^\alpha(\mathbf{x}')\,, \tag{4.1.92}$$

and, respectively,

$$\begin{aligned} \delta_* u_i^\alpha(\mathbf{x}) &= \frac{\partial}{\partial x_i} \delta_* u^\alpha(\mathbf{x})\,, \\ \delta u_i^\alpha(\mathbf{x}) &= \frac{\partial}{\partial x_i} \delta_* u^\alpha(\mathbf{x}) + u_{ij}^\alpha \delta x_j\,. \end{aligned}$$

Now, the variation of the functional $\mathcal{F}$ becomes

$$\delta\mathcal{F} = \int_\Omega \left[\frac{d}{dx_i} (\mathcal{L}\, \delta x_i) + \frac{\partial \mathcal{L}}{\partial u^\alpha} \delta_* u^\alpha + \frac{\partial \mathcal{L}}{\partial u_i^\alpha} \delta_* u_i^\alpha \right] dx\,.$$

Here, we have introduced the total differentiation operator

$$\frac{d}{dx_i} = \frac{\partial}{\partial x_i} + u_i^\alpha \frac{\partial}{\partial u^\alpha} + u_{ji}^\alpha \frac{\partial}{\partial u_j^\alpha}\,, \quad i = 1, 2, \ldots, n\,. \tag{4.1.93}$$

The last term in $\delta\mathcal{F}$ can be written in the form

$$\frac{\partial \mathcal{L}}{\partial u_i^\alpha} \delta_* u_i^\alpha = \frac{d}{dx_i} \left(\frac{\partial \mathcal{L}}{\partial u_i^\alpha} \delta_* u^\alpha \right) - \frac{d}{dx_i} \left(\frac{\partial \mathcal{L}}{\partial u_i^\alpha} \right) \delta_* u^\alpha\,,$$

and we obtain

$$\delta\mathcal{F} = \int_\Omega \left\{ \frac{d}{dx_i}\left(\mathcal{L}\,\delta x_i + \frac{\partial\mathcal{L}}{\partial u_i^\alpha}\,\delta_* u^\alpha\right) + [\mathcal{L}]^\alpha\,\delta_* u^\alpha \right\} dx\,,$$

where

$$[\mathcal{L}]^\alpha = \frac{\partial\mathcal{L}}{\partial u^\alpha} - \frac{d}{dx_i}\left(\frac{\partial\mathcal{L}}{\partial u_i^\alpha}\right), \quad \alpha = 1, 2, \ldots, m\,, \tag{4.1.94}$$

is the *Euler-Lagrange derivative* (also known as the variational derivative). Finally, by means of Eqs.(4.1.91) we obtain the variation of the functional $\mathcal{F}$ in the form

$$\delta\mathcal{F} = \int_\Omega \left\{ \frac{d}{dx_i}\left(\mathcal{L}\,\delta x_i - \frac{\partial\mathcal{L}}{\partial u_i^\alpha}\,u_j^\alpha\,\delta x_j + \frac{\partial\mathcal{L}}{\partial u_i^\alpha}\,\delta u^\alpha\right) \right.$$
$$\left. + [\mathcal{L}]^\alpha\left(\delta u^\alpha - u_j^\alpha\,\delta x_j\right) \right\} dx\,. \tag{4.1.95}$$

To obtain the equations of motion, under the conditions of Hamilton's principle, we assume that the variations δx_i vanish inside and on the boundary of the domain Ω (i.e., the integration domain remains unchanged), and the variations of the functions of state (u^α) are identically zero on the boundary of Ω. Thus, by applying the flux-divergence formula (the divergence theorem of Gauss) we show that the first term in (4.1.95) vanishes and, from $\delta\mathcal{F} = 0$, we obtain the equations of motion in the differential form

$$[\mathcal{L}]^\alpha = 0\,, \quad \alpha = 1, 2, \ldots, m\,. \tag{4.1.96}$$

These equations are invariant if we replace the function $\mathcal{L}$ by $\lambda\mathcal{L}$, where λ is an arbitrary non-zero constant (*scale transformation*), or by $\mathcal{L} + df_j/dx_j$, where $f_j(x_i, u^\alpha)$ are arbitrary functions of class C^2 (*divergence transformation*). It is easy to show that $[df_j/dx_j]^\alpha = 0$, which proves the invariance of (4.1.96) with respect to divergence transformations. In particular, when the only independent variable is the time (t) we obtain the gauge transformation discussed in Chap. 4 Sec. 1.1.

By *symmetry transformations* we denote the transformations of independent variables under which the form of the equations of motion (4.1.96) remains invariant. The general form of a transformation of independent variables is

$$x_i' = \varphi_i(\mathbf{x})\,, \quad i = 1, 2, \ldots, n\,, \tag{4.1.97}$$

and it changes the functions of state into

$$u'^\alpha(\mathbf{x}') = \Phi^\alpha(x_i, u^\alpha)\,, \quad \alpha = 1, 2, \ldots, m\,. \tag{4.1.98}$$

The functional (4.1.83) is invariant with respect to the transformation (4.1.97) if the relation

$$\mathcal{L}'(x'_i, u'^\alpha, u'^\alpha_i)\, dx' = \mathcal{L}(x_i, u^\alpha, u^\alpha_i)\, dx \tag{4.1.99}$$

holds. On the other hand, the equations of motion (4.1.96) are invariant if

$$\mathcal{L}'(x'_i, u'^\alpha, u'^\alpha_i) = \mathcal{L}(x'_i, u'^\alpha, u'^\alpha_i) + \frac{df_j}{dx'_j}\,. \tag{4.1.100}$$

It follows that the transformations (4.1.97) are symmetry transformations for the considered physical system if and only if both conditions (4.1.99) and (4.1.100) are satisfied. In the case of the infinitesimal transformations (4.1.84) these conditions become

$$\begin{aligned}
&\mathcal{L}'(x_i + \delta x_i, u^\alpha + \delta u^\alpha, u^\alpha_i + \delta u^\alpha_i)\, dx' = \mathcal{L}(x_i, u^\alpha, u^\alpha_i)\, dx\,,\\
&\mathcal{L}'(x_i + \delta x_i, u^\alpha + \delta u^\alpha, u^\alpha_i + \delta u^\alpha_i)\, dx'\\
&\quad = \mathcal{L}(x_i + \delta x_i, u^\alpha + \delta u^\alpha, u^\alpha_i + \delta u^\alpha_i)\, dx' + \frac{d}{dx_j}(\delta f_j)\, dx\,;
\end{aligned}$$

in the last term we have used the relation $\delta f_j(\mathbf{x}') = \delta f_j(\mathbf{x})$, because δf_j are of order ε, dx' differs from dx by terms of order ε, and we neglect all the terms of order ε^2 and higher. It follows that

$$\mathcal{L}(x_i + \delta x_i, u^\alpha + \delta u^\alpha, u^\alpha_i + \delta u^\alpha_i) = \left[\mathcal{L}(x_i, u^\alpha, u^\alpha_i) - \frac{d}{dx_j}\,\delta f_j\right] J^{-1}\,.$$

Finally, substituting (4.1.88) and neglecting terms of order ε^2 and higher, we obtain the equation

$$\left(\delta x_i\,\frac{\partial}{\partial x_i} + \delta u^\alpha\,\frac{\partial}{\partial u^\alpha} + \delta u^\alpha_i\,\frac{\partial}{\partial u^\alpha_i} + \frac{\partial}{\partial x_i}\,\delta x_i\right)\mathcal{L} = -\frac{d}{dx_j}\,\delta f_j(x_i, u^\alpha)\,. \tag{4.1.101}$$

Hence, an infinitesimal transformation of the form (4.1.84) is a symmetry transformation if, for a given $\mathcal{L}$, there exist functions $f_j(x_i, u^\alpha)$ of class C^2, so that the equation (4.1.101) is satisfied. It can be shown that, for a given $\mathcal{L}$, the symmetry transformations form a group, which represents the *symmetry group* of the considered physical system.

From (4.1.86) and (4.1.101) we have

$$\delta\mathcal{F} + \int_\Omega \frac{d}{dx_i}(\delta f_i)\, dx = 0\,,$$

so that (4.1.95) can be written in the form

$$\begin{aligned}
\int_\Omega \Bigg\{ \frac{d}{dx_i}\left(\mathcal{L}\,\delta x_i - \frac{\partial\mathcal{L}}{\partial u^\alpha_i}\, u^\alpha_j\,\delta x_j + \frac{\partial\mathcal{L}}{\partial u^\alpha_i}\,\delta u^\alpha + \delta f_i\right)&\\
+ [\mathcal{L}]^\alpha\left(\delta u^\alpha - u^\alpha_j\,\delta x_j\right)\Bigg\} dx = 0\,.&
\end{aligned}$$

This integral has to vanish on any domain Ω. If we assume that the integrand is a continuous function, it must vanish at any point $\mathbf{x} \in \Omega$. Consequently, when the equations of motion (4.1.96) are satisfied, we obtain an *equation of conservation* of the form

$$\frac{d}{dx_i}\left(\mathcal{L}\,\delta x_i - \frac{\partial \mathcal{L}}{\partial u_i^\alpha}\,u_j^\alpha\,\delta x_j + \frac{\partial \mathcal{L}}{\partial u_i^\alpha}\,\delta u^\alpha + \delta f_i\right) = 0\,. \qquad (4.1.102)$$

It follows that a symmetry transformation of a physical system is associated with a certain equation of conservation. As the symmetry transformations form a group the equation of conservation (4.1.102) establishes the connection between the symmetry group of a physical system and a certain conservation law. The general connection between the symmetry properties of a physical system and the conservation laws is given by Noether's theorem

THEOREM 4.1.5 (NOETHER) *If the Lagrangian of a physical system is invariant with respect to a continuous group of transformations with* p *parameters then there exist* p *quantities which are conserved during the evolution of the system.*

1.4.2 Lagrangians that do not explicitly depend on time

Let be a mechanical system described by a Lagrangian $\mathcal{L}(\mathbf{q},\dot{\mathbf{q}})$ which does not explicitly depend on time (for instance, a mechanical system subject to scleronomic constraints). We also assume that nor the Hamiltonian $H(\mathbf{q},\mathbf{p})$ (which is a consequence in the case of scleronomic constraints), neither the integrals of motion of the system, do not explicitly depend on time. For an infinitesimal transformation of the generalized coordinates

$$q_k' = q_k + \delta q_k\,, \quad k = 1, 2, \ldots, s\,, \qquad (4.1.103)$$

we have

$$\delta \dot{q}_k = \frac{d}{dt}\,\delta q_k\,.$$

We obtain the relation between the Lagrangian $\mathcal{L}(\mathbf{q},\dot{\mathbf{q}})$ and the Hamiltonian $H(\mathbf{q},\mathbf{p})$ by means of the generalized momenta defined in Chap. 4 Sec. 1.2. Thus, to go from H to $\mathcal{L}$ we use the transformation

$$\dot{q}_k = \frac{\partial H}{\partial p_k}\,, \quad k = 1, 2, \ldots, s\,.$$

If $C(\mathbf{q},\mathbf{p})$ is an integral of motion in the Hamiltonian formulation, then in the Lagrangian formulation this integral of motion will take the form $D(\mathbf{q},\dot{\mathbf{q}}) =$

$C(\mathbf{q}, \mathbf{p})$. Also, along the trajectory of the system ($[\mathcal{L}]^k = 0,\ k = 1, 2, \ldots, s$) this integral of motion will satisfy the equations

$$\frac{d}{dt} C(\mathbf{q}, \mathbf{p}) = \frac{d}{dt} D(\mathbf{q}, \dot{\mathbf{q}}) = 0\,. \tag{4.1.104}$$

The variation of the Lagrangian, induced by the transformations (4.1.103), can be written in the form (we use the summation convention for dummy indices from 1 to s)

$$\delta\mathcal{L} = \frac{d}{dt}\left(\frac{\partial\mathcal{L}}{\partial\dot{q}_k}\,\delta q_k\right) - [\mathcal{L}]^k\delta q_k\,,$$

which, along the system trajectory, becomes

$$\delta\mathcal{L} = \frac{d}{dt}\left(\frac{\partial\mathcal{L}}{\partial\dot{q}_k}\,\delta q_k\right). \tag{4.1.105}$$

There are only three cases when the equation (4.1.105) leads to conserved quantities (integrals of motion), i.e.:

1 If $\delta\mathcal{L} = 0$, the corresponding integral of motion is

$$D(\mathbf{q}, \dot{\mathbf{q}})\delta\tau = \frac{\partial\mathcal{L}}{\partial\dot{q}_k}\,\delta q_k = p_k\,\delta q_k\,, \tag{4.1.106}$$

where p_k are the generalized momenta in the Hamiltonian formulation, and $\delta\tau$ is a constant factor.

2 If there is a function $f(\mathbf{q})$ such that $\delta\mathcal{L} = d\,[\delta f(\mathbf{q})]/dt$, we obtain the integral of motion

$$D(\mathbf{q}, \dot{\mathbf{q}})\delta\tau = \left(\frac{\partial\mathcal{L}}{\partial\dot{q}_k} - \frac{\partial f}{\partial q_k}\right)\delta q_k = \left(p_k - \frac{\partial f}{\partial q_k}\right)\delta q_k\,. \tag{4.1.107}$$

In this case, we may assume that the Lagrangian has the form $\mathcal{L}' = \mathcal{L} - df/dt$ so that $\delta\mathcal{L}' = 0$. Note that the two Lagrangians ($\mathcal{L}$ and $\mathcal{L}'$) are related by a divergence transformation. Consequently, they are equivalent in the sense that the form of the equations of motion is preserved.

3 If

$$\delta\mathcal{L} = \frac{d}{dt}\,g(\mathbf{q}, \dot{\mathbf{q}})\,, \tag{4.1.108}$$

the associated integral of motion is

$$D(\mathbf{q}, \dot{\mathbf{q}})\delta\tau = \frac{\partial\mathcal{L}}{\partial\dot{q}_k}\,\delta q_k - g(\mathbf{q}, \dot{\mathbf{q}}) = p_k\,\delta q_k - g(\mathbf{q}, \dot{\mathbf{q}})\,. \tag{4.1.109}$$

In this case there is no Lagrangian $\mathcal{L}'$ which is equivalent to $\mathcal{L}$ and satisfies the equation $\delta\mathcal{L}' = 0$.

It is evident that the first and the second case are particular instances of the third case. Actually, there is no other case (i.e., $\delta\mathcal{L} \neq d\,g(\mathbf{q},\dot{\mathbf{q}})/dt$) which could lead to an integral of motion associated with the infinitesimal transformation (4.1.103).

Note that not all the three cases considered above correspond to symmetry transformations. The only independent variable is the time (t), and the Lagrangian $\mathcal{L}$ does not explicitly depend on t. Hence, the defining equation (4.1.101) takes the form

$$\delta\mathcal{L} = \left(\delta q_k \frac{\partial}{\partial q_k} + \delta\dot{q}_k \frac{\partial}{\partial \dot{q}_k}\right)\mathcal{L} = -\frac{d}{dt}\,\delta f(t,\mathbf{q})\,. \tag{4.1.110}$$

It follows that only the first and the second case correspond to symmetry transformations.

THEOREM 4.1.6 (NOETHER) *To every infinitesimal transformation of the form (4.1.103), inducing a variation of the Lagrangian of the form (4.1.108), it corresponds a conserved quantity defined by (4.1.109).*

If the infinitesimal transformation (4.1.103) is a symmetry transformation (i.e., $g(\mathbf{q},\dot{\mathbf{q}}) = \delta f(\mathbf{q})$), then there is an invariant Lagrangian

$$\mathcal{L}' = \mathcal{L} - \frac{d}{dt}\,f(\mathbf{q})\,. \tag{4.1.111}$$

1.4.3 The reciprocal of Noether's theorem

The Hessian of a given Lagrangian, with respect to the generalized velocities, has the elements

$$\mathcal{H}_{jk} = \mathcal{H}_{kj} = \frac{\partial^2\mathcal{L}}{\partial\dot{q}_j\partial\dot{q}_k} = \frac{\partial p_j}{\partial\dot{q}_k} = \frac{\partial p_k}{\partial\dot{q}_j}\,, \quad j,k,=1,2,\ldots,s\,, \tag{4.1.112}$$

and $\det\left[\mathcal{H}_{jk}\right] \neq 0$ is the necessary and sufficient condition for the integrability of Lagrange's equations. The inverse of the matrix $[\mathcal{H}]$ has the elements

$$\mathcal{H}^{-1}_{jk} = \mathcal{H}^{-1}_{kj} = \frac{\partial\dot{q}_j}{\partial p_k} = \frac{\partial\dot{q}_k}{\partial p_j}\,, \quad j,k,=1,2,\ldots,s\,. \tag{4.1.113}$$

The relation (4.1.109) represents the condition which must be satisfied by the most general integral of motion of a mechanical system whose evolution is described by Lagrange's equations. An equivalent form of this condition is given by (4.1.104). Thus, we have

$$\frac{d}{dt}D(\mathbf{q},\dot{\mathbf{q}})=\frac{\partial D}{\partial q_k}\dot{q}_k+\frac{\partial D}{\partial \dot{q}_k}\ddot{q}_k=0\,. \tag{4.1.114}$$

In our case $\mathcal{L}=\mathcal{L}(\mathbf{q},\dot{\mathbf{q}})$, $H=H(\mathbf{q},\mathbf{p})$, and $\dot{q}_k=\dot{q}_k(\mathbf{q},\mathbf{p})$, so that

$$\frac{\partial \dot{q}_k}{\partial q_j}=\frac{\partial}{\partial q_j}\left(\frac{\partial H}{\partial p_k}\right)=\frac{\partial}{\partial p_k}\left(\frac{\partial H}{\partial q_j}\right)=-\frac{\partial}{\partial p_k}\left(\frac{\partial \mathcal{L}}{\partial q_j}\right)=-\frac{\partial^2\mathcal{L}}{\partial \dot{q}_l\partial q_j}\frac{\partial \dot{q}_l}{\partial p_k}\,,$$

and we obtain the generalized accelerations in the form

$$\begin{aligned}\ddot{q}_k=\frac{d}{dt}\dot{q}_k&=\frac{\partial \dot{q}_k}{\partial q_j}\dot{q}_j+\frac{\partial \dot{q}_k}{\partial p_j}\dot{p}_j=-\frac{\partial^2\mathcal{L}}{\partial \dot{q}_l\partial q_j}\frac{\partial \dot{q}_l}{\partial p_k}\dot{q}_j+\frac{\partial \dot{q}_k}{\partial p_j}\frac{\partial \mathcal{L}}{\partial q_j}\\&=\left(\frac{\partial \mathcal{L}}{\partial q_j}-\frac{\partial^2\mathcal{L}}{\partial \dot{q}_j\partial q_l}\dot{q}_l\right)\frac{\partial \dot{q}_k}{\partial p_j}\,.\end{aligned} \tag{4.1.115}$$

In terms of the elements of the inverse Hessian (4.1.113), and by means of (4.1.115), the condition (4.1.114) becomes

$$\dot{q}_k\frac{\partial D}{\partial q_k}=\left(\frac{\partial^2\mathcal{L}}{\partial \dot{q}_k\partial q_l}\dot{q}_l-\frac{\partial \mathcal{L}}{\partial q_k}\right)\mathcal{H}_{kj}^{-1}\frac{\partial D}{\partial \dot{q}_j}\,, \tag{4.1.116}$$

where $D(\mathbf{q},\dot{\mathbf{q}})$ is an integral of motion determined by (4.1.109). The condition (4.1.116) can also be expressed by means of Poisson's bracket (note that $\partial C(\mathbf{q},\mathbf{p})/\partial t=0$)

$$[C(\mathbf{q},\mathbf{p}),H]=0\,, \tag{4.1.117}$$

this being a necessary and sufficient condition for $C(\mathbf{q},\mathbf{p})$ to be an integral of motion of the canonical system.

For $K=C$ in (4.1.82), and by means of (4.1.113), we may rewrite the variations of the generalized coordinates in the form

$$\delta q_k=[q_k,C]\,\delta\tau=\frac{\partial C}{\partial p_k}\delta\tau=\frac{\partial D}{\partial \dot{q}_j}\frac{\partial \dot{q}_j}{\partial p_k}\delta\tau=\frac{\partial D}{\partial \dot{q}_j}\mathcal{H}_{jk}^{-1}\delta\tau\,; \tag{4.1.118}$$

the conditions (4.1.116) and (4.1.118) are equivalent to (4.1.109). It follows that an integral of motion $D(\mathbf{q},\dot{\mathbf{q}})$ is associated with an infinitesimal transformation defined by (4.1.103) and (4.1.118), such that the condition (4.1.109) is satisfied, and the function $g(\mathbf{q},\dot{\mathbf{q}})$ is determined by

$$g(\mathbf{q},\dot{\mathbf{q}})=\left(\frac{\partial \mathcal{L}}{\partial \dot{q}_k}\frac{\partial D}{\partial \dot{q}_j}\mathcal{H}_{jk}^{-1}-D\right)\delta\tau\,. \tag{4.1.119}$$

The variation of the Lagrangian, induced by this infinitesimal transformation, is given by (4.1.108). The condition (4.1.116) is a necessary condition which has to be satisfied by the most general integral of motion given by Noether's theorem. Because (4.1.116) is also a homogeneous differential equation of the first order, in $2s$ independent variables $(\mathbf{q}, \dot{\mathbf{q}})$, it has $2s - 1$ independent solutions $D(\mathbf{q}, \dot{\mathbf{q}})$. Hence, (4.1.116) yields all the integrals of motion $D(\mathbf{q},\dot{\mathbf{q}})$ which do not explicitly depend on time. The condition (4.1.116) is also a sufficient condition for the integrals of motion and we may rewrite it in the form

$$\frac{d}{dt} D(\mathbf{q}, \dot{\mathbf{q}}) = \left[\frac{d}{dt} \left(\frac{\partial \mathcal{L}}{\partial \dot{q}_k} \right) - \frac{\partial \mathcal{L}}{\partial q_k} \right] \frac{\partial D}{\partial \dot{q}_j} \mathcal{H}_{jk}^{-1} = - [\mathcal{L}]^k \frac{\partial D}{\partial \dot{q}_j} \mathcal{H}_{jk}^{-1} , \quad (4.1.120)$$

to prove that $dD(\mathbf{q}, \dot{\mathbf{q}})/dt = 0$ along the system trajectory.

In the case of a symmetry transformation we have

$$g(\mathbf{q}, \dot{\mathbf{q}}) = \delta f(\mathbf{q}) = \frac{\partial f}{\partial q_k} \delta q_k = \frac{\partial f}{\partial q_k} \frac{\partial D}{\partial \dot{q}_j} \mathcal{H}_{jk}^{-1} \delta \tau , \quad (4.1.121)$$

and by means of (4.1.109) and (4.1.119) we obtain

$$D(\mathbf{q}, \dot{\mathbf{q}}) = E(\mathbf{q}, \mathbf{p}') = p'_k \frac{\partial E}{\partial p'_k} , \quad (4.1.122)$$

where

$$p'_k = \frac{\partial \mathcal{L}}{\partial \dot{q}_k} - \frac{\partial f}{\partial q_k} = \frac{\partial \mathcal{L}'}{\partial q_k} , \quad k = 1, 2, \ldots, s . \quad (4.1.123)$$

It follows that the integral of motion $D(\mathbf{q},\dot{\mathbf{q}})$ has to be a homogeneous function of the generalized momenta (4.1.123). This condition is equivalent to the condition which has to be satisfied by the generating function of infinitesimal, homogeneous, canonical transformations, and (4.1.118) shows how to obtain the related infinitesimal canonical transformation starting with a given integral of motion. Also, the condition (4.1.122), or the equivalent condition

$$D(\mathbf{q}, \dot{\mathbf{q}}) = \left(\frac{\partial \mathcal{L}}{\partial \dot{q}_k} - \frac{\partial f}{\partial q_k} \right) \frac{\partial D}{\partial \dot{q}_j} \mathcal{H}_{jk}^{-1} \quad (4.1.124)$$

[derived from (4.1.107) and (4.1.118)], represent sufficient conditions for the second case discussed in Chap. 4 Sec. 1.4.2.

All these results can be summarized in the form of the reciprocal of Noether's theorem:

THEOREM 4.1.7 (NOETHER; RECIPROCAL) *Suppose, $D(\mathbf{q}, \dot{\mathbf{q}})$ is an integral of motion. To this integral of motion we associate an infinitesimal variation (4.1.118) of the generalized coordinates, which induces a variation of the Lagrangian*

$$\delta \mathcal{L} = \frac{d}{dt} \left[\left(\frac{\partial \mathcal{L}}{\partial \dot{q}_k} \frac{\partial D}{\partial \dot{q}_j} \mathcal{H}_{jk}^{-1} - D \right) \delta \tau \right] . \quad (4.1.125)$$

If and only if there is a Lagrangian $\mathcal{L}'$ equivalent to the Lagrangian $\mathcal{L}$, as defined in (4.1.111), and $D(\mathbf{q},\dot{\mathbf{q}})$ is a homogeneous function of the first degree in the generalized momenta (4.1.123), then the integral of motion $D(\mathbf{q},\dot{\mathbf{q}})$ can be correlated with an invariance property of the Lagrangian ($\delta\mathcal{L} = 0$).

2. Symmetry laws and applications

2.1 Lie groups with one parameter and with *m* parameters

2.1.1 One-parameter Lie groups

We consider the transformations

$$x_i' = f_i(\mathbf{x}; a) = f_i(x_1, x_2, \ldots, x_{2n}; a)\,, \quad i = 1, 2, \ldots, 2n\,, \tag{4.2.1}$$

which depend on one parameter (a), and with the identity transformation corresponding to the zero value of the parameter ($a = 0$), that is

$$x_i = f_i(x_1, x_2, \ldots, x_{2n}; 0)\,. \tag{4.2.2}$$

Let

$$x_i'' = f_i(x_1', x_2', \ldots, x_{2n}'; b)\,, \quad i = 1, 2, \ldots, 2n\,, \tag{4.2.3}$$

be some other transformation of the form (4.2.1). The transformations of the form (4.2.1) constitute a *finite one-parameter topological group* if there is a value c of the parameter such that

$$x_i'' = f_i(x_1, x_2, \ldots, x_{2n}; c)\,, \quad i = 1, 2, \ldots, 2n\,. \tag{4.2.4}$$

According to (1.2.8) c is a function of a and b,

$$c = \varphi(a; b)\,, \quad \varphi(a; 0) = a\,.$$

The group of transformations (4.2.1) is a Lie group if the functions f_i are analytic functions in the parameter a, and the function φ is analytic in both arguments.

In terms of the general method for infinitesimal transformations, described in Chap. 1 Sec. 2, we have

$$\begin{aligned} x_i'' &= x_i' + dx_i' = f_i(\mathbf{x}'; \delta a) = x_i' + \left.\frac{\partial f_i(\mathbf{x}'; a)}{\partial a}\right|_{a=0} \delta a = x_i' + u_i(\mathbf{x}')\delta a\,, \\ c &= a + da = \varphi(a; \delta a) = a + \left.\frac{\partial \varphi(a; b)}{\partial b}\right|_{b=0} \delta a = a + \mu(a)\delta a\,, \end{aligned} \tag{4.2.5}$$

where c is related with the transformation

$$x_i' + dx_i' = f_i(\mathbf{x}; a + da)\,.$$

Now, the system of differential equations (1.2.15) becomes

$$\frac{dx_i'}{da} = u_i(\mathbf{x}')\,\lambda(a)\,, \quad i = 1, 2, \ldots, 2n\,, \tag{4.2.6}$$

where the functions $u_i(\mathbf{x})$ form a velocity field. By introducing the parameter

$$\alpha = \int_0^a \lambda(\widetilde{a})\,d\widetilde{a}\,, \tag{4.2.7}$$

we rewrite (4.1.30) in the form

$$\frac{dx_i'}{d\alpha} = u_i(\mathbf{x}')\,, \quad i = 1, 2, \ldots, 2n\,. \tag{4.2.8}$$

The new parameter α is an additive quantity with respect to the composition of infinitesimal transformations, and we have

$$f_i(f_1(\mathbf{x};\alpha), f_2(\mathbf{x};\alpha), \ldots, f_{2n}(\mathbf{x};\alpha);\beta) = f_i(\mathbf{x};\alpha+\beta)\,. \tag{4.2.9}$$

From (1.2.19) we obtain the infinitesimal generator of the transformation (4.2.1)

$$\mathsf{X} = \sum_{i=1}^{2n} u_i \frac{\partial}{\partial x_i}\,. \tag{4.2.10}$$

Hence, the solution of the system of differential equations (4.2.8) can be expressed in terms of the power series

$$x_i' = x_i + \alpha \mathsf{X} x_i + \frac{\alpha^2}{2!}\mathsf{X}^2 x_i + \ldots\,, \quad i = 1, 2, \ldots, 2n\,, \tag{4.2.11}$$

which, for infinitesimal transformations, reduces to the simple form

$$x_i' = x_i + \alpha \mathsf{X} x_i = x_i + \alpha u_i(\mathbf{x})\,, \quad i = 1, 2, \ldots, 2n\,. \tag{4.2.12}$$

From (4.2.11) it follows

$$x_i' = f_i(\mathbf{x};\alpha) = e^{\alpha \mathsf{X}} x_i\,,$$

and we are led to the equations

$$\frac{\partial f_i}{\partial \alpha} = \mathsf{X} f_i\,, \quad i = 1, 2, \ldots, 2n\,. \tag{4.2.13}$$

We are now in a position to establish the connection between the canonical transformations (4.1.60-4.1.63), and the one-parameter Lie group, formed by the transformations (4.2.1). This will allow us to make use of Noether's theorem (Theorem 4.1.5), to find the corresponding conserved quantities.

First of all, it is worth-while to note that a special form of (4.2.13) is obtained if we set $f_i(\mathbf{x};\alpha) = x_i(\alpha)$ and $\alpha = t$, such that (4.2.13) become the canonical equations

$$\frac{dx_i}{dt} = \mathsf{X}_t\, x_i\,, \quad i = 1, 2, \ldots, 2n\,, \tag{4.2.14}$$

where we have attached the label t to the infinitesimal generator to emphasize the associated parameter.

Now, we assume that the transformations (4.2.1) constitute a Lie group, and note that in the phase space (Γ_{2n}), these transformations can be written in a form similar to (4.1.60), that is

$$q_i' \equiv Q_i = f_i(\mathbf{q}, \mathbf{p}; \alpha)\,, \quad p_i' \equiv P_i = g_i(\mathbf{q}, \mathbf{p}; \alpha)\,, \quad i = 1, 2, \ldots, n\,. \tag{4.2.15}$$

We sought the conditions that have to be satisfied by these transformations such that they are also canonical transformations. A sufficient condition, consisting in the existence of a function $W = W(\mathbf{Q}, \mathbf{P})$ whose total differential is the differential one-form (4.1.65), has been derived in Chap. 4 Sec. 1.3.2. Thus, if we denote by

$$\lambda_i = \frac{\partial f_i}{\partial \alpha}\,, \quad \mu_i = \frac{\partial g_i}{\partial \alpha}\,, \tag{4.2.16}$$

from (4.2.15), we obtain

$$\left.\frac{dQ_i}{d\alpha}\right|_{\alpha=0} = \lambda_i(\mathbf{Q}, \mathbf{P})\,, \quad \left.\frac{dP_i}{d\alpha}\right|_{\alpha=0} = \mu_i(\mathbf{Q}, \mathbf{P})\,.$$

Also, the infinitesimal generator (4.2.10) becomes

$$\mathsf{X} = \lambda_i\, \frac{\partial}{\partial q_i} + \mu_i\, \frac{\partial}{\partial p_i}\,, \tag{4.2.17}$$

where we have used the summation convention for dummy indices from 1 to n.

By means of the power series solution (4.2.11), and assuming the series expansion

$$W(\mathbf{Q}, \mathbf{P}) = \sum_{k=0}^{\infty} \alpha^k W_k(\mathbf{Q}, \mathbf{P})\,,$$

we obtain the equation

$$p_i dq_i - \left(p_i + \alpha\mu_i + \frac{\alpha^2}{2!}\,\mathsf{X}^2\mu_i + \ldots\right)\left(dq_i + \alpha d\lambda_i + \frac{\alpha^2}{2!}\, d\mathsf{X}^2\lambda_i + \ldots\right)$$
$$= \sum_{k=0}^{\infty} \alpha^k dW_k(\mathbf{Q}, \mathbf{P})\,,$$

which has to be identically satisfied. Hence, equating coefficients of equal powers of α yields the relations

$$dW_0 = 0\,, \quad dW_1 = -\left(\mu_i dq_i + p_i d\lambda_i\right)\,, \quad \ldots\,.$$

All these relations are simultaneously satisfied if we assume that the function W has the form

$$W = \alpha W_1 + \frac{\alpha^2}{2!}\mathsf{X}\, W_1 + \frac{\alpha^3}{3!}\mathsf{X}^2\, W_1 + \ldots\,.$$

To find the functions λ_i and μ_i, we introduce the function

$$U(\mathbf{q},\mathbf{p}) = -W_1 - p_i\,\lambda_i\,, \tag{4.2.18}$$

such that

$$dU = \frac{\partial U}{\partial q_i}\,dq_i + \frac{\partial U}{\partial p_i}\,dp_i = \mu_i dq_i - \lambda_i dp_i\,.$$

The last equality is identically satisfied in the variables q_i and p_i, if we choose

$$\lambda_i = -\frac{\partial U}{\partial p_i}\,, \quad \mu_i = \frac{\partial U}{\partial q_i}\,, \quad i = 1, 2, \ldots, n\,. \tag{4.2.19}$$

Thus, we obtain the condition which has to be satisfied such that the transformations (4.2.15) are canonical transformations. Consequently, in terms of Poisson's bracket, the infinitesimal generator (4.2.17) takes the form

$$\mathsf{X} = \frac{\partial U}{\partial q_i}\frac{\partial}{\partial p_i} - \frac{\partial U}{\partial p_i}\frac{\partial}{\partial q_i} \equiv [U,\ ]\,. \tag{4.2.20}$$

Finally, by applying the operator (4.2.20) to the functions $\partial U/\partial q_k$ and $\partial U/\partial p_k$, we obtain the whole group of canonical transformations, that is

$$\begin{aligned} Q_i &= q_i - \alpha\,\frac{\partial U}{\partial p_i} + \frac{\alpha^2}{2!}\mathsf{X}\,\frac{\partial U}{\partial p_i} - \ldots\,, \\ P_i &= p_i + \alpha\,\frac{\partial U}{\partial q_i} + \frac{\alpha^2}{2!}\mathsf{X}\,\frac{\partial U}{\partial q_i} + \ldots\,, \quad i = 1, 2, \ldots, n\,. \end{aligned} \tag{4.2.21}$$

Note that we are led to the same result by integrating the system of partial differential equations (4.2.13)

$$\frac{\partial f_i}{\partial \alpha} = [U, f_i]\,, \quad i = 1, 2, \ldots, 2n\,, \tag{4.2.22}$$

or, by integrating the system of differential equations (4.2.16), written in the from

$$\frac{dQ_i}{d\alpha} = \frac{\partial \widetilde{U}}{\partial P_i}\,, \quad \frac{dP_i}{d\alpha} = -\frac{\partial \widetilde{U}}{\partial Q_i}\,, \quad i = 1, 2, \ldots, n\,, \tag{4.2.23}$$

where $\widetilde{U}(\mathbf{Q},\mathbf{P}) = -U(\mathbf{q},\mathbf{p})$. All these three methods used to determine the Lie group of canonical transformations are equivalent. The function U is the generating function of canonical transformations (see Chap. 4 Sec. 1.3.2), while the operator $\mathsf{X} = [U, \ ]$ is the infinitesimal generator of the Lie group. It is also easy to see that, for $\alpha = t$ and $\widetilde{U} = \widetilde{H}(\mathbf{Q},\mathbf{P})$, the system of differential equations (4.2.23) is identical with Hamilton's equations (4.1.62).

We denote by $\mathbf{x}$ the set of canonical variables $\{\mathbf{q},\mathbf{p}\}$. Then, a function $f(\mathbf{x})$ is invariant with respect to the Lie group of canonical transformations, if the equation

$$f(\mathbf{x}') = \sum_{k=0}^{\infty} \frac{\alpha^k}{k!} \mathsf{X}^k f(\mathbf{x}) = f(\mathbf{x}) \tag{4.2.24}$$

is identically satisfied. This can be accomplished if and only if $\mathsf{X} f(\mathbf{x}) = 0$, $\forall \mathbf{x} \in \mathbb{R}^{2n}$. Explicitly, by using the expression (4.2.20) for the infinitesimal generator X, we may rewrite this condition in terms of canonical variables

$$\frac{\partial U}{\partial q_i}\frac{\partial f}{\partial p_i} - \frac{\partial U}{\partial p_i}\frac{\partial f}{\partial q_i} = [U, f] = 0\,. \tag{4.2.25}$$

It follows that the conserved quantities are invariants of the group of canonical transformations, and also solutions of (4.2.25). If we find m independent invariants of this kind, then we may choose them as generalized coordinates (cyclic coordinates in this case), and there remain to integrate only $2n - m$ equations of motion.

2.1.2 Lie groups with m parameters

Let

$$x_i' = f_i(x_1, x_2, \ldots, x_{2n}; a_1, a_2, \ldots, a_m)\,, \quad i = 1, 2, \ldots, 2n\,, \tag{4.2.26}$$

be a set of general transformations, which depend on m real parameters, and are of the form (1.2.4). We assume that $a_1, a_2, \ldots, a_m$ are essential parameters, and therefore the transformations (4.2.26) form a Lie group with m-parameters. In this case the functions (1.2.9) are linearly independent, the differential equations of the group have the form (1.2.15), and the infinitesimal generators $\{\mathsf{X}_k\}$ $(k = 1, 2, \ldots, m)$ are given by (1.2.19). The formulation outlined in the previous section can easily be generalized by introducing m generating functions $U_k(\mathbf{q},\mathbf{p})$, such that (4.2.20) becomes

$$\mathsf{X}_k = \frac{\partial U_k}{\partial q_i}\frac{\partial}{\partial p_i} - \frac{\partial U_k}{\partial p_i}\frac{\partial}{\partial q_i} \equiv [U_k, \ ]\,. \tag{4.2.27}$$

The commutator of two infinitesimal generators can be evaluated by applying it to an arbitrary function $f(\mathbf{q}, \mathbf{p})$. Thus, we have

$$[\mathsf{X}_j, \mathsf{X}_k] f = (\mathsf{X}_j \mathsf{X}_k - \mathsf{X}_k \mathsf{X}_j) f = [U_j, [U_k, f]] - [U_k, [U_j, f]] \,. \tag{4.2.28}$$

By means of (4.2.28), we prove that the infinitesimal generators (4.2.27) satisfy the relations (1.2.24) and (1.2.26), that is Poisson's brackets are skew-symmetric

$$[U_j, U_k] = -[U_k, U_j] \,, \tag{4.2.29}$$

and verify the Poisson-Jacobi identity

$$[[U_i, U_j], U_k] + [[U_j, U_k], U_i] + [[U_k, U_i], U_j] = 0 \,. \tag{4.2.30}$$

If we set one of the generating functions $U_k = H$, we are led to the Jacobi-Poisson theorem (Theorem 4.1.2) for the case when the time does not appear explicitly: *If U_i and U_j are two conserved quantities (i.e., $[U_i, H] = [U_j, H] = 0$), then $[U_i, U_j]$ is also a conserved quantity.*

Finally, from (4.2.28) and (1.2.23), and by means of (4.2.29) and (4.2.30), we obtain the equation for the structure constants of the group

$$[U_i, U_j] = \sum_{k=1}^{m} c_{ij}^k \, U_k \,, \quad i, j = 1, 2, \ldots, m \,. \tag{4.2.31}$$

2.2 The Symplectic and Euclidean groups

2.2.1 The Symplectic group

DEFINITION 4.2.1 *The* **real Symplectic group of order** $2n$ *is the group of unimodular linear transformations, acting in the linear space of even dimension $\mathbb{R}^{2n}$ ($n = 1, 2, \ldots$), that leave invariant a non-degenerate skew-symmetric bilinear form $\varphi(\mathbf{x}, \mathbf{y})$, $\mathbf{x}, \mathbf{y} \in \mathbb{R}^{2n}$.*

The real Symplectic group of order $2n$, denoted by $\mathsf{Sp}(2n, \mathbb{R})$, is a subgroup of the special linear group $\mathsf{SL}(2n, \mathbb{R})$, defined in Chap. 2 Sec. 3.3.1. It has been shown that the transformations in the Symplectic group $\mathsf{Sp}(2n, \mathbb{R})$ preserve the classical Poisson brackets from classical mechanics, as well as the commutation relations among n pairs of canonical variables in quantum mechanics. We may also apply the Definition 4.2.1 to the case of a linear complex space of even dimension, the corresponding Symplectic group being the *complex Symplectic group of order* $2n$, denoted by $\mathsf{Sp}(2n, \mathbb{C}) \subset \mathsf{SL}(2n, \mathbb{C})$. In this case, the general real linear group $\mathsf{GL}(n, \mathbb{R})$ is a subgroup of the Symplectic group $\mathsf{Sp}(2n, \mathbb{C})$.

To determine the form of the matrices corresponding to the elements of the real Symplectic group, we choose an orthonormal basis in the space $\mathbb{R}^{2n}$ and denote the components of two vectors $\mathbf{x}, \mathbf{y} \in \mathbb{R}^{2n}$ by $(x_1, x_2, \ldots, x_{2n})$ and

$(y_1, y_2, \ldots, y_{2n})$, respectively. We also assume that the scalar product of these two vectors has the form

$$< \mathbf{x}, \mathbf{y} > = \sum_{k=1}^{2n} x_k \, y_k \,. \tag{4.2.32}$$

With these notations the skew-symmetric bilinear form $\varphi(\mathbf{x}, \mathbf{y})$, defined on $\mathbb{R}^{2n}$, has the expression

$$\varphi(\mathbf{x}, \mathbf{y}) = x_1 y_{2n} + x_2 y_{2n-1} + \ldots + x_n y_{n+1} - x_{n+1} y_n - x_{n+2} y_{n-1} - \ldots - x_{2n} y_1 \,, \tag{4.2.33}$$

and the associated matrix

$$\mathbf{s_0} = \begin{bmatrix} \mathbf{0} & -\mathbf{s_1} \\ \mathbf{s_1} & \mathbf{0} \end{bmatrix} , \tag{4.2.34}$$

where $\mathbf{s_1}$ is a square matrix of order n, which contains only once the number 1 in each row and each column, all the other elements being zero

$$\mathbf{s_1} = \begin{bmatrix} 0 \; 0 & \ldots & 0 \; 1 \\ 0 \; 0 & \ldots & 1 \; 0 \\ \vdots \; \vdots & & \vdots \; \vdots \\ 0 \; 1 & \ldots & 0 \; 0 \\ 1 \; 0 & \ldots & 0 \; 0 \end{bmatrix} .$$

Now, in terms of the scalar product (4.2.32), the bilinear form (4.2.33) becomes

$$\varphi(\mathbf{x}, \mathbf{y}) = < \mathbf{s_0} \mathbf{x}, \mathbf{y} > \,. \tag{4.2.35}$$

The invariance of the bilinear form $\varphi(\mathbf{x}, \mathbf{y})$ with respect to a transformation $g \in \mathsf{Sp}(2n, \mathbb{R})$ can be expressed in the form

$$< \mathbf{s_0} \, \mathbf{g} \, \mathbf{x}, \mathbf{g} \, \mathbf{y} > = < \mathbf{s_0} \, \mathbf{x}, \mathbf{y} > , \tag{4.2.36}$$

where $\mathbf{g}$ is the matrix corresponding to the transformation g. Therefore, we have

$$\mathbf{g}^{\mathrm{T}} \, \mathbf{s_0} \, \mathbf{g} = \mathbf{s_0} \,, \tag{4.2.37}$$

where the superscript T denotes a transposed matrix. The determinants of the matrices in (4.2.37) satisfy the relation $\det \, (\mathbf{g}^{\mathrm{T}} \, \mathbf{s_0} \, \mathbf{g}) = |\det \mathbf{g}|^2 \det \mathbf{s_0}$, hence

$$|\det \mathbf{g}|^2 = 1 \,. \tag{4.2.38}$$

The relation (4.2.37) determines the form of the matrices associated with the elements of the Symplectic group $\mathsf{Sp}(2n, \mathbb{R})$, and reciprocally, every matrix which satisfies (4.2.37) is an element of the group of Symplectic matrices.

Let

$$\mathbf{u} = \begin{bmatrix} \mathbf{e} & \mathbf{0} \\ \mathbf{0} & \mathbf{s}_1 \end{bmatrix} \tag{4.2.39}$$

be the matrix of an orthogonal transformation of the basis. From $\mathbf{s}_1{}^{\mathrm{T}} = \mathbf{s}_1$ and $\mathbf{s}_1\mathbf{s}_1 = \mathbf{e}$, it follows that

$$\mathbf{u}\mathbf{u}^{\mathrm{T}} = \mathbf{u}^{\mathrm{T}}\mathbf{u} = \begin{bmatrix} \mathbf{e} & \mathbf{0} \\ \mathbf{0} & \mathbf{e} \end{bmatrix}.$$

Consequently, the bilinear form $\varphi(\mathbf{x}, \mathbf{y})$ becomes

$$\varphi'(\mathbf{x}, \mathbf{y}) = < \mathbf{s}_0\, \mathbf{u}\,\mathbf{x}, \mathbf{u}\,\mathbf{y} > = < \mathbf{u}^{\mathrm{T}}\,\mathbf{s}_0\,\mathbf{u}\,\mathbf{x}, \mathbf{y} > = < \mathbf{s}\,\mathbf{x}, \mathbf{y} >, \tag{4.2.40}$$

where $\mathbf{s}$ is the matrix associated with the bilinear form $\varphi'(\mathbf{x}, \mathbf{y})$. By employing the explicit expressions (4.2.34) and (4.2.39), we obtain the matrix associated with the bilinear form $\varphi'(\mathbf{x}, \mathbf{y})$

$$\mathbf{s} = \mathbf{u}^{\mathrm{T}}\,\mathbf{s}_0\,\mathbf{u} = \begin{bmatrix} \mathbf{0} & -\mathbf{e} \\ \mathbf{e} & \mathbf{0} \end{bmatrix}. \tag{4.2.41}$$

The transformation (4.2.39) is orthogonal such that it leaves invariant the scalar product (4.2.32), therefore we may define the matrices associated with the elements of the Symplectic group in terms of the condition

$$\mathbf{g}^{\mathrm{T}}\,\mathbf{s}\,\mathbf{g} = \mathbf{s}\,. \tag{4.2.42}$$

Note that $\mathbf{s}\,\mathbf{g}^{\mathrm{T}}\,\mathbf{s}\,\mathbf{g} = \mathbf{s}^2 = -\mathbf{e}$, and so the inverse matrices are given by $\mathbf{g}^{-1} = -\mathbf{s}\,\mathbf{g}^{\mathrm{T}}\,\mathbf{s}$.

THEOREM 4.2.1 *Every irreducible representation, of finite dimension, of the group* $\mathsf{Sp}(2n, \mathbb{R})$ *is completely determined by two ordered sets of integers* $\alpha_1 \geq \alpha_2 \geq \ldots \geq \alpha_n \geq 0$ *and* $\beta_1 \geq \beta_2 \geq \ldots \geq \beta_n \geq 0$.

We shown in Chap. 4 Sec. 1.3.2 that the necessary and sufficient condition for the transformation (4.1.60) to be a canonical transformation, is represented by (4.1.78), which is identical to (4.2.42). If we denote by

$$\mathbf{x} = (x_1, x_2, \ldots, x_{2n}) \equiv (q_1, q_2, \ldots, q_n, p_1, p_2, \ldots, p_n) = (\mathbf{q}, \mathbf{p}) \tag{4.2.43}$$

a vector in the phase space Γ_{2n}, then the Poisson bracket (4.1.37) can be written in the form

$$[f, g] = \sum_{j,k=1}^{2n} s_{jk}\, \frac{\partial f}{\partial x_j}\, \frac{\partial g}{\partial x_k}\,. \tag{4.2.44}$$

Also, the transformation (4.1.60) becomes

$$X_i = X_i(x_1, x_2, \dots, x_{2n}), \quad i = 1, 2, \dots, 2n, \tag{4.2.45}$$

with the Jacobian

$$J = \det [J_{jk}] = \det \left[\frac{\partial x_j}{\partial X_k} \right],$$

and the condition (4.1.78) takes the form

$$\mathbf{J}^{\mathrm{T}} \mathbf{s} \mathbf{J} = \mathbf{s}. \tag{4.2.46}$$

It follows that, *a transformation of the form* (4.1.60), *or* (4.2.45), *is a canonical transformation if and only if the Jacobi matrix associated to this transformation is Symplectic*. Consequently, the canonical transformations form a group isomorphic to the real Symplectic group $\mathsf{Sp}(2n, \mathbb{R})$.

We shown in Chap. 4 Sec. 1.3.2 that Poisson's bracket is invariant with respect to the canonical transformations. Now, in terms of the variables X_i, defined in (4.2.45), we obtain

$$[f, g]_{\mathbf{P},\mathbf{Q}} = \sum_{j,k=1}^{2n} s_{jk} \frac{\partial f}{\partial X_j} \frac{\partial g}{\partial X_k} = \sum_{j,k=1}^{2n} \sum_{l,m=1}^{2n} s_{jk} \frac{\partial f}{\partial x_l} \frac{\partial x_l}{\partial X_j} \frac{\partial g}{\partial x_m} \frac{\partial x_m}{\partial X_k}$$

$$= \sum_{l,m=1}^{2n} \left(\sum_{j,k=1}^{2n} J_{lj} s_{jk} J_{km}^{\mathrm{T}} \right) \frac{\partial f}{\partial x_l} \frac{\partial g}{\partial x_m} = \sum_{l,m=1}^{2n} s_{lm} \frac{\partial f}{\partial x_l} \frac{\partial g}{\partial x_m} = [f, g]_{\mathbf{p},\mathbf{q}},$$

where we have used the relation

$$\mathbf{J}\mathbf{s}\mathbf{J}^{\mathrm{T}} = \mathbf{s}\mathbf{s}\mathbf{J}\mathbf{s}\mathbf{J}^{\mathrm{T}}\mathbf{s}\mathbf{s} = \mathbf{s}\mathbf{s}\mathbf{J}(\mathbf{s}\mathbf{J}^{\mathrm{T}}\mathbf{s})\mathbf{s} = -\mathbf{s}\mathbf{s}\mathbf{s} = \mathbf{s} = \mathbf{J}^{\mathrm{T}}\mathbf{s}\mathbf{J}.$$

An analogous proof demonstrates the invariance of Lagrange's bracket with respect to the canonical transformations.

Also, the conditions (4.1.69) are *necessary and sufficient* to determine if a transformation of the form (4.1.60) is a canonical transformation. In terms of the variables X_i, these conditions become

$$\{Q_j, Q_k\} = \{P_j, P_k\} = 0, \quad \{Q_j, P_k\} = \delta_{jk}, \quad j, k = 1, 2, \dots, n, \tag{4.2.47}$$

where

$$\{f, g\} = \sum_{i=1}^{n} \left(\frac{\partial q_i}{\partial f} \frac{\partial p_i}{\partial g} - \frac{\partial q_i}{\partial g} \frac{\partial p_i}{\partial f} \right) = \sum_{j,k=1}^{2n} s_{jk} \frac{\partial x_j}{\partial f} \frac{\partial x_k}{\partial g} \tag{4.2.48}$$

is the Lagrange bracket corresponding to the parameters f and g. It is easy to show that the Poisson and Lagrange brackets satisfy the equation

$$\sum_{i=1}^{2n} [f_i, f_j] \{f_i, f_k\} = \delta_{jk}, \quad j, k = 1, 2, \dots, 2n, \tag{4.2.49}$$

where $\{f_1, f_2, \ldots, f_{2n}\}$ is a set of $2n$ linearly independent functions.

We will now concentrate on the changes induced in the differential one-form (4.1.65) by a transformation of the form (4.1.60). If the differential one-form (4.1.65) is a total differential, then, in the phase space Γ_{2n}, we will also have a differential form

$$\sum_{i=1}^{n} [(\pm \eta_i d\xi_i) - (\pm Y_i dX_i)] \,, \tag{4.2.50}$$

where $\boldsymbol{\xi} \equiv \{\xi_1, \xi_2, \ldots, \xi_n\}$ and $\mathbf{X} \equiv \{X_1, X_2, \ldots, X_n\}$ are components of an arbitrary partition into n-dimensional subsets, of the $2n$-dimensional sets of canonical variables $\{\mathbf{q}, \mathbf{p}\}$, and $\{\mathbf{Q}, \mathbf{P}\}$, respectively, while $\boldsymbol{\eta} \equiv \{\eta_1, \eta_2, \ldots, \eta_n\}$ and $\mathbf{Y} \equiv \{Y_1, Y_2, \ldots, Y_n\}$ are the corresponding conjugate variables. We have to pick the upper or lower sign if $\boldsymbol{\xi}$ and $\mathbf{X}$ are, respectively, generalized coordinates or generalized momenta. In this context, we say that the transformation (4.1.60) is a complete canonical transformation, if there is a function $V(\boldsymbol{\xi}, \mathbf{X})$ such that the relations

$$\eta_i = \pm \frac{\partial V}{\partial \xi_i} \,, \quad Y_i = \pm \frac{\partial V}{\partial X_i} \,, \quad i = 1, 2, \ldots, n \,, \tag{4.2.51}$$

and the transformation equations (4.1.60), are identical. If this is the case, we say that the canonical transformation is generated by the function V.

Let us consider the transformations (4.2.1) and (4.2.3), having the Jacobi matrices $\mathbf{J}$ and $\mathbf{J}'$, respectively. If these two transformations are canonical transformations, then the matrices $\mathbf{J}$ and $\mathbf{J}'$ are Symplectic matrices. Hence, the matrix $\mathbf{J}'' = \mathbf{J}\mathbf{J}'$, corresponding to the transformation

$$x_i'' = f_i''(x_1, x_2, \ldots, x_{2n}) = f_i'(f_1(\mathbf{x}), f_2(\mathbf{x}), \ldots, f_{2n}(\mathbf{x})) \,, \; i = 1, 2, \ldots, 2n \,, \tag{4.2.52}$$

is also a Symplectic matrix, and therefore, the transformation (4.2.52) is also a canonical transformation. Now, we assume that the canonical transformations of the form (4.2.1) and (4.2.3) are generated by the functions $V(\boldsymbol{\xi}, \boldsymbol{\xi}')$ and $V'(\boldsymbol{\xi}', \boldsymbol{\xi}'')$, that is

$$\begin{aligned} &\sum_{i=1}^{n} \left[(\pm \eta_i d\xi_i) - (\pm \eta_i' d\xi_i')\right] = dV(\boldsymbol{\xi}, \boldsymbol{\xi}') \,, \\ &\sum_{i=1}^{n} \left[(\pm \eta_i' d\xi_i') - (\pm \eta_i'' d\xi_i'')\right] = dV'(\boldsymbol{\xi}', \boldsymbol{\xi}'') \,. \end{aligned} \tag{4.2.53}$$

The transformation (4.2.52) is also a canonical transformation such that we may assume that it is generated by a function $V''(\boldsymbol{\xi}, \boldsymbol{\xi}'')$, i.e.,

$$\sum_{i=1}^{n} \left[(\pm \eta_i d\xi_i) - (\pm \eta_i'' d\xi_i'') \right] = dV''(\boldsymbol{\xi}, \boldsymbol{\xi}'') . \tag{4.2.54}$$

From (4.2.53) we deduce

$$V''(\boldsymbol{\xi}, \boldsymbol{\xi}'') = V(\boldsymbol{\xi}, \boldsymbol{\xi}') + V'(\boldsymbol{\xi}', \boldsymbol{\xi}'') , \tag{4.2.55}$$

where, according to (4.2.51), the function V'' is independent of $\boldsymbol{\xi}'$. Indeed, from (4.2.55) and (4.2.53) we have

$$\frac{\partial V''}{\partial \xi_i'} = \frac{\partial}{\partial \xi_i'} \left[V(\boldsymbol{\xi}, \boldsymbol{\xi}') + V'(\boldsymbol{\xi}', \boldsymbol{\xi}'') \right] = \mp \eta_i' \pm \eta_i' = 0 , \quad i = 1, 2, \ldots, n .$$

Consequently, the generating functions exhibit the property of additivity with respect to the composition of canonical transformations, defined by (4.2.52).

2.2.2 The Euclidean group

DEFINITION 4.2.2 *It is called* **rigid motion** *of the n-dimensional Euclidean space $\mathcal{E}_n$, an inhomogeneous linear transformation of the form*

$$x_i' = \sum_{j=1}^{n} R_{ij} x_j + a_i , \quad i = 1, 2, \ldots, n , \tag{4.2.56}$$

where $\mathbf{x} \equiv (x_1, x_2, \ldots, x_n)$, $\mathbf{x}' \equiv (x_1', x_2', \ldots, x_n')$, and $\mathbf{a} \equiv (a_1, a_2, \ldots, a_n)$, are vectors of $\mathcal{E}_n$, while the matrix $\mathbf{R}$ represents a rotation of $\mathcal{E}_n$ ($\mathbf{R} \in \mathsf{SO}(n)$).

Note that, a transformation of the form (4.2.56), which we will denote by $g = g(\mathbf{a}, \mathbf{R})$, preserves the distance between any two points in $\mathcal{E}_n$, and the orientation of $\mathcal{E}_n \subset \mathcal{E}_{n+1}$. The set of all rigid motions (4.2.56) forms the Euclidean group $\mathsf{E}(n)$, which is isomorphic to the group of matrices

$$\mathbf{t}(g) = \begin{bmatrix} \mathbf{R} & \mathbf{a} \\ \mathbf{0} & \mathbf{1} \end{bmatrix} , \tag{4.2.57}$$

of order $n + 1$, where $\mathbf{R}$ is an orthogonal matrix, $\mathbf{a}$ is an n-dimensional column vector, and $\mathbf{0}$ is an n-dimensional row vector of zero elements. Due to this isomorphism the set of matrices (4.2.57) form a faithful representation of the Euclidean group $\mathsf{E}(n)$. The elements of $\mathsf{E}(n)$ depend on $n(n+1)/2$ parameters, which are

- $n(n-1)/2$ Euler angles

$$\theta_{1j} \in [0, 2\pi), \quad \theta_{ij} \in [0, \pi), \quad 1 < i \leq j, \ j = 1, 2, \ldots, n-1,$$

  for the rotation $\mathbf{R}$, and

- n spherical coordinates

$$r \in [0, \infty), \quad \varphi_1 \in [0, 2\pi), \quad \varphi_k \in [0, \pi), \quad k = 2, 3, \ldots, n-1,$$

  for the translation defined by the vector $\mathbf{a}$.

The product of two elements from $\mathsf{E}(n)$ has the form

$$g(\mathbf{a}_1, \mathbf{R}_1) g(\mathbf{a}_2, \mathbf{R}_2) = g(\mathbf{a}_1 + \mathbf{R}_1 \mathbf{a}_2, \mathbf{R}_1 \mathbf{R}_2), \tag{4.2.58}$$

and corresponds to the successive application (composition) of the transformations (4.2.56) associated to the group elements. Note that the following relations hold

$$g(\mathbf{a}, \mathbf{R}) = g(\mathbf{a}, \mathbf{e}) g(\mathbf{0}, \mathbf{R}) = g(\mathbf{0}, \mathbf{R}) g(\mathbf{R}^{-1}\mathbf{a}, \mathbf{e}),$$

where $\mathbf{e} \in \mathsf{SO}(n)$ is the identity element.

DEFINITION 4.2.3 *The* **direct product** *of the groups $G_1, G_2, \ldots, G_n$, is defined as the set of all systems $g = \{g_1, g_2, \ldots, g_n\}$, formed by $g_1 \in G_1, g_2 \in G_2, \ldots, g_n \in G_n$, endowed with the composition law*

$$gg' = \{g_1 g_1', g_2 g_2', \ldots, g_n g_n'\}, \tag{4.2.59}$$

where $g' = \{g_1', g_2', \ldots, g_n'\}$.

The direct product of groups, denoted by

$$G = G_1 \times G_2 \times \ldots \times G_n, \tag{4.2.60}$$

is a group having the identity element $e = \{e_1, e_2, \ldots, e_n\}$, where $e_1 \in G_1, e_2 \in G_2, \ldots, e_n \in G_n$, represent the identity elements in each component group of G. The mapping $g_1 \to \{g_1, e_2, \ldots, e_n\}$ shows that G_1 is an invariant subgroup of G. The elements of the group G can be written in the form $g = g_1 g_2 \ldots g_n$ and, by taking into account the commutation relations $g_j g_k = g_k g_j$ ($g_j \in \mathsf{G}_j$, $g_k \in G_k$, $j \neq k$), we deduce that $G_2, G_3, \ldots, G_n$ are also invariant subgroups of G. Reciprocally, if a group G has n invariant subgroups $G_1, G_2, \ldots, G_n$ such that $g = g_1 g_2 \ldots g_n$, $g_i \in G_i$ ($i = 1, 2, \ldots, n$), $\forall g \in G$, and $g_j g_k = g_k g_j$ ($j \neq k$), then we say that G is the direct product of the subgroups $G_1, G_2, \ldots, G_n$. If we pick only $m < n$ invariant subgroups of G, then we have

$$G_1 \times G_2 \times \ldots \times G_m \lhd G,$$

that is, the direct product of m invariant subgroups of G is an invariant subgroup of G.

A particular form of the direct product of groups arises in relation to the group of automorphisms of a given group. Let G be a group, and let $\mathcal{A}$ be the group formed by all the automorphisms of G. We denote by $\alpha(g) \in G$ the image of the element $g \in G$ under the automorphism $\alpha \in \mathcal{A}$.

DEFINITION 4.2.4 *Suppose A is a subgroup of $\mathcal{A}$. It is called* **semidirect product**, *denoted by $G \wedge A$, the group formed by the ordered pairs (g, α), satisfying the composition law*

$$(g, \alpha)(h, \beta) = (g\,\alpha(h), \alpha\beta) \quad \forall g, h \in G\,,\; \alpha, \beta \in A\,, \tag{4.2.61}$$

the identity element being (e, e_α), and the inverse $(g, \alpha)^{-1} = (\alpha^{-1}(g^{-1}), \alpha^{-1})$.

If $G = H \wedge A \equiv \{(h, \alpha)\}$, where $H \equiv \{(h, e_\alpha)\}$ and $A \equiv \{(e, \alpha)\}$, it can be shown that H is an invariant subgroup of G, and

$$G = HA\,, \quad H \cap A = \{(e, e_\alpha)\}\,, \quad G/H = A\,. \tag{4.2.62}$$

The n-dimensional Euclidean space $\mathcal{E}_n$ can be regarded as an Abelian group with respect to the composition of vectors, while $\mathsf{SO}(n)$ is its group of automorphisms. In this case, by comparing (4.2.58) and (4.2.61), we show that the Euclidean group is the semidirect product

$$\mathsf{E}(n) = \mathcal{E}_n \wedge \mathsf{SO}(n)\,. \tag{4.2.63}$$

We will examine now two particular instances of the Euclidean group, namely the two-, and three-dimensional Euclidean groups. Thus, for $n = 2$ we obtain the *group of rigid motions of the plane*, denoted by $\mathsf{E}(2)$, which consists of the group of plane translations and the group of rotations about an axis normal to the plane ($\mathsf{SO}(2)$). If we define an orthogonal system of coordinates in the plane, then a rigid motion $g \in \mathsf{E}(2)$, which moves a point $P(x, y)$ into the point $P'(x', y')$, is expressed by the equations

$$x' = x\,\cos\theta - y\,\sin\theta + a\,, \quad y' = x\,\cos\theta + y\,\sin\theta + b\,, \tag{4.2.64}$$

where $a, b \in (-\infty, \infty)$, and $\theta \in [0, 2\pi)$ are the parameters of the $\mathsf{E}(2)$ group. The mapping

$$g(a, b, \theta) \to \mathbf{t}(g) = \begin{bmatrix} \cos\theta & -\sin\theta & a \\ \sin\theta & \cos\theta & b \\ 0 & 0 & 1 \end{bmatrix}, \tag{4.2.65}$$

provides a faithful representation of $\mathsf{E}(2)$. The product of two elements $g_1, g_2 \in \mathsf{E}(2)$ has the form

$$g(a, b, \theta) = g_1(a_1, b_1, \theta_1)\, g_2(a_2, b_2, \theta_2)\,, \tag{4.2.66}$$

and, from the composition law (4.2.58), or, equally well from the matrix product $\mathbf{t}(g) = \mathbf{t}(g_1)\mathbf{t}(g_2)$, we determine the parameters of g:

$$\begin{aligned} a &= a_2 \cos\theta_1 - b_2 \sin\theta_1 + a_1\,, \\ b &= a_2 \sin\theta_1 + b_2 \cos\theta_1 + b_1\,, \\ \theta &= \theta_1 + \theta_2\,. \end{aligned} \tag{4.2.67}$$

The matrices $\mathbf{t}(g)$ in (4.2.65) correspond to the linear operators, which are associated to the group elements $g \in \mathsf{E}(2)$, and act in the linear space of the vectors $(x, y, 1)$. Now, the decomposition (4.2.63) takes the form

$$\mathsf{E}(2) = \mathcal{E}_2 \wedge \mathsf{SO}(2)\,. \tag{4.2.68}$$

Note that the two-dimensional Euclidean group $\mathsf{E}(2)$ is isomorphic to the set of complex square matrices, of the second order. This isomorphism is defined by the mapping

$$g(a, b, \theta) \to \mathbf{q}(g) = \begin{bmatrix} e^{i\theta} & a + ib \\ 0 & 1 \end{bmatrix}, \tag{4.2.69}$$

which also led us to the relations (4.2.67) for the parameters of the product (4.2.66).

The infinitesimal generators of $\mathsf{E}(2)$ follow from the analytic expression of the representation matrices (4.2.65), by means of (1.2.31). Hence, we have

$$\mathbf{a}_1 = \left.\frac{d\mathbf{t}(g)}{da}\right|_{a=b=\theta=0} = \begin{bmatrix} 0 & 0 & 1 \\ 0 & 0 & 0 \\ 0 & 0 & 0 \end{bmatrix},$$

$$\mathbf{a}_2 = \left.\frac{d\mathbf{t}(g)}{db}\right|_{a=b=\theta=0} = \begin{bmatrix} 0 & 0 & 0 \\ 0 & 0 & 1 \\ 0 & 0 & 0 \end{bmatrix},$$

$$\mathbf{a}_3 = \left.\frac{d\mathbf{t}(g)}{d\theta}\right|_{a=b=\theta=0} = \begin{bmatrix} 0 & -1 & 0 \\ 1 & 0 & 0 \\ 0 & 0 & 0 \end{bmatrix}.$$

The matrices $\mathbf{a}_1$, $\mathbf{a}_2$, and $\mathbf{a}_3$ are linearly independent, and satisfy the commutation relations

$$[\mathbf{a}_1, \mathbf{a}_2] = \mathbf{0}\,, \quad [\mathbf{a}_2, \mathbf{a}_3] = \mathbf{a}_1\,, \quad [\mathbf{a}_3, \mathbf{a}_1] = \mathbf{a}_2\,. \tag{4.2.70}$$

Consequently, these matrices constitute a basis of the Lie algebra associated to $\mathsf{E}(2)$.

For $n = 3$, the group $\mathsf{E}(3)$ is isomorphic to the group of real matrices

$$\mathbf{t}(g) = \begin{bmatrix} \mathbf{R} & \mathbf{a} \\ \mathbf{0} & \mathbf{1} \end{bmatrix}, \tag{4.2.71}$$

where $\mathbf{R} \in \mathsf{SO}(3)$ and $\mathbf{a} \in \mathcal{E}_3$. The group $\mathsf{E}(3)$ is the semidirect product

$$\mathsf{E}(3) = \mathcal{E}_3 \wedge \mathsf{SO}(3), \tag{4.2.72}$$

and its elements are determined by 6 parameters (three Euler angles for the rotation, and three parameters for the translation). The infinitesimal generators of $\mathsf{E}(3)$ have the form

$$\mathbf{a}_1 = \begin{bmatrix} 0 & 0 & 0 & 1 \\ 0 & 0 & 0 & 0 \\ 0 & 0 & 0 & 0 \\ 0 & 0 & 0 & 0 \end{bmatrix}, \quad \mathbf{a}_2 = \begin{bmatrix} 0 & 0 & 0 & 0 \\ 0 & 0 & 0 & 1 \\ 0 & 0 & 0 & 0 \\ 0 & 0 & 0 & 0 \end{bmatrix}, \quad \mathbf{a}_3 = \begin{bmatrix} 0 & 0 & 0 & 0 \\ 0 & 0 & 0 & 0 \\ 0 & 0 & 0 & 1 \\ 0 & 0 & 0 & 0 \end{bmatrix},$$

$$\mathbf{a}_4 = \begin{bmatrix} 0 & -1 & 0 & 0 \\ 1 & 0 & 0 & 0 \\ 0 & 0 & 0 & 0 \\ 0 & 0 & 0 & 0 \end{bmatrix}, \quad \mathbf{a}_5 = \begin{bmatrix} 0 & 0 & -1 & 0 \\ 0 & 0 & 0 & 0 \\ 1 & 0 & 0 & 0 \\ 0 & 0 & 0 & 0 \end{bmatrix}, \quad \mathbf{a}_6 = \begin{bmatrix} 0 & 0 & 0 & 0 \\ 0 & 0 & -1 & 0 \\ 0 & 1 & 0 & 0 \\ 0 & 0 & 0 & 0 \end{bmatrix},$$

and satisfy the commutation relations

[,]	$\mathbf{a}_1$	$\mathbf{a}_2$	$\mathbf{a}_3$	$\mathbf{a}_4$	$\mathbf{a}_5$	$\mathbf{a}_6$
$\mathbf{a}_1$	0	0	0	$-\mathbf{a}_2$	$-\mathbf{a}_3$	0
$\mathbf{a}_2$		0	0	$\mathbf{a}_1$	0	$-\mathbf{a}_3$
$\mathbf{a}_3$			0	0	$\mathbf{a}_1$	$\mathbf{a}_2$
$\mathbf{a}_4$				0	$\mathbf{a}_6$	$-\mathbf{a}_5$
$\mathbf{a}_5$					0	$\mathbf{a}_4$
$\mathbf{a}_6$						0

Note that the infinitesimal generators $\mathbf{a}_1$, $\mathbf{a}_2$, and $\mathbf{a}_4$, satisfy commutation relations identical to (4.2.70). Therefore, these infinitesimal generators constitute the basis of an algebra isomorphic to the algebra associated to the $\mathsf{E}(2)$ group, a consequence of the fact that $\mathsf{E}(2)$ is a subgroup of $\mathsf{E}(3)$.

3. Space-time symmetries. Conservation laws

3.1 Particular groups. Noether's theorem

3.1.1 Conservation laws

Let be a discrete mechanical system, formed by n particles of masses m_α, $\alpha = 1, 2, \ldots, n$, described by the Lagrangian $\mathcal{L}(\mathbf{q}, \dot{\mathbf{q}}; t)$, where $\mathbf{q}$ is a vector of generalized coordinates $q_k(t)$, $k = 1, 2, \ldots, s$, in the configuration space Λ_s. Note that, the only independent variable is the time t, such that the generalized coordinates play the rôle of state functions. When $s = 3n$, the relation between the generalized coordinates and the Cartesian coordinates of the particle α, denoted by x_j^α, $j = 1, 2, 3$, is given by the relation $k = 3(\alpha - 1) + j$. In terms of the Cartesian coordinates, the condition (4.1.101) which has to be satisfied by a transformation in order to be a symmetry transformation has the form

$$\left[\delta t\,\frac{\partial}{\partial t} + \delta x_j^\alpha\,\frac{\partial}{\partial x_j^\alpha} + \delta \dot{x}_j^\alpha\,\frac{\partial}{\partial \dot{x}_j^\alpha} + \frac{\partial(\delta t)}{\partial t}\right]\mathcal{L} = -\frac{d(\delta f)}{dt}\,, \tag{4.3.1}$$

where we have used the summation convention from 1 to 3, and from 1 to n, for Latin, and Greek indices, respectively. Consequently, the left side in (4.3.1) has to be the total derivative, with respect to the time variable, of a function depending on x_j^α and t. Also, the equation of conservation (4.1.102), corresponding to the invariance of the Lagrangian with respect to a symmetry transformation, becomes

$$\frac{d}{dt}\left[\left(\mathcal{L} - \frac{\partial \mathcal{L}}{\partial \dot{x}_j^\alpha}\dot{x}_j^\alpha\right)\delta t + \frac{\partial \mathcal{L}}{\partial \dot{x}_j^\alpha}\delta x_j^\alpha + \delta f\right] = 0\,, \tag{4.3.2}$$

which is equivalent to

$$\left(\mathcal{L} - \frac{\partial \mathcal{L}}{\partial \dot{x}_j^\alpha}\dot{x}_j^\alpha\right)\delta t + \frac{\partial \mathcal{L}}{\partial \dot{x}_j^\alpha}\delta x_j^\alpha + \delta f = \text{const}\,. \tag{4.3.3}$$

The equations (4.3.1) and (4.3.3) represent the analytic form of Noether's theorem in classical mechanics (see Theorem 4.1.5). Thus, to every symmetry transformation of a mechanical system, that is an infinitesimal transformation which satisfies the condition (4.3.1), it corresponds a conservation law expressed by (4.3.3).

3.1.2 Translation group. Conservation of total generalized momentum

The simplest symmetry transformations are the *translations of the origin of Cartesian coordinates* by a time-independent vector $(\delta x_1, \delta x_2, \delta x_3)$. These transformations are defined by the relations

$$\begin{aligned} &t' = t\,, \quad \delta t = 0\,, \\ &x'^\alpha_j = x_j^\alpha + \delta x_j\,, \quad j = 1, 2, 3\,, \quad \alpha = 1, 2, \ldots, n\,, \\ &\dot{x}'^\alpha_j = \dot{x}_j^\alpha\,. \end{aligned} \tag{4.3.4}$$

If we choose $\delta f = 0$, then (4.3.1) takes a simpler form

$$\sum_{j=1}^{3}\left(\sum_{\alpha=1}^{n}\frac{\partial \mathcal{L}}{\partial x_j^\alpha}\right)\delta x_j = 0\,,$$

and, because the infinitesimal displacements δx_1, δx_2, and δx_3 are independent, it follows that

$$\sum_{\alpha=1}^{n} \frac{\partial \mathcal{L}}{\partial x_j^\alpha} = 0\,, \quad j = 1,2,3\,. \tag{4.3.5}$$

Now, from (4.3.2) we obtain

$$\frac{d}{dt}\sum_{\alpha=1}^{n} \frac{\partial \mathcal{L}}{\partial \dot{x}_j^\alpha}\delta x_j = \delta x_j \frac{d}{dt}\sum_{\alpha=1}^{n} \frac{\partial \mathcal{L}}{\partial \dot{x}_j^\alpha} = 0\,;$$

consequently, the total generalized momentum $\mathbf{P}$ of components

$$P_j = \sum_{\alpha=1}^{n} p_j^\alpha = \sum_{\alpha=1}^{n} \frac{\partial \mathcal{L}}{\partial \dot{x}_j^\alpha}\,, \quad j = 1,2,3\,, \tag{4.3.6}$$

is conserved, if the condition (4.3.5) is satisfied. Relations (4.3.5) represent not only a *sufficient* condition but also a *necessary* one, because taking into account Lagrange's equations of motion (4.1.5), the hypothesis of conservation of the total generalized momentum leads to

$$0 = \frac{dP_j}{dt} = \frac{d}{dt}\sum_{\alpha=1}^{n} \frac{\partial \mathcal{L}}{\partial \dot{x}_j^\alpha} = \sum_{\alpha=1}^{n} \frac{d}{dt}\left(\frac{\partial \mathcal{L}}{\partial \dot{x}_j^\alpha}\right) = \sum_{\alpha=1}^{n} \frac{\partial \mathcal{L}}{\partial x_j^\alpha}\,, \quad j = 1,2,3\,.$$

3.1.3 Rotation group. Conservation of total generalized angular momentum

Let us consider now a rotation of the system of orthogonal axes, defined by the relations

$$x_i' = a_{ij}x_j\,, \quad i = 1,2,3\,, \tag{4.3.7}$$

where $\mathbf{a} \in \mathsf{SO}(3)$ is an orthogonal matrix. In case of an infinitesimal rotation, this matrix has the form

$$\mathbf{a} = \mathbf{e} + \mathbf{a}_k \delta\xi_k\,, \tag{4.3.8}$$

where $\mathbf{e}$ is the identity matrix, $\mathbf{a}_k$ are the matrices of the infinitesimal generators (2.1.43), and $\delta\xi_k$ (k = 1, 2, 3) represent the infinitesimal variations of the rotation parameters. Hence, we have

$$a_{ij} = \delta_{ij} + (\mathbf{a}_k\delta\xi_k)_{ij} = \delta_{ij} + \delta a_{ij}\,, \quad \delta a_{ij} = -\delta a_{ji}\,, \quad i,j = 1,2,3\,, \tag{4.3.9}$$

with δ_{ij} being the Kronecker symbol. We assume that the transformation (4.3.7) is of the form (4.1.84), such that, by substituting (4.3.9) into (4.3.7), we have

$$\delta x_i = x_j \delta a_{ij} = -x_j \delta a_{ji}\,.$$

It follows that the variations of the particle coordinates are of the form

$$\begin{aligned} \delta x_i^\alpha &= -x_j^\alpha \delta a_{ji}\,, \\ & \qquad\qquad i = 1,2,3\,, \quad \alpha = 1,2,\dots,n\,, \\ \delta \dot{x}_i^\alpha &= -\dot{x}_j^\alpha \delta a_{ji}\,, \end{aligned} \tag{4.3.10}$$

while the independent variable t remains unchanged, that is $\delta t = 0$. Thus, from the condition (4.3.1), with $\delta f = 0$, we obtain

$$\left(x_j^\alpha \frac{\partial \mathcal{L}}{\partial x_k^\alpha} + \dot{x}_j^\alpha \frac{\partial \mathcal{L}}{\partial \dot{x}_k^\alpha} \right) \delta a_{jk} = 0\,. \tag{4.3.11}$$

The matrix $\delta\mathbf{a}$ is skew-symmetric, and the variations $\delta\xi_k$ are independent, such that (4.3.11) can be rewritten as a system of equations

$$x_j^\alpha \frac{\partial \mathcal{L}}{\partial x_k^\alpha} - x_k^\alpha \frac{\partial \mathcal{L}}{\partial x_j^\alpha} + \dot{x}_j^\alpha \frac{\partial \mathcal{L}}{\partial \dot{x}_k^\alpha} - \dot{x}_k^\alpha \frac{\partial \mathcal{L}}{\partial \dot{x}_j^\alpha} = 0\,, \quad j,k = 1,2,3\,. \tag{4.3.12}$$

If these equations are satisfied, then the transformation (4.3.7) is a symmetry transformation, and (4.3.2) yields the equation

$$\frac{d}{dt}\left(\frac{\partial \mathcal{L}}{\partial \dot{x}_k^\alpha} x_j^\alpha \delta a_{jk} \right) = \frac{d}{dt}\left[\frac{1}{2}\left(\frac{\partial \mathcal{L}}{\partial \dot{x}_k^\alpha} x_j^\alpha - \frac{\partial \mathcal{L}}{\partial \dot{x}_j^\alpha} x_k^\alpha \right) \right] \delta a_{jk} = 0\,,$$

which is equivalent to the system of equations

$$x_j^\alpha \frac{\partial \mathcal{L}}{\partial \dot{x}_k^\alpha} - x_k^\alpha \frac{\partial \mathcal{L}}{\partial \dot{x}_j^\alpha} = x_j^\alpha p_k^\alpha - x_k^\alpha p_j^\alpha = \text{const.}\,, \quad j,k = 1,2,3\,.$$

Therefore, the conserved quantity is the *total generalized angular momentum* $\mathbf{K}$, which is a skew-symmetric tensor of rank two, of components

$$K_{jk} = x_j^\alpha p_k^\alpha - x_k^\alpha p_j^\alpha\,, \quad j,k = 1,2,3\,. \tag{4.3.13}$$

Conversely, assuming the conservation of the total generalized angular momentum, and taking into account Lagrange's equations of motion (4.1.5), one can write

$$\begin{aligned} 0 = \frac{dK_{jk}}{dt} &= \frac{d}{dt}\left(x_j^\alpha \frac{\partial \mathcal{L}}{\partial \dot{x}_k^\alpha} - x_k^\alpha \frac{\partial \mathcal{L}}{\partial \dot{x}_j^\alpha} \right) \\ &= \dot{x}_j^\alpha \frac{\partial \mathcal{L}}{\partial \dot{x}_k^\alpha} + x_j^\alpha \frac{d}{dt}\left(\frac{\partial \mathcal{L}}{\partial \dot{x}_k^\alpha} \right) - \dot{x}_k^\alpha \frac{\partial \mathcal{L}}{\partial \dot{x}_j^\alpha} - x_k^\alpha \frac{d}{dt}\left(\frac{\partial \mathcal{L}}{\partial \dot{x}_j^\alpha} \right) \\ &= \dot{x}_j^\alpha \frac{\partial \mathcal{L}}{\partial \dot{x}_k^\alpha} - \dot{x}_k^\alpha \frac{\partial \mathcal{L}}{\partial \dot{x}_j^\alpha} + x_j^\alpha \frac{\partial \mathcal{L}}{\partial x_k^\alpha} - x_k^\alpha \frac{\partial \mathcal{L}}{\partial x_j^\alpha}\,, \quad j,k = 1,2,3\,. \end{aligned}$$

Therefore, the relations (4.3.12) represent *necessary and sufficient* conditions for the conservation of total generalized angular momentum.

3.1.4 Time translations. Conservation of total energy

The *time translations* are defined by the equations

$$\begin{aligned} & t' = t + \delta t\,, \\ & x'^{\alpha}_j = x^{\alpha}_j\,, \quad \delta x^{\alpha}_j = 0\,, \\ & \dot{x}'^{\alpha}_j = \dot{x}^{\alpha}_j\,, \quad \delta \dot{x}^{\alpha}_j = 0\,, \quad j = 1, 2, 3\,,\ \alpha = 1, 2, \ldots, n\,, \end{aligned} \tag{4.3.14}$$

and form a one-parameter Abelian group. If we choose $\delta f = 0$, for $\delta t = \text{const.}$, the condition (4.3.1) reduces to

$$\frac{\partial \mathcal{L}}{\partial t} = 0\,, \tag{4.3.15}$$

that is, *the Lagrangian does not explicitly depend on time*. Also, the equation of conservation (4.3.2) takes the form

$$\frac{d}{dt}\left(\frac{\partial \mathcal{L}}{\partial \dot{x}^{\alpha}_j}\dot{x}^{\alpha}_j - \mathcal{L}\right) = 0\,. \tag{4.3.16}$$

According to (4.1.25), in the case of scleronomic constraints the first term in the parenthesis, in (4.3.16), represents the double of the total kinetic energy of the mechanical system. It follows that the *total mechanical energy*

$$E = \frac{\partial \mathcal{L}}{\partial \dot{x}^{\alpha}_j}\,\dot{x}^{\alpha}_j - \mathcal{L} = p^{\alpha}_j\,\dot{x}^{\alpha}_j - \mathcal{L} = H \tag{4.3.17}$$

is a conserved quantity. Reciprocally, if the conservation equation (4.3.16) holds, we can write

$$\begin{aligned} \frac{dE}{dt} &= \frac{d}{dt}\left(\frac{\partial \mathcal{L}}{\partial \dot{x}^{\alpha}_k}\dot{x}^{\alpha}_k - \mathcal{L}\right) \\ &= \frac{d}{dt}\left(\frac{\partial \mathcal{L}}{\partial \dot{x}^{\alpha}_k}\right)\dot{x}^{\alpha}_k + \frac{\partial \mathcal{L}}{\partial \dot{x}^{\alpha}_k}\,\ddot{x}^{\alpha}_k - \frac{d\mathcal{L}}{dt} = -\frac{\partial \mathcal{L}}{\partial t}\,. \end{aligned} \tag{4.3.18}$$

Here, we have used the relation

$$\frac{d\mathcal{L}}{dt} = \frac{\partial \mathcal{L}}{\partial t} + \frac{\partial \mathcal{L}}{\partial x^{\alpha}_k}\,\dot{x}^{\alpha}_k + \frac{\partial \mathcal{L}}{\partial \dot{x}^{\alpha}_k}\,\ddot{x}^{\alpha}_k\,,$$

and the Lagrange equations of motion (4.1.5). Hence, the condition (4.3.15) is not only a sufficient condition, but it is also a necessary one for the conservation of total mechanical energy.

3.1.5 The Galilei group. The theorem of centre of mass motion

Let be two reference frames: one considered to be a fixed frame of reference, and the other moving uniformly in a straight line relative to the first one, with velocity $\mathbf{v} = (v_1, v_2, v_3)$. The transformation connecting the coordinates of a particle in the two reference frames are of the form

$$t' = t\,, \quad x'_k = x_k + v_k t\,, \quad k = 1,2,3\,, \tag{4.3.19}$$

where the primed coordinates correspond to the fixed reference frame. The operator associated to this transformation acts in the space of vectors $(\mathbf{r}, t) = (x_1, x_2, x_3, t)$ and has the associated matrix

$$\mathbf{t}(\mathbf{v}) = \begin{bmatrix} 1 & 0 & 0 & v_1 \\ 0 & 1 & 0 & v_2 \\ 0 & 0 & 1 & v_3 \\ 0 & 0 & 0 & 1 \end{bmatrix}. \tag{4.3.20}$$

The set of transformations (4.3.19) forms an Abelian group with three parameters (the components of the velocity), called the *Galilei group* (denoted here by Γ). The matrices (4.3.20) satisfy the relation

$$\mathbf{t}(\mathbf{u} + \mathbf{v}) = \mathbf{t}(\mathbf{u})\mathbf{t}(\mathbf{v})\,,$$

where $\mathbf{u}$ and $\mathbf{v}$ are two arbitrary velocities. Therefore, the set of matrices (4.3.20) forms a representation of Γ. The infinitesimal transformation, corresponding to (4.3.19) has the form

$$\begin{aligned} t' &= t\,, \\ x'_k &= x_k + t\,\delta v_k\,, \quad k = 1,2,3\,, \\ \dot{x}'_k &= \dot{x}_k + \delta v_k\,, \end{aligned} \tag{4.3.21}$$

while the variations of the coordinates of the system of particles are given by the relations

$$\begin{aligned} \delta t &= 0\,, \\ \delta x_k^\alpha &= t\,\delta v_k\,, \quad k = 1,2,3\,, \quad \alpha = 1,2,\ldots,n\,, \\ \delta \dot{x}_k^\alpha &= \delta v_k\,. \end{aligned} \tag{4.3.22}$$

With these notations, the condition (4.3.1) becomes

$$\sum_{\alpha=1}^{n} \left(t\,\frac{\partial \mathcal{L}}{\partial x_k^\alpha} + \frac{\partial \mathcal{L}}{\partial \dot{x}_k^\alpha} \right) \delta v_k = -\frac{d(\delta f)}{dt}\,, \tag{4.3.23}$$

and assuming that (4.3.5) is satisfied (i.e., the total generalized momentum is conserved), we have

$$\frac{d(\delta f)}{dt} = -\sum_{\alpha=1}^{n} \frac{\partial \mathcal{L}}{\partial \dot{x}_k^\alpha} \delta v_k = -P_k \delta v_k \,. \tag{4.3.24}$$

The *centre of mass* of the discrete mechanical system has the coordinates

$$\xi_k = \frac{1}{M} m_\alpha x_k^\alpha \,, \quad k = 1, 2, 3 \,, \tag{4.3.25}$$

where $M = \sum_{\alpha=1}^{n} m_\alpha$ is the total mass of the mechanical system. Therefore, taking into account the definition (4.3.6) we obtain

$$M\dot{\xi}_k = m_\alpha \dot{x}_k^\alpha = P_k \,, \quad k = 1, 2, 3 \,. \tag{4.3.26}$$

Hence, up to an additive constant, we may choose

$$\delta f = -M\xi_k \delta v_k \,, \tag{4.3.27}$$

such that the equation of conservation (4.3.2) takes the form

$$\frac{d}{dt}\left[\left(t\sum_{\alpha=1}^{n} \frac{\partial \mathcal{L}}{\partial \dot{x}_k^\alpha} - M\xi_k\right) \delta v_k\right] = 0 \,,$$

and, by means of (4.3.6) we may write

$$\frac{d}{dt}\left[(P_k t - M\xi_k)\, \delta v_k\right] = 0 \,.$$

Because the variations δv_k are independent we also have

$$\frac{d}{dt}(P_k t - M\xi_k) = 0 \,, \quad k = 1, 2, 3 \,,$$

or $P_k t - M\xi_k = \text{const.}$, where the constant is determined by the initial conditions

$$(P_k t - M\xi_k)\Big|_{t=0} = -M\xi_k^0 \,, \quad k = 1, 2, 3 \,.$$

Finally, we obtain

$$\xi_k^0 = \xi_k - \frac{1}{M} P_k\, t \,, \quad k = 1, 2, 3 \,. \tag{4.3.28}$$

Taking into account the conservation of the total generalized momentum, it follows that *the centre of mass of the considered mechanical system has a uniform rectilinear motion relative to both reference frames*. It is a characteristic of classical mechanics that, in the case of a Lagrangian invariant with respect to space translations, not only the total generalized momentum is conserved but the equation (4.3.28) also holds.

3.2 The reciprocal of Noether's theorem

3.2.1 Lagrangians with certain symmetry properties

We will focus now on implications of the reciprocal of Noether's theorem in classical mechanics. We restrict our considerations to two-particle interactions, thus neglecting higher order interactions, and make use of the superposition principle. Therefore, we assume that the Lagrangian is a sum of Lagrangians, some of them associated with single particle motion, while the others correspond to two-particle interactions, that is

$$\mathcal{L} = \sum_{\alpha=1}^{n} \mathcal{L}_\alpha(\mathbf{x}_\alpha, \dot{\mathbf{x}}_\alpha; t) + \sum_{\alpha,\beta=1}^{n} \mathcal{L}_{\alpha\beta}(\mathbf{x}_\alpha, \mathbf{x}_\beta, \dot{\mathbf{x}}_\alpha, \dot{\mathbf{x}}_\beta; t)\,, \tag{4.3.29}$$

where $\mathbf{x}_\alpha = (x_1^\alpha, x_2^\alpha, x_3^\alpha)$ are the Cartesian coordinates of particle α, and similarly for $\mathbf{x}_\beta$. On this general form of the Lagrangian we impose the conditions (4.3.5), (4.3.12), and (4.3.15), that is we want the Lagrangian (4.3.29) to be invariant with respect to spatial translations, spatial rotations, and time translations. In other words, we consider that the space, in which we study the evolution of our discrete mechanical system, is homogeneous (in both spatial and time variables) and isotropic. In fact, the conditions (4.3.5), (4.3.12), and (4.3.15) represent restrictions with respect to the functional form of the Lagrangian in variables t, x_k^α, $\dot{x}_k^\alpha$, etc.

Firstly, the condition (4.3.15) leads to

$$\mathcal{L} = \sum_{\alpha=1}^{n} \mathcal{L}_\alpha(\mathbf{x}_\alpha, \dot{\mathbf{x}}_\alpha) + \sum_{\alpha,\beta=1}^{n} \mathcal{L}_{\alpha\beta}(\mathbf{x}_\alpha, \mathbf{x}_\beta, \dot{\mathbf{x}}_\alpha, \dot{\mathbf{x}}_\beta)\,. \tag{4.3.30}$$

Then, for every index $k = 1, 2, 3$, the condition (4.3.5) can be regarded as a partial differential equation, the solutions of this equation being the first integrals of the system of differential equations

$$\frac{dx_k^1}{1} = \frac{dx_k^2}{1} = \ldots = \frac{dx_k^n}{1} = d\lambda\,,$$

where λ is an arbitrary parameter. Thus, we obtain $n(n-1)/2$ first integrals of the form $\varphi_k^{\alpha\beta} = x_k^\alpha - x_k^\beta = \text{const.}$, $\alpha > \beta$ $(\alpha, \beta = 1, 2, \ldots, n)$, but only $n-1$ are linearly independent. Consequently, if the Lagrangian is of the form (4.3.30), then $\mathcal{L}_\alpha$ has to be independent of $\mathbf{x}_\alpha$, while $\mathcal{L}_{\alpha\beta}$ has to depend on x_k^α and x_k^β only by the intermediate of functions $\varphi_k^{\alpha\beta}$. From the condition (4.3.12), it follows that the Lagrangian has to depend only on the dot products $|\mathbf{x}_\alpha|^2$, $|\dot{\mathbf{x}}_\alpha|^2$, $|\mathbf{x}_\alpha - \mathbf{x}_\beta|^2$, $|\dot{\mathbf{x}}_\alpha - \dot{\mathbf{x}}_\beta|^2$, $(\mathbf{x}_\alpha - \mathbf{x}_\beta) \cdot (\dot{\mathbf{x}}_\alpha - \dot{\mathbf{x}}_\beta)$, which are invariant with respect to the spatial rotations. Hence, this particular Lagrangian has to be of the form

$$\mathcal{L} = \sum_{\alpha=1}^{n} T_\alpha \left(|\dot{\mathbf{x}}_\alpha|^2 \right) - \sum_{\alpha,\beta=1}^{n} V_{\alpha\beta} \left(|\mathbf{x}_\alpha - \mathbf{x}_\beta|^2 , \dot{\mathbf{x}}_\alpha, \dot{\mathbf{x}}_\beta \right) , \qquad (4.3.31)$$

where we have introduced the notations $T_\alpha = \mathcal{L}_\alpha$ and $V_{\alpha\beta} = -\mathcal{L}_{\alpha\beta}$.

In classical mechanics, the equations of motion are invariant with respect to the Galilei group (*Galilean invariance of non-relativistic mechanics*), such that, by assuming that $V_{\alpha\beta}$ does not depend on the velocities $\dot{\mathbf{x}}_\alpha$ and $\dot{\mathbf{x}}_\beta$, the condition (4.3.24) becomes

$$\frac{d(\delta f)}{dt} = -\frac{\partial T_\alpha}{\partial \dot{x}_k^\alpha} \, \delta v_k \, . \qquad (4.3.32)$$

Here, f is a function of x_k^α ($k = 1, 2, 3$, $\alpha = 1, 2, \ldots, n$), and the right side of (4.3.32) is a total time derivative of a function of this kind, only if the functions T_α are proportional to $|\dot{\mathbf{x}}_\alpha|^2$. Finally, the Lagrangian which describes the motion of a system of n particles, invariant with respect to the group of space translations, the group of space rotations, the group of time translations, and also conforms to the Galilean invariance, has to be of the form

$$\mathcal{L} = \frac{1}{2} \sum_{\alpha=1}^{n} m_\alpha \, |\dot{\mathbf{x}}_\alpha|^2 - \sum_{\alpha,\beta=1}^{n} V_{\alpha\beta} \left(|\mathbf{x}_\alpha - \mathbf{x}_\beta|^2 \right) , \qquad (4.3.33)$$

where we have introduced the proportionality factors $m_\alpha/2$ ($\alpha = 1, 2, \ldots, n$) to set up the dependence of the functions T_α on $|\dot{\mathbf{x}}_\alpha|^2$. Actually, m_α is the mass of the particle of index α. In the case of such a Lagrangian, the generalized momentum of this particle has the components

$$p_k^\alpha = \frac{\partial \mathcal{L}}{\partial \dot{x}_k^\alpha} = m_\alpha \dot{x}_k^\alpha \, , \quad k = 1, 2, 3 \, , \qquad (4.3.34)$$

while the components of the vector associated to the generalized angular momentum of this particle, are

$$K_i^\alpha = \epsilon_{ijk} \, x_j^\alpha \, p_k^\alpha = \frac{1}{2} \, \epsilon_{ijk} \, K_{jk}^\alpha \, , \quad j = 1, 2, 3 \, . \qquad (4.3.35)$$

3.2.2 The reciprocal of Noether's theorem

The example in the previous section shows in a very explicit manner, the close relation existing between the motion laws in classical mechanics, inferred directly from the Lagrangian, and the geometrical properties of the space-time, which impose the analytic form of the Lagrangian. Now, we will use the form (4.3.33) of the Lagrangian to illustrate how to find the symmetry group corresponding to a conservation law, that is how to apply the reciprocal of Noether's theorem.

To simplify the formulation we will consider the motion of a single particle, described by the Lagrangian

$$\mathcal{L} = \frac{m}{2}\left(\dot{x}_1^2 + \dot{x}_2^2 + \dot{x}_3^2\right) - V(\mathbf{x}), \tag{4.3.36}$$

where $V(\mathbf{x})$ is the potential energy of the field of forces acting upon the particle. The Hessian of this Lagrangian with respect to velocities is proportional to the identity matrix $\mathbf{e}$ of rank three

$$\mathcal{H} = m \begin{bmatrix} 1 & 0 & 0 \\ 0 & 1 & 0 \\ 0 & 0 & 1 \end{bmatrix} = m\,\mathbf{e},$$

and has the inverse

$$\mathcal{H}^{-1} = \frac{1}{m} \begin{bmatrix} 1 & 0 & 0 \\ 0 & 1 & 0 \\ 0 & 0 & 1 \end{bmatrix} = \frac{1}{m}\,\mathbf{e}. \tag{4.3.37}$$

Under the assumption that the momentum of the particle is conserved, we obtain three first integrals (integrals of motion)

$$D_k(\mathbf{x}, \dot{\mathbf{x}}) = m\,\dot{x}_k = \text{const.}, \quad k = 1, 2, 3, \tag{4.3.38}$$

and from (4.1.118) it follows

$$\delta x_k = \frac{\partial D_l}{\partial \dot{x}_j}\,\mathcal{H}_{jk}^{-1}\,\delta\tau_l = \delta\tau_k, \quad k = 1, 2, 3. \tag{4.3.39}$$

Hence, the corresponding infinitesimal transformation with respect to which the Lagrangian remains invariant is of the form

$$x_k' = x_k + \delta\tau_k, \quad k = 1, 2, 3, \tag{4.3.40}$$

that is an infinitesimal translation of the form (4.3.4). Consequently, the conservation of the momentum leads to the invariance of the Lagrangian with respect to the translations of the Euclidean space $\mathcal{E}_3$.

The conservation of the angular momentum provides three first integrals

$$D_k(\mathbf{x}, \dot{\mathbf{x}}) = m\,\epsilon_{ijk}\,x_i\dot{x}_j = \text{const.}, \quad k = 1, 2, 3. \tag{4.3.41}$$

The first integral $D_3(\mathbf{x}, \dot{\mathbf{x}})$ leads to variations of coordinates of the form

$$\delta x_1 = -x_2\delta\tau_3, \quad \delta x_2 = x_1\delta\tau_3, \quad \delta x_3 = 0,$$

respectively to the infinitesimal transformation

$$x_1' = x_1 - x_2 \delta\tau_3 , \quad x_2' = x_2 + x_1 \delta\tau_3 , \quad x_3' = x_3 .$$

In matrix notation, this transformation takes the form

$$\begin{bmatrix} x_1' \\ x_2' \\ x_3' \end{bmatrix} = \begin{bmatrix} 1 & 0 & 0 \\ 0 & 1 & 0 \\ 0 & 0 & 1 \end{bmatrix} \begin{bmatrix} x_1 \\ x_2 \\ x_3 \end{bmatrix} + \begin{bmatrix} 0 & -1 & 0 \\ 1 & 0 & 0 \\ 0 & 0 & 0 \end{bmatrix} \delta\tau_3 , \tag{4.3.42}$$

or

$$\mathbf{x}' = \mathbf{e}\,\mathbf{x} + \widetilde{\mathbf{X}}_3 \,\delta\tau_3 ,$$

where $\mathbf{e}$ represents the identity matrix, and $\widetilde{\mathbf{X}}_3$ is the infinitesimal generator of the group of transformations associated to the first integral $D_3(\mathbf{x}, \dot{\mathbf{x}})$. By the same method we obtain for the first integrals $D_1(\mathbf{x}, \dot{\mathbf{x}})$ and $D_2(\mathbf{x}, \dot{\mathbf{x}})$ the infinitesimal generators

$$\widetilde{\mathbf{X}}_1 = \begin{bmatrix} 0 & 0 & 0 \\ 0 & 0 & -1 \\ 0 & 1 & 0 \end{bmatrix} , \quad \widetilde{\mathbf{X}}_2 = \begin{bmatrix} 0 & 0 & 1 \\ 0 & 0 & 0 \\ -1 & 0 & 0 \end{bmatrix} . \tag{4.3.43}$$

The product of the three transformations obtained above leads to an infinitesimal transformation defined by the matrix

$$\mathbf{u}(\delta\tau_1, \delta\tau_2, \delta\tau_3) = \mathbf{e} + \widetilde{\mathbf{X}}_k \,\delta\tau_k , \tag{4.3.44}$$

which is a square orthogonal matrix $\mathbf{u} \in \mathsf{SO}(3)$. The matrices $\widetilde{\mathbf{X}}_i$ $(i = 1, 2, 3)$ satisfy the commutation relations (1.2.34) and can be identified with the matrices of the infinitesimal generators of $\mathsf{SO}(3)$ group in exponential parametrization, defined in (2.1.43). Consequently, the transformations defined by matrices of the form (4.3.43) form a group isomorphic to $\mathsf{SO}(3)$, and the relations expressing the conservation of the angular momentum lead to the invariance of the Lagrangian with respect to the proper rotations of the Euclidean space $\mathcal{E}_3$. At the same time, this property of invariance imposes a restriction on the form of the potential energy $V(\mathbf{x})$ in (4.3.36). This function has to depend on the variables x_1, x_2, and x_3 in the form $V(x_1^2 + x_2^2 + x_3^2)$, the norm of the vector $\mathbf{x}$ (i.e., $\mathbf{x} \cdot \mathbf{x} = x_1^2 + x_2^2 + x_3^2$) being invariant with respect to the proper rotations of $\mathcal{E}_3$. Indeed, from the conditions (4.1.116) it follows that $V(\mathbf{x})$ is a solution of the system of differential equations

$$\frac{\partial D_k}{\partial \dot{x}_1} \frac{\partial V}{\partial x_1} + \frac{\partial D_k}{\partial \dot{x}_2} \frac{\partial V}{\partial x_2} + \frac{\partial D_k}{\partial \dot{x}_3} \frac{\partial V}{\partial x_3} = 0 , \quad k = 1, 2, 3,$$

respectively

$$\epsilon_{ijk}\, x_j \frac{\partial V}{\partial x_k} = 0 , \quad i = 1, 2, 3,$$

which admits the general solution $V = V(\varphi)$, where $\varphi = \mathbf{x} \cdot \mathbf{x}$.

The method described above can be generalized to systems of n particles because for $\partial \mathcal{L}/\partial t = 0$ and $\partial V/\partial \dot{x}_k^\alpha = 0$, in the case $\mathcal{L} = T - V$, the function T is a homogeneous function of the second degree, which can be reduced to its canonical form (i.e., a sum of squares) by means of a point transformation. Hence, the Hessian of the transformed Lagrangian with respect to the generalized velocities, is a diagonal matrix, and the relations of conservation of the total generalized angular momentum lead to the infinitesimal generators of the $\mathsf{SO}(3)$ group.

We have shown in (4.3.18) that whenever the Lagrangian does not explicitly depend on time ($\partial \mathcal{L}/\partial t = 0$), along a system trajectory (i.e., when the Lagrange equations of motion (4.1.5) are satisfied), the total energy

$$D(\mathbf{x}, \dot{\mathbf{x}}) = E = \frac{\partial \mathcal{L}}{\partial \dot{x}_j}\, \dot{x}_j - \mathcal{L} \tag{4.3.45}$$

is conserved ($dE/dt = 0$). In our case, the only independent variable is the time t and we consider an infinitesimal transformation of the form $t' = t + \delta t$, where δt is an arbitrary infinitesimal function of class C^2. If we choose $\delta f = 0$, then the equation of conservation (4.3.2) becomes

$$-E\, \frac{d(\delta t)}{dt} + \frac{d}{dt}\left(\frac{\partial \mathcal{L}}{\partial \dot{x}_j}\, \delta x_j \right) = 0\,,$$

which is identically satisfied for $\delta t = \mathrm{const.}$, that is the considered transformation is a time translation. Note that, for time translations

$$\delta x_k(t) = x'_k(t') - x_k(t) = 0\,, \quad k = 1, 2, 3, \tag{4.3.46}$$

because $x'(t')$ and $x(t)$ have to represent the same point in the space of coordinates. Consequently, we also have $\delta \dot{x}_k = 0$. It follows that the symmetry transformation corresponding to the conservation of energy is the time translation (4.3.14) for an one-particle system

$$\begin{aligned} t' &= t + \delta t\,, \\ x'_k &= x_k\,, \quad \delta x_k = 0\,, \\ \dot{x}'_k &= \dot{x}_k\,, \quad \delta \dot{x}_k = 0\,, \quad k = 1, 2, 3\,. \end{aligned} \tag{4.3.47}$$

Note that the method used here to determine the symmetry transformation corresponding to the conservation of energy does not rely on (4.1.118) which can be applied directly only in the case of point transformations. What we actually obtain from (4.1.118), (4.3.36), (4.3.37), and (4.3.45) is the synchronous variation

$$\delta_* x_k(t) \equiv x'_k(t) - x_k(t) = \dot{x}_k(t)\delta\tau\,, \quad k = 1, 2, 3,$$

such that for an arbitrary infinitesimal parameter $\delta\tau$ we have

$$x_k'(t) = x_k(t) + \dot{x}_k(t)\delta\tau = x_k(t+\delta\tau)\,, \quad k = 1,2,3\,.$$

Now, by changing t into $t' = t+\delta t$, and defining the new infinitesimal parameter $\delta t = -\delta\tau$, we are lead successively to $x_k'(t') = x_k(t)$, that is (4.3.46), and finally to (4.3.47).

3.2.3 The Galilei-Newton group

Let $\mathcal{E}_4$ be the Euclidean space spanned by the vectors $\mathbf{x} = (x_1, x_2, x_3, t) \equiv (\mathbf{r}, t)$. The *Galilei-Newton group* G is the group of endomorphisms of $\mathcal{E}_4$, of the form

$$t' = t + \tau\,, \quad \mathbf{r}' = \mathbf{R}\,\mathbf{r} + \mathbf{v}\,t + \mathbf{a}\,, \tag{4.3.48}$$

and it consists of a time translation, defined by τ, a translation in $\mathcal{E}_3 \subset \mathcal{E}_4$, defined by $\mathbf{a} \in \mathcal{E}_3$, a transformation of the form (4.3.19) from the Galilei group Γ, and a rotation $\mathbf{R} \in \mathsf{SO}(3)$.The Galilei-Newton group is a subgroup of the affine group in four dimensions or, in homogeneous coordinates, a subgroup of the general linear group $\mathsf{GL}(5, \mathbb{R})$.

The time interval between two events in the space $\mathcal{E}_4$

$$\Delta t = t_2 - t_1\,, \tag{4.3.49}$$

as well as the spatial distance between two simultaneous events

$$\Delta r = |\mathbf{r}_2 - \mathbf{r}_1|\,, \quad t_2 = t_1\,, \tag{4.3.50}$$

are invariant with respect to the transformations of G. Reciprocally, the Galilei-Newton group is the most general group of linear transformations in $\mathcal{E}_4$, which preserve the intervals Δt and Δr. The elements of the Galilei-Newton group depend on ten parameters, and are denoted by $g(\tau, \mathbf{a}, \mathbf{v}, \mathbf{R})$. The internal composition law is defined as

$$g(\tau', \mathbf{a}', \mathbf{v}', \mathbf{R}')\, g(\tau, \mathbf{a}, \mathbf{v}, \mathbf{R}) = g(\tau' + \tau, \mathbf{a}' + \mathbf{R}'\mathbf{a} + \mathbf{v}'\tau, \mathbf{v}' + \mathbf{R}'\mathbf{v}, \mathbf{R}'\mathbf{R})\,. \tag{4.3.51}$$

The identity element in G has the form $e = g(0, \mathbf{0}, \mathbf{0}, \mathbf{e}_3)$, where $\mathbf{e}_3$ is the identity element in $\mathsf{SO}(3)$, while the inverse of a given element is $g^{-1} = g(\tau, \mathbf{a}, \mathbf{v}, \mathbf{R})^{-1} = g(-\tau, \mathbf{R}^{-1}(\mathbf{a} - \mathbf{v}\tau), -\mathbf{R}^{-1}\mathbf{v}, \mathbf{R}^{-1})$. The Galilei-Newton group is isomorphic to the group of matrices

$$g \to \mathbf{t}(g) = \begin{bmatrix} & & & v_1 & a_1 \\ & \mathbf{R} & & v_2 & a_2 \\ & & & v_3 & a_3 \\ 0 & 0 & 0 & 1 & \tau \\ 0 & 0 & 0 & 0 & 1 \end{bmatrix}, \tag{4.3.52}$$

of order five, which depend on ten parameters and satisfy the relation

$$\mathbf{t}(g_1 g_2) = \mathbf{t}(g_1)\,\mathbf{t}(g_2)\,,$$

identical to the internal composition law (4.3.51).

We will outline now some results from the theory of the Galilei-Newton group. Thus, it can be shown that the Galilei-Newton group is a non-compact Lie group. The most important subgroups of the Galilei-Newton group are:

- $\mathcal{T} = \{g(\tau, \mathbf{0}, \mathbf{0}, \mathbf{e}_3)\}$ - the group of time translations,
- $T = \{g(0, \mathbf{a}, \mathbf{0}, \mathbf{e}_3)\}$ - the group of space translations in $\mathcal{E}_3$,
- $\Gamma = \{g(0, \mathbf{0}, \mathbf{v}, \mathbf{e}_3)\}$ - the Galilei group,
- $\mathsf{SO}(3) = \{g(0, \mathbf{0}, \mathbf{0}, \mathbf{R})\}$ - the group of proper rotations in $\mathcal{E}_3$.

The direct product of groups $\mathcal{T} \times T$ form an invariant Abelian subgroup of G, and the corresponding factor group is the semidirect product $\Gamma \wedge \mathsf{SO}(3)$, that leads to the decomposition

$$\mathsf{G} = (\mathcal{T} \times T) \wedge (\Gamma \wedge \mathsf{SO}(3))\,. \tag{4.3.53}$$

Note that there also are other decompositions of G, like this one

$$\mathsf{G} = (T \times \Gamma) \wedge (\mathcal{T} \times \mathsf{SO}(3))\,, \tag{4.3.54}$$

where $T \times \Gamma$ is the maximal invariant Abelian subgroup of G.

The basis of the Lie algebra associated to G comprises ten elements corresponding to the subgroups $\mathcal{T}, T, \Gamma$, and $\mathsf{SO}(3)$. Consequently, the first integrals obtained from the invariance of the Lagrangian with respect to these subgroups, determine the infinitesimal generators (4.2.27) of the Lie algebra. Thus, we obtain

- H - the infinitesimal generator of $\mathcal{T}$,
- $\mathsf{P}_i,\ i = 1, 2, 3$ - the infinitesimal generators of T,
- $\mathsf{K}_i,\ i = 1, 2, 3$ - the infinitesimal generators of Γ,
- $\mathsf{M}_i = \frac{1}{2}\,\epsilon_{ijk}\,\mathsf{M}_{jk},\ i = 1, 2, 3$ - the infinitesimal generators of $\mathsf{SO}(3)$.

Hence, the infinitesimal generator of G has the form

$$\mathsf{X} = \mathsf{H}\delta\tau + \mathsf{P}_i\delta a_i + \mathsf{K}_i\delta v_i + \frac{1}{2}\mathsf{M}_{ij}\delta\kappa_{ij} = \mathsf{H}\delta\tau + \mathsf{P}_i\delta a_i + \mathsf{K}_i\delta v_i + \mathsf{M}_i\delta\theta_i\,. \tag{4.3.55}$$

The matrix form of these infinitesimal generators follows from the representation (4.3.52). We can also obtain these infinitesimal generators in the form of differential operators. Thus, we consider the space of square-integrable functions $f(\mathbf{r}, t)$ defined on $\mathcal{E}_4$. Then, according to (3.2.1) we obtain the regular representation of G as the set of differential operators u acting in the space of the functions f, defined by the equation

$$\mathsf{u}(\tau, \mathbf{a}, \mathbf{v}, \mathbf{R})\, f(\mathbf{r}, t) = f(\mathbf{R}^{-1}(\mathbf{r} - \mathbf{v}t - \mathbf{a} + \mathbf{v}\tau), t - \tau)\,. \tag{4.3.56}$$

Finally, we obtain the basis of the Lie algebra associated to G, as the set of infinitesimal generators

$$\mathsf{H} = \frac{\partial}{\partial t}\,, \quad \mathbf{P} = \nabla\,, \quad \mathbf{K} = t\,\nabla\,, \quad \mathbf{M} = \mathbf{r} \times \nabla\,, \tag{4.3.57}$$

that satisfy the commutation relations

$$\begin{aligned}
&[\mathsf{M}_i, \mathsf{M}_j] = \epsilon_{ijk}\,\mathsf{M}_k\,, \quad [\mathsf{M}_i, \mathsf{K}_j] = \epsilon_{ijk}\,\mathsf{K}_k\,, \quad [\mathsf{M}_i, \mathsf{P}_j] = \epsilon_{ijk}\,\mathsf{P}_k\,,\\
&[\mathsf{K}_i, \mathsf{H}] = \mathsf{P}_i\,,\\
&[\mathsf{M}_i, \mathsf{H}] = [\mathsf{K}_i, \mathsf{K}_j] = [\mathsf{K}_i, \mathsf{P}_j] = [\mathsf{P}_i, \mathsf{P}_j] = [\mathsf{P}_i, \mathsf{H}] = [\mathsf{H}, \mathsf{H}] = 0\,.
\end{aligned} \tag{4.3.58}$$

The subgroup structure of G can be determined by means of the Lie subalgebras which can be identified on the basis of these commutation relations. From (4.3.58) we can also obtain the two invariants of the Galilei-Newton group

$$P^2 = \mathbf{P}^2\,, \quad N^2 = (\mathbf{K} \times \mathbf{P})^2\,. \tag{4.3.59}$$

The invariance of the Lagrangian of a mechanical system with respect to the Galilei-Newton group is the necessary and sufficient condition for the conservation of the total generalized momentum, of the total generalized angular momentum, of the total mechanical energy, and of the uniform rectilinear motion of the centre of mass. The corresponding Lagrangian has the form (4.3.33), and it is easy to show that, under the transformations of the Galilei-Newton group, the mechanical energy and the momentum of a free particle are transformed according to the relations

$$E' = E + \mathbf{v} \cdot (\mathbf{R}\,\mathbf{p}) + \frac{1}{2}\,m\,\mathbf{v}^2\,, \quad \mathbf{p}' = \mathbf{R}\,\mathbf{p} + m\,\mathbf{v}\,. \tag{4.3.60}$$

It follows that the quantity

$$U = E' - \frac{1}{2m}\,\mathbf{p}'^2 = E - \frac{1}{2m}\,\mathbf{p}^2\,, \tag{4.3.61}$$

is invariant with respect to the transformations of G. To prove (4.3.61) we used the invariance of the scalar product with respect to $\mathsf{SO}(3)$, that is $(\mathbf{R}\,\mathbf{p})\cdot(\mathbf{R}\,\mathbf{p}) = \mathbf{p} \cdot \mathbf{p}$.

From a physical point of view, U represents the difference between the total mechanical energy and the kinetic energy of the particle, that is the energy of the particle at rest ($\mathbf{p} = 0$). Hence, we may say that U represents a sort of "intrinsical energy" of the particle. It is worth-while to mention that, while in the relativistic mechanics the existence of the rest energy of a particle is an obvious result, a similar energy appears here in the context of classical mechanics, from considerations on the symmetry properties of the Lagrangian in a four dimensional Euclidean space-time.

3.3 The Hamilton-Jacobi equation for a free particle

3.3.1 The case of the three-dimensional space

There is a maximal symmetry group, in classical physics, consisting of all the symmetry transformations which leave invariant the Hamilton-Jacobi equation for a *free particle* of mass m

$$\frac{\partial S}{\partial t} + \frac{1}{2m}(\nabla S)^2 = 0\,, \tag{4.3.62}$$

where S is the action defined in (4.1.44). To find this symmetry group we will use the method described in Chap. 3 Sec. 4. Thus, according to (3.4.58) the infinitesimal generator of the symmetry group can be written in the form

$$\mathsf{L} = \xi\partial_x + \eta\partial_y + \zeta\partial_z + \tau\partial_t + \varphi\partial_S\,, \tag{4.3.63}$$

where the coefficients ξ, η, ζ, τ, and φ are analytic functions in the variables x, y, z, t, and S, considered as independent variables. Then, the first order prolongation of the infinitesimal generator L is

$$\mathsf{pr}^{(1)}\mathsf{L} = \mathsf{L} + \varphi^x\frac{\partial}{\partial S_x} + \varphi^y\frac{\partial}{\partial S_y} + \varphi^z\frac{\partial}{\partial S_z} + \varphi^t\frac{\partial}{\partial S_t}\,, \tag{4.3.64}$$

with the coefficients given by (3.4.60), that is

$$\begin{aligned}
\varphi^x &= D^x\,\psi + S_{xx}\,\xi + S_{xy}\,\eta + S_{xz}\,\zeta + S_{xt}\,\tau\,,\\
\varphi^y &= D^y\,\psi + S_{yx}\,\xi + S_{yy}\,\eta + S_{yz}\,\zeta + S_{yt}\,\tau\,,\\
\varphi^z &= D^z\,\psi + S_{zx}\,\xi + S_{zy}\,\eta + S_{zz}\,\zeta + S_{zt}\,\tau\,,\\
\varphi^t &= D^t\,\psi + S_{tx}\,\xi + S_{ty}\,\eta + S_{tz}\,\zeta + S_{tt}\,\tau\,,
\end{aligned} \tag{4.3.65}$$

where $\psi = \varphi - S_x\,\xi - S_y\,\eta - S_z\,\zeta - S_t\,\tau$. Hence, in the case of equation (4.3.62) the condition (3.4.55) becomes

$$\varphi^t + \frac{1}{m}\left(\varphi^x\,S_x + \varphi^y\,S_y + \varphi^z\,S_z\right) = 0\,. \tag{4.3.66}$$

This condition is identically satisfied when S is a solution of (4.3.62). It follows that after substituting (4.3.65) into (4.3.66), we have to take into account the relation $S_x^2 + S_y^2 + S_z^2 = -2mS_t$. Then, by cancelling the coefficients of different derivatives of the function S we obtain the system of differential equations

$$\begin{aligned} &\tau_S = 0\,, \quad \tau_x = m\,\xi_S\,, \quad \tau_y = m\,\eta_S\,, \quad \tau_z = m\,\zeta_S\,, \\ &\eta_x + \xi_y = \zeta_y + \eta_z = \xi_z + \zeta_x = 0\,, \\ &\varphi_x = m\,\xi_t\,, \quad \varphi_y = m\,\eta_t\,, \quad \varphi_z = m\,\zeta_t\,, \quad \varphi_t = 0\,, \\ &\varphi_S + \tau_t = 2\xi_x = 2\eta_y = 2\zeta_z\,. \end{aligned} \tag{4.3.67}$$

The general solution of this system can be written in the form

$$\begin{aligned} &\tau(\mathbf{r}, t) = m\alpha_8^i x_i t + \tfrac{1}{2} m\alpha_9 r^2 + m\alpha_{10}^i x_i + \alpha_3 t^2 + \alpha_2 t + \alpha_1\,, \\ &\rho_i(\mathbf{r}, t, S) = \alpha_8^j \left[\left(tS - \tfrac{1}{2} m r^2 \right) \delta_{ij} + m x_i x_j \right] + \alpha_9 x_i S + \alpha_3 t x_i \\ &\quad + \tfrac{1}{2}\alpha_2 x_i + \tfrac{1}{2}\alpha_7 x_i + \alpha_5^i t + \alpha_4^i + \alpha_{10} S + R_{ij} x_j\,, \quad i = 1, 2, 3, \\ &\varphi(\mathbf{r}, t, S) = m\alpha_8^i x_i S + \tfrac{1}{2} m\alpha_3 r^2 + m\alpha_5^i x_i + \alpha_9 S^2 + \alpha_7 S + \alpha_6\,, \end{aligned} \tag{4.3.68}$$

where we have introduced the notations $\boldsymbol{\rho} = (\rho_1, \rho_2, \rho_3) \equiv (\xi, \eta, \zeta)$, $\mathbf{r} = (x_1, x_2, x_3) \equiv (x, y, z)$, and $r^2 = x^2 + y^2 + z^2$. Here, we have also used the summation convention for dummy indices from 1 to 3. The constants R_{ij} satisfy the antisymmetry relation $R_{ij} = -R_{ji}$, and consequently the general solution (4.3.68) depends on 21 parameters (α_i, R_{jk}), which represent the parameters of the symmetry group of the Hamilton-Jacobi equation for a free particle.

3.3.2 The case of the n-dimensional space

The motion of a free particle in an n-dimensional space, spanned by the vectors $\mathbf{r} = (x_1, x_2, \ldots, x_n)$, is described by a Hamilton-Jacobi equation that has the same form as (4.3.62) by exception of the differential operator ∇ which takes the form

$$\nabla = (\partial_1, \partial_2, \ldots, \partial_n)\,, \quad \partial_i = \partial/\partial x_i\,, i = 1, 2, \ldots, n\,.$$

The general solution of the differential system, corresponding to (4.3.67), will depend now on $N = (n+3)(n+4)/2$ parameters, and by means of relations generalizing (4.3.63) and (4.3.68) to the n-dimensional case, we obtain the following set of differential operators

$$\begin{aligned} &\mathsf{X}_1 = \partial_t,\ \mathsf{X}_2 = t\partial_t + \frac{1}{2}\mathbf{r}\cdot\nabla,\ \mathsf{X}_3 = t^2\partial_t + t\,\mathbf{r}\cdot\nabla + \frac{1}{2} m r^2 \partial_S,\ \mathbf{X}_4 = \nabla, \\ &\mathbf{X}_5 = t\nabla + m\mathbf{r}\,\partial_S,\ \mathsf{X}_6 = \partial_S,\ \mathsf{X}_7 = \frac{1}{2}\mathbf{r}\cdot\nabla + S\,\partial_S, \\ &\mathbf{X}_8 = m\mathbf{r}\,(\mathbf{r}\cdot\nabla) + \left(tS - \frac{1}{2} m r^2 \right) \nabla + mt\mathbf{r}\,\partial_t + mS\mathbf{r}\,\partial_S, \end{aligned} \tag{4.3.69}$$

$$\mathsf{X}_9 = S\,\mathbf{r}\cdot\nabla + \frac{1}{2}mr^2\partial_t + S^2\partial_S,$$
$$\mathbf{X}_{10} = S\,\nabla + m\mathbf{r}\partial_t,\ \mathsf{X}_{11}^{ij} = x_j\partial_i - x_i\partial_j,\ i,j,=1,2,\ldots,n.$$

These operators form the basis of a Lie algebra isomorphic to the Lie algebra of the conformal group $\mathsf{O}(n+2,2)$. The elements of $\mathsf{O}(n+2,2)$ are transformations acting in an $(n+2)$-dimensional Minkowskian space, denoted by M_{n+2}, which leave invariant the quadratic form

$$\varphi(\mathbf{x},\mathbf{x}) = x_1^2 + x_2^2 + \ldots + x_n^2 - x_{n+1}^2 - x_{n+2}^2\,.$$

Note that, generally, we denote by $\mathsf{O}(p+q,q)$ the group of transformations that leave invariant the quadratic form

$$\varphi(\mathbf{x},\mathbf{x}) = x_1^2 + x_2^2 + \ldots + x_p^2 - x_{p+1}^2 - x_{p+2}^2 - \ldots - x_{p+q}^2\,,$$

defined on the Minkowskian space M_{p+q}.

To prove the isomorphism between the Lie algebra of the symmetry group of Hamilton-Jacobi equation and the Lie algebra of the conformal group $\mathsf{O}(n+2,2)$, we introduce the variables

$$x_0 = \frac{1}{\sqrt{2}}\left(t + \frac{S}{m}\right), \quad x_{n+1} = \frac{1}{\sqrt{2}}\left(t - \frac{S}{m}\right),$$

and the metric tensor

$$g_{00} = -g_{ii} = g_{n+1,n+1} = 1\,, \quad i = 1,2,\ldots,n\,,$$
$$g_{\mu\nu} = 0\,, \quad \mu \neq \nu\,, \quad \mu,\nu = 0,1,2,\ldots,n+1\,.$$

Then, from the set of infinitesimal generators (4.3.69) we form the operators

$$\begin{aligned}
&\mathsf{M}_{i0} = -\frac{1}{\sqrt{2}}\left(\mathsf{X}_5^i + \frac{1}{m}\mathsf{X}_{10}^i\right), \quad \mathsf{M}_{n+1,0} = \mathsf{X}_2 - \mathsf{X}_7\,,\\
&\mathsf{M}_{n+1,i} = -\frac{1}{\sqrt{2}}\left(\mathsf{X}_5^i - \frac{1}{m}\mathsf{X}_{10}^i\right),\\
&\mathsf{M}_{ij} = \mathsf{X}_{11}^{ij}\,, \quad \mathsf{D} = \mathsf{X}_2 + \mathsf{X}_7\,,\\
&\mathsf{P}_i = \mathsf{X}_4^i\,, \quad \mathsf{P}_0 = \frac{1}{\sqrt{2}}(\mathsf{X}_1 + m\mathsf{X}_6)\,, \quad \mathsf{P}_{n+1} = \frac{1}{\sqrt{2}}(\mathsf{X}_1 - m\mathsf{X}_6)\,,\\
&\mathsf{K}_i = -\frac{2}{m}\mathsf{X}_8^i,\ \mathsf{K}_0 = \sqrt{2}\left(\mathsf{X}_3 + \frac{1}{m}\mathsf{X}_9\right),\ \mathsf{K}_{n+1} = -\sqrt{2}\left(\mathsf{X}_3 - \frac{1}{m}\mathsf{X}_9\right),
\end{aligned}\tag{4.3.70}$$

for $i,j = 1,2,\ldots,n$. Here, we have also used the components of the vector operators

$$\mathbf{X}_k = \left(\mathsf{X}_k^1, \mathsf{X}_k^2, \ldots, \mathsf{X}_k^n\right)\,, \quad k = 4,5,8,10.$$

The expressions (4.3.70) can be written in a more compact form if we introduce the notation $\mathbf{x} = (x_0, x_1, x_2, \ldots, x_{n+1}) \in M_{n+2}$, such that the norm of $\mathbf{x}$ is given by $x^2 = \mathbf{x} \cdot \mathbf{x} = g_{\mu\nu} x_\mu x_\nu$. Thus, we have

$$\begin{aligned} \mathsf{M}_{\mu\nu} &= x_\mu \partial_\nu - x_\nu \partial_\mu \,, \quad \mathsf{P}_\mu = \partial_\mu \,, \\ \mathsf{D} &= \mathbf{x} \cdot \nabla = g_{\mu\nu} x_\mu \partial_\nu \,, \\ \mathsf{K}_\mu &= 2\, x_\mu (\mathbf{x} \cdot \nabla) - x^2 \partial_\mu \,, \end{aligned} \tag{4.3.71}$$

where $\partial_\mu = \partial/\partial x_\mu$, and $\mu, \nu = 0, 1, 2, \ldots, n+1$. The operators (4.3.71) satisfy the commutation relations

$$\begin{aligned} &[\mathsf{M}_{\mu\nu}, \mathsf{M}_{\sigma\rho}] = g_{\mu\rho} \mathsf{M}_{\nu\sigma} + g_{\nu\sigma} \mathsf{M}_{\mu\rho} - g_{\mu\sigma} \mathsf{M}_{\nu\rho} - g_{\nu\rho} \mathsf{M}_{\mu\sigma} \,, \\ &[\mathsf{M}_{\mu\nu}, \mathsf{P}_\sigma] = g_{\nu\sigma} \mathsf{P}_\mu - g_{\mu\sigma} \mathsf{P}_\nu \,, \\ &[\mathsf{M}_{\mu\nu}, \mathsf{K}_\sigma] = g_{\nu\sigma} \mathsf{K}_\mu - g_{\mu\sigma} \mathsf{K}_\nu \,, \\ &[\mathsf{M}_{\mu\nu}, \mathsf{D}] = [\mathsf{P}_\mu, \mathsf{P}_\nu] = [\mathsf{K}_\mu, \mathsf{K}_\nu] = [\mathsf{D}, \mathsf{D}] = 0 \,, \\ &[\mathsf{P}_\mu, \mathsf{D}] = \mathsf{P}_\mu \,, \\ &[\mathsf{D}, \mathsf{K}_\mu] = \mathsf{K}_\mu \,, \\ &[\mathsf{K}_\mu, \mathsf{P}_\nu] = -2 \left(\mathsf{M}_{\mu\nu} + g_{\mu\nu} \mathsf{D} \right) \,, \end{aligned}$$

which are identical with the commutation relations satisfied by the infinitesimal generators of the conformal group $\mathsf{O}(n+2, 2)$. The transformations of the coordinates, in the Minkowskian space M_{n+2}, associated to the elements of the conformal group are determined by the subgroup structure of $\mathsf{O}(n+2, 2)$. These subgroups comprise the following transformations:

- *homogeneous Lorentz transformations* generated by $\mathsf{M}_{\mu\nu}$

$$x'_\mu = \Lambda^\nu_\mu x_\nu \,, \quad \Lambda^\nu_\mu \in \mathsf{O}(n+1, 1) \,,$$

- *general translations* generated by P_μ

$$x'_\mu = x_\mu + a_\mu \,, \quad a_\mu \in \mathbb{R} \,,$$

- *dilatations* generated by D

$$x'_\mu = s\, x_\mu \,, \quad s \in \mathbb{R} \,, \quad s > 0 \,,$$

- *special conformal transformations* generated by K_μ

$$x'_\mu = \frac{x_\mu - a_\mu x^2}{1 - 2\mathbf{a} \cdot \mathbf{x} + a^2 x^2} \,, \quad a_\mu \in \mathbb{R} \,,$$

where $\mathbf{a} \cdot \mathbf{x} = g_{\mu\nu} a_\mu x_\nu$, $a^2 = g_{\mu\nu} a_\mu a_\nu$, and $x^2 = g_{\mu\nu} x_\mu x_\nu$.

For $n = 3$ the Lie algebra of the infinitesimal generators (4.3.71) is isomorphic to the Lie algebra associated to the conformal group $\mathsf{O}(5,2)$. This group contains the *Galilei-Newton group* of classical mechanics, the *Poincaré group* of relativistic mechanics, the *conformal group* $\mathsf{O}(4,2)$ of electrodynamics, and the *symmetry group of the Schrödinger equation* of quantum mechanics. Consequently, the conformal group $\mathsf{O}(5,2)$, that is the symmetry group of the Hamilton-Jacobi equation for a free particle, can be regarded as the maximal symmetry group, in classical physics.

Finally, we point out that the mathematical modelling of different phenomena in theoretical physics is based on variational principles starting with a Lagrangian, or equivalently a Hamiltonian, and leading to a Hamilton-Jacobi type equation, or a canonical system. The free particle represents the simplest physical system, and it also evolves in an ideal homogeneous and isotropic space-time. Hence, the corresponding mathematical model exhibits all the possible symmetries assembled into a maximal symmetry group, while the mathematical model of any other physical system, subject to some constraints, inherits only a subgroup of this maximal symmetry group, by a symmetry breaking mechanism generated by the constraints. Such subgroups will be discussed in detail in the next chapters.

4. Applications in the theory of vibrations

4.1 General considerations

4.1.1 Lagrange's equations. Normal modes

Let be a discrete mechanical system of particles having the kinetic and potential energies, respectively

$$T = \frac{1}{2}\,\dot{q}_i a_{ij} \dot{q}_j\,, \quad V = \frac{1}{2}\,q_i b_{ij} q_j\,, \tag{4.4.1}$$

where q_i represents the generalized coordinates, $\dot{q}_i$ are the generalized velocities, while a_{ij} and b_{ij} are real constants. We assume that q_i and $\dot{q}_i$ are expressed as column vectors, such that (4.4.1) can be written in matrix form as

$$T = \frac{1}{2}\,\dot{\mathbf{q}}^{\mathrm{T}}\mathbf{A}\dot{\mathbf{q}}\,, \quad V = \frac{1}{2}\,\mathbf{q}^{\mathrm{T}}\mathbf{B}\mathbf{q}\,, \tag{4.4.2}$$

where the square matrices $\mathbf{A} = [a_{ij}]$ and $\mathbf{B} = [b_{ij}]$ are symmetric matrices. Note that a skew-symmetric part will have a zero contribution to T and V. The kinetic energy is a positively defined quadratic form (that is, $T = 0$ only when all the generalized velocities are equal to zero), therefore $\mathbf{A}$ is also a positively defined matrix. The Lagrangian of this mechanical system has the form

$$\mathcal{L} = T - V = \frac{1}{2}\,\dot{\mathbf{q}}^{\mathrm{T}}\mathbf{A}\dot{\mathbf{q}} - \frac{1}{2}\,\mathbf{q}^{\mathrm{T}}\mathbf{B}\mathbf{q}\,, \tag{4.4.3}$$

and the Lagrange equations are

$$\mathbf{A}\ddot{\mathbf{q}} + \mathbf{B}\mathbf{q} = \mathbf{0}\,, \tag{4.4.4}$$

corresponding to *small oscillations of the discrete mechanical system, subject to holonomic and scleronomic constraints, and conservative forces, about a stable position of equilibrium.*

By definition, the *normal mode* of oscillation has the form

$$\mathbf{q}(t) = \mathbf{q}_0\, e^{-i\omega t}\,, \tag{4.4.5}$$

and, by substituting (4.4.5) into (4.4.4), we obtain the homogeneous linear system

$$\left(\mathbf{B} - \omega^2\mathbf{A}\right)\mathbf{q}_0 = \mathbf{0}\,, \tag{4.4.6}$$

which admits non-trivial solutions only when the determinant of coefficients vanishes, i.e.,

$$\det\left(\mathbf{B} - \omega^2\mathbf{A}\right) = 0\,. \tag{4.4.7}$$

The equation (4.4.7) is called the *secular equation* of the mechanical system, and the roots $\{\omega_k^2\}$ are the squares of the *angular frequencies* of oscillation ($\omega_k = 2\pi\nu_k$, ν_k being the *frequencies* of oscillation). The solutions of the secular equation are also called *resonant frequencies* of the mechanical system. If $\mathbf{A}$ and $\mathbf{B}$ are $n \times n$ matrices, the secular equation is of degree n in ω^2, and there are at most n distinct frequencies. If $\omega_k^2 < 0$, then the corresponding normal mode is not an oscillatory one (the generalized coordinates q increase or decrease exponentially in time), and if $\omega_k^2 = 0$, then q_i are linear functions of time.

4.1.2 Normal coordinates

THEOREM 4.4.1 (J. J. SYLVESTER) *If $\mathbf{A}$ and $\mathbf{B}$ are symmetric real matrices, and $\mathbf{A}$ is also a positively defined matrix, then there is a non-singular real matrix $\mathbf{S}$ such that*

$$\mathbf{S}^{\mathrm{T}}\mathbf{A}\mathbf{S} = \mathbf{E}\,, \quad \mathbf{S}^{\mathrm{T}}\mathbf{B}\mathbf{S} = \mathbf{\Lambda}\,, \tag{4.4.8}$$

where $\mathbf{E}$ represents the identity matrix, and $\mathbf{\Lambda}$ is a diagonal matrix with the diagonal elements being the roots of the equation

$$\det\left(\mathbf{B} - \lambda\mathbf{A}\right) = 0\,.$$

By means of this theorem we may transform the Lagrange equations (4.4.4) into

$$\ddot{\mathbf{q}}' + \mathbf{\Lambda}\mathbf{q}' = \mathbf{0}\,, \tag{4.4.9}$$

where $\mathbf{q}' = \mathbf{S}^{-1}\mathbf{q}$. The secular equations associated to (4.4.4) and (4.4.9) have the same roots since

$$\det(\mathbf{\Lambda} - \lambda\mathbf{E}) = \det\left[\mathbf{S}^{\mathrm{T}}(\mathbf{B} - \lambda\mathbf{A})\mathbf{S}\right] = (\det\mathbf{S})^2 \det(\mathbf{B} - \lambda\mathbf{A}), \quad (4.4.10)$$

and the matrix $\mathbf{S}$ is non-singular, that is $\det\mathbf{S} \neq 0$. By comparing (4.4.7) and (4.4.10) we obtain $\nu_k^2 = \lambda_k/4\pi^2$, hence the squares of the oscillation frequencies of the normal mode are proportional to the eigenvalues of the diagonal matrix $\mathbf{\Lambda}$. Also, note that the Lagrange equations are invariant with respect to the point transformation $\mathbf{q}' = \mathbf{S}^{-1}\mathbf{q}$.

The generalized coordinates, which satisfy the equation of motion (4.4.9), have the general form

$$q_j'(t) = \alpha_j\, e^{-i\omega_j t} + \beta_j\, e^{i\omega_j t},$$

such that each of them is a simply periodic function involving only one resonant frequency. The new generalized coordinates q_j' are called *normal coordinates* of the mechanical system. The matrices $\mathbf{A}$ and $\mathbf{B}$ take simultaneously the diagonal form when all the generalized coordinates are normal, therefore in normal coordinates the equations of motion become independent equations.

THEOREM 4.4.2 (DANIEL BERNOULLI) *The small permanent oscillations of a discrete mechanical system, subject to holonomic and scleronomic constraints, and conservative forces, about a stable position of equilibrium, can be obtained as a superposition of a finite number of independent harmonic vibrations.*

4.2 Transformations of normal coordinates

4.2.1 The general case

There are two main problems in the theory of oscillations: finding out the frequencies of the normal mode, and determining the normal coordinates as functions of the generalized coordinates used to formulate the initial problem. The frequencies are given by the secular equation (4.4.7), but the problem of determining the normal coordinates is much more difficult since we have to find out the matrix $\mathbf{S}$ in the transformation (4.4.8). Note that, if the matrix $\mathbf{S}$ is determined, we also have the frequencies of the normal mode directly from $\mathbf{\Lambda} = \mathbf{S}^{\mathrm{T}}\mathbf{B}\mathbf{S}$, where the superscript T denotes a transposed matrix. Both problems require the precise form of the matrices $\mathbf{A}$ and $\mathbf{B}$, in other words, we have to know the masses of the particles in the mechanical system and the constants which characterize the forces. Even if it cannot produce a solution, the group theory can be used to simplify these problems. Thus, on the basis of symmetry considerations we can determine the limits of the number of frequencies, an upper limit for the number of constants required to specify the kinetic and potential energies, a partial factorization of the secular equation, and the system of coordinates to reduce the matrices $\mathbf{A}$ and $\mathbf{B}$ (not necessarily to a diagonal form).

Suppose, we have determined a matrix $\mathbf{D} \neq \mathbf{E}$, of the same dimension as the vector of generalized coordinates $\mathbf{q}$, such that

$$V(\mathbf{Dq}) = V(\mathbf{q}), \quad \forall \mathbf{q}. \tag{4.4.11}$$

In this case, the configurations $\mathbf{q}$ and $\mathbf{Dq}$ have the same potential energy, and from

$$V(\mathbf{Dq}) = \frac{1}{2}(\mathbf{Dq})^{\mathrm{T}}\mathbf{B}(\mathbf{Dq}) = \frac{1}{2}\mathbf{q}^{\mathrm{T}}(\mathbf{D}^{\mathrm{T}}\mathbf{BD})\mathbf{q} = V(\mathbf{q}),$$

it follows that

$$\mathbf{D}^{\mathrm{T}}\mathbf{BD} = \mathbf{B}. \tag{4.4.12}$$

Note that we also have

$$V(\mathbf{D}^2\mathbf{q}) = V(\mathbf{D}(\mathbf{Dq})) = V(\mathbf{Dq}) = V(\mathbf{q}),$$

and, substituting $\mathbf{q}$ by $\mathbf{D}^{-1}\mathbf{q}$ into (4.4.11), we obtain

$$V(\mathbf{D}^{-1}\mathbf{q}) = V(\mathbf{q}).$$

Consequently, the set of all matrices $\mathbf{D}$ which satisfy (4.4.11) form a representation of a group G, that leave invariant the potential V.

Suppose, that the relation $\mathbf{D}^{\mathrm{T}}\mathbf{AD} = \mathbf{A}$ is also satisfied (i.e., the Lagrangian (4.4.3) is invariant with respect to the transformations of G). Then (4.4.12) becomes the commutation relation

$$[\mathbf{B}, \mathbf{D}] = \mathbf{0}. \tag{4.4.13}$$

If the representation D is orthogonal and fully reducible, and each irreducible component is orthogonal, then there is a similarity transformation $\mathbf{S}$ which transforms each of the matrices in D into a direct sum of matrices

$$\mathbf{D}' = \mathbf{S}^{-1}\mathbf{DS} = \begin{bmatrix} \mathbf{D}^{(1)} & & & & & \\ & \mathbf{D}^{(1)} & & & & \\ & & \ddots & & & \\ & & & \mathbf{D}^{(2)} & & \\ & & & & \mathbf{D}^{(2)} & \\ & & & & & \ddots \end{bmatrix}, \tag{4.4.14}$$

corresponding to the decomposition of D into the direct sum irreducible representations

$$\mathsf{D} = \sum_{k}{}_{\oplus} c^{(k)}\mathsf{D}^{(k)}. \tag{4.4.15}$$

Now, we apply the similarity transformation $\mathbf{S}$ to the commutation relations (4.4.13). This gives

$$\mathbf{0} = \mathbf{S}^{-1}[\mathbf{B}, \mathbf{D}]\mathbf{S} = [\mathbf{S}^{-1}\mathbf{B}\mathbf{S}, \mathbf{D}'] .$$

It follows that the transformed matrix $\mathbf{B}' = \mathbf{S}^{-1}\mathbf{B}\mathbf{S}$ commutes with all the matrices in the representation D and, according to Schur's lemma (1.1.2), it has to be of the form

$$\mathbf{B}' = \mathbf{S}^{-1}\mathbf{B}\mathbf{S} = \begin{bmatrix} \mathbf{B}^{(1)} & & \\ & \mathbf{B}^{(2)} & \\ & & \ddots \end{bmatrix}, \tag{4.4.16}$$

where $\mathbf{B}^{(k)}$ is a submatrix of the same dimension with the block of all the matrices $\mathbf{D}^{(k)}$ in (4.4.14), defined by the relation

$$\mathbf{B}^{(k)} = \left(\mathbf{B}^{(k)}\right)^{\mathrm{T}} = \begin{bmatrix} \beta_{11}^{(k)}\mathbf{E}_k & \beta_{12}^{(k)}\mathbf{E}_k & \dots \\ \beta_{12}^{(k)}\mathbf{E}_k & \beta_{22}^{(k)}\mathbf{E}_k & \dots \\ \dots & \dots & \dots \end{bmatrix} = \boldsymbol{\beta}^{(k)} \times \mathbf{E}_k . \tag{4.4.17}$$

Here, we have pointed out the Cartesian product of the matrix $\boldsymbol{\beta}^{(k)}$ and the identity matrix of dimension $\dim \mathbf{E}_k = \dim \mathbf{D}^{(k)} = n_k$. The matrix $\boldsymbol{\beta}^{(k)}$ is symmetric and has the dimension $\dim \boldsymbol{\beta}^{(k)} = c^{(k)}$, where $c^{(k)}$ is the multiplicity of $\mathbf{D}^{(k)}$ in the decomposition (4.4.15). The matrix $\mathbf{A}$ takes a similar form

$$\mathbf{A}' = \begin{bmatrix} \mathbf{A}^{(1)} & & \\ & \mathbf{A}^{(2)} & \\ & & \ddots \end{bmatrix}, \quad \mathbf{A}^{(k)} = \boldsymbol{\alpha}^{(k)} \times \mathbf{E}_k , \tag{4.4.18}$$

$\boldsymbol{\alpha}^{(k)}$ being a matrix of dimension $c^{(k)}$. It follows that the secular equation can be written in the factorized form

$$\det\left(\mathbf{B} - \lambda\mathbf{A}\right) = \prod_k \left[\det\left(\boldsymbol{\beta}^{(k)} - \lambda\boldsymbol{\alpha}^{(k)}\right)\right]^{n_k} = 0 . \tag{4.4.19}$$

Consequently, the number of distinct frequencies is at most equal to the number of irreducible components in the representation D, and the degeneracy of a given frequency is at least equal to the dimension of the corresponding irreducible component. Since the representation D can be reduced to the block diagonal form by means of a real transformation, we may assume that the transformed matrices $\mathbf{A}'$ and $\mathbf{B}'$ are also real. Hence, the number of real constants required to specify the matrices $\mathbf{A}$ and $\mathbf{B}$ is given by the formula

$$\frac{1}{2}\sum_k c^{(k)}\left(c^{(k)} + 1\right) . \tag{4.4.20}$$

If an irreducible representation $\mathsf{D}^{(p)}$ appears only once in (4.4.15), then $c^{(p)} = 1$, and $\dim \alpha^{(p)} = \dim \beta^{(p)} = 1$. Hence, $\mathbf{A}^{(p)}$ and $\mathbf{B}^{(p)}$ are diagonal, and the corresponding symmetry coordinates (in which the representation D is completely reduced) are normal coordinates.

Let $\mathbf{q}'$ and $\mathbf{q}''$ be two systems of normal coordinates. Suppose, that in the coordinates $\mathbf{q}'$ the matrices $\mathbf{A}$ and $\mathbf{B}$ have the form

$$\mathbf{A} = \mathbf{E}\,, \quad \mathbf{B} = \mathbf{\Lambda} = \begin{bmatrix} \mathbf{\Lambda}_1 & & \\ & \mathbf{\Lambda}_2 & \\ & & \ddots \end{bmatrix}, \tag{4.4.21}$$

where $\mathbf{\Lambda}_k = \lambda_k \mathbf{E}_k$, λ_k are scalars, and $\lambda_i \neq \lambda_j$ for $i \neq j$. If $\mathbf{q}' = \mathbf{R}\mathbf{q}''$, then in the new coordinates $\mathbf{q}''$, the matrices $\mathbf{A}$ and $\mathbf{B}$ will be given by the relations

$$\mathbf{A}' = \mathbf{R}^{\mathrm{T}}\mathbf{E}\mathbf{R} = \mathbf{R}^{\mathrm{T}}\mathbf{R}\,, \quad \mathbf{B}' = \mathbf{R}^{\mathrm{T}}\mathbf{\Lambda}\mathbf{R}\,.$$

We have assumed that the coordinates $\mathbf{q}''$ are also normal, therefore

$$\mathbf{A}' = \mathbf{E}\,, \quad \mathbf{B}' = \mathbf{\Lambda}\,,$$

and the matrix $\mathbf{R}$ has to satisfy the relations

$$\mathbf{R}^{\mathrm{T}}\mathbf{R} = \mathbf{E}\,, \quad \mathbf{R}^{\mathrm{T}}\mathbf{\Lambda}\mathbf{R} = \mathbf{\Lambda}\,, \tag{4.4.22}$$

that is $[\mathbf{R}, \mathbf{\Lambda}] = \mathbf{0}$. It follows that the most general matrix $\mathbf{R}$, which transforms a system of normal coordinates into another system of normal coordinates, preserving the order of eigenvalues in the matrix $\mathbf{\Lambda}$, is an orthogonal matrix of the form

$$\mathbf{R} = \begin{bmatrix} \mathbf{R}_1 & & \\ & \mathbf{R}_2 & \\ & & \ddots \end{bmatrix}. \tag{4.4.23}$$

If the representation D in (4.4.15) has been real before being reduced to the block diagonal form, we may assume that it has been reduced by means of a real similarity transformation. Thus, the matrices $\boldsymbol{\alpha}^{(k)}$ and $\boldsymbol{\beta}^{(k)}$, in (4.4.17) and (4.4.18), are also real. Since these two matrices are symmetric, and $\boldsymbol{\alpha}^{(k)}$ is positively defined, then, from Sylvester's theorem (Theorem 4.4.1) it follows that there is a non-singular real matrix $\mathbf{S}_k$, such that

$$\mathbf{S}_k^{\mathrm{T}}\boldsymbol{\alpha}^{(k)}\mathbf{S}_k = \mathbf{E}^{(k)}\,, \quad \mathbf{S}_k^{\mathrm{T}}\boldsymbol{\beta}^{(k)}\mathbf{S}_k = \mathbf{b}^{(k)}\,,$$

$\mathbf{E}^{(k)}$ being the identity matrix of the same dimension as $\boldsymbol{\alpha}^{(k)}$, and $\mathbf{b}^{(k)}$ a diagonal matrix. Hence, if the lines of the Cartesian products are properly ordered, we have

$$\begin{aligned} (\mathbf{S}_k \times \mathbf{E}_k)^{\mathrm{T}}\mathbf{A}^{(k)}(\mathbf{S}_k \times \mathbf{E}_k) &= \mathbf{E}^{(k)} \times \mathbf{E}_k\,, \\ (\mathbf{S}_k \times \mathbf{E}_k)^{\mathrm{T}}\mathbf{B}^{(k)}(\mathbf{S}_k \times \mathbf{E}_k) &= \mathbf{b}^{(k)} \times \mathbf{E}_k\,. \end{aligned} \tag{4.4.24}$$

Finally, by combining all the matrices $\mathbf{S}_k$ into the direct sum

$$\mathbf{S} = \sum_{k}{}_{\oplus} \mathbf{S}_k \times \mathbf{E}_k \,, \tag{4.4.25}$$

we obtain a block diagonal matrix of the form (4.4.23). This is the matrix for the similarity transformation $\mathbf{S}^{-1}\mathbf{D}\mathbf{S} = \mathbf{D}'$, which also determines the normal system of coordinates $\mathbf{Q} = \mathbf{S}^{-1}\mathbf{q}$. Therefore, in the system of normal coordinates $\mathbf{Q}$, the representation D is completely reduced, that is the normal coordinates $\mathbf{Q}$ are also symmetry coordinates. The matrix (4.4.25) transforms each irreducible component of D into an equivalent irreducible component. It follows that all the systems of normal coordinates, in which $\mathbf{\Lambda}$ has the form (4.4.21), are systems of symmetry coordinates.

The normal coordinates corresponding to different frequencies cannot be transformed one into another because, if Q_j is a normal coordinate, then $\ddot{Q}_j + \omega_j^2 Q_j = 0$, and if Q'_j is also a normal coordinate, obtained by a transformation, then $\ddot{Q}'_j + \omega_j^2 Q'_j = 0$ and the only angular frequency which can appear in the expression of Q'_j is ω_j.

4.2.2 Examples of simple vibrating systems

To illustrate the usefulness of the group theory in the study of vibrating systems we will consider two examples: a completely symmetric system (see Fig. 4.1), and a system showing a reduced symmetry (see Fig. 4.2). The symmetry groups used in these two examples are S_3 and $\mathsf{G}_1 \subset \mathsf{S}_3$. Here, we focus on the method of using the symmetry properties of the mechanical systems to find the oscillation frequencies of the normal mode, and refer the reader to Chap. 1 Sec. 1.4 for details about different representations of S_3 and G_1, and their specific properties.

In the first case, the particles in the mechanical system have the same mass $m_1 = m_2 = m_3 \equiv m$, and are located at the vertices of an equilateral triangle. They are also connected by identical springs of constant k. The problem we have to solve is to find the normal modes of this mechanical system, assuming that the particles oscillate only in the xy-plane.

Let (x_1, y_1) be the coordinates (displacements) of the particle of mass m_1 relative to its equilibrium position, and similarly for the other two particles. Thus, the vector of the generalized coordinates is $\mathbf{q} = (q_1, q_2, \ldots, q_6) \equiv (x_1, y_1, x_2, y_2, x_3, y_3)$, and the kinetic and potential energies of the mechanical system are, respectively,

$$T = \frac{m}{2} \sum_{i=1}^{6} \dot{q}_i^2 \,, \tag{4.4.26}$$

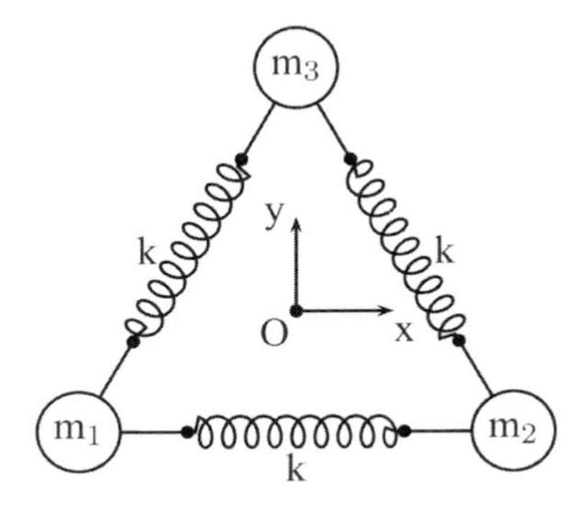

Figure 4.1. An S_3 symmetric system consisting of three equal masses ($m_1 = m_2 = m_3$) connected by identical springs.

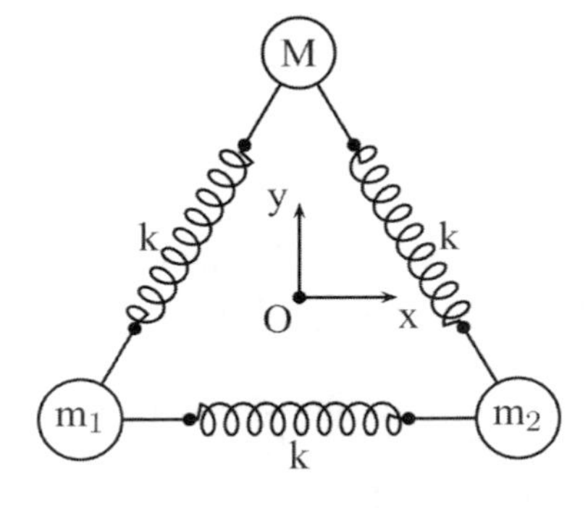

Figure 4.2. An example of symmetry breaking. The springs are identical, but one mass is larger than the other two ($m_1 = m_2 \neq M$).

and

$$V = \frac{k}{2}\left\{(x_2 - x_1)^2 + \left[-\frac{1}{2}(x_3 - x_2) + \frac{\sqrt{3}}{2}(y_3 - y_2)\right]^2 + \left[\frac{1}{2}(x_1 - x_3) + \frac{\sqrt{3}}{2}(y_1 - y_3)\right]^2\right\}. \tag{4.4.27}$$

Here, we have expressed the variation of distances between the particles according to the geometrical constraints in our problem. By comparing (4.4.26) and (4.4.27) with (4.4.1) we obtain the Lagrangian of the mechanical system in the form

$$\mathcal{L} = T - V = \frac{1}{2}\dot{\mathbf{q}}^{\mathrm{T}}\mathbf{A}\dot{\mathbf{q}} - \frac{1}{2}\mathbf{q}^{\mathrm{T}}\mathbf{B}\mathbf{q}, \tag{4.4.28}$$

where the matrices $\mathbf{A}$ and $\mathbf{B}$ are

$$\mathbf{A} = m\,\mathbf{E}, \quad \mathbf{B} = \frac{k}{4}\begin{bmatrix} 5 & \sqrt{3} & -4 & 0 & -1 & -\sqrt{3} \\ \sqrt{3} & 3 & 0 & 0 & -\sqrt{3} & -3 \\ -4 & 0 & 5 & -\sqrt{3} & -1 & \sqrt{3} \\ 0 & 0 & -\sqrt{3} & 3 & \sqrt{3} & -3 \\ -1 & -\sqrt{3} & -1 & \sqrt{3} & 2 & 0 \\ -\sqrt{3} & -3 & \sqrt{3} & -3 & 0 & 6 \end{bmatrix}. \tag{4.4.29}$$

The considered mechanical system has the symmetry of the equilateral triangle (see Fig. 4.1), the corresponding symmetry group being the group S_3. In a two-dimensional space a representation of S_3 is formed by rotations of an equilateral triangle about its centre by $2\pi/3$, and reflections about the axes passing through the centre of the triangle and one of its vertices. All these operations are represented by the set of matrices

$$\mathsf{T}_2 = \left\{ \begin{bmatrix} 1 & 0 \\ 0 & 1 \end{bmatrix}, \begin{bmatrix} -c & -s \\ s & -c \end{bmatrix}, \begin{bmatrix} -c & s \\ -s & -c \end{bmatrix}, \begin{bmatrix} -1 & 0 \\ 0 & 1 \end{bmatrix}, \begin{bmatrix} c & s \\ s & -c \end{bmatrix}, \begin{bmatrix} c & -s \\ -s & -c \end{bmatrix} \right\}, \tag{4.4.30}$$

acting in the space of vectors $\mathbf{r} = (x, y)$. Here, we have introduced the notations $\cos(2\pi/3) = -\cos(\pi/3) = -c$, and $\sin(2\pi/3) = \sin(\pi/3) = s$, that is $c = 1/2$ and $s = \sqrt{3}/2$. According to the notations used in Chap. 1 Sec. 1.4.2, these matrices are identified by the elements in the set

$$\mathsf{T}_2 = \{\mathbf{t}_2(\sigma_0), \mathbf{t}_2(\sigma_1), \mathbf{t}_2(\sigma_2), \mathbf{t}_2(\sigma_3), \mathbf{t}_2(\sigma_4), \mathbf{t}_2(\sigma_5)\}\ . \tag{4.4.31}$$

Note that, leaving out the identity matrix $\mathbf{t}_2(\sigma_0)$, the essential elements of S_3 are $\mathbf{t}_2(\sigma_1)$ (rotation about the origin of coordinates by $2\pi/3$) and $\mathbf{t}_2(\sigma_3)$ (reflection about the y-axis), because all the other elements are given by the relations $\mathbf{t}_2(\sigma_2) = \mathbf{t}_2(\sigma_1)^2$, $\mathbf{t}_2(\sigma_4) = \mathbf{t}_2(\sigma_3)\mathbf{t}_2(\sigma_1)$, and $\mathbf{t}_2(\sigma_5) = \mathbf{t}_2(\sigma_3)\mathbf{t}_2(\sigma_2)$, as it can be seen in the Cayley table of S_3. Hence, we may analyse the symmetry properties of the mechanical system only in terms of the matrices associated to the group elements σ_0, σ_1, and σ_3. Note that the group S_3 has three conjugacy classes, and each of these classes is determined by one of the elements in the subset $\{\sigma_0, \sigma_1, \sigma_3\}$.

The vibrations of a particle in the mechanical system are described by the corresponding coordinates (x, y) relative to its equilibrium position. Therefore, the configuration space of the mechanical system is a six-dimensional space, corresponding to three two-dimensional systems of coordinates, one for each particle. In this space, the action of the rotation operator associated to the element σ_1, comprises a rotation of each two-dimensional system of coordinates by $2\pi/3$, followed by a rotation of $2\pi/3$ of the particles about the centre of the triangle. The rotation of the two-dimensional systems is described by the matrix $\mathbf{t}_2(\sigma_1)$, while the rotation about the centre of the triangle is equivalent to an even permutation of the particles, which is described by the matrix

$$\mathbf{t}_3(\sigma_1) = \begin{bmatrix} 0 & 0 & 1 \\ 1 & 0 & 0 \\ 0 & 1 & 0 \end{bmatrix},$$

from the three-dimensional representation T_3 of S_3 (here we have denoted by $\mathbf{t}_3(\sigma)$ the matrices $\mathbf{s}(\sigma)$ defined in (1.1.42)). Accordingly, the action of σ_1 in the configuration space is given by the Cartesian product

$$\mathbf{t}(\sigma_1) = \mathbf{t}_2(\sigma_1) \times \mathbf{t}_3(\sigma_1) = \begin{bmatrix} 0 & 0 & \mathbf{t}_2(\sigma_1) \\ \mathbf{t}_2(\sigma_1) & 0 & 0 \\ 0 & \mathbf{t}_2(\sigma_1) & 0 \end{bmatrix}, \tag{4.4.32}$$

which is a 6×6 matrix from the direct product of representations $\mathsf{T}_2 \otimes \mathsf{T}_3$. Similarly, we obtain

$$\mathbf{t}(\sigma_3) = \mathbf{t}_2(\sigma_3) \times \mathbf{t}_3(\sigma_3) = \begin{bmatrix} 0 & \mathbf{t}_2(\sigma_3) & 0 \\ \mathbf{t}_2(\sigma_3) & 0 & 0 \\ 0 & 0 & \mathbf{t}_2(\sigma_3) \end{bmatrix}, \tag{4.4.33}$$

and $\mathbf{t}(\sigma_0) = \mathbf{E}$, that is the identity matrix. Then, by means of Cayley's table we can find all the matrices in the representation of S_3 in the configuration space

$$\mathsf{T} = \{\mathbf{t}(\sigma_0), \mathbf{t}(\sigma_1), \mathbf{t}(\sigma_2), \mathbf{t}(\sigma_3), \mathbf{t}(\sigma_4), \mathbf{t}(\sigma_5)\} \,. \tag{4.4.34}$$

The explicit form of the matrices in the representation T, and of the matrix $\mathbf{B}$ given in (4.4.29), can be used now to verify directly the commutation relations (4.4.13)

$$[\mathbf{B}, \mathbf{t}(\sigma)] = \mathbf{0}\,, \quad \forall\, \sigma \in S_3 \,. \tag{4.4.35}$$

In our case, the matrix $\mathbf{A}$, defined in (4.4.29), is proportional to the identity matrix. Therefore, $[\mathbf{A}, \mathbf{t}(\sigma_i)] = \mathbf{0}$ $(i = 0, 1, \ldots, 5)$, and consequently the Lagrangian (4.4.28) is invariant with respect to the transformations of S_3.

The group S_3 has three conjugacy classes, and therefore three irreducible representations, two of dimension one, and one of dimension two. It can be shown that the representation T is equivalent to the regular representation R of S_3, defined in (1.1.41). Hence, T is fully reducible and can be decomposed into the direct sum of irreducible representations

$$\mathsf{T} = \mathsf{T}_1^{\mathrm{s}} \oplus \mathsf{T}_1^{\mathrm{a}} \oplus 2\mathsf{T}_2 \,, \tag{4.4.36}$$

by means of a similarity transformation. Here,

$$\begin{aligned} \mathsf{T}_1^{\mathrm{s}} &= \{[1], [1], [1], [1], [1], [1]\}\,, \\ \mathsf{T}_1^{\mathrm{a}} &= \{[1], [1], [1], [-1], [-1], [-1]\}\,, \end{aligned}$$

and T_2 has been defined in (4.4.30). It is only a coincidence that in this example, the configuration space has the same dimension as the space of the regular representation of S_3. Therefore, the action of the operators associated to the elements of S_3 in the configuration space is described by the regular representation (4.4.34). This will simplify our analysis, as we already know the decomposition (4.4.36), otherwise, in the case of a configuration space of a larger (or lower) dimension we have to determine the corresponding decomposition. With these notations, the irreducible form of the representation T comprises the block diagonal matrices

$$\mathbf{t}'(\sigma) = \begin{bmatrix} \mathbf{t}_1^{\mathrm{s}}(\sigma) & & & \\ & \mathbf{t}_1^{\mathrm{a}}(\sigma) & & \\ & & \mathbf{t}_2(\sigma) & \\ & & & \mathbf{t}_2(\sigma) \end{bmatrix}, \quad \sigma \in S_3 \,. \tag{4.4.37}$$

Note that, the block diagonal form (4.4.37), and consequently the decomposition (4.4.36), are obtained from the regular representation (4.4.34) by means of a certain orthogonal similarity transformation. If we denote by $\mathbf{S}$ this transformation, then we have

$$\mathbf{t}'(\sigma) = \mathbf{S}^{\mathrm{T}}\mathbf{t}(\sigma)\mathbf{S}\,, \quad \sigma \in \mathsf{S}_3\,.$$

Now, if we apply the similarity transformation $\mathbf{S}$ to the commutation relations (4.4.35), that is

$$\mathbf{0} = \mathbf{S}^{\mathrm{T}}\left[\mathbf{B}, \mathbf{t}(\sigma)\right]\mathbf{S} = \left[\mathbf{S}^{\mathrm{T}}\mathbf{B}\mathbf{S}, \mathbf{t}'(\sigma)\right]\,, \quad \sigma \in \mathsf{S}_3\,, \tag{4.4.38}$$

it follows that a transformed matrix $\mathbf{B}' = \mathbf{S}^{\mathrm{T}}\mathbf{B}\mathbf{S}$ has to commute with all the matrices (4.4.37). According to the analysis in the previous section, this transformed matrix has necessarily to be of a block diagonal form, similar to the form of matrices in (4.4.37). Furthermore, according to Schur's lemma (Lemma 1.1.2), the blocks on the diagonal have to be multiples of the corresponding identity matrices. Hence, in the basis in which the matrices of the representation T have the form (4.4.37) the matrix $\mathbf{B}$ becomes

$$\mathbf{B}' = \begin{bmatrix} \lambda_1 & & & & & \\ & \lambda_2 & & & & \\ & & \lambda_3 & & & \\ & & & \lambda_3 & & \\ & & & & \lambda_4 & \\ & & & & & \lambda_4 \end{bmatrix}\,. \tag{4.4.39}$$

The trace of a matrix is not affected by a similarity transformation, and we may use it to relate the two forms of the matrix $\mathbf{B}$, associated respectively to the representations (4.4.34) and (4.4.37), by means of different matrix products. Consequently, we are lead to numerical relations between the eigenvalues of $\mathbf{B}$ which, in turn, can be used to determine at least some of these eigenvalues without solving the secular equation. Thus, by using (4.4.29) and (4.4.34) we have

$$\mathsf{Tr}\left[\mathbf{t}(\sigma_0)\mathbf{B}\right] = 6k\,, \quad \mathsf{Tr}\left[\mathbf{t}(\sigma_1)\mathbf{B}\right] = \frac{3}{2}\,k\,, \quad \mathsf{Tr}\left[\mathbf{t}(\sigma_3)\mathbf{B}\right] = 3k\,, \tag{4.4.40}$$

while (4.4.39) and (4.4.37) yield

$$\begin{aligned} \mathsf{Tr}\left[\mathbf{t}'(\sigma_0)\mathbf{B}'\right] &= \lambda_1 + \lambda_2 + 2(\lambda_3 + \lambda_4)\,, \\ \mathsf{Tr}\left[\mathbf{t}'(\sigma_1)\mathbf{B}'\right] &= \lambda_1 + \lambda_2 - (\lambda_3 + \lambda_4)\,, \\ \mathsf{Tr}\left[\mathbf{t}'(\sigma_3)\mathbf{B}'\right] &= \lambda_1 - \lambda_2\,. \end{aligned} \tag{4.4.41}$$

The cyclic property of the trace, that is $\mathsf{Tr}(\mathbf{ABC}) = \mathsf{Tr}(\mathbf{CAB}) = \mathsf{Tr}(\mathbf{BCA})$, allows us to equate (4.4.40) and (4.4.41). Thus, without specifying the form of the matrix $\mathbf{S}$ we obtain a linear system of equations for the unknowns λ_j, which can be reduced to the following relations

$$\lambda_1 = 3k\,, \quad \lambda_2 = 0\,, \quad \lambda_3 + \lambda_4 = \frac{3}{2}\,k\,. \tag{4.4.42}$$

Another relation between the eigenvalues of $\mathbf{B}$ can be given for instance by

$$\mathsf{Tr}\,\mathbf{B}'^2 = \mathsf{Tr}\,\mathbf{B}^2\,,$$

that is

$$\lambda_1^2 + \lambda_2^2 + 2(\lambda_3^2 + \lambda_4^2) = \frac{27}{2}\,k^2\,.$$

Hence, the eigenvalues of $\mathbf{B}$ are $\lambda \equiv \{3k, 0, 0, 0, 3k/2, 3k/2\}$, and the oscillation frequencies of the normal mode can be found from the relations

$$\nu_j^2 = \frac{1}{4\pi^2 m}\,\lambda_j\,, \quad j = 1, 2, \ldots, 6.$$

Note that the Lagrangian (4.4.28) does not directly depend on time, i.e., it satisfies the condition (4.3.15), which ensures the conservation of mechanical energy.

In the second example of simple vibrating systems, we will study the same mechanical system, but changing one of the masses of particles in the mechanical system ($m_1 = m_2 \equiv m$, $m_3 = M \neq m$), while leaving the springs unaltered (see Fig. 4.2). Now, the Lagrangian remains of the form (4.4.28),

$$\mathcal{L} = \frac{1}{2}\,\dot{\mathbf{q}}^{\mathrm{T}}\mathbf{A}\dot{\mathbf{q}} - \frac{1}{2}\,\mathbf{q}^{\mathrm{T}}\mathbf{B}\mathbf{q}\,, \tag{4.4.43}$$

with the matrix $\mathbf{B}$ unchanged, but $\mathbf{A} = \mathbf{ME}$, where

$$\mathbf{M} = \begin{bmatrix} m_1 & & & & & \\ & m_1 & & & & \\ & & m_2 & & & \\ & & & m_2 & & \\ & & & & M & \\ & & & & & M \end{bmatrix}. \tag{4.4.44}$$

We can simplify the problem by rescaling the generalized coordinates $\mathbf{Q} = \mathbf{M}^{1/2}\mathbf{q}$, and writing the Lagrangian (4.4.43) in the form

$$\mathcal{L} = \frac{1}{2}\,\dot{\mathbf{Q}}^{\mathrm{T}}\dot{\mathbf{Q}} - \frac{1}{2}\,\mathbf{Q}^{\mathrm{T}}\mathbf{D}\mathbf{Q}\,, \tag{4.4.45}$$

with $\mathbf{D} = \mathbf{M}^{-1/2}\mathbf{B}\mathbf{M}^{-1/2}$.

It is easy to observe in Fig. 4.2 that the considered mechanical system is invariant only with respect to the permutations of equal masses (m_1 and m_2), which represent reflections about the y-axis. Consequently, this mechanical system exhibits a reduced symmetry corresponding to the subgroup $\mathsf{G}_2 = \{\sigma_0, \sigma_3\} \subset \mathsf{S}_3$. Each element in G_2 forms a conjugacy class by itself, and consequently the group G_2 has two one-dimensional irreducible representations; one of them is the trivial representation

$$\mathsf{T}_1^{(G)} = \{[1], [1]\}, \tag{4.4.46}$$

while the second is obtained by associating $+1$ to the even permutation (σ_0), and -1 to the odd permutation (σ_3)

$$\mathsf{T}_2^{(G)} = \{[1], [-1]\}. \tag{4.4.47}$$

There are no other irreducible representations of G_2, such that the characters table reads

	$\chi^{(1)}$	$\chi^{(2)}$	p
C_0	1	1	1
C_3	1	-1	1

where p represents the number of group elements in each conjugacy class.

In the configuration space spanned by the vectors $\mathbf{Q}$ we have a six-dimensional representation of G_2 formed by the matrices $\mathbf{t}(\sigma_0) = \mathbf{E}$ and $\mathbf{t}(\sigma_3)$ defined in (4.4.33). In contrast to the previous example, this representation is no longer the regular representation of G_2 and we have to determine its decomposition into a direct sum of irreducible representations. Let

$$\mathsf{T}^{(G)} = \{\mathbf{t}(\sigma_0), \mathbf{t}(\sigma_3)\} \tag{4.4.48}$$

be the representation of G_2 in the configuration space. Then, the conjugacy classes of $\mathsf{T}^{(G)}$ have the characters

	χ	p
C_0	6	1
C_3	0	1

Now, we may use the orthogonality properties of the characters

$$\mu_k = \frac{1}{n} \sum_{\sigma \in \{\sigma_0, \sigma_3\}} p(C_\sigma)\, \overline{\chi^{(k)}(C_\sigma)}\, \chi(C_\sigma), \quad k = 1, 2, \tag{4.4.49}$$

where $n = 2$ is the number of elements in G_2, and μ_k is the multiplicity of the irreducible representation $\mathsf{T}_k^{(G)}$ in the decomposition of $\mathsf{T}^{(G)}$. Thus, we obtain

$$\mathsf{T}^{(G)} = \mu_1 \mathsf{T}_1^{(G)} \oplus \mu_2 \mathsf{T}_2^{(G)} = 3\,\mathsf{T}_1^{(G)} \oplus 3\,\mathsf{T}_2^{(G)}, \tag{4.4.50}$$

the corresponding matrices being of diagonal form

$$\mathbf{t}'(\sigma_0) = \mathbf{E}\,, \quad \mathbf{t}'(\sigma_3) = \begin{bmatrix} 1 & & & & & \\ & 1 & & & & \\ & & 1 & & & \\ & & & -1 & & \\ & & & & -1 & \\ & & & & & -1 \end{bmatrix} . \tag{4.4.51}$$

The matrix $\mathbf{D}$ defined in (4.4.45) commutes with the matrices (4.4.48), while the transformed matrix

$$\mathbf{D}' = \begin{bmatrix} \lambda_1 & & & & & \\ & \lambda_2 & & & & \\ & & \lambda_3 & & & \\ & & & \lambda_4 & & \\ & & & & \lambda_5 & \\ & & & & & \lambda_6 \end{bmatrix} \tag{4.4.52}$$

commutes with the matrices (4.4.51). Consequently, by the same method as the one used in the previous example we obtain

$$\begin{aligned} \operatorname{Tr}\left[\mathbf{t}(\sigma_0)\mathbf{D}\right] &= 2k\left(\frac{2}{m} + \frac{1}{M}\right), \\ \operatorname{Tr}\left[\mathbf{t}(\sigma_3)\mathbf{D}\right] &= k\left(\frac{2}{m} + \frac{1}{M}\right), \end{aligned}$$

and

$$\begin{aligned} \operatorname{Tr}\left[\mathbf{t}'(\sigma_0)\mathbf{D}'\right] &= \lambda_1 + \lambda_2 + \lambda_3 + \lambda_4 + \lambda_5 + \lambda_6\,, \\ \operatorname{Tr}\left[\mathbf{t}'(\sigma_3)\mathbf{D}'\right] &= \lambda_1 + \lambda_2 + \lambda_3 - \lambda_4 - \lambda_5 - \lambda_6\,. \end{aligned}$$

Compared with the first example, this is a case of symmetry breaking, when a modified mechanical system is invariant with respect to a subgroup of the symmetry group associated to the original mechanical system. Thus, we may conclude only that the eigenvalues of the matrix $\mathbf{D}$ satisfy the relations

$$\begin{aligned} \lambda_1 + \lambda_2 + \lambda_3 &= 3k\left(\frac{1}{m} + \frac{1}{2M}\right), \\ \lambda_4 + \lambda_5 + \lambda_6 &= k\left(\frac{1}{m} + \frac{1}{2M}\right), \end{aligned}$$

and we have to use other relations to determine the set of all eigenvalues. However, in this particular case, the secular equation

$$\begin{aligned} 4m^3M^2\lambda^6 - 8km^2M(m+2M)\lambda^5 + 3k^2m(m^2 + 8mM + 6M^2)\lambda^4 \\ -3k^3(2m^2 + 5mM + 2M^2)\lambda^3 = 0 \end{aligned}$$

shows that three eigenvalues are equal to zero, and we may use the relation

$$\operatorname{Tr}\mathbf{D}'^2 = \operatorname{Tr}\mathbf{D}^2 ,$$

which yields

$$\lambda_1^2 + \lambda_2^2 + \lambda_3^2 + \lambda_4^2 + \lambda_5^2 + \lambda_6^2 = k^2 \left(\frac{7}{m^2} + \frac{5}{2M^2} + \frac{4}{mM} \right)$$

to find all the eigenvalues. Finally, by setting $\lambda_3 = \lambda_4 = \lambda_5 = 0$, we get the eigenvalues

$$\lambda = \left\{ \frac{3k}{2m} + \frac{3k}{4M} + \frac{\sqrt{3}\,k}{4mM} \sqrt{3m^2 - 4mM + 4M^2} \right.,$$

$$\left. \frac{3k}{2m} + \frac{3k}{4M} - \frac{\sqrt{3}\,k}{4mM} \sqrt{3m^2 - 4mM + 4M^2} , 0 , 0 , 0 , \frac{k}{m} + \frac{k}{2M} \right\},$$

and the oscillation frequencies of the normal mode

$$\nu_j^2 = \frac{1}{4\pi^2} \lambda_j , \quad j = 1, 2, \ldots, 6.$$

Note that, for $M = m$, the eigenvalues of $\mathbf{D}$ can be written in the form

$$\left\{ \frac{3k}{m} , \frac{3k}{2m} , 0 , 0 , 0 , \frac{3k}{2m} \right\},$$

that are the eigenvalues of the matrix $\mathbf{B}$ in the first example, divided by m and reordered.

Chapter 5

APPLICATIONS IN THE THEORY OF RELATIVITY AND THEORY OF CLASSICAL FIELDS

1. Theory of Special Relativity

1.1 Preliminary considerations

1.1.1 Rigid motions of the Lobachevskian plane

Let us consider in the Euclidean three-dimensional space $\mathcal{E}_3$ a surface determined by the parametric equations

$$x = x(u,v)\,, \quad y = y(u,v)\,, \quad z = z(u,v)\,. \tag{5.1.1}$$

The *distance* between two points on this surface is defined in terms of the element of arc length (the *first fundamental form*)

$$ds^2 = dx^2 + dy^2 + dz^2 = Edu^2 + 2Fdudv + Gdv^2\,, \tag{5.1.2}$$

where

$$\begin{aligned} E(u,v) &= \left(\frac{\partial x}{\partial u}\right)^2 + \left(\frac{\partial y}{\partial u}\right)^2 + \left(\frac{\partial z}{\partial u}\right)^2, \\ F(u,v) &= \frac{\partial x}{\partial u}\frac{\partial x}{\partial v} + \frac{\partial y}{\partial u}\frac{\partial y}{\partial v} + \frac{\partial z}{\partial u}\frac{\partial z}{\partial v}, \\ G(u,v) &= \left(\frac{\partial x}{\partial v}\right)^2 + \left(\frac{\partial y}{\partial v}\right)^2 + \left(\frac{\partial z}{\partial v}\right)^2. \end{aligned}$$

When $E = G = 1$ and $F = 0$ we say that the surface (5.1.1) is characterized by an *Euclidean geometry* (or *parabolic geometry*), otherwise we have a *non-Euclidean geometry* (or *Riemannian geometry*). In both cases (5.1.2) determines the metric of the geometry. By a suitable transformation of the parameters u and v we may reduce the expression (5.1.2) to the form

$$ds^2 = a^2du^2 + dv^2\,, \tag{5.1.3}$$

such that the *total surface curvature* is given by the formula

$$K = -\frac{1}{a}\frac{d^2 a}{dv^2}\,. \tag{5.1.4}$$

When $K = 0$ the surface (5.1.1) is an Euclidean plane, and a is a linear function of v. The rigid motions in this Euclidean plane form the Euclidean group $\mathsf{E}(3)$, which allows to overlay two identical forms by a translation and a rotation. The Euclidean group $\mathsf{E}(3)$ preserves the distances between two points in the plane as well as the angle between two directions in the plane (see Chap. 4 Sec. 2.2). If $K \neq 0$ (*Riemannian plane*) we can find a similar property only when the total surface curvature K is constant. In this case we have an *elliptical geometry* ($K > 0$) or a *hyperbolic geometry* ($K < 0$). The only three-dimensional surface having a constant and positive total curvature ($K = 1/R^2$) is the *sphere*, determined by the parametric equations

$$x = R\cos u \sin v\,, \quad y = R\sin u \sin v\,, \quad z = R\cos v\,, \tag{5.1.5}$$

with the parameters varying in the domains $u \in [0, 2\pi)$ and $v \in [0, \pi]$. Correspondingly, the three-dimensional surface having a constant and negative total curvature ($K = -1/R^2$) is called *pseudo-sphere* (or *Lobachevskian plane*), and is determined by the parametric equations

$$x = R\cos u/\cosh v\,, \quad y = R\sin u/\cosh v\,, \quad z = R(v - \tanh v)\,, \tag{5.1.6}$$

where $u \in [0, 2\pi)$ and $v \in (-\infty, \infty)$. Since $x^2 + y^2 = R^2/\cosh^2 v$, we may regard the pseudo-sphere as the surface generated by the motion of a circle, of variable radius, parallel to the plane $z = 0$ and having all the time the centre on the Oz axis. Likewise, the intersection of the surface (5.1.6) with the $x = 0$ plane ($u = \pi/2$) is the parametric curve

$$y = R/\cosh v\,, \quad z = R(v - \tanh v)\,, \tag{5.1.7}$$

called *tractrix*, and the pseudo-sphere can be obtained by rotating this curve about its asymptote (the Oz axis).

A unified model of these three geometries (Euclidean, elliptical and hyperbolic) has been introduced by Cayley. In the two-dimensional case, Cayley considered the homogeneous coordinates x_1, x_2, x_3 in the projective plane, and defined a non-degenerate conic having the canonical form

$$x_1^2 + x_2^2 + \frac{1}{K}\,x_3^2 = 0\,, \tag{5.1.8}$$

called the *Absolute* of the geometry. In this model, the *distance* between two points A and B in the projective plane is defined in terms of the intersection points of the line AB with the Absolute of the geometry.

We mention that Cayley's model can be generalized to the case of three-dimensional projective space by considering that the Absolute of the geometry is the quadric

$$x_1^2 + x_2^2 + x_3^2 + \frac{1}{K}\,x_4^2 = 0\,, \tag{5.1.9}$$

in the homogeneous coordinates x_1, x_2, x_3, x_4. Thus, the rigid motions of *Cayley's space* can be expressed in the form of homogeneous linear transformations

$$x'_\mu = a_{\mu\nu} x_\nu\,, \quad \mu = \overline{1,4}\,, \tag{5.1.10}$$

which leave invariant the quadric (5.1.9). Generally, these transformations depend on six parameters and form a group that corresponds to the group of rigid motions in the three-dimensional Euclidean space $\mathcal{E}_3$ (see Chap. 4 Sec. 2.2). When $K = -1/R^2$ we obtain the model of *Lobachevskian space*, the Absolute of the geometry being

$$x_1^2 + x_2^2 + x_3^2 - R^2 x_4^2 = 0\,. \tag{5.1.11}$$

The rigid motions in the Lobachevskian space (i.e., the transformations (5.1.10) that leave invariant the quadric (5.1.11) and preserve the orientation of space objects) form a group isomorphic to the group of proper Lorentz transformations (see Chap. 2 Sec. 4.1).

In a Lobachevskian plane the distance between two points $A(x_1, y_1)$ and $B(x_2, y_2)$ is given by the formula

$$d^2 = (x_1 - x_2)^2 - (y_1 - y_2)^2\,. \tag{5.1.12}$$

Note that the sign of the distance d is not defined and d can be real or purely imaginary. Even if the two points A and B are distinct, the distance d can be equal to zero when

$$(x_1 - x_2) = \pm(y_1 - y_2)\,. \tag{5.1.13}$$

When the point B is placed at the origin of coordinates ($x_2 = y_2 = 0$) the lines $y_1 = \pm x_1$ divide the plane into four quadrants. Note that all the points on these lines are located at a zero distance from the origin. Then, the rigid motions of the Lobachevskian plane are defined as the inhomogeneous and linear transformations that do not change orientations, preserve the distances between the points in the plane, and leave invariant the quadrants one by one. It can be shown that such transformations have the form

$$\begin{aligned} x' &= x\cosh\varphi + y\sinh\varphi + a_1\,, \\ y' &= x\sinh\varphi + y\cosh\varphi + a_2\,, \end{aligned} \tag{5.1.14}$$

where $a_1, a_2, \varphi \in (-\infty, \infty)$. Therefore, each rigid motion g is determined by three real parameters. The set of all rigid motions $g(a_1, a_2, \varphi)$ form a subgroup of the Poincaré group, denoted by $\mathsf{O}(2,1)$[1]. The composition law of this group can be obtained directly from its definition (5.1.14). Thus, we have

$$g(a_1, a_2, \varphi) g(b_1, b_2, \psi) = g(c_1, c_2, \theta)\,, \tag{5.1.15}$$

where

$$\begin{aligned} c_1 &= b_1 \cosh\varphi + b_2 \sinh\varphi + a_1\,, \\ c_2 &= b_1 \sinh\varphi + b_2 \cosh\varphi + a_2\,, \\ \theta &= \varphi + \psi\,. \end{aligned}$$

Let us consider two subgroups of $\mathsf{O}(2,1)$: the subgroup of *hyperbolic rotations* $\Omega = \{g(0,0,\varphi)\}$, which leave the origin fixed, and the subgroup of *translations* $A = \{g(a_1, a_2, 0)\}$. The elements of the subgroup A are in one-to-one correspondence with the points in the Lobachevskian plane, while the elements of the subgroup Ω can be regarded as automorphisms of the subgroup A. Consequently, the group $\mathsf{O}(2,1)$ is obtained as the semidirect product of the subgroup of translations A and the group of its automorphisms

$$\mathsf{O}(2,1) = A \wedge \Omega\,. \tag{5.1.16}$$

From the definition (5.1.14) it follows that the group $\mathsf{O}(2,1)$ is isomorphic to the group of matrices

$$g \to \mathbf{t}(g) = \begin{bmatrix} \cosh\varphi & \sinh\varphi & a_1 \\ \sinh\varphi & \cosh\varphi & a_2 \\ 0 & 0 & 1 \end{bmatrix}. \tag{5.1.17}$$

The matrices in (5.1.17) satisfy the relation $\mathbf{t}(g_1)\mathbf{t}(g_2) = \mathbf{t}(g_1 g_2)$, therefore they form a representation of $\mathsf{O}(2,1)$. It can be shown that this representation is reducible, but not complete reducible. The infinitesimal generators of the representation (5.1.17) have the form

$$\begin{aligned} \mathbf{a}_1 &= \left.\frac{\partial \mathbf{t}}{\partial a_1}\right|_{a_1=a_2=\varphi=0} = \begin{bmatrix} 0&0&1 \\ 0&0&0 \\ 0&0&0 \end{bmatrix}, \\ \mathbf{a}_2 &= \left.\frac{\partial \mathbf{t}}{\partial a_2}\right|_{a_1=a_2=\varphi=0} = \begin{bmatrix} 0&0&0 \\ 0&0&1 \\ 0&0&0 \end{bmatrix}, \\ \mathbf{a}_3 &= \left.\frac{\partial \mathbf{t}}{\partial \varphi}\right|_{a_1=a_2=\varphi=0} = \begin{bmatrix} 0&1&0 \\ 1&0&0 \\ 0&0&0 \end{bmatrix}, \end{aligned} \tag{5.1.18}$$

[1] Generally, $\mathsf{O}(m+n,n)$ denotes the *conformal group*, which is the group of linear transformations (in the Minkowski space $\mathcal{M}_{m+n}$) that leave invariant the quadratic form $x_1^2 + \cdots + x_m^2 - x_{m+1}^2 - \cdots - x_{m+n}^2$.

and satisfy the commutation relations

$$[\mathbf{a}_1, \mathbf{a}_2] = \mathbf{0}\,, \quad [\mathbf{a}_2, \mathbf{a}_3] = -\mathbf{a}_1\,, \quad [\mathbf{a}_3, \mathbf{a}_1] = \mathbf{a}_2\,. \tag{5.1.19}$$

1.1.2 Elements of tensor analysis

In the case of a Riemannian n-dimensional space the element of arc length (5.1.1), also called *interval*, takes the general form

$$ds^2 = g_{\mu\nu} dx^\mu dx^\nu\,, \tag{5.1.20}$$

where $g_{\mu\nu}(\mathbf{x})$ represents the *metric tensor*, which determines the geometry of the considered space. Note that, in (5.1.20) we have used the summation convention for dummy Greek indices from 1 to n. In what follows we will focus on the case $n = 4$, which is of particular interest in the theory of relativity, but all the results presented in this section are also valid in the general case of a Riemannian n-dimensional space.

Let

$$x'^\mu = f^\mu(\mathbf{x})\,, \quad \mu = \overline{1,4}\,, \tag{5.1.21}$$

be a transformation of coordinates, where $\mathbf{x} = (x^1, x^2, x^3, x^4)$, and f^μ are real, independent functions, having the Jacobian

$$\det \left[\frac{\partial(f^\mu)}{\partial(x^\nu)} \right] \neq 0\,. \tag{5.1.22}$$

Hence, there is an inverse transformation

$$x^\nu = h^\nu(\mathbf{x}')\,, \quad \nu = \overline{1,4}\,. \tag{5.1.23}$$

By differentiating (5.1.21) we obtain the relations

$$dx'^\mu = \frac{\partial x'^\mu}{\partial x^\nu} dx^\nu = \frac{\partial f^\mu}{\partial x^\nu} dx^\nu\,, \tag{5.1.24}$$

where we have used the summation convention for dummy Greek indices from 1 to 4. If the transformations (5.1.21) form a group, then the differentials of the coordinates generate a representation of this group.

DEFINITION 5.1.1 *The set of functions* $\{V^\mu(\mathbf{x})\}_{\mu=\overline{1,4}}$ *form the components of a* **contravariant vector** *in the system of coordinates* x^μ, *if they satisfy the same transformation relations as the differentials of the coordinates.*

Thus, in the system of coordinates x'^μ the components of a contravariant vector are given by the formulae

$$V'^\mu(\mathbf{x}') = \frac{\partial x'^\mu}{\partial x^\nu} V^\nu(\mathbf{x}) = \frac{\partial f^\mu}{\partial x^\nu} V^\nu(\mathbf{x})\,. \tag{5.1.25}$$

Note that we use upper indices on all coordinates, since the differentials dx^μ form a contravariant vector.

Let $\varphi(\mathbf{x})$ be a scalar function. Under a transformation of the form (5.1.23), the derivatives of this function (the components of the *gradient*) become

$$\frac{\partial \varphi}{\partial x'^\mu} = \frac{\partial \varphi}{\partial x^\nu}\frac{\partial x^\nu}{\partial x'^\mu} = \frac{\partial \varphi}{\partial x^\nu}\frac{\partial h^\nu}{\partial x'^\mu}\,. \tag{5.1.26}$$

DEFINITION 5.1.2 *The set of functions* $\{V_\mu(\mathbf{x})\}_{\mu=\overline{1,4}}$ *form the components of a* **covariant vector** *in the system of coordinates* x^μ, *if they satisfy the same transformation relations as the gradient of a scalar function.*

It follows that

$$V'_\mu(\mathbf{x}') = \frac{\partial x^\nu}{\partial x'^\nu}\,V_\nu(\mathbf{x})\,. \tag{5.1.27}$$

DEFINITION 5.1.3 *The set of functions* $\{T^{\mu_1\mu_2\ldots\mu_p}_{\lambda_1\lambda_2\ldots\lambda_q}(\mathbf{x})\}_{1\leq\lambda_k,\mu_k\leq 4}$ *form the components of a* **tensor** *of rank* $p+q$, p *times* **contravariant** *and* q *times* **covariant**, *in the system of coordinates* x^μ, *if these functions satisfy the transformation relations*

$$T'^{\mu_1\mu_2\ldots\mu_p}_{\lambda_1\lambda_2\ldots\lambda_q}(\mathbf{x}') = \frac{\partial x^{\rho_1}}{\partial x'^{\lambda_1}}\frac{\partial x^{\rho_2}}{\partial x'^{\lambda_2}}\cdots\frac{\partial x^{\rho_q}}{\partial x'^{\lambda_q}}\frac{\partial x'^{\mu_1}}{\partial x^{\nu_1}}\frac{\partial x'^{\mu_2}}{\partial x^{\nu_2}}\cdots\frac{\partial x'^{\mu_p}}{\partial x^{\nu_p}} \times T^{\nu_1\nu_2\ldots\nu_p}_{\rho_1\rho_2\ldots\rho_q}(\mathbf{x})\,. \tag{5.1.28}$$

Consequently, the covariant components of the metric tensor satisfy the transformation relations

$$g'_{\mu\nu}(\mathbf{x}') = \frac{\partial x^\lambda}{\partial x'^\mu}\frac{\partial x^\rho}{\partial x'^\nu}\,g_{\lambda\rho}(\mathbf{x})\,. \tag{5.1.29}$$

The quantities

$$g'_{\mu\nu}V'^\nu = \frac{\partial x^\lambda}{\partial x'^\mu}\frac{\partial x^\rho}{\partial x'^\nu}\frac{\partial x'^\nu}{\partial x^\sigma}\,g_{\lambda\rho}V^\sigma = \frac{\partial x^\lambda}{\partial x'^\mu}\,\delta^\rho_\sigma\,g_{\lambda\rho}V^\sigma = \frac{\partial x^\lambda}{\partial x'^\mu}g_{\lambda\sigma}V^\sigma\,,$$

that transform according to the relations (5.1.27), constitute the components of a covariant vector, and the relations

$$V_\nu(\mathbf{x}) = g_{\nu\mu}(\mathbf{x})V^\mu(\mathbf{x})\,, \tag{5.1.30}$$

set up a one-to-one correspondence between contravariant and covariant components of the same vector $\mathbf{V}$. The inverse of (5.1.30) is obtained by means of the inverse of the metric tensor, having the components $g^{\mu\nu}$, which satisfy the relations $g^{\mu\rho}g_{\nu\rho} = \delta^{\mu}_{\nu}$. Thus, we have

$$V^{\mu}(\mathbf{x}) = g^{\mu\nu}(\mathbf{x})V_{\nu}(\mathbf{x}) . \tag{5.1.31}$$

By means of (5.1.30) and (5.1.31) the dot product of two vectors can be expressed in one of the forms

$$g_{\mu\nu}V^{\mu}W^{\nu} = V_{\nu}W^{\nu} = V^{\mu}W_{\mu} = g^{\mu\nu}V_{\mu}W_{\nu} . \tag{5.1.32}$$

The derivatives of a scalar function (tensor of rank zero) $\varphi(\mathbf{x})$ transform according to (5.1.26), and represent the components of a covariant vector. This is the only case when, in a general system of coordinates, the derivative of a tensor is also a tensor. It can be shown that in the case of a contravariant vector (contravariant tensor of rank one) whose components satisfy the transformation relations

$$V^{\mu}(\mathbf{x}) = \frac{\partial x^{\mu}}{\partial x'^{\nu}}\, V'(\mathbf{x}') ,$$

the quantities

$$V^{\mu}_{;\sigma} = \frac{\partial V^{\mu}}{\partial x^{\sigma}} + \left\{ \begin{matrix} \mu \\ \rho \;\; \sigma \end{matrix} \right\} V^{\rho} , \tag{5.1.33}$$

where

$$\left\{ \begin{matrix} \mu \\ \rho \;\; \sigma \end{matrix} \right\} = \frac{1}{2}\, g^{\mu\nu} \left(\frac{\partial g_{\sigma\nu}}{\partial x^{\rho}} + \frac{\partial g_{\rho\nu}}{\partial x^{\sigma}} - \frac{\partial g_{\sigma\rho}}{\partial x^{\nu}} \right) \tag{5.1.34}$$

are the *Christoffel symbols of the second kind* (also known as *affine connections* or *connection coefficients*), constitute the components of a mixed tensor of rank two. In the case of a covariant vector (covariant tensor of rank one), the quantities

$$V_{\mu;\nu} = \frac{\partial V_{\mu}}{\partial x^{\nu}} - \left\{ \begin{matrix} \rho \\ \mu \;\; \nu \end{matrix} \right\} V_{\rho} \tag{5.1.35}$$

form the components of a covariant tensor of rank two, being related to the components (5.1.33) by the formula

$$V_{\mu;\nu} = g_{\mu\sigma}V^{\sigma}_{;\nu} . \tag{5.1.36}$$

The equations (5.1.33) and (5.1.35) define the *covariant derivative* of a contravariant and a covariant vector, respectively.

Other notations for the Christoffel symbol of the second kind are

$$\left\{ \begin{matrix} \mu \\ \rho \;\; \sigma \end{matrix} \right\} \equiv \{\rho\,\sigma, \mu\} \equiv \Gamma^{\mu}_{\rho\sigma} .$$

The latter symbol suggests however a tensor character, which is not true in general. The Christoffel symbol can be thought of as describing how a coordinate system twists and stretches as you move about on the manifold. It thus is about a particular coordinate system in two different ways: it describes a particular coordinate system, and it is expressed in that same coordinate system. Also, the Christoffel symbols of the second kind are related to the Christoffel symbols of the first kind $[\rho\,\sigma,\nu]$, defined in (4.1.15), by

$$\left\{ {\mu \atop \rho\ \sigma} \right\} = g^{\mu\nu}\,[\rho\,\sigma,\nu]\,.$$

The covariant derivative of the tensors of rank two is given by the formulae

$$\begin{aligned} T_{\mu\nu;\rho} &= \frac{\partial T_{\mu\nu}}{\partial x^\rho} - \left\{ {\lambda \atop \mu\ \rho} \right\} T_{\lambda\nu} - \left\{ {\lambda \atop \nu\ \rho} \right\} T_{\mu\lambda}\,, \\ T^{\mu\nu}_{;\rho} &= \frac{\partial T^{\mu\nu}}{\partial x^\rho} + \left\{ {\mu \atop \lambda\ \rho} \right\} T^{\lambda\nu} + \left\{ {\nu \atop \lambda\ \rho} \right\} T^{\mu\lambda}\,, \\ T^{\mu}_{\nu;\rho} &= \frac{\partial T^{\mu}_{\nu}}{\partial x^\rho} + \left\{ {\mu \atop \lambda\ \rho} \right\} T^{\lambda}_{\nu} - \left\{ {\lambda \atop \nu\ \rho} \right\} T^{\mu}_{\lambda}\,. \end{aligned} \tag{5.1.37}$$

Also, the covariant derivative of tensors of arbitrary rank can be obtained in a similar way.

THEOREM 5.1.1 (RICCI) *The covariant derivative of the metric tensor is equal to zero*

$$g^{\mu\nu}_{;\lambda} = g_{\mu\nu;\lambda} = 0\,. \tag{5.1.38}$$

Differentiating the covariant tensor $V_{\mu;\nu}$ we obtain

$$V_{\mu;\nu\rho} - V_{\mu;\rho\nu} = R^{\sigma}_{\mu\rho\nu} V_\sigma\,, \tag{5.1.39}$$

where **R** is a tensor of rank four, called the *curvature tensor* of the Riemannian space (or the *Riemann-Christoffel tensor*). By means of the relations (5.1.35), (5.1.37) and (5.1.39) we can find the components of the curvature tensor

$$\begin{aligned} R^{\sigma}_{\mu\rho\nu} &= \frac{\partial}{\partial x^\rho} \left\{ {\sigma \atop \mu\ \nu} \right\} - \frac{\partial}{\partial x^\nu} \left\{ {\sigma \atop \mu\ \rho} \right\} \\ &\quad + \left\{ {\lambda \atop \mu\ \nu} \right\} \left\{ {\sigma \atop \lambda\ \rho} \right\} - \left\{ {\lambda \atop \mu\ \rho} \right\} \left\{ {\sigma \atop \lambda\ \nu} \right\}. \end{aligned} \tag{5.1.40}$$

Then, we may *lower* the upper index to obtain the components of the associated covariant tensor

$$R_{\mu\rho\nu\sigma} = g_{\sigma\lambda} R^{\lambda}_{\mu\rho\nu}\,.$$

These covariant components satisfy the relations

$$\begin{aligned} &R_{\mu\rho\nu\sigma} = -R_{\rho\mu\nu\sigma} = -R_{\mu\rho\sigma\nu} = R_{\nu\sigma\mu\rho} , \\ &R_{\rho\lambda\mu\nu} + R_{\rho\mu\nu\lambda} + R_{\rho\nu\lambda\mu} = 0 , \\ &R_{\rho\sigma\lambda\mu;\nu} + R_{\rho\sigma\mu\nu;\lambda} + R_{\rho\sigma\nu\lambda;\mu} = 0 , \end{aligned} \tag{5.1.41}$$

the last two being called *Bianchi's identities*. Therefore, in a Riemannian four-dimensional space the curvature tensor comprises only 20 non-zero independent components.

The contracted tensor

$$R_{\mu\nu} = R^{\sigma}_{\mu\sigma\nu} \tag{5.1.42}$$

is called *Ricci's tensor*, while the invariant

$$R = g^{\mu\nu} R_{\mu\nu} = R^{\mu}_{\mu} \tag{5.1.43}$$

is called *Ricci's scalar* (the *scalar Riemannian curvature*). The quantities (5.1.42) and (5.1.43) are used to build *Einstein's tensor*

$$G_{\mu\nu} = R_{\mu\nu} - \frac{1}{2} g_{\mu\nu} R . \tag{5.1.44}$$

1.1.3 Transformations of the metric tensor

In all mathematical models, classical or quantic, non-relativistic or relativistic, the coordinate transformations that leave invariant the functional form of the metric tensor, are of a particular importance. Such transformations emphasize certain symmetry properties of the geometry of the space, which in turn lead to conservation laws characteristic to these symmetry properties. Thus, in the case of the Riemannian four-dimensional space, the first fundamental form is

$$ds^2 = g_{\mu\nu}(\mathbf{x}) dx^{\mu} dx^{\nu} . \tag{5.1.45}$$

Under a transformation of coordinates of the form (5.1.21), the components of the metric tensor in the two systems of coordinates are related by the formula (5.1.29). In the case of an infinitesimal transformation of coordinates, the transformation (5.1.21) takes the form

$$x'^{\mu} = x^{\mu} + \delta x^{\mu} = x^{\mu} + \varepsilon\, \xi^{\mu}(\mathbf{x}) , \quad \mu = \overline{1,4} , \tag{5.1.46}$$

where ε represents a positive infinitesimal constant, and $\xi^{\mu}(\mathbf{x})$ are functions of class C^1. By substituting (5.1.46) into (5.1.29), and neglecting terms of order ε^2 (such that we may identify the differentiation with respect to x'^{μ} and the differentiation with respect to x^μ) we obtain

$$g'_{\mu\nu}(\mathbf{x}') = \left(\delta^\lambda_\mu - \varepsilon\,\frac{\partial\xi^\lambda}{\partial x'^\mu}\right)\left(\delta^\rho_\nu - \varepsilon\,\frac{\partial\xi^\rho}{\partial x'^\nu}\right) g_{\lambda\rho}(\mathbf{x})$$
$$= g_{\mu\nu}(\mathbf{x}) - \varepsilon\, g_{\lambda\nu}(\mathbf{x})\,\frac{\partial\xi^\lambda}{\partial x^\mu} - \varepsilon\, g_{\mu\rho}(\mathbf{x})\,\frac{\partial\xi^\rho}{\partial x^\nu} + \mathcal{O}(\varepsilon^2)\,. \quad (5.1.47)$$

Now, we substitute into (5.1.47) the Taylor series

$$g'_{\mu\nu}(\mathbf{x}') = g'_{\mu\nu}(\mathbf{x} + \varepsilon\,\boldsymbol{\xi}(\mathbf{x})) = g'_{\mu\nu}(\mathbf{x}) + \frac{\partial g_{\mu\nu}}{\partial x^\lambda}\,\varepsilon\,\xi^\lambda(\mathbf{x}) + \mathcal{O}(\varepsilon^2)$$

where, in the correction terms, we used the relations

$$\left.\frac{\partial g'_{\mu\nu}}{\partial x'^\kappa}\right|_{\varepsilon=0} = \left\{\frac{\partial}{\partial x'^\kappa}\left[\left(\delta^\lambda_\mu - \varepsilon\,\frac{\partial\xi^\lambda}{\partial x'^\mu}\right)\left(\delta^\rho_\nu - \varepsilon\,\frac{\partial\xi^\rho}{\partial x'^\nu}\right) g_{\lambda\rho}(\mathbf{x})\right]\right\}_{\varepsilon=0} = \frac{\partial g_{\mu\nu}}{\partial x^\kappa}\,.$$

Finally, by neglecting terms of order ε^2 or higher, we obtain the variation of the components of the metric tensor in terms of the initial coordinates

$$\delta_* g_{\mu\nu}(\mathbf{x}) = g'_{\mu\nu}(\mathbf{x}) - g_{\mu\nu}(\mathbf{x}) = -\varepsilon\left(\frac{\partial g_{\mu\nu}}{\partial x^\lambda}\,\xi^\lambda + g_{\lambda\nu}\,\frac{\partial\xi^\lambda}{\partial x^\mu} + g_{\mu\rho}\,\frac{\partial\xi^\rho}{\partial x^\nu}\right). \quad (5.1.48)$$

The quantity $\delta_* g_{\mu\nu}(\mathbf{x})$ is called the *Lie differential* of the tensor $\mathbf{g}$, while the quantity

$$\mathsf{L}_{\boldsymbol{\xi}}\, g_{\mu\nu} = \frac{\partial g_{\mu\nu}}{\partial x^\lambda}\,\xi^\lambda + g_{\lambda\nu}\,\frac{\partial\xi^\lambda}{\partial x^\mu} + g_{\mu\rho}\,\frac{\partial\xi^\rho}{\partial x^\nu} \quad (5.1.49)$$

is called the *Lie derivative* of the tensor $\mathbf{g}$ with respect to the vector $\boldsymbol{\xi}$. It can be shown that the operator $\mathsf{L}_{\boldsymbol{\xi}}$ commutes with the operators of partial differentiation and satisfies the Leibnitz formula for product differentiation.

The transformations of coordinates which preserve the functional form of the metric tensor (i.e., $\delta_* g_{\mu\nu}(\mathbf{x}) = 0$ or $\mathsf{L}_{\boldsymbol{\xi}}\, g_{\mu\nu} = 0$) constitute the *group of rigid motions* of the Riemannian space, with the metric tensor $g_{\mu\nu}(\mathbf{x})$. Such transformations that transform the space into itself are also called *isometries* of the considered space. If $\delta_* g_{\mu\nu}(\mathbf{x}) = 0$ then $g_{\mu\nu}(\mathbf{x}) = g'_{\mu\nu}(\mathbf{x})$, or $g_{\mu\nu}(\mathbf{x}') = g'_{\mu\nu}(\mathbf{x}')$, and the vector $\boldsymbol{\xi}$ has to be a solution of the equation

$$\frac{\partial g_{\mu\nu}}{\partial x^\lambda}\,\xi^\lambda + g_{\lambda\nu}\,\frac{\partial\xi^\lambda}{\partial x^\mu} + g_{\mu\rho}\,\frac{\partial\xi^\rho}{\partial x^\nu} = 0\,. \quad (5.1.50)$$

By applying Ricci's theorem (see Theorem 5.1.1) we obtain the equations

$$\xi_{\mu;\nu} + \xi_{\nu;\mu} = 0\,, \quad (5.1.51)$$

called *Killing's equations*. The solutions of these equations are called *Killing's vectors*. The equations (5.1.51) represent the necessary and sufficient conditions for the infinitesimal transformations (5.1.46) to be rigid motions (or isometries) of the Riemannian space with the metric tensor $g_{\mu\nu}(\mathbf{x})$.

1.2 Applications in the theory of Special Relativity

1.2.1 The model of the theory of Special Relativity

The theory of *Special (Restricted) Relativity*, founded by A. Einstein, reformulates the basic principles of classical mechanics and physics, making them adequate to describe phenomena involving velocities comparable with the speed of light. It gives a unified account of the laws of mechanics and electromagnetism. This theory is called *Special* (or *Restricted*) because it is limited to inertial frames of reference (i.e., reference frames in which the first principle of classical mechanics holds) defined in a flat space-time, and in a uniform and rectilinear motion to each other. The elaboration of the theory of Special Relativity was based on a series of theoretical and experimental studies concerning the electrodynamics of moving bodies. In the frame of pre-relativity physics such phenomena were explained in terms of the existence of a hypothetical medium called *ether*. This medium was supposed to fill all space and to support the propagation of electromagnetic waves. Another theory considered the electromagnetic field in vacuum as a special state of the ether. Both interpretations proved to be wrong on the basis of experimental evidence (the Michelson-Morley experiments). Thus, it has been shown that the ether theories of the electromagnetic field, led to contradictory properties of the ether in the same physical conditions. It has also been shown the impossibility of detecting experimentally the uniform motion of an inertial reference frame relative to another inertial reference frame. Consequently, the Michelson-Morley experiments can be explained only by assuming that the speed of light is the same independently of the relative motion of the source, or observer. These experiments have also shown the necessity of considering that the physical laws are invariant with respect to a group of transformations different from the Galilei group. Note that Maxwell equations are not invariant with respect to the transformations in the Galilei group. Therefore, a unified theory of mechanics and electromagnetism should rely on a different symmetry group.

The theory of Special Relativity is based on two postulates:

1. The laws of physics must take the same form with respect to all inertial frames of reference.

2. The speed of light in vacuum has the same constant value with respect to all inertial frames of reference, independently of the relative motions of sources and observers.

Einstein showed that these two postulates were equivalent to the requirement that the coordinates of different inertial frames of reference should be related by a certain group of transformations, namely the *Lorentz group*. In addition, the theory of Special Relativity satisfies the *correspondence principle*. That is, the laws of non-relativistic classical physics have to follow from the corresponding laws of relativistic physics, as a limiting case when all the velocities are much smaller than the speed of light.

The equivalence of mass and energy, time dilatation, and length contraction are the main outcomes of the theory of Special Relativity. Thus, the space and the time are no longer absolute, and become relative and interdependent quantities. Consequently, this new conception about space and time requires a new geometry, in accordance to the theory of Special Relativity. Such a geometry has been founded by H. Minkowski, on the basis of a fictitious four-dimensional space denoted by $\mathcal{M}_4$. This space is formed by the set of all *world points* (or *events*), which have a determined location in space, and take place at a given instant. Hence, the position of a *world point* is defined by four coordinates: three spatial coordinates (x, y, and z), and a temporal coordinate (t). In the Minkowskian geometry these coordinates are denoted by $x^1 = x$, $x^2 = y$, $x^3 = z$, and $x^4 = ict$, where $i = \sqrt{-1}$, and c represents the speed of light. By introducing four coordinates, the interdependency between space and time has been established in terms of separated coordinates that is, separately preserving the physical properties of space and time.

In what follows we will use the real coordinate $x^4 = ct$, instead of the imaginary coordinate, and the metric tensor given by (2.4.4). Thus, the position of a *world point* is determined by the contravariant four-vector

$$(x^\mu) = (x^1, x^2, x^3, x^4) = (\mathbf{r}, ct) = (x, y, z, ct)\,,$$

or by the associated covariant four-vector

$$(x_\mu) = (g_{\mu\nu} x^\nu) = (-\mathbf{r}, ct) = (-x, -y, -z, ct)\,,$$

where $\mathbf{r} = (x, y, z)$ is the spatial (three-dimensional) radius vector. We will also adopt the summation convention for dummy Greek indices from 1 to 4, and for dummy Latin indices from 1 to 3. Consequently, the inner product of two arbitrary four-vectors, $(a^\mu) = (\mathbf{a}, a^4)$ and $(b^\mu) = (\mathbf{b}, b^4)$, can be written in one of the forms

$$g_{\mu\nu} a^\mu b^\nu = a^\mu b_\mu = a_\nu b^\nu = a^k b_k + a^4 b_4 = a_l b^l + a_4 b^4 = -\mathbf{a}\cdot\mathbf{b} + a^4 b_4\,.$$

Note that, $\mathbf{a}$ and $\mathbf{b}$ are the spatial Cartesian components of the two four-vectors.

The *distance* between two world points is defined by the square of the *line element* (also called *interval*)

$$ds^2 = -dx^2 - dy^2 - dz^2 + c^2 dt^2 = g_{\mu\nu} dx^\mu dx^\nu\,, \tag{5.1.52}$$

which, according to (2.4.12), is invariant with respect to the Lorentz transformations.

In classical mechanics, the equations of motion have been obtained by means of Hamilton's principle. In the case of relativistic mechanics, we may use the same variational principle but now, the functional to be minimized has to refer to the world lines. Hence, for a free particle we may assume that the action is proportional to the distance between two given world points a and b, and has the form

$$\mathcal{A} = -\alpha \int_a^b ds\,, \tag{5.1.53}$$

invariant with respect to the Lorentz transformations. Here, α is a constant which characterizes the particle, and the integration is carried out along a world line connecting the two world points a and b. In the definition (5.1.52), we may express the line element in terms of the spatial (three-dimensional) velocity $\mathbf{v}$, which characterizes the propagation of the event between the world points. Consequently, we have

$$ds^2 = c^2\left(1 - \frac{v^2}{c^2}\right)dt^2 = c^2\gamma^2 dt^2\,, \tag{5.1.54}$$

where

$$v^2 = \left(\frac{dx}{dt}\right)^2 + \left(\frac{dy}{dt}\right)^2 + \left(\frac{dz}{dt}\right)^2 = \left(\frac{dx^1}{dt}\right)^2 + \left(\frac{dx^2}{dt}\right)^2 + \left(\frac{dx^3}{dt}\right)^2$$

and

$$\gamma = \sqrt{1 - \frac{v^2}{c^2}} = \sqrt{1 + \frac{1}{c^2}\,v^k v_k}\,,$$

such that (5.1.53) becomes

$$\mathcal{A} = -\alpha c \int_{t_0}^{t_1} \gamma dt\,, \tag{5.1.55}$$

and by comparing with (4.1.4) we obtain the relativistic Lagrangian for a free particle

$$\mathcal{L} = -\alpha c\gamma\,. \tag{5.1.56}$$

To determine the constant α, we employ the correspondence principle, which states that in the limit $v \ll c$ the relativistic Lagrangian (5.1.56) has to take the form of the classical Lagrangian for a free particle $\mathcal{L}_c = m_0 v^2/2$, where m_0 is the mass of the particle in classical mechanics. Thus, we consider the Taylor series of the Lagrangian (5.1.56)

$$\mathcal{L} = -\alpha c + \frac{\alpha v^2}{2c} - \ldots \tag{5.1.57}$$

In classical mechanics, the Lagrangian is invariant with respect to gauge transformations (see Chap. 4 Sec. 1.1.3), that is it is defined up to the total time derivative of an arbitrary function. Therefore, we may omit the constant term

αc in (5.1.57), and by comparing $\mathcal{L}$ with the classical form $\mathcal{L}_c$ it follows that $\alpha = m_0 c$. In conclusion, the relativistic action corresponding to a free particle is

$$\mathcal{A} = -m_0 c \int_a^b ds \,, \tag{5.1.58}$$

and the relativistic Lagrangian has the form

$$\mathcal{L} = -m_0 c^2 \gamma = -m_0 c^2 \sqrt{1 + \frac{1}{c^2}\, v^k v_k} \,. \tag{5.1.59}$$

The spatial momentum of the particle has the components

$$p_k = \frac{\partial \mathcal{L}}{\partial v^k} = -\frac{m_0 v_k}{\gamma} = -m v_k = -m g_{kl} v^l = m v^k, \quad k = 1,2,3 \tag{5.1.60}$$

or, simply

$$\mathbf{p} = \frac{m_0 \mathbf{v}}{\gamma} = m\, \mathbf{v} \,, \tag{5.1.61}$$

where $v^k = dx^k/dt$, and

$$m = \frac{m_0}{\gamma} \tag{5.1.62}$$

is the mass of the particle as measured by an observer moving with a speed v with respect to it. Now, the *rest mass* m_0 (the mass of the particle in classical mechanics) can be defined as being the mass of the particle measured by an observer at rest with respect to the particle ($v = 0$). Note that (5.1.60) takes the classical form $p_k = m_0 v^k$ when $v \ll c$, or equally well when the relative velocity tends to zero ($v \to 0$). Also, the *total mechanical energy* of the particle is given by the well-known relativistic formula

$$E = \frac{\partial \mathcal{L}}{\partial v^k}\, v^k - \mathcal{L} = \mathbf{p} \cdot \mathbf{v} - \mathcal{L} = \frac{m_0 c^2}{\gamma} = m c^2 \,, \tag{5.1.63}$$

from which we deduce the *rest mechanical energy* of the particle

$$E_0 = E \Big|_{v=0} = m_0 c^2 \,. \tag{5.1.64}$$

Finally, by squaring (5.1.61) and (5.1.63), and comparing the results, we obtain the energy-momentum formula

$$E^2 = \mathbf{p}^2 c^2 + m_0^2 c^4 \,. \tag{5.1.65}$$

The Hamiltonian is given by the total mechanical energy expressed in terms of the momenta, that is

$$H = \sqrt{\mathbf{p}^2 c^2 + m_0^2 c^4} \,. \tag{5.1.66}$$

The Minkowskian space is a Riemannian four-dimensional space with the metric tensor (2.4.4). In this space, the variation of the action integral (5.1.53) can be written in the form

$$\begin{aligned}\delta\mathcal{A} &= -m_0 c\delta\int_a^b \sqrt{g_{\mu\nu}dx^\mu dx^\nu} = -m_0 c\int_a^b \frac{g_{\mu\nu}dx^\mu\delta dx^\nu}{\sqrt{g_{\mu\nu}dx^\mu dx^\nu}}\\ &= -m_0 c\int_a^b g_{\mu\nu}\frac{dx^\mu}{ds}\delta dx^\nu \qquad (5.1.67)\end{aligned}$$

and Hamilton's principle implies $\delta\mathcal{A} = 0$. The motion of a particle in the three-dimensional space is determined by the parametric equations

$$x = x(t)\,, \quad y = y(t)\,, \quad z = z(t)\,, \qquad (5.1.68)$$

and to each pair position-time it corresponds a world point $(\mathbf{r}, ct) \equiv (x^\mu) \in \mathcal{M}_4$. When t varies in an interval $t \in [t_0, t_1]$, the world point (5.1.68) follows a curve in $\mathcal{M}_4$, called the *world line* of the considered particle. The parametric equations of the world line can be obtained by means of a parameter τ, which is a relativistic invariant, and depends on time. Thus we have

$$\mathbf{r} = \mathbf{r}(\tau)\,, \quad x^4 = x^4(\tau)\,, \qquad (5.1.69)$$

and we may choose the parameter τ as the solution of $d\tau^2 = ds^2/c^2 > 0$, which is real, and satisfies the condition of invariance. Hence, for a given mechanical system we may write

$$\tau = \int_{t_0}^{t_1} \gamma dt\,.$$

In such a case, the four-vector

$$u^\mu = \frac{dx^\mu}{d\tau}\,, \quad \mu = \overline{1,4}\,, \qquad (5.1.70)$$

is tangent to the world line and represents a generalisation of the velocity in $\mathcal{M}_4$. From $ds^2 = c^2 d\tau^2$ it follows the property $g_{\mu\nu}u^\mu u^\nu = c^2$.

Taking into account the proportionality between $d\tau$ and ds we may also choose the arc length as parameter in the equations of the world line, which become

$$\mathbf{r} = \mathbf{r}(s)\,, \quad x^4 = x^4(s)\,. \qquad (5.1.71)$$

Now, the generalized velocity has the components

$$u^\mu = \frac{dx^\mu}{ds}\,, \quad \mu = \overline{1,4}\,, \qquad (5.1.72)$$

and satisfies the relation $g_{\mu\nu}u^\mu u^\nu = 1$. Then, by substituting (5.1.72) into (5.1.67) the variation of the action takes the form

$$\delta\mathcal{A} = -m_0 c \int_a^b g_{\mu\nu} u^\mu d\delta x^\nu \,, \tag{5.1.73}$$

such that an integration by parts gives

$$\delta\mathcal{A} = -m_0 c g_{\mu\nu} u^\mu \delta x^\nu \Big|_a^b + m_0 c \int_a^b g_{\mu\nu} \delta x^\nu \frac{du^\mu}{ds} ds \,. \tag{5.1.74}$$

The arbitrary world lines for the variation $\delta\mathcal{A}$ have fixed ends, that is $(\delta x^\nu)_a = (\delta x^\nu)_b = 0$, and the condition $\delta\mathcal{A} = 0$ leads us to the motion equations $du^\mu/ds = 0$. Hence, the world point associated to a free particle has a uniform motion in $\mathcal{M}_4$.

Note that the action $\mathcal{A}$ can be expressed in a form similar to (4.1.45)

$$\mathcal{A} = S\Big|_a^b \,, \tag{5.1.75}$$

where S is a primitive of the integrand in (5.1.73), such that from (5.1.74) and the equations of motion ($du^\mu/ds = 0$) we obtain

$$\delta S = -m_0 c g_{\mu\nu} u^\mu \delta x^\nu = m_0 c \left(u^1 \delta x^1 + u^2 \delta x^2 + u^3 \delta x^3 - u^4 \delta x^4\right) . \tag{5.1.76}$$

Then, by generalizing the relation $\mathbf{p} = \nabla S$, which follows from (4.1.46), we write the components of the *four-momentum* vector as

$$p_\nu = \frac{\partial S}{\partial x^\nu} = -m_0 c g_{\mu\nu} u^\mu \,, \quad \nu = \overline{1,4} \,, \tag{5.1.77}$$

or, separating the three-dimensional components

$$p_k = m_0 c u^k = m_0 c \frac{dx^k}{c\gamma\, dt} = \frac{m_0 v^k}{\gamma} \,, \tag{5.1.78}$$

$$p_4 = -m_0 c u^4 = -\frac{m_0 c}{\gamma} = -\frac{1}{c} E \,, \tag{5.1.79}$$

where $k = 1, 2, 3$. Thus, it follows that the momentum and the total mechanical energy constitute the components of a four-dimensional vector. The fourth coordinate in $\mathcal{M}_4$ is $x^4 = ct$, such that

$$\frac{\partial S}{\partial x^4} = \frac{1}{c}\frac{\partial S}{\partial t} = -\frac{1}{c} E \,,$$

and we obtain

$$\frac{\partial S}{\partial t} = -E \,. \tag{5.1.80}$$

Finally, by substituting (5.1.78) into (5.1.66), and replacing $H = E = -\partial S/\partial t$, we obtain the relativistic Hamilton-Jacobi equation

$$g_{\mu\nu}\frac{\partial S}{\partial x^{\mu}}\frac{\partial S}{\partial x^{\nu}} - m_0^2 c^2 = 0\,. \tag{5.1.81}$$

1.2.2 Conservation laws inferred from Lorentz invariance

In the theory of Special Relativity the time is a coordinate in the usual sense, while in the classical physics it is used as a parameter. Thus, some of the conserved quantities associated to the invariance of a physical system with respect to Lorentz transformations are not, in fact, *conserved* because they depend on the space-time coordinates. It is also interesting to note that such physical quantities associated to the Lorentz transformations have not been given a name, like in the case of similar quantities which have an equivalent in classical mechanics.

The relativistic Lagrangian for a free particle (5.1.59) is invariant with respect to the Lorentz transformations (see Chap. 2 Sec. 4.1). This Lagrangian does not directly depend on the coordinates, that is

$$\frac{\partial \mathcal{L}}{\partial x^{\mu}} = 0\,, \quad \mu = \overline{1,4}\,, \tag{5.1.82}$$

and this results in the conservation of the four-momentum vector, respectively the conservation of the three-dimensional momentum and mechanical energy. We are led to the same result if we make $\delta\mathcal{A} = 0$ in (5.1.74). Thus, from the equations of motion we obtain $u^{\mu} = \text{constant}$, such that $p_{\nu} = -mcg_{\mu\nu}u^{\mu} = \text{constant}$.

The group L^{+} (the group of proper Lorentz transformations) comprises rotations of $\mathcal{M}_4$, which have the infinitesimal form

$$x'^{\mu} = x^{\mu} + x^{\nu}\delta\theta^{\mu}_{\nu}\,, \quad \mu = \overline{1,4}\,, \tag{5.1.83}$$

depending on the infinitesimal coefficients $\delta\theta^{\mu}_{\nu}$. From the condition of invariance of intervals with respect to the Lorentz transformations ($g_{\mu\nu}x'^{\mu}x'^{\nu} = g_{\mu\nu}x^{\mu}x^{\nu}$), it follows that these coefficients represent the components of a skew-symmetric tensor

$$\delta\theta^{\mu}_{\nu} = -\delta\theta^{\nu}_{\mu}\,. \tag{5.1.84}$$

Now, an infinitesimal transformation of the form (5.1.83) induces a change of the action S in (5.1.76) of the form

$$\delta S = p_{\mu}\delta x^{\mu} = p_{\mu}x^{\nu}\delta\theta^{\mu}_{\nu}\,. \tag{5.1.85}$$

Here, $p_{\mu}x^{\nu}$ form the components of a mixed tensor which can be decomposed into a sum of a symmetric and a skew-symmetric tensor. As the product of a symmetric tensor and a skew-symmetric tensor vanishes, we may keep only the skew-symmetric part of $p_\mu x^\nu$ that is

$$\delta S = \frac{1}{2}\left(p_\mu x^\nu - p_\nu x^\mu\right)\delta\theta^\mu_\nu\,. \tag{5.1.86}$$

The quantities

$$M^\nu_\mu = x^\nu p_\mu - x^\mu p_\nu \tag{5.1.87}$$

represent the components of a skew-symmetric tensor called the *angular four-momentum* tensor. The Lagrangian (5.1.59) is invariant with respect to the Lorentz transformations, hence the infinitesimal coefficients $\delta\theta^\mu_\nu$ are cyclic coordinates, and the quantities $\delta S/\delta\theta^\mu_\nu = M^\nu_\mu$ are conserved. The spatial components of M^ν_μ are identical to the components of the three-dimensional angular momentum $\mathbf{K} = \mathbf{r} \times \mathbf{p}$, while the components M^k_4 $(k = 1, 2, 3)$ form a vector

$$c\left(t\,\mathbf{p} - \frac{1}{c^2}\,E\,\mathbf{r}\right) = \mathbf{const}\,. \tag{5.1.88}$$

In the case of a discrete mechanical system of non-interacting particles, the momentum and mechanical energy associated to each particle are independent of time, and accordingly the four-momentum associated to each particle is conserved. It follows that the total four-momentum of the mechanical system is also conserved. The total angular four-momentum becomes

$$M^\nu_\mu = \sum\left(x^\nu p_\mu - x^\mu p_\nu\right)\,,$$

where the summation extends over all the particles in the mechanical system. On the basis of the invariance of the Lagrangian with respect to the Lorentz transformations, it can be shown that the total angular four-momentum is conserved during the evolution of the mechanical system. Now, the components M^k_4 $(k = 1, 2, 3)$ take the form

$$c\left[t\sum\mathbf{p} - \frac{1}{c^2}\sum(E\,\mathbf{r})\right] = \mathbf{const}\,. \tag{5.1.89}$$

As the total mechanical energy of the discrete system $(\sum E)$ is conserved, we may write (5.1.89) in the form

$$\frac{\sum(E\mathbf{r})}{\sum E} - \frac{c^2\sum\mathbf{p}}{\sum E}\,t = \mathbf{const}\,, \tag{5.1.90}$$

which shows that a "particle" having the position

$$\boldsymbol{\xi} = \frac{\sum(E\mathbf{r})}{\sum E}\,, \tag{5.1.91}$$

experiences a uniform motion with the velocity

$$\mathbf{v} = \frac{c^2 \sum \mathbf{p}}{\sum E} . \tag{5.1.92}$$

The relation (5.1.91) represents the relativistic definition of the *centre of mass of the discrete mechanical system*, because in the case when all the velocities are much smaller than the speed of light $E \cong m_0 c^2$, and we are led to the classical definition of the centre of mass

$$\boldsymbol{\xi} \to \frac{\sum (m_0 \mathbf{r})}{\sum m_0} . \tag{5.1.93}$$

In this limit the relation (5.1.92) also becomes

$$\mathbf{v} \to \frac{\sum \mathbf{p}}{\sum m_0} = \frac{\mathbf{P}}{M_0} , \tag{5.1.94}$$

where $\mathbf{P}$ and M_0 represent, respectively, the total momentum and the total mass of the mechanical system. Finally, the relation (5.1.90) takes the form

$$\boldsymbol{\xi} - \frac{\mathbf{P}}{M_0} t = \boldsymbol{\xi}_0 = \textbf{const} , \tag{5.1.95}$$

that is the *theorem of the centre of mass motion* (4.3.28).

1.2.3 General forms of conservation laws for a free particle

The results in the previous section can also be obtained from a general method based on the invariance of the metric tensor with respect to rigid motions (or isometries) of the Riemannian space (see Chap 5 Sec. 1.1.3). For the sake of clarity we will begin by presenting the use of this method in the classical case of a free particle in the Euclidean three-dimensional space. Hence, the geometric space is $\mathcal{E}_3$, and the time is an absolute scalar, considered to be a parameter (independent variable) which describes the evolution of the mechanical system. We consider a system of curvilinear coordinates in $\mathcal{E}_3$, such that the line element is given by the formula

$$ds^2 = g_{ij}(\mathbf{x}) dx^i dx^j , \tag{5.1.96}$$

where we used the summation convention for dummy Latin indices from 1 to 3. Note that the metric tensor determines the geometry in the considered system of coordinates, while the motion of the particle is described by the Lagrange equations

$$\frac{d}{dt}\left(\frac{\partial \mathcal{L}}{\partial \dot{x}^k}\right) - \frac{\partial \mathcal{L}}{\partial x^k} = 0 , \quad k = 1, 2, 3. \tag{5.1.97}$$

According to Noether's theorem (Theorem 4.1.5) the invariance of the functional form of the Lagrangian $\mathcal{L}(x^k, \dot{x}^k)$ with respect to a group of transformations determines the maximum number of conservation laws, which in turn are directly related to the first integrals of Lagrange's equations.

Let

$$x'^i = x^i + \varepsilon\,\xi^i(\mathbf{x})\,, \quad i = 1, 2, 3, \tag{5.1.98}$$

be an infinitesimal transformation of the form (5.1.46), that leaves invariant the functional form of the Lagrangian (symmetry transformation). The time is assumed to be an absolute scalar ($\delta t = 0$), and by means of Lagrange's equations (5.1.97), we obtain

$$\begin{aligned}\delta\mathcal{L} &= \frac{\partial \mathcal{L}}{\partial x^k}\,\delta x^k + \frac{\partial \mathcal{L}}{\partial \dot{x}^k}\,\delta \dot{x}^k = \frac{d}{dt}\left(\frac{\partial \mathcal{L}}{\partial \dot{x}^k}\right)\delta x^k + \frac{\partial \mathcal{L}}{\partial \dot{x}^k}\,\delta \dot{x}^k \\ &= \frac{d}{dt}\left(\frac{\partial \mathcal{L}}{\partial \dot{x}^k}\,\delta x^k\right) = \frac{d}{dt}\left(\frac{\partial \mathcal{L}}{\partial \dot{x}^k}\,\varepsilon\,\xi^k\right).\end{aligned} \tag{5.1.99}$$

Note that this expression is identical with the variation of the Lagrangian (4.1.105) if we set $\delta q_k = \varepsilon \xi^k$.

Let us now assume that we have found r independent vectors $\boldsymbol{\xi}_{(\alpha)}$ ($\alpha = 1, 2, \ldots, r$), which determine the variations $\delta x^k_{(\alpha)} = \varepsilon \xi^k_{(\alpha)}$ associated to the invariance of the functional form of the Lagrangian under the transformations (5.1.98), that is $\delta\mathcal{L} = 0$. Consequently, we obtain r conservation laws

$$\frac{d}{dt}\left(\frac{\partial \mathcal{L}}{\partial \dot{x}^k}\,\xi^k_{(\alpha)}\right) = 0\,, \quad \alpha = 1, 2, \ldots, r\,, \tag{5.1.100}$$

corresponding to the integrals of motion

$$D_{(\alpha)}(x^k, \dot{x}^k) = \frac{\partial \mathcal{L}}{\partial \dot{x}^k}\,\xi^k_{(\alpha)} = p_k\,\xi^k_{(\alpha)} = C_{(\alpha)} = \text{const}\,, \tag{5.1.101}$$

which are similar to the integrals of motion (4.1.106). The quantity $C_{(\alpha)}$ in (5.1.101) is called the *generator of the conservation law*.

In the considered curvilinear system of coordinates the velocity of a particle is $v = ds/dt$, and so the Lagrangian has the form

$$\mathcal{L} = T - V = \frac{m_0}{2}\left(\frac{ds}{dt}\right)^2 - V = \frac{m_0}{2}\,g_{ij}(\mathbf{x})\,\frac{dx^i}{dt}\,\frac{dx^j}{dt} - V\,. \tag{5.1.102}$$

Then, if the function V does not depend on velocities, the relation (5.1.101) becomes

$$m_0\,g_{ij}(\mathbf{x})\,\frac{dx^i}{dt}\,\xi^j_{(\alpha)} = C_{(\alpha)}\,, \quad \alpha = 1, 2, \ldots, r\,. \tag{5.1.103}$$

The metric tensor is symmetric, and for a free particle $V(\mathbf{x}) = 0$ (therefore $\ddot{x}^k = 0$). Hence, by differentiating (5.1.103) with respect to the time variable we obtain the equation

$$\frac{d}{dt}\left(m_0\, g_{ij}\,\frac{dx^i}{dt}\,\xi^j\right) = \frac{1}{2}\,m_0\,(\xi_{i;j} + \xi_{j;i})\,\frac{dx^i}{dt}\,\frac{dx^j}{dt} = 0\,, \tag{5.1.104}$$

where we have omitted the index α. Note the use of the covariant derivative that is required in a system of curvilinear coordinates. It follows that $p_k\xi^k_{(\alpha)}$ is an integral of motion if $\xi^k_{(\alpha)}$ is a solution of the Killing equation (5.1.51). In this case the symmetry transformations of the Lagrangian are isometries that leave invariant the metric tensor, and reciprocally. In Cartesian coordinates $g_{ij} = \delta_{ij}$, and so there is no difference between covariant and contravariant components. Thus, the general solution of the Killing equation is the linear function

$$\xi_i = g_{ij}\xi^j = \omega_{ij}x^j + \kappa_i\,, \quad i = 1, 2, 3, \tag{5.1.105}$$

where $\omega_{ij} = -\omega_{ji}$ and κ_i are integration constants, which can be regarded as independent parameters. By introducing the pseudovector $\boldsymbol{\omega}$ of components $\omega_i = \epsilon_{ijk}\omega_{jk}/2$ $(i = 1, 2, 3)$, we may write the Killing vectors in the form

$$\xi_i = \epsilon_{ijk}x^j\omega_k + \kappa_i\,, \tag{5.1.106}$$

which depends on six independent parameters ω_k and κ_i. Thus, by setting to zero all the parameters except one of them, we obtain six independent vectors $\boldsymbol{\xi}_{(\alpha)}$, corresponding to six integrals of motion

$$m_0\, g_{ij}\,\frac{dx^i}{dt}\,\xi^j_{(\alpha)} = C_{(\alpha)}\,, \quad \alpha = 1, 2, \ldots, 6,. \tag{5.1.107}$$

The first three integrals of motion

$$m_0\,\epsilon_{ijk}\,x^j\,\frac{dx^i}{dt} = C_{(k)}\,, \quad \omega_k \neq 0\,, \quad k = 1, 2, 3, \tag{5.1.108}$$

represent the conservation of the angular momentum, while the remaining three integrals of motion

$$m_0\,\frac{dx^i}{dt} = C_{(3+i)}\,, \quad \kappa_i \neq 0\,, \quad i = 1, 2, 3, \tag{5.1.109}$$

correspond to the conservation of the linear momentum.

Note that this method is restricted to symmetry transformations that leave invariant both the Lagrangian and the metric tensor (rigid motions of $\mathcal{E}_3$). Hence, we have obtained only the conservation laws associated to geometrical properties of the Euclidean three-dimensional space. We cannot derive, for instance, the conservation of total energy or the theorem of the centre of mass motion.

However, this method provides the complete set of conservation laws corresponding to the space-time geometrical properties, in the frame of relativistic theories.

Now, we will consider the problem of a free particle in the theory of Special Relativity. In terms of the parametrization (5.1.71) the motion of the particle in $\mathcal{M}_4$ is described by the Lagrange equations

$$\frac{d}{ds}\left(\frac{\partial \mathcal{L}}{\partial u^\mu}\right) - \frac{\partial \mathcal{L}}{\partial x^\mu} = 0\,, \quad \mu = \overline{1,4}\,, \tag{5.1.110}$$

with the Lagrangian

$$\mathcal{L} = \mathcal{L}(s, x^\mu, u^\mu).$$

Under an infinitesimal transformation of coordinates of the form

$$x'^\mu = x^\mu + \varepsilon\,\xi^\mu(\mathbf{x})\,, \quad \mu = \overline{1,4}\,, \tag{5.1.111}$$

assuming $\delta s = 0$ (rigid motions of $\mathcal{M}_4$), the variation of the Lagrangian can be written as

$$\begin{aligned}\delta\mathcal{L} &= \frac{d}{ds}\left(\frac{\partial \mathcal{L}}{\partial u^\mu}\,\varepsilon\,\xi^\mu\right) = \frac{d}{ds}\,(p_\mu\,\varepsilon\,\xi^\mu) = -\frac{d}{ds}\,(m_0 c\, g_{\nu\mu} u^\nu\,\varepsilon\,\xi^\mu) \\ &= -\frac{d}{ds}\left(m_0 c\, g_{\nu\mu}\,\frac{dx^\nu}{ds}\,\varepsilon\,\xi^\mu\right),\end{aligned} \tag{5.1.112}$$

where we have used the expression (5.1.77) for the four-momentum of the free particle. Hence, the generic equation of conservation given by $\delta\mathcal{L} = 0$ reads

$$\frac{d}{dt}\left(m_0 c\, g_{\mu\nu}\,\frac{dx^\mu}{ds}\,\xi^\nu\right) = \frac{1}{2}\,m_0 c\,(\xi_{\mu;\nu} + \xi_{\nu;\mu})\,\frac{dx^\mu}{ds}\,\frac{dx^\nu}{ds} = 0\,. \tag{5.1.113}$$

It follows that, like in the case of classical mechanics, the Lagrangian and the metric tensor are both invariant relative to the rigid motions of $\mathcal{M}_4$, defined by the solutions of the Killing equation (5.1.51). For a system of Cartesian coordinates, the Killing vectors are given by the relations

$$\xi_\mu = \omega_{\mu\nu} x^\nu + \kappa_\mu\,, \quad \mu = \overline{1,4}\,, \tag{5.1.114}$$

depending on ten independent parameters. In the same way as in the three-dimensional case we may replace the parameters $\omega_{\mu\nu} = -\omega_{\nu\mu}$ by a set of six parameters ω_λ according to the relations

$$\begin{bmatrix} 0 & \omega_{12} & \omega_{13} & \omega_{14} \\ -\omega_{12} & 0 & \omega_{23} & \omega_{24} \\ -\omega_{13} & -\omega_{23} & 0 & \omega_{34} \\ -\omega_{14} & -\omega_{24} & -\omega_{34} & 0 \end{bmatrix} = \begin{bmatrix} 0 & \omega_3 & -\omega_2 & -\omega_4 \\ -\omega_3 & 0 & \omega_1 & -\omega_5 \\ \omega_2 & -\omega_1 & 0 & -\omega_6 \\ \omega_4 & \omega_5 & \omega_6 & 0 \end{bmatrix}. \tag{5.1.115}$$

To simplify the notations, we also replace the parameters κ_μ by $\omega_{\mu+6} = \kappa_\mu$ $(\mu = \overline{1,4})$. The ten linearly independent integrals of motion are obtained by setting to zero all the parameters except one of them, in the generic equation of conservation (5.1.113). Replacing $x^4 = ct$ and observing that

$$\frac{dx^\mu}{ds} = \frac{dx^\mu}{dt}\frac{dt}{ds} = \frac{1}{c\gamma}\frac{dx^\mu}{dt}, \qquad \gamma = \sqrt{1-\frac{v^2}{c^2}},$$

one obtains ten integrals of motion in the form

$$\begin{aligned} \frac{m_0}{c\gamma}\,\epsilon_{ijk}\,x^j\,v^k &= C_{(i)}\,, \\ \frac{m_0}{\gamma}(x^i - v^i t) &= C_{(i+3)}\,, \\ \frac{m_0}{c\gamma}\,v^i &= C_{(i+6)}\,, \end{aligned} \tag{5.1.116}$$

for $i = 1,2,3$, and

$$\frac{m_0}{\gamma} = C_{(10)}\,. \tag{5.1.117}$$

From the corresponding relativistic definitions, it follows that $c\,C_{(1)}$, $c\,C_{(2)}$, and $c\,C_{(3)}$ are the components of the angular momentum, while $c\,C_{(7)}$, $c\,C_{(8)}$, and $c\,C_{(9)}$ are the components of the linear momentum, for a free particle of mass m_0. Hence these integrals of motion are the relativistic equivalents of (5.1.108) and (5.1.109). In concordance with $\gamma =$ const., corresponding to $C_{(10)}$, the second group of integrals of motion in (5.1.116) indicates that the particle has a rectilinear motion. These integrals of motion represent a finite Lorentz transformation

$$x'^i = \frac{x^i - v^i t}{\gamma}, \qquad i = 1,2,3, \tag{5.1.118}$$

which relates the motion of a particle as seen in the reference frames x'^i and x^i; in the reference frame x'^i, attached to the particle (reference frame that is moving with a constant speed $\mathbf{v}$ with respect to the reference frame x^i) the motion of the particle is uniform.

The last integral of motion (5.1.117) has a special meaning. In the limit $v \ll c$, all the terms of order higher than $(v/c)^2$ in the series expansion of $C_{(10)}$ are negligible, and the integral of motion (5.1.117) leads to

$$m_0\left(1 + \frac{1}{2}\frac{v^2}{c^2}\right) \cong C_{(10)}\,, \tag{5.1.119}$$

wherefrom

$$m_0 c^2 + \frac{1}{2}m_0 v^2 \cong c^2\,C_{(10)}\,. \tag{5.1.120}$$

Consequently, *the total energy of a free particle is the sum of the rest energy, contained in the structure of the particle itself, and the classical kinetic energy.* If we denote $C_{(10)} = m$, then

$$m = \frac{m_0}{\gamma} = \frac{m_0}{\sqrt{1 - v^2/c^2}},$$

can be regarded as the mass of the moving particle, such that

$$m_0 c^2 + \frac{1}{2} m v^2 \cong m c^2.$$

Therefore, the mass energy relation $E = mc^2$ can be obtained as a linear integral of motion for the free particle, derived from the symmetry properties of the four-dimensional space-time.

2. Theory of electromagnetic field

2.1 Noether's theorem for the electromagnetic field

2.1.1 Lagrangian formulation of Maxwell's equations

The theory of Special Relativity is essentially a theory of the electromagnetic field generated by moving bodies, that carry an electric charge. This electromagnetic field is characterized by two vectors: $\mathbf{E}$ (the *intensity of the electric field*) and $\mathbf{B}$ (the *density of the magnetic flux*), which depend on the spatial position and time, and are determined by the Maxwell equations which, in vacuum, have the form

$$\begin{aligned} &\nabla \times \mathbf{B} = \mu_0 \mathbf{j} + \mu_0 \varepsilon_0 \frac{\partial \mathbf{E}}{\partial t}, \quad \nabla \cdot \mathbf{E} = \frac{\rho}{\varepsilon_0}, \\ &\nabla \times \mathbf{E} = -\frac{\partial \mathbf{B}}{\partial t}, \quad \nabla \cdot \mathbf{B} = 0, \end{aligned} \tag{5.2.1}$$

$$\mathbf{H} = \frac{1}{\mu_0} \mathbf{B}, \quad \mathbf{D} = \varepsilon_0 \mathbf{E}. \tag{5.2.2}$$

Here, $\mathbf{j}$ represents the *current density*, ρ is the *volume density of the electric charge*, ε_0 and μ_0 are, respectively, the *electric permittivity* and the *magnetic permeability* of the vacuum, $\mathbf{H}$ is the intensity of the magnetic field, and $\mathbf{D}$ is the *electric displacement*. Note that the speed of light in vacuum is given by $c = 1/\sqrt{\varepsilon_0 \mu_0}$. The two equations in (5.2.2) are called *material equations*.

Similarly to the case of Lagrange's equations in classical mechanics, the Maxwell equations (5.2.1) can be obtained from Hamilton's principle, by means of the Lagrangian action. We thus consider the motion of a charged particle in an electrostatic field. The force acting upon the particle has the form

$$\mathbf{F} = q \mathbf{E} = -q \nabla \varphi(\mathbf{r}),$$

where q is the electric charge of the particle, and $\varphi(\mathbf{r})$ represents the *electrostatic potential*. Also, the motion of the particle in this force field is described by the equation

$$\frac{d\mathbf{p}}{dt} = -q\,\nabla\varphi(\mathbf{r}) \,, \tag{5.2.3}$$

$\mathbf{p}$ being the linear momentum of the particle. It is easy to show that the Lagrangian action for the equation of motion (5.2.3) is

$$\mathcal{A} = \mathcal{A}_0 + \mathcal{A}_{\text{int}} \,, \tag{5.2.4}$$

where

$$\mathcal{A}_0 = -m_0\,c \int_a^b ds \,, \tag{5.2.5}$$

with m_0 being the rest mass of the particle, and

$$\mathcal{A}_{\text{int}} = -\int_{t_a}^{t_b} q\,\varphi(\mathbf{r})\,dt = -\frac{1}{c}\,q \int_a^b \varphi(\mathbf{x})\,dx^4 \,, \tag{5.2.6}$$

which brings out the Lagrangian action corresponding to the interaction of the charged particle with the electromagnetic field. From (5.2.1) it follows that the vectors $\mathbf{E}$ and $\mathbf{B}$ can be expressed in terms of a *scalar potential* $\varphi(\mathbf{r};t)$ and a vector potential $\mathbf{A}(\mathbf{r};t)$, according to the relations

$$\begin{aligned} \mathbf{E} &= -\nabla\,\varphi(\mathbf{r};t) - \frac{\partial}{\partial t}\,\mathbf{A}(\mathbf{r};t)\,, \\ \mathbf{B} &= \nabla \times \mathbf{A}(\mathbf{r};t)\,. \end{aligned} \tag{5.2.7}$$

If we introduce the four-vector potential (A_μ), $\mu = \overline{1,4}$, having the spatial components equal to the components of the vector potential $-\mathbf{A}(\mathbf{r};t)$, the fourth component being $A_4 = \varphi(\mathbf{r};t)/c$ then, by generalizing (5.2.6), we may write

$$\mathcal{A}_{\text{int}} = -q \int_a^b A_\mu\,dx^\mu \,. \tag{5.2.8}$$

Hence, the Lagrangian action corresponding to the motion of a charged particle in a time-dependent electromagnetic field has the general form

$$\mathcal{A} = -\int_a^b (m_0\,c\,ds + q\,A_\mu\,dx^\mu)\,, \tag{5.2.9}$$

or

$$\begin{aligned} \mathcal{A} &= \int_a^b (-m_0\,c\,ds + q\,\mathbf{A}\cdot d\mathbf{r} - q\,\varphi\,dt) \\ &= \int_{t_a}^{t_b} \left[-m_0\,c^2\gamma + q(\mathbf{A}\cdot\mathbf{v} - \varphi)\right] dt\,. \end{aligned} \tag{5.2.10}$$

Consequently, the Lagrangian of this physical system is

$$\mathcal{L} = -m_0\, c^2 \gamma + q(\mathbf{A} \cdot \mathbf{v} - \varphi)\,, \tag{5.2.11}$$

where $\gamma = \sqrt{1 - v^2/c^2}$.

We may now obtain the generalized momentum of the particle

$$\mathbf{P} = \left(\frac{\partial \mathcal{L}}{\partial v^1}, \frac{\partial \mathcal{L}}{\partial v^2}, \frac{\partial \mathcal{L}}{\partial v^3}\right) = \frac{m_0 \mathbf{v}}{\gamma} + q\,\mathbf{A} = \mathbf{p} + q\,\mathbf{A}\,, \tag{5.2.12}$$

where $\mathbf{p}$ is the linear momentum of the free particle (5.1.61). It follows that the total energy of the physical system has the form

$$E = \mathbf{P} \cdot \mathbf{v} - \mathcal{L} = \frac{m_0\, c^2}{\gamma} + q\,\varphi\,. \tag{5.2.13}$$

It is thus evident that, by substituting $\mathbf{p} = \mathbf{P} - q\mathbf{A}$ into (5.1.66) and adding the electrostatic energy $q\varphi$, we may write the Hamiltonian of the physical system as

$$H = \sqrt{(\mathbf{P} - q\,\mathbf{A})^2 c^2 + m_0^2 c^4} + q\,\varphi\,. \tag{5.2.14}$$

Finally, replacing into (5.2.14) the generalized momenta P_k by $\partial S/\partial x^k$, $k = 1, 2, 3$, and the Hamiltonian H by $-\partial S/\partial t$, we get the relativistic Hamilton-Jacobi equation

$$(\nabla S - q\,\mathbf{A})^2 - \frac{1}{c^2}\left(\frac{\partial S}{\partial t} + q\,\varphi\right)^2 + m_0^2\, c^2 = 0\,. \tag{5.2.15}$$

Hamilton's principle for the Lagrangian action (5.2.9) can be expressed in the form

$$\delta\,\mathcal{A} = -\delta \int_a^b (m_0\, c\, ds + q\, A_\mu\, dx^\mu) = 0 \tag{5.2.16}$$

and, by means of $ds^2 = g_{\mu\nu} dx^\mu dx^\nu$, we obtain

$$\begin{aligned}
\delta\,\mathcal{A} &= -\int_a^b [m_0\, c\,\delta\, ds + q\,\delta(A_\mu\, dx^\mu)] \\
&= -\int_a^b \left(m_0\, c\, g_{\mu\nu}\, \frac{dx^\mu}{ds}\, d\delta x^\nu + q\,\delta A_\mu\, dx^\mu + q\, A_\mu\, d\delta x^\mu\right) \\
&= -\int_a^b [(m_0\, c\, g_{\mu\nu}\, u^\mu + q\, A_\nu) d\delta x^\nu + e\,\delta A_\mu\, dx^\mu]\,.
\end{aligned}$$

Then, we integrate by parts the first term

$$\begin{aligned}
\delta\,\mathcal{A} = \int_a^b &(m_0\, c\, du_\mu \delta x^\mu + q\, dA_\mu\, \delta x^\mu - q\,\delta A_\mu\, dx^\mu) \\
&- (m_0\, c\, u_\mu + q\, A_\mu)\, \delta x^\mu \Big|_a^b = 0\,,
\end{aligned} \tag{5.2.17}$$

and substitute

$$\delta A_\mu = \frac{\partial A_\mu}{\partial x^\nu}\,\delta x^\nu\,, \quad dA_\mu = \frac{\partial A_\mu}{\partial x^\nu}\,dx^\nu\,.$$

The last term in (5.2.17) vanishes because $(\delta x^\mu)_a = (\delta x^\mu)_b = 0$, so that we have

$$\int_a^b \left[m_0\,c\,du_\mu - q\left(\frac{\partial A_\nu}{\partial x^\mu} - \frac{\partial A_\mu}{\partial x^\nu}\right)dx^\nu\right]\delta x^\mu = 0\,,$$

where δx^μ are independent arbitrary variations. Consequently, the integrand vanishes and we thus obtain the motion equations

$$m_0\,c\,\frac{du_\mu}{ds} = q\left(\frac{\partial A_\nu}{\partial x^\mu} - \frac{\partial A_\mu}{\partial x^\nu}\right)u^\nu\,. \tag{5.2.18}$$

Now, we may introduce the *tensor of the electromagnetic field* defined by the relations

$$F_{\mu\nu} = \frac{\partial A_\nu}{\partial x^\mu} - \frac{\partial A_\mu}{\partial x^\nu}\,, \quad \mu,\nu = \overline{1,4}\,. \tag{5.2.19}$$

Taking into account the definition of the components of the four-vector potential A_μ, and the relations (5.2.7), it follows that this tensor can be represented in the form of the skew-symmetric matrix

$$[F_{\mu\nu}] = \begin{bmatrix} 0 & B_3 & -B_2 & E_1/c \\ -B_3 & 0 & B_1 & E_2/c \\ B_2 & -B_1 & 0 & E_3/c \\ -E_1/c & -E_2/c & -E_3/c & 0 \end{bmatrix}. \tag{5.2.20}$$

The components of the vectors $\mathbf{E}$ and $\mathbf{B}$ are thus related to the components of the electromagnetic field tensor by the formulae

$$\begin{aligned} E_j &= c\,F_{j4} = -c\,F_{4j}\,, \quad j = 1,2,3, \\ B^j &= \frac{1}{2}\,\epsilon^{jkl}\,F_{kl}\,, \quad F_{kl} = \epsilon_{jkl}\,B^j\,, \quad j,k,l = 1,2,3. \end{aligned} \tag{5.2.21}$$

One can use the components $F_{\mu\nu}$ to form quantities that are invariant with respect to the Lorentz transformations. This invariants are the coefficients of the secular equation $\det\left(F_{\mu\nu} - \lambda\,\delta_{\mu\nu}\right) = 0$, which has the explicit form

$$\lambda^4 + \left(\mathbf{B}^2 - \frac{1}{c^2}\,\mathbf{E}^2\right)\lambda^2 - \frac{1}{c^2}\,(\mathbf{B}\cdot\mathbf{E})^2 = 0\,. \tag{5.2.22}$$

Hence, the tensor (5.2.19) has only two independent invariants

$$\mathbf{B}^2 - \frac{1}{c^2}\,\mathbf{E}^2 = \frac{1}{2}\,F_{\mu\nu}F^{\mu\nu}\,, \quad \frac{1}{c^2}\,(\mathbf{B}\cdot\mathbf{E})^2 = \epsilon^{\mu\nu\rho\sigma}\,F_{\mu\nu}F_{\rho\sigma}\,, \tag{5.2.23}$$

where $\epsilon^{\mu\nu\rho\sigma}$ is Ricci's symbol of the fourth order.

Note that the Lagrangian action $\mathcal{A}$ can be written in the form (5.1.75) such that, by assuming that the motion equations (5.2.18) are satisfied, we have

$$\delta S = -\left(m_0\, c\, u_\mu + q\, A_\mu\right) \delta x^\mu\,. \tag{5.2.24}$$

It follows that the generalized momenta are

$$P_\mu = \frac{\partial S}{\partial x^\mu} = -m_0\, c\, u_\mu - q\, A_\mu\,, \tag{5.2.25}$$

corresponding to (5.2.12) for $\mu = 1, 2, 3$, and to (5.2.13) for $\mu = 4$;

$$P_4 = \frac{1}{c}\frac{\partial S}{\partial t} = -\frac{1}{c}\left(E + q\,\varphi\right) = -\frac{1}{c}\, H\,.$$

Generally, in the case of a system of charged particles, subjected to an electromagnetic field, the Lagrangian action comprises three terms

$$\mathcal{A} = \mathcal{A}_0 + \mathcal{A}_{\text{int}} + \mathcal{A}_e\,, \tag{5.2.26}$$

where $\mathcal{A}_0$ and $\mathcal{A}_{\text{int}}$ are given, respectively by (5.2.5) and (5.2.8), while $\mathcal{A}_e$ represents the Lagrangian action of the electromagnetic field in the absence of any charge, that is the Lagrangian action of the electromagnetic field in vacuum. It can be shown that the action S_e, corresponding to $\mathcal{A}_e$, has the form

$$S_e = \frac{\alpha}{c}\int F_{\mu\nu}F^{\mu\nu}\, d\omega\,, \tag{5.2.27}$$

where $d\omega = dx^1 dx^2 dx^3 dx^4$, and α is a constant which depends on the system of units. In SI units, the Lagrangian determined by (5.2.27) is

$$\mathcal{L}_e = -\frac{1}{4\mu_0}\int F_{\mu\nu}F^{\mu\nu}\, dx^1 dx^2 dx^3\,,$$

and the quantity

$$L_e = -\frac{1}{4\mu_0}\, F_{\mu\nu}F^{\mu\nu} \tag{5.2.28}$$

is called the *Lagrangian density*. In this case, the coordinates x^μ represent the parameters of the field, that is they play the same rôle as the time t in classical mechanics, while the components of the four-vector potential A_μ have a rôle similar to the state functions $q_k(t)$. The equations of motion (the field equations in our case) follow from Hamilton's principle applied to the action S_e of the electromagnetic field in vacuum

$$\delta S_e = \frac{1}{c}\,\delta\int L_e\, d\omega = 0\,. \tag{5.2.29}$$

Thus, we obtain the field equations in the absence of charges

$$\frac{\partial}{\partial x^\mu}\left(\frac{\partial L_e}{\partial A_{\nu,\mu}}\right) - \frac{\partial L_e}{\partial A_\nu} = 0\,, \quad \nu = \overline{1,4}\,, \tag{5.2.30}$$

where $A_{\nu,\mu} = \partial A_\nu / \partial x^\mu$. From

$$\begin{aligned}\delta L_e &= -\frac{1}{4\mu_0}\,\delta\left(F_{\mu\nu}F^{\mu\nu}\right) = -\frac{1}{2\mu_0}\,F^{\mu\nu}\delta F_{\mu\nu} = -\frac{1}{2\mu_0}\,F^{\mu\nu}\left(\delta A_{\nu,\mu} - \delta A_{\mu,\nu}\right)\\ &= -\frac{1}{2\mu_0}\left(F^{\mu\nu}\delta A_{\nu,\mu} + F^{\nu\mu}\delta A_{\mu,\nu}\right) = -\frac{1}{\mu_0}\,F^{\mu\nu}\delta A_{\nu,\mu}\,,\end{aligned} \tag{5.2.31}$$

we have

$$\frac{\partial L_e}{\partial A_{\nu,\mu}} = -\frac{1}{\mu_0}\,F^{\mu\nu} \tag{5.2.32}$$

and, because $\partial L_e / \partial A_\nu = 0$, the field equations (5.2.30) take the form

$$F^{\nu\mu}_{\ ,\mu} \equiv \frac{\partial F^{\nu\mu}}{\partial x^\mu} = 0\,, \quad \nu = \overline{1,4}\,. \tag{5.2.33}$$

Taking into account the expression (5.2.20), it is easy to show that the field equations (5.2.33) correspond to the curl equations in (5.2.1) for $\mathbf{j} = \mathbf{0}$ and $\rho = 0$. Note that, in this case the divergence equations appear as a direct consequence of the curl equations.

In the presence of electric charges, the Lagrangian action becomes

$$\mathcal{A} = \mathcal{A}_{\text{int}} + \mathcal{A}_e\,. \tag{5.2.34}$$

If the total electric charge is Q we may introduce the volume density of the electric charge ρ defined by the relation

$$Q = \int \rho\, dx^1 dx^2 dx^3\,.$$

Thus, on account of (5.2.8) we may write

$$\begin{aligned}S_{\text{int}} &= -Q\int A_\mu\, dx^\mu = -\int \rho\left(A_\mu\, dx^\mu\right) dx^1 dx^2 dx^3\\ &= \int \rho\left(\mathbf{A}\cdot d\mathbf{r} - \varphi\, dt\right) dx^1 dx^2 dx^3\\ &= \int \left[\mathbf{A}\cdot(\rho\mathbf{v}) - \rho\varphi\right] dx^1 dx^2 dx^3 dt = -\frac{1}{c}\int A_\mu j^\mu\, d\omega\,,\end{aligned} \tag{5.2.35}$$

where $(j^\mu) = (\mathbf{j}, c\rho)$ is the four-vector current density, the conduction current (due to the motion of electric charges) being $\mathbf{j} = \rho\mathbf{v}$. Consequently, from (5.2.34), (5.2.35), and (5.2.27) we may infer

$$S = -\frac{1}{c}\int\left(A_\mu j^\mu - \frac{1}{4\mu_0}\,F_{\mu\nu}F^{\mu\nu}\right) d\omega\,, \tag{5.2.36}$$

corresponding to the Lagrangian density

$$L = A_\mu j^\mu - \frac{1}{4\mu_0} F_{\mu\nu} F^{\mu\nu} . \tag{5.2.37}$$

Now, the field equations (5.2.30) for the Lagrangian density L lead us to the field equations

$$F^{\nu\mu}_{\ ,\mu} = \mu_0 j^\nu , \quad \nu = \overline{1,4} , \tag{5.2.38}$$

which represent the four-dimensional form of the curl equations in (5.2.1). Also, from the definition (5.2.19) of the electromagnetic field tensor $F_{\mu\nu}$, we deduce the relations

$$\frac{\partial F_{\mu\nu}}{\partial x^\sigma} + \frac{\partial F_{\nu\sigma}}{\partial x^\mu} + \frac{\partial F_{\sigma\mu}}{\partial x^\nu} = 0 , \tag{5.2.39}$$

which can also be written in the form

$$\epsilon^{\rho\sigma\mu\nu} \frac{\partial F_{\mu\nu}}{\partial x^\sigma} = 0 , \quad \rho = \overline{1,4} . \tag{5.2.40}$$

By substituting the components of the electromagnetic field tensor from (5.2.20) into (5.2.40) we obtain the divergence equations of (5.2.1). Thus, in the theory of Special Relativity the Maxwell equations (5.2.1) are used in their four-dimensional form (5.2.38), (5.2.39).

2.1.2 Symmetry transformations

In the case of the electromagnetic field, the Lagrangian density (5.2.37) depends only on the four-potential (A_μ) and the derivatives of its components, contained in the electromagnetic field tensor $F_{\mu\nu}$. Therefore, the action has the general form

$$S = \int L\,(A_\mu, A_{\mu,\nu})\, d\omega . \tag{5.2.41}$$

Also, the components A_μ play the rôle of the state functions, and the coordinates x^μ are the independent parameters. Thus, using the same method as in Chap. 4 Sec. 1.4.1 for an infinitesimal transformation

$$x'^\mu = x^\mu + \delta x^\mu , \quad \mu = \overline{1,4}, \tag{5.2.42}$$

we obtain (assuming that the field equations are satisfied) the variation of the action

$$\delta S = \int \frac{\partial}{\partial x^\rho} \left[(L\, \delta^\rho_\nu - L^{\mu\rho}\, A_{\mu,\nu})\, \delta x^\nu + L^{\mu\rho}\, \delta A_\mu \right] d\omega , \tag{5.2.43}$$

where $L^{\mu\rho} = \partial L / \partial A_{\mu,\rho}$. If we denote by

$$\psi^\rho = (L\, \delta^\rho_\nu - L^{\mu\rho}\, A_{\mu,\nu})\, \delta x^\nu + L^{\mu\rho}\, \delta A_\mu , \tag{5.2.44}$$

then (5.2.43) leads to a continuity equation for the functions ψ^ρ, that is

$$\psi^\rho_{,\rho} \equiv \frac{\partial \psi^\rho}{\partial x^\rho} = 0\,. \tag{5.2.45}$$

We assume that the domain of integration in (5.2.43) is a finite volume $\Omega \subset \mathcal{M}_4$ bounded by a three-dimensional manifold $\Sigma \subset \mathcal{M}_4$. Hence, by separating the spatial and temporal components we have

$$\begin{aligned}\int_\Omega \frac{\partial \psi^\rho}{\partial x^\rho}\, d\omega &= \int dx^4 \int_{D_3} \frac{\partial \psi^k}{\partial x^k}\, dx^1 dx^2 dx^3 + \frac{1}{c}\int dx^4 \int_{D_3} \frac{\partial \psi^4}{\partial t}\, dx^1 dx^2 dx^3 \\ &= \int dx^4 \oint_{S_3} \psi^k\, d\sigma_k + \frac{1}{c}\int dx^4 \frac{d}{dt}\int_{D_3} \psi^4\, dx^1 dx^2 dx^3\,,\end{aligned}$$

where D_3 represents the spatial part of Ω and S_3 is the boundary of D_3. If we extend the domain D_3 such that it covers the whole three-dimensional space, and take into account that the surface integral vanishes for a sphere of infinite radius, we obtain

$$\int \frac{\partial \psi^\rho}{\partial x^\rho}\, d\omega = \frac{1}{c}\int dx^4 \frac{d}{dt}\int \psi^4\, dx^1 dx^2 dx^3 = 0\,. \tag{5.2.46}$$

Consequently, the conserved quantity, associated to a transformation of the form (5.2.42), is

$$C_{(4)} = \frac{1}{c}\int \psi^4\, dx^1 dx^2 dx^3\,, \tag{5.2.47}$$

satisfying $dC_{(4)}/dt = 0$. This result represents the extension of Noether's theorem, from classical mechanics, to the case of the electromagnetic field. Note that, the conserved quantities (5.2.47) are constant in time like in the case of classical mechanics.

In the general case, when the Lagrangian density $L(x^\mu, A_\mu, A_{\mu,\nu})$ depends on the coordinates x^μ, a transformation of the form (5.2.42) is a symmetry transformation if the condition

$$\left[\delta x^\nu \frac{\partial}{\partial x^\nu} + \delta A_\nu \frac{\partial}{\partial A_\nu} + \delta A_{\nu,\mu} \frac{\partial}{\partial A_{\nu,\mu}} + \frac{\partial(\delta x^\nu)}{\partial x^\nu}\right] L = -d_\mu\left(\delta f^\mu\right) \tag{5.2.48}$$

is satisfied. Here, f^μ, $\mu = \overline{1,4}$, are arbitrary functions of class C^2 which depend on x^μ and A_ν (not on $A_{\nu,\mu}$), and the operator in the right hand side of (5.2.48) has the form

$$d_\mu = \frac{\partial}{\partial x^\mu} + A_{\nu,\mu}\frac{\partial}{\partial A_\nu} + A_{\nu,\mu\rho}\frac{\partial}{\partial A_{\nu,\rho}}\,, \tag{5.2.49}$$

with $A_{\nu,\mu\rho} = \partial^2 A_\nu/\partial x^\mu \partial x^\rho$. Consequently, the conservation equation corresponding to a transformation (5.2.42) is of the form of a generalized divergence

$$d_\mu\left[\left(L\,\delta^\mu_\rho - L^{\mu\nu} A_{\nu,\rho}\right)\delta x^\rho + L^{\mu\nu}\,\delta A_\nu + \delta f^\mu\right] = 0\,. \tag{5.2.50}$$

Also, in terms of the function ψ defined by (5.2.44), the conservation equation (5.2.50) can be written as

$$d_\mu \left(\psi^\mu + \delta f^\mu\right) = 0\,,$$

which essentially represents Noether's theorem in the frame of the theory of Special Relativity.

2.1.3 Translations in $\mathcal{M}_4$

The infinitesimal translations in $\mathcal{M}_4$ are defined by the relations

$$x'^\mu = x^\mu + \varepsilon^\mu\,, \quad \mu = \overline{1,4}, \tag{5.2.51}$$

where $\delta x^\mu = \varepsilon^\mu$ are infinitesimal parameters (independent of x^μ), while $\delta A_\mu = 0$. If we choose $\delta f^\mu = 0$, then the transformations (5.2.51) are symmetry transformations, that is the condition (5.2.48) is satisfied, only when

$$\frac{\partial L}{\partial x^\nu}\,\varepsilon^\nu = 0\,. \tag{5.2.52}$$

Since the infinitesimal variations ε^ν are independent, from (5.2.52) we obtain the conditions

$$\frac{\partial L}{\partial x^\nu} = 0\,, \quad \nu = \overline{1,4}, \tag{5.2.53}$$

and the equation of conservation (5.2.50) becomes

$$d_\mu \left[\left(L\,\delta^\mu_\rho - L^{\mu\nu}\,A_{\nu,\rho}\right)\varepsilon^\rho\right] = \varepsilon^\rho\, d_\mu \left(L\,\delta^\mu_\rho - L^{\mu\nu}\,A_{\nu,\rho}\right) = 0\,. \tag{5.2.54}$$

Now, the function ψ^4 in (5.2.47) has the form

$$\psi^4 = \left(L\,\delta^4_\nu - L^{\mu 4}\,A_{\mu,\nu}\right)\varepsilon^\nu\,. \tag{5.2.55}$$

For time translations (along the axis x^4) the transformation (5.2.51) has the parameters

$$\varepsilon^1 = \varepsilon^2 = \varepsilon^3 = 0\,, \quad \varepsilon^4 \neq 0\,,$$

and (5.2.47) becomes

$$\begin{aligned} C_{(4)} &= \frac{1}{c}\int\left(L - L^{\mu 4}\,\frac{\partial A_\mu}{\partial x^4}\right)\varepsilon^4\, dx^1 dx^2 dx^3 \\ &= -\,\frac{1}{c}\int\left(\frac{\partial L}{\partial \dot{A}_\mu}\,\dot{A}_\mu - L\right)\varepsilon^4\, dx^1 dx^2 dx^3\,. \end{aligned} \tag{5.2.56}$$

By analogy with the nomenclature used in classical mechanics, the quantities $\partial L/\partial \dot{A}_\mu = \Pi^\mu$ are called *densities of generalized momenta*, and the *Hamiltonian density* is $H = \Pi^\mu \dot{A}_\mu - L$. Thus, we may rewrite (5.2.56) in the form

$$C_{(4)} = -\frac{1}{c}\int \left(\Pi^\mu \dot{A}_\mu - L\right) \varepsilon^4 \, dx^1 dx^2 dx^3$$
$$= -\frac{1}{c}\,\varepsilon^4 \int H \, dx^1 dx^2 dx^3 = -\frac{1}{c}\,\mathcal{H}\,\varepsilon^4 , \tag{5.2.57}$$

where $\mathcal{H}$ is the Hamiltonian of the electromagnetic field. Thus, the invariance of the Lagrangian density with respect to time translations leads us to the conservation of the total energy.

For translations along the spatial axes the transformation (5.2.51) has the parameters

$$\varepsilon^1 \neq 0 , \quad \varepsilon^2 \neq 0 , \quad \varepsilon^3 \neq 0 , \quad \varepsilon^4 = 0 ,$$

and the conserved quantity is

$$\frac{1}{c}\int L^{\mu 4}\,\frac{\partial A_\mu}{\partial x^k}\,\varepsilon^k \, dx^1 dx^2 dx^3 = \frac{1}{c}\int c\,\frac{\partial L}{\partial \dot{A}_\mu}\,\frac{\partial A_\mu}{\partial x^k}\,\varepsilon^k \, dx^1 dx^2 dx^3 = \frac{1}{c}\,G_k\,\varepsilon^k , \tag{5.2.58}$$

where G_k, $k = 1,2,3$, represent the components of the electromagnetic momentum. Since the parameters ε^k are independent, from

$$\frac{d}{dt}\left(G_k\,\varepsilon^k\right) = \varepsilon^k\,\dot{G}_k = 0$$

it results that the components of the vector $\mathbf{G}$ are conserved. Therefore, from the invariance of the Lagrangian density with respect to the symmetry transformations (5.2.51), when the Lagrangian density satisfies the conditions (5.2.53), we obtain the conservation of the components of the *energy-momentum four-vector* $(G_\mu) = (-\mathbf{G}, \mathcal{H}/c)$ for the electromagnetic field.

Note that the quantity ψ^ρ, defined in (5.2.44), can be written as

$$\psi^\rho = T^{\rho\nu}\,\delta x_\nu + L^{\mu\rho}\,\delta A_\mu ,$$

and the translations (5.2.51) correspond to $\delta A_\mu = 0$ such that

$$\psi^\rho = T^{\rho\nu}\,\varepsilon_\nu . \tag{5.2.59}$$

The tensor $T^{\rho\nu}$ is called the *energy-momentum density tensor* (or the *canonical tensor*) of the electromagnetic field and is related to the four-vector (G_μ) by the formulae

$$G_\mu = -\frac{1}{c}\int T_{\mu 4}\, dx^1 dx^2 dx^3 , \quad \mu = \overline{1,4}. \tag{5.2.60}$$

In terms of the canonical tensor components the conservation equation (5.2.54) takes the form

$$d_\mu \left(T^{\rho\mu}\, \varepsilon_\rho \right) = \varepsilon_\rho\, d_\mu\, T^{\rho\mu} = 0\,, \tag{5.2.61}$$

such that we have

$$d_\mu\, T^{\rho\mu} = 0\,, \quad \rho = \overline{1,4}. \tag{5.2.62}$$

2.1.4 Rotations in $\mathcal{M}_4$

The rotations in $\mathcal{M}_4$ are homogeneous proper Lorentz transformations, defined by the relations (see Chap. 2 Sec. 4.1)

$$x'^\mu = \Lambda^\mu_\nu\, x^\nu\,, \quad \Lambda^\mu_\nu \Lambda^\nu_\rho = \delta^\mu_\rho\,, \quad \mu, \rho = \overline{1,4}\,. \tag{5.2.63}$$

For infinitesimal rotations, by expanding x'^μ into a Taylor series, we get

$$x'^\mu = x^\mu + g^{\mu\nu}\,\delta x_\nu = x^\mu + \delta\omega^\mu_\nu\, x^\nu\,, \quad \delta x_\mu = \delta\omega_{\mu\nu} x^\nu\,, \quad \mu = \overline{1,4}\,, \tag{5.2.64}$$

such that, the condition $\delta(x^\mu x_\mu) = 0$ for the conservation of distances leads us to the relation

$$\delta\omega_{\mu\nu} = -\delta\omega_{\nu\mu},\,. \tag{5.2.65}$$

Now, from (5.2.64) we obtain the variations of coordinates

$$\delta x^\mu = g^{\mu\nu}\,\delta\omega_{\nu\rho} x^\rho\,, \quad \delta\omega_{\nu\rho} = g_{\mu\nu}\,\frac{\partial(\delta x^\mu)}{\partial x^\rho} = \frac{\partial(\delta x_\nu)}{\partial x^\rho}\,. \tag{5.2.66}$$

The components of the four-vector potential are transformed according to the formulae

$$A'_\mu(\mathbf{x}') = \Lambda^\nu_\mu(\delta\omega) A_\nu(\mathbf{x}) \tag{5.2.67}$$

and, by expanding into a Taylor series, we are lead to the transformation relations

$$A'_\mu(\mathbf{x}') = A_\mu(\mathbf{x}) + \delta\omega_{\rho\sigma} \left[\frac{\partial\Lambda^\nu_\mu}{\partial(\delta\omega_{\rho\sigma})} \right]_{\delta\omega=0} A_\nu(\mathbf{x})\,.$$

We denote by

$$X^{\nu\rho\sigma}_\mu = \left[\frac{\partial\Lambda^\nu_\mu}{\partial(\delta\omega_{\rho\sigma})} \right]_{\delta\omega=0} = -X^{\nu\sigma\rho}_\mu\,,$$

so that the variations δA_μ take the form

$$\delta A_\mu = \frac{1}{2}\, X^{\nu\rho\sigma}_\mu\, A_\nu\, \delta\omega_{\rho\sigma}\,. \tag{5.2.68}$$

The transformation (5.2.63) is a symmetry transformation for the Lagrangian density (5.2.37) because

$$\begin{aligned} L(A_\mu, A_{\nu,\sigma}) &= A_\mu\, j^\mu - \frac{1}{4\mu_0}\, F_{\nu\sigma} F^{\nu\sigma} \\ &= \Lambda^\rho_\mu\, \Lambda^\mu_\kappa\, A'_\rho\, j'^\kappa - \frac{1}{4\mu_0}\, \Lambda^\alpha_\nu\, \Lambda^\beta_\sigma\, \Lambda^\nu_\gamma\, \Lambda^\sigma_\delta\, F'_{\alpha\beta}\, F'^{\gamma\delta} \\ &= \delta^\rho_\kappa\, A'_\rho\, j'^\kappa - \frac{1}{4\mu_0}\, \delta^\alpha_\gamma\, \delta^\beta_\delta\, F'_{\alpha\beta}\, F'^{\gamma\delta} \\ &= A'_\rho\, j'^\rho - \frac{1}{4\mu_0}\, F'_{\alpha\beta}\, F'^{\alpha\beta} = L(A'_\mu, A'_{\nu,\sigma}) , \end{aligned}$$

the Lagrangian density in the new variables resulting from L simply replacing the old variables by the new variables (the functional form of the Lagrangian density is invariant), that is

$$L'(A'_\rho, A'_{\alpha\beta}) = L(A'_\rho, A'_{\alpha\beta}) . \tag{5.2.69}$$

In this case, with $\delta f^\mu = 0$, the conservation equation (5.2.50) becomes

$$\begin{aligned} & d_\lambda \left[x^\nu\, T^{\rho\lambda}\, \delta\omega_{\rho\nu} + \frac{1}{2}\, L^{\mu\lambda}\, X^{\sigma\rho\nu}_\mu\, A_\sigma\, \delta\omega_{\rho\nu} \right] \\ & \qquad = \delta\omega_{\rho\nu}\, d_\lambda \left[x^\nu\, T^{\rho\lambda} + \frac{1}{2}\, Y^{\lambda\rho\nu} \right] = 0 , \end{aligned} \tag{5.2.70}$$

where the tensor

$$Y^{\lambda\rho\nu} = L^{\mu\lambda}\, X^{\sigma\rho\nu}_\mu\, A_\sigma \tag{5.2.71}$$

is skew-symmetric relative to the indexes ρ and ν. Since the variations $\delta_{\rho\nu}$ are independent, and the tensor $\delta_{\rho\nu}$ is skew-symmetric, the equation (5.2.70) leads us to the system of equations

$$d_\lambda \left[x^\nu\, T^{\rho\lambda} - x^\rho\, T^{\nu\lambda} + \frac{1}{2} \left(Y^{\lambda\rho\nu} - Y^{\lambda\nu\rho} \right) \right] = 0 . \tag{5.2.72}$$

If we assume that the conservation equation (5.2.62) is satisfied, then (5.2.72) shows that the tensor

$$Q^{\mu\nu} = T^{\mu\nu} + \frac{1}{2}\, d_\lambda\, Y^{\lambda\mu\nu} \tag{5.2.73}$$

is symmetric. We may add to the tensor $Q^{\mu\nu}$ the divergence

$$\frac{1}{2}\, d_\lambda \left(Y^{\mu\nu\lambda} + Y^{\nu\mu\lambda} \right)$$

to obtain

$$Q^{\mu\nu} = T^{\mu\nu} + \frac{1}{2}\, d_\lambda \left(Y^{\lambda\mu\nu} + Y^{\mu\nu\lambda} + Y^{\nu\mu\lambda} \right) = T^{\mu\nu} + d_\lambda\, V^{\lambda\mu\nu} ,$$

where the tensor $V^{\lambda\mu\nu}$ is skew-symmetric relative to the first two indexes, and satisfies the relation

$$Y^{\lambda\mu\nu} = V^{\lambda\mu\nu} - V^{\lambda\nu\mu}\,. \tag{5.2.74}$$

Thus, the conserved quantity in (5.2.72) becomes

$$\begin{aligned}\mathcal{M}^{\nu\lambda\rho} &= x^\rho\, T^{\nu\lambda} - x^\nu\, T^{\rho\lambda} - Y^{\lambda\rho\nu}\\ &= x^\rho\, T^{\nu\lambda} - x^\nu\, T^{\rho\lambda} - V^{\lambda\rho\nu} + V^{\lambda\nu\rho}\\ &= x^\rho\, \mathcal{T}^{\nu\lambda} - x^\nu\, \mathcal{T}^{\rho\lambda} + d_\sigma\left(x^\nu\, V^{\lambda\sigma\rho} - x^\rho\, V^{\lambda\sigma\nu}\right),\end{aligned}$$

where we have used the notation

$$\mathcal{T}^{\nu\lambda} = T^{\nu\lambda} + d_\sigma\, V^{\lambda\sigma\nu}\,. \tag{5.2.75}$$

Also, the system of equations (5.2.72) can be written in the form

$$d_\lambda\, \mathcal{M}^{\nu\lambda\rho} = d_\lambda\left(x^\rho\, \mathcal{T}^{\nu\lambda} - x^\nu\, \mathcal{T}^{\rho\lambda}\right) = 0\,. \tag{5.2.76}$$

Note that here we employed the relation

$$\begin{aligned}&d_\lambda d_\sigma(x^\nu\, V^{\lambda\sigma\rho} - x^\rho\, V^{\lambda\sigma\nu})\\ &\qquad = d_\lambda(\delta^\nu_\sigma\, V^{\lambda\sigma\rho} - \delta^\rho_\sigma\, V^{\lambda\sigma\nu} + x^\nu\, d_\sigma\, V^{\lambda\sigma\rho} - x^\rho\, d_\sigma\, V^{\lambda\sigma\nu})\\ &\qquad = d_\lambda(V^{\lambda\nu\rho} - V^{\lambda\rho\nu} + x^\nu\, d_\sigma\, V^{\lambda\sigma\rho} - x^\rho\, d_\sigma\, V^{\lambda\sigma\nu})\\ &\qquad = x^\nu\, d_\lambda\, d_\sigma\, V^{\lambda\sigma\rho} - x^\rho\, d_\lambda\, d_\sigma\, V^{\lambda\sigma\nu} = 0\,,\end{aligned}$$

and took into account the fact that all expressions of the form $d_\lambda d_\sigma V^{\lambda\sigma\rho}$ vanish due to the skew-symmetry of the tensor $V^{\lambda\sigma\rho}$ relative to the indexes λ and σ. The quantity

$$\mathcal{M}^{\mu\nu\lambda} = x^\lambda\, \mathcal{T}^{\mu\nu} - x^\mu\, \mathcal{T}^{\lambda\nu} \tag{5.2.77}$$

is called the *angular momentum density tensor*, and the conserved quantities are

$$M^{\mu\lambda} = \frac{1}{c}\int \mathcal{M}^{\mu 4\lambda}\, dx^1\, dx^2\, dx^3 = \frac{1}{c}\int (x^\lambda\, \mathcal{T}^{\mu 4} - x^\mu\, \mathcal{T}^{\lambda 4})\, dx^1\, dx^2\, dx^3\,. \tag{5.2.78}$$

Because we have assumed that the Lagrangian density is invariant with respect to translations ($\partial L/\partial x^\mu = 0$), and the last term in the relation

$$d_\mu\, \mathcal{T}^{\lambda\mu} = d_\mu\, T^{\lambda\mu} + d_\mu\, d_\sigma\, V^{\mu\sigma\lambda}$$

vanishes, it follows that the tensor $\mathcal{T}^{\lambda\mu}$ satisfies the same conservation equation as the canonical tensor $T^{\lambda\mu}$. Also, the relations (5.2.60) remain unchanged if we substitute $T^{\lambda\mu}$ by $\mathcal{T}^{\lambda\mu}$, that is

$$G_\mu = -\frac{1}{c}\int \mathcal{T}_{\mu 4}\, dx^1\, dx^2\, dx^3\,, \tag{5.2.79}$$

since the last term in the definition (5.2.75) does not contribute to the integral. Consequently, we may consider that the tensor $\mathcal{T}^{\lambda\mu}$ is the *energy-momentum density tensor*.

2.1.5 The Lorentz gauge

The potentials $\mathbf{A}$ and φ are not uniquely defined by the relations (5.2.7), and there exists a set of such functions for any given pair of fields $\mathbf{E}$ and $\mathbf{B}$. Indeed, the form of the relations (5.2.7) remains unchanged if we transform the potentials according to the formulae

$$\begin{aligned} \mathbf{A} \to \mathbf{A}' &= \mathbf{A} + \nabla h \,, \\ \varphi \to \varphi' &= \varphi - \frac{\partial h}{\partial t} \,, \end{aligned} \tag{5.2.80}$$

where $h = h(\mathbf{r}, t)$ is an arbitrary function. Also, taking into account that $(A_\mu) = (-\mathbf{A}, \varphi/c)$ and $(\partial_\mu) = (\nabla, \partial/\partial x^4)$, the four-dimensional form of the transformation (5.2.80) is

$$A_\mu \to A'_\mu = A_\mu - \partial_\mu h \,, \quad \mu = \overline{1,4} \,. \tag{5.2.81}$$

The transformations (5.2.80) are called *gauge transformations of the second kind*, while the invariance of the fields $\mathbf{E}$ and $\mathbf{B}$ with respect to the transformations (5.2.80) is called *gauge invariance*.

The set of transformations (5.2.80) can be reduced by imposing certain conditions to the function f. Thus, when the function f satisfies the wave equation

$$\left(\Delta - \frac{1}{c^2} \frac{\partial^2}{\partial t^2} \right) f = 0 \,, \tag{5.2.82}$$

it follows that the expression

$$W = \nabla \cdot \mathbf{A} + \frac{1}{c^2} \frac{\partial \varphi}{\partial t} = \nabla \cdot \mathbf{A}' + \frac{1}{c^2} \frac{\partial \varphi'}{\partial t} \tag{5.2.83}$$

is invariant with respect to the gauge transformations (5.2.80). Reciprocally, the invariance condition (5.2.83) implies (5.2.82). The value of the invariant W is arbitrary and we may choose $W = 0$, that is

$$\nabla \cdot \mathbf{A} + \frac{1}{c^2} \frac{\partial \varphi}{\partial t} = 0 \,, \tag{5.2.84}$$

or, in a Lorentz invariant form,

$$\partial_\mu A^\mu = 0 \,. \tag{5.2.85}$$

Note that $(A^\mu) = (g^{\mu\nu} A_\nu) = (\mathbf{A}, \varphi/c)$, with $g^{\mu\nu} = g_{\mu\nu}$ being the metric tensor defined in (2.4.4). The condition (5.2.84), called the *Lorentz gauge*, has been introduced only to simplify the calculations. When we substitute the potentials $\mathbf{A}$ and φ, defined by the relations (5.2.7), into the Maxwell equations (5.2.1) we obtain a system of coupled equations for $\mathbf{A}$ and φ. If we admit that the potentials satisfy the condition (5.2.84), then this system can be decoupled into two independent equations

$$\left(\Delta - \frac{1}{c^2}\frac{\partial^2}{\partial t^2}\right)\mathbf{A} = -\mu_0\,\mathbf{j}\,,$$
$$\left(\Delta - \frac{1}{c^2}\frac{\partial^2}{\partial t^2}\right)\varphi = -\frac{\rho}{\varepsilon_0}\,. \tag{5.2.86}$$

Because the Maxwell equations (5.2.1) are identically satisfied when the fields $\mathbf{E}$ and $\mathbf{B}$ are replaced according to (5.2.7) and (5.2.84), the potentials $\mathbf{A}$ and φ being solutions of (5.2.86), it follows that the system of equations (5.2.86) is equivalent to (5.2.1). Consequently, the system of equations (5.2.86) represents the Maxwell equations in the Lorentz gauge.

Generally, the Lagrangian density for the Maxwell equations (5.2.1), contains the components of the four-potential (A_μ) as dependent variables (functions of state), the independent variables being the coordinates in the Minkowski space $\mathcal{M}_4$. Because the equations (5.2.86) satisfied by the components A_μ are partial differential equations of the second order, the Lagrangian has to contain only the first order derivatives of A_μ and moreover, the Lagrangian has to be invariant with respect to the Lorentz transformations. It follows that the Lagrangian density may have the form

$$L_1 = -\frac{1}{2\mu_0}\,\partial_\mu A_\nu \partial^\mu A^\nu + A_\mu j^\mu\,. \tag{5.2.87}$$

Then, on the ground of Hamilton's principle, we may write the Euler-Lagrange equations

$$\partial_\nu\,\frac{\partial L_1}{\partial(\partial_\nu A_\mu)} - \frac{\partial L_1}{\partial A_\mu} = 0\,, \tag{5.2.88}$$

being thus led to the equations (5.2.86) in terms of A_μ

$$\left(\Delta - \frac{1}{c^2}\frac{\partial^2}{\partial t^2}\right)A_\mu = -\mu_0\,j_\mu\,. \tag{5.2.89}$$

The equations (5.2.89), supplemented by the condition (5.2.85) and the definition (5.2.19), result in the form (5.2.38) of the Maxwell equations.

Another choice of the Lagrangian density for the Maxwell equations is the expression (5.2.37), which leads us directly to the Maxwell equations (5.2.38) without any supplementary condition. Thus, this choice is more appropriate. The difference between the two expressions of the Lagrangian densities (5.2.37) and (5.2.87) is given by

$$\begin{aligned} L - L_1 &= -\frac{1}{4\mu_0} F_{\mu\nu}F^{\mu\nu} + \frac{1}{2\mu_0} \partial_\mu A_\nu \partial^\mu A^\nu = \frac{1}{2\mu_0} (\partial_\nu A_\mu)(\partial^\mu A^\nu) \\ &= \frac{1}{2\mu_0} \partial_\nu (A_\mu \partial^\mu A^\nu) - \frac{1}{2\mu_0} A_\mu \partial_\nu \partial^\mu A^\nu \\ &= \frac{1}{2\mu_0} \partial_\nu (A_\mu \partial^\mu A^\nu) - \frac{1}{2\mu_0} A_\mu \partial^\mu (\partial_\nu A^\nu) , \end{aligned}$$

and the Lorentz gauge (5.2.85) leads us to

$$L - L_1 = \frac{1}{2\mu_0} \partial_\nu (A_\mu \partial^\mu A^\nu) . \tag{5.2.90}$$

Therefore, L and L_1 differ only by a divergence such that the Euler-Lagrange equations obtained, using any of the forms (5.2.37) or (5.2.87), are identical.

However, the form (5.2.37) of the Lagrangian density is not invariant with respect to the gauge transformation (5.2.81), because the first term in (5.2.37) contains the expression

$$-j^\mu \partial_\mu h = -\partial_\mu (j^\mu h) + h\, \partial_\mu j^\mu . \tag{5.2.91}$$

The gauge transformation (5.2.81) is a symmetry transformation (i.e., we have $f^\mu \to j^\mu h$ in (5.2.48)) if the condition $\partial_\mu j^\mu = 0$, or

$$\frac{\partial \rho}{\partial t} + \nabla \cdot \mathbf{j} = 0 , \tag{5.2.92}$$

is fulfilled, and in such cases the two forms of the Lagrangian density differ only by a divergence. Note that (5.2.92) represents the differential form of the conservation law of the electric charge. Also, if the Lagrangian density is invariant with respect to the gauge transformations, then the field equations will be invariant with respect to these transformations. Due to the skew-symmetry of the tensor $F_{\mu\nu}$, the equation (5.2.92) results directly from (5.2.38).

In the case of the Lagrangian density (5.2.87) the action is invariant with respect to the gauge transformations if the field equations (5.2.89) are satisfied. Note that the equivalence of these equations with the Maxwell equations implies the existence of the condition (5.2.85). Thus, the conservation law of the electric charge (5.2.92) follows as a consequence of the field equations together with the Lorentz gauge.

2.2 Conformal transformations in four dimensions

2.2.1 The infinitesimal generators of the Poincaré group

In Chap. 2 Sec. 4.1.2 we shown that the group of inhomogeneous Lorentz transformations, called the Poincaré group, preserves the norm of the vectors in $\mathcal{M}_4$. Thus, we may regard this group as the group of rigid motions in $\mathcal{M}_4$. Here, we will detail some of the properties of Poincaré's group.

We begin by considering a Lie transformation of the form (1.2.4)

$$x'^{\mu} = f^{\mu}(x^{\nu}; a^i) = f^{\mu}(x^{\nu}; a_0^i + \delta a^i), \tag{5.2.93}$$

where a_0^i, $i = 1, 2, \ldots, m$, are the parameters corresponding to the identity transformation. Then, we expand x'^{μ} into the Taylor series

$$x'^{\mu} = f^{\mu}(x^{\nu}; a_0^i) + \left.\frac{\partial f^{\mu}}{\partial a^i}\right|_{\mathbf{a}=\mathbf{a}_0} \delta a^i + \ldots,$$

with the generalized velocities (1.2.9) being of the form

$$u_i^{\mu} = \left.\frac{\partial f^{\mu}}{\partial a^i}\right|_{\mathbf{a}=\mathbf{a}_0},$$

and the notation $\mathbf{a} = \mathbf{a}_0$ standing for $a^i = a_0^i$, $i = 1, 2, \ldots, m$. We thus obtain the infinitesimal form of the transformation (5.2.93)

$$x'^{\mu} = x^{\mu} + u_i^{\mu}\, \delta a^i. \tag{5.2.94}$$

Correspondingly, the infinitesimal generators defined by (1.2.20) can be expressed as differential operators

$$\mathsf{X}_i = u_i^{\mu} \frac{\partial}{\partial x^{\mu}}, \quad i = 1, 2, \ldots, m, \tag{5.2.95}$$

that satisfy the commutation relations (1.2.23). Also, according to the general theory of Lie groups, if the explicit form of the infinitesimal transformation of coordinates (5.2.94) is known, then the form of the corresponding finite transformation is

$$x'^{\mu} = \exp\left(\sum_{i=1}^{m} \mathsf{X}_i\, \delta a^i\right) x^{\mu}. \tag{5.2.96}$$

In the case of the rigid motions in $\mathcal{M}_4$, the infinitesimal generators X_i of the Poincaré group are obtained by means of the Killing vector components ξ^{λ}, which are solutions of Killing's equations (5.1.51). In Cartesian coordinates, and using the metric tensor (2.4.4), the Killing equations are

$$g_{\lambda\nu} \frac{\partial \xi^{\lambda}}{\partial x^{\mu}} + g_{\mu\lambda} \frac{\partial \xi^{\lambda}}{\partial x^{\nu}} = 0, \quad \mu, \nu = \overline{1, 4}. \tag{5.2.97}$$

The components of the metric tensor are constant and $g_{\mu\nu} = g_{\nu\mu}$ such that, with $\xi_{\nu} = g_{\lambda\nu}\xi^{\lambda}$, we have

$$\frac{\partial \xi_{\nu}}{\partial x^{\mu}} + \frac{\partial \xi_{\mu}}{\partial x^{\nu}} = 0.$$

These equations are identically satisfied if

$$\frac{\partial \xi_\nu}{\partial x^\mu} = \omega_{\nu\mu} = -\omega_{\mu\nu}\,,$$

where $\omega_{\mu\nu}$ are constants. Therefore, we may write

$$\xi_\nu = g_{\lambda\nu}\,\xi^\lambda = \omega_{\nu\mu}\,x^\mu + \kappa_\nu\,, \tag{5.2.98}$$

the quantities κ_ν being integration constants. To obtain the components ξ^λ we multiply (5.2.98) by $g^{\nu\rho}$ and sum with respect to ν

$$g^{\nu\rho} g_{\lambda\nu}\,\xi^\lambda = \delta^\rho_\lambda \xi^\lambda = \xi^\rho = g^{\rho\nu}\omega_{\nu\mu}\,x^\mu + g^{\rho\nu}\kappa_\nu\,.$$

Thus, the solutions of the Killing equations (5.2.97) have the form

$$\xi^\lambda = \omega^\lambda_\mu\,x^\mu + \kappa^\lambda\,, \quad \omega^\lambda_\mu = -\omega^\mu_\lambda\,. \tag{5.2.99}$$

Now, we may introduce the new parameters ω_i, $i = \overline{1,10}$, defined by the relations

$$\begin{aligned}
&\omega_1 = \omega^3_2 = -\omega^2_3\,, \quad \omega_2 = \omega^1_3 = -\omega^3_1\,, \quad \omega_3 = \omega^2_1 = -\omega^1_2\,,\\
&\omega_4 = \omega^1_4 = -\omega^4_1\,, \quad \omega_5 = \omega^2_4 = -\omega^4_2\,, \quad \omega_6 = \omega^3_4 = -\omega^4_3\,,\\
&\omega_7 = \kappa^1\,, \quad \omega_8 = \kappa^2\,, \quad \omega_9 = \kappa^3\,, \quad \omega_{10} = \kappa^4\,.
\end{aligned} \tag{5.2.100}$$

Then, by comparing (5.1.46) and (5.2.94), we may write

$$\varepsilon\,\xi^\mu = \sum_{i=1}^{10} u^\mu_i\,\delta a^i$$

or, multiplying by $\partial/\partial x^\mu$ and summing over μ,

$$\varepsilon\,\xi^\mu\,\frac{\partial}{\partial x^\mu} = \left(u^\mu_i\,\frac{\partial}{\partial x^\mu}\right)\delta a^i = \mathsf{X}_i\,\delta a^i\,.$$

Also, by setting the variations of the parameters $\delta a^i = -\varepsilon\,\omega_i$, $i = \overline{1,10}$, we obtain

$$\begin{aligned}
\xi^\mu\,\frac{\partial}{\partial x^\mu} &= g^{\mu\nu}\,\xi_\nu\,\frac{\partial}{\partial x^\mu}\\
&= -\xi_1\,\frac{\partial}{\partial x^1} - \xi_2\,\frac{\partial}{\partial x^2} - \xi_3\,\frac{\partial}{\partial x^3} + \xi_4\,\frac{\partial}{\partial x^4} = \sum_{i=1}^{10}\mathsf{X}_i\,\omega_i
\end{aligned}$$

and, from (5.2.99) and (5.2.100) it results the differential form of infinitesimal generators X_i

$$\mathsf{X}_1 = x^2\frac{\partial}{\partial x^3} - x^3\frac{\partial}{\partial x^2}, \quad \mathsf{X}_2 = x^3\frac{\partial}{\partial x^1} - x^1\frac{\partial}{\partial x^3}, \quad \mathsf{X}_3 = x^1\frac{\partial}{\partial x^2} - x^2\frac{\partial}{\partial x^1},$$

(5.2.101)

$$\mathsf{X}_4 = x^1 \frac{\partial}{\partial x^4} - x^4 \frac{\partial}{\partial x^1}, \quad \mathsf{X}_5 = x^2 \frac{\partial}{\partial x^4} - x^4 \frac{\partial}{\partial x^2}, \quad \mathsf{X}_6 = x^3 \frac{\partial}{\partial x^4} - x^4 \frac{\partial}{\partial x^3}, \tag{5.2.102}$$

$$\mathsf{X}_7 = \frac{\partial}{\partial x^1}, \quad \mathsf{X}_8 = \frac{\partial}{\partial x^2}, \quad \mathsf{X}_9 = \frac{\partial}{\partial x^3}, \quad \mathsf{X}_{10} = \frac{\partial}{\partial x^4}. \tag{5.2.103}$$

From the commutation relations satisfied by these infinitesimal generators we find the structure constants of the *Poincaré group*. Also, it is now clear that the Poincaré group is a Lie group, depending on ten parameters, and comprising the rigid motions of the Minkowski space $\mathcal{M}_4$.

Let us consider separately the one-parameter subgroups of the Poincaré group, which contain finite transformations of the form (5.2.96). For small variations of the parameters the series expansion of the exponential is

$$x'^\mu = x^\mu + \left(\mathsf{X}_i\, \delta a^i\right) x^\mu + \frac{1}{2!} \left(\mathsf{X}_i\, \delta a^i\right)^2 x^\mu + \dots . \tag{5.2.104}$$

Let us now assume that all the variations δa^i vanish except δa^1, and substitute into (5.2.104) the expression of X_1 from (5.2.101). We thus obtain

$$x'^\mu = x^\mu + \left(x^2\, \delta_3^\mu - x^3\, \delta_2^\mu\right) \delta a^1 - \frac{1}{2!} \left(x^2\, \delta_2^\mu + x^3\, \delta_3^\mu\right) (\delta a^1)^2 + \dots , \quad \mu = \overline{1,4} ,$$

that is the coordinate transformation

$$\begin{aligned}
x'^1 &= x^1 , \\
x'^2 &= \left[1 - \frac{1}{2!} (\delta a^1)^2 + \dots\right] x^2 - \left[\delta a^1 - \frac{1}{3!} (\delta a^1)^3 + \dots\right] x^3 \\
&= x^2 \cos (\delta a^1) - x^3 \sin (\delta a^1) , \\
x'^3 &= \left[\delta a^1 - \frac{1}{3!} (\delta a^1)^3 + \dots\right] x^2 + \left[1 - \frac{1}{2!} (\delta a^1)^2 + \dots\right] x^3 \\
&= x^2 \sin (\delta a^1) + x^3 \cos (\delta a^1) , \\
x'^4 &= x^4 .
\end{aligned}$$

Consequently, the transformations determined by the infinitesimal generators of the form (5.2.101) represent rotations of the three-dimensional space. For instance, the transformations determined by X_1 are rotations of angle δa^1 about the x^1-axis. In a similar manner we find that the infinitesimal generators (5.2.102) determine special Lorentz transformations. Thus, the transformation associated to the infinitesimal generator X_6 has the form

$$\begin{aligned}
x'^1 &= x^1 , \\
x'^2 &= x^2 ,
\end{aligned}$$

$$x'^3 = x^3 \cosh(\delta a^6) + x^4 \sinh(\delta a^6),$$
$$x'^4 = x^3 \sinh(\delta a^6) + x^4 \cosh(\delta a^6).$$

This transformation is identical to a transformation of the form (5.1.14) in the hyperplane x^3x^4, for $a_1 = a_2 = 0$ and $\varphi = \delta a^6$. Also, if we set $\cosh(\delta a^6) = 1/\gamma$, $\sinh(\delta a^6) = -\beta/\gamma$, with $\gamma = 1/\sqrt{1-\beta^2}$, then we obtain a special Lorentz transformation corresponding to a matrix of the form $\mathbf{\Lambda}_{34}(\beta)$ (see Chap. 2 Sec. 4.1.1). The infinitesimal generators $\mathsf{X}_7, \ldots \mathsf{X}_{10}$ are associated to the infinitesimal transformations

$$x'^\mu = x^\mu + \delta a^\mu, \quad \mu = \overline{1,4},$$

which represent translations along the coordinate axes.

2.2.2 Conformal transformations

In 1910, H. Bateman and E. Cunnigham shown that the Maxwell equations are invariant with respect to the *inversion*

$$x'^\mu = k^2 \frac{x^\mu}{x_\nu x^\nu}, \quad k \in \mathbb{R}. \tag{5.2.105}$$

By combining this symmetry property with the invariance with respect to the Poincaré group and the invariance relative to dilatations, we obtain the symmetry group of the Maxwell equations in the form of a group depending on 15 parameters, called the *conformal group*. The conformal group comprises the following subgroups:

1 the *homogeneous Lorentz transformations* (six parameters)

$$x'^\mu = \omega^\mu_\nu x^\nu, \quad \omega^\mu_\nu = -\omega^\nu_\mu, \quad \omega^\mu_\nu \in \mathbb{R}, \tag{5.2.106}$$

2 the *translations in* $\mathcal{M}_4$ (four parameters)

$$x'^\mu = x^\mu + \kappa^\mu, \quad \kappa^\mu \in \mathbb{R}, \tag{5.2.107}$$

3 the *Weyl-Zeemann dilatations* (one parameter)

$$x'^\mu = s\, x^\mu, \quad s \in \mathbb{R}, \tag{5.2.108}$$

4 the *special conformal transformations* (four parameters)

$$x'^\mu = \frac{x^\mu + a^\mu \mathbf{x}^2}{1 + 2a^\nu x_\nu + \mathbf{a}^2\mathbf{x}^2}, \quad a^\mu \in \mathbb{R}, \tag{5.2.109}$$

where $a^\nu x_\nu = g^{\mu\nu} a_\mu x_\nu$, $\mathbf{a}^2 = a_\nu a^\nu$, and $\mathbf{x}^2 = x_\nu x^\nu$.

The special conformal transformations (5.2.109) consist of a sequence of three transformations: inversion, translation, inversion, where the inversion is of the form (5.2.105). We thus have (for $k = 1$)

$$x^\mu \to \frac{x^\mu}{\mathbf{x}^2} \to \frac{x^\mu}{\mathbf{x}^2} + a^\mu \to \frac{x^\mu/\mathbf{x}^2 + a^\mu}{(x^\nu/\mathbf{x}^2 + a^\nu)(x_\nu/\mathbf{x}^2 + a_\nu)}$$
$$= \frac{x^\mu + a^\mu\,\mathbf{x}^2}{1 + 2a^\nu x_\nu + \mathbf{a}^2\mathbf{x}^2}\,. \tag{5.2.110}$$

We also mention that the dilatations (5.2.108) can be written in the form

$$x'^\mu = e^\lambda\, x^\mu\,, \quad \lambda \in \mathbb{R}\,. \tag{5.2.111}$$

The linear representations of the conformal group can be obtained in a six-dimensional space, being of the form of linear transformations that leave invariant a bilinear form with the metric tensor

$$\mathbf{G} = [g_{ab}] = \begin{bmatrix} 1 & & & & & \\ & -1 & & & & \\ & & -1 & & & \\ & & & -1 & & \\ & & & & -1 & \\ & & & & & 1 \end{bmatrix}.$$

Note that in the four-dimensional Minkowski space the representations of the conformal group are not linear. The algebra associated to the conformal group is the algebra $\mathsf{O}(6,2)$, isomorphic to the spinor algebra $\mathsf{SU}(2,2)$. The infinitesimal generators of the $\mathsf{O}(6,2)$ algebra form a skew-symmetric tensor having the components

$$\mathsf{J}_{\mu\nu} = \mathsf{M}_{\mu\nu},\ \mathsf{J}_{5\mu} = \frac{1}{2}\,(\mathsf{P}_\mu - \mathsf{K}_\mu)\,,\ \mathsf{J}_{6\mu} = \frac{1}{2}\,(\mathsf{P}_\mu + \mathsf{K}_\mu)\,,\ \mathsf{J}_{65} = \mathsf{D},\ \mu,\nu = \overline{1,4},$$

where $\mathsf{M}_{\mu\nu}$ are the infinitesimal generators of the homogeneous Lorentz transformations, P_μ are the infinitesimal generators of the translations, D is the infinitesimal generator of dilatations, and K_μ represent the infinitesimal generators of the special conformal transformations. Note that P_μ and K_μ are components of four-vectors. The commutation relations have the form

$$[\mathsf{J}_{ab}, \mathsf{J}_{cd}] = i\,(g_{ad}\,\mathsf{J}_{bc} + g_{bc}\,\mathsf{J}_{ad} - g_{ac}\,\mathsf{J}_{bd} - g_{bd}\,\mathsf{J}_{ac})\,, \quad a,b,c,d = \overline{1,6}\,.$$

Thus, it can be seen that the conformal group contains two Abelian subgroups: one generated by P_μ, and the other generated by K_μ. Also, the algebra of the conformal group contains two Poincaré subalgebras: $(\mathsf{M}_{\mu\nu}, \mathsf{P}_\mu)$ and $(\mathsf{M}_{\mu\nu}, \mathsf{K}_\mu)$.

The transformations (5.2.106) and (5.2.107), considered in the form (5.1.46), will contain the Killing vectors (5.2.99). In the case of dilatations (5.2.108) the associated infinitesimal transformations are

$$x'^{\mu} = x^{\mu} + \varepsilon\, \eta_5\, x^{\mu}\,, \tag{5.2.112}$$

where $\varepsilon\eta_5 = \delta s$. Also, the infinitesimal transformations corresponding to (5.2.109) have the form

$$\begin{aligned} x'^{\mu} &= x^{\mu} + \left(\frac{\partial x^{\mu}}{\partial a^{\nu}}\right)_{\mathbf{a}=\mathbf{0}} \delta a^{\nu} = x^{\mu} + \left[\frac{1}{2}\,\delta^{\mu}_{\nu}\,\mathbf{x}^2 - \delta^{\rho}_{\nu}\,x^{\mu}\,x_{\rho}\right]\delta a^{\nu} \\ &= x^{\mu} + \varepsilon\left[x^{\mu}\,x_{\nu} - \frac{1}{2}\,\delta^{\mu}_{\nu}\,\mathbf{x}^2\right]\eta^{\nu}\,, \end{aligned}$$

where $\delta a^{\nu} = -\varepsilon\eta^{\nu}$, and the associated Killing vectors are

$$\begin{aligned} \xi^{\mu} &= \eta^{\nu} x_{\nu} x^{\mu} - \frac{1}{2}\,\eta^{\mu}\mathbf{x}^2 = \eta^{\nu} g_{\nu\rho} x^{\rho} x^{\mu} - \frac{1}{2}\,\eta^{\mu}\mathbf{x}^2 \\ &= \eta_{\rho} x^{\rho} x^{\mu} - \frac{1}{2}\,\eta^{\mu} g_{\nu\rho} x^{\nu} x^{\rho}\,. \end{aligned} \tag{5.2.113}$$

Written in the form (5.1.46) the most general transformation of the conformal group is associated to the Killing vector

$$\xi^{\mu} = \eta_{\nu} x^{\nu} x^{\mu} - \frac{1}{2}\,\eta^{\mu} g_{\nu\rho} x^{\nu} x^{\rho} + \eta_5 x^{\mu} + \omega^{\mu}_{\nu} x^{\nu} + \kappa^{\mu}\,, \quad \mu = \overline{1,4}\,. \tag{5.2.114}$$

Here, $\eta_1, \ldots, \eta_5$ are parameters of the transformations (5.2.108) and (5.2.109), in addition to the ten parameters ($\omega^{\mu}_{\nu} = -\omega^{\nu}_{\mu}$ and κ^{μ}) of the Poincaré group. The Lie derivative of the metric tensor of the Minkowski space $\mathcal{M}_4$, relative to ξ^{μ}, is given by

$$\mathsf{L}_{\boldsymbol{\xi}}\, g_{\mu\nu} = 2\left(\sum_{\rho=1}^{4} \eta_{\rho} x^{\rho} + \eta_5\right) g_{\mu\nu}\,; \tag{5.2.115}$$

therefore, the transformations (5.2.108) and (5.2.109) do not represent rigid motions of $\mathcal{M}_4$. Because the transformations (5.2.106 - 5.2.109) entail a relation of the form $[\sigma(\mathbf{x})]^2 ds^2 = [\sigma(\mathbf{x}')]^2 ds'^2$, it follows that the element of arc does not have an absolute meaning since a comparison of length at two different points implies two different functions $\sigma(\mathbf{x})$, but the ratio of two arc elements at the same point is well defined. Also, the angles at a given point are well defined, the expression

$$\cos\theta = \frac{g_{\mu\nu}\,dx^{\mu}\,dx^{\nu}}{\sqrt{g_{\alpha\beta}\,dx^{\alpha}\,dx^{\beta}}\sqrt{g_{\rho\sigma}\,dx^{\rho}\,dx^{\sigma}}}$$

being invariant relative to the transformations (5.2.106 - 5.2.109).

The invariance of the physical laws relative to the conformal transformations, respectively their invariance when the arc element ds^2 is multiplied by an arbitrary function of coordinates, is equivalent to the independence of these laws with respect to the choice of units. This invariance has the same importance as the invariance with respect to the transformations of coordinates. In the frame of the theory of Special Relativity the group of conformal transformations is associated to the uniformly accelerated motion of reference frames, while the Lorentz transformations (which form a subgroup) correspond to the motion of inertial frames of reference (i.e., reference frames in a uniform motion).

3. Theory of gravitational field

3.1 General equations

3.1.1 The equation of geodesics

The *general theory of relativity* is essentially a theory of the gravitational field. While the theory of Special Relativity is concerned with the relative motion in different nonaccelerated (inertial) frames of reference, the general theory of relativity deals with the relative motion in different accelerated (noninertial) frames of reference, and is based on two fundamental principles:

1 *The equivalence principle.* In a small spatial domain, the gravitational field (homogeneous and uniform) is equivalent, regarding its effects, to an acceleration field (the *field of inertial forces*).

2 *The covariance principle.* The physical laws are covariant in any system of coordinates.

The physical space of the general theory of relativity is a Riemannian four-dimensional space, which differs from the Minkowskian space. In such spaces the motion of a free particle is represented by a line called *geodesic*, which is no longer straight in the Euclidean sense, but is the line corresponding to the minimum action. Actually, the motion of a free particle is represented by a zero length geodesic. The element of the arc is given by (5.1.45) where $g_{\mu\nu}(\mathbf{x}) = g_{\nu\mu}(\mathbf{x})$ are the components of the metric tensor. Only in the particular case of an inertial reference frame (*Galilean reference frame*), the matrix of the metric tensor is diagonal and consists of constant elements (see (2.4.4)).

The geodesics are given by the integral trajectories of minimum length, along which the action integral for a free particle

$$\mathcal{A} = -m_0\, c \int_a^b ds \tag{5.3.1}$$

is stationary. Thus, by applying the Hamilton principle ($\delta\mathcal{A} = 0$) we have

$$-\delta\mathcal{A} = m_0\, c \int_a^b \delta\, ds = m_0\, c \int_a^b \frac{\delta(g_{\mu\nu} dx^\mu dx^\nu)}{2ds} = 0\,, \tag{5.3.2}$$

where a and b are two points on the geodesic. If we introduce the components of the generalized velocity (5.1.72), and take into account the relation

$$\delta(g_{\mu\nu}dx^\mu dx^\nu) = (\delta g_{\mu\nu})dx^\mu dx^\nu + g_{\mu\nu}\delta(dx^\mu)dx^\nu + g_{\mu\nu}dx^\mu\delta(dx^\nu)\,,$$

then we may rewrite (5.3.2) in the form

$$\begin{aligned} -\delta\mathcal{A} &= m_0 c \int_a^b \left[\frac{1}{2}\frac{\partial g_{\mu\nu}}{\partial x^\rho}\,u^\mu u^\nu \delta x^\rho + g_{\mu\nu}u^\mu\,\frac{d(\delta x^\nu)}{ds}\right] ds \\ &= m_0 c \int_a^b \left[\frac{1}{2}\frac{\partial g_{\mu\nu}}{\partial x^\rho}\,u^\mu u^\nu \delta x^\rho - \frac{d}{ds}(g_{\mu\nu}u^\mu)\delta x^\nu + \frac{d}{ds}(g_{\mu\nu}u^\mu\delta x^\nu)\right] ds \\ &= m_0 c \int_a^b \left[\frac{1}{2}\frac{\partial g_{\mu\nu}}{\partial x^\rho}\,u^\mu u^\nu - \frac{d}{ds}(g_{\mu\rho}u^\mu)\right] \delta x^\rho ds + m_0 c g_{\mu\nu}u^\mu \delta x^\nu \Big|_a^b . \end{aligned}$$

By admitting that $(\delta x^\nu)_a = (\delta x^\nu)_b = 0$ we obtain the equation of geodesics

$$\frac{1}{2}\frac{\partial g_{\mu\nu}}{\partial x^\rho}\,u^\mu u^\nu - \frac{d}{ds}(g_{\mu\rho}u^\mu) = 0\,, \quad \rho = \overline{1,4}\,. \tag{5.3.3}$$

The second term in (5.3.3) can be written as

$$\begin{aligned} \frac{d}{ds}(g_{\mu\rho}u^\mu) &= \frac{\partial g_{\mu\rho}}{\partial x^\lambda}\,u^\mu u^\lambda + g_{\mu\rho}\,\frac{du^\mu}{ds} \\ &= \frac{1}{2}\left(\frac{\partial g_{\mu\rho}}{\partial x^\lambda} + \frac{\partial g_{\lambda\rho}}{\partial x^\mu}\right) u^\mu u^\lambda + g_{\mu\rho}\,\frac{du^\mu}{ds}\,, \end{aligned}$$

and substituting into (5.3.3) it results

$$g_{\mu\rho}\,\frac{du^\mu}{ds} + [\mu\,\lambda,\,\rho]\,u^\mu u^\lambda = 0\,, \quad \rho = \overline{1,4}\,,$$

where we have used the Christoffel symbols of the first kind

$$[\mu\,\lambda,\,\rho] = \frac{1}{2}\left(\frac{\partial g_{\mu\rho}}{\partial x^\lambda} + \frac{\partial g_{\lambda\rho}}{\partial x^\mu} - \frac{\partial g_{\mu\lambda}}{\partial x^\rho}\right). \tag{5.3.4}$$

In terms of the Christoffel symbols of the second kind (see the definition of the covariant derivative (5.1.33))

$$\left\{ {\nu \atop \mu\ \lambda} \right\} = g^{\nu\rho}\,[\mu\,\lambda,\,\rho]\,, \tag{5.3.5}$$

taking into account that $g^{\nu\rho}g_{\rho\mu} = \delta^\nu_\mu$, we obtain the explicit covariant form of the geodesic equations

$$\frac{du^\nu}{ds} + \left\{ {\nu \atop \mu\ \lambda} \right\} u^\mu u^\lambda = 0\,, \quad \nu = \overline{1,4}\,, \tag{5.3.6}$$

where $du^\nu/ds = d^2x^\nu/ds^2$ are the components of the particle *four-acceleration*. The equation (5.3.6) describes the motion of a free particle in a Riemannian four-dimensional space with the metric tensor $g_{\mu\nu}(\mathbf{x})$.

Now, according to (5.1.76) the variation of the action along the geodesics is given by

$$\delta S = -m_0\, c\, g_{\mu\nu} u^\mu \delta x^\nu \,, \tag{5.3.7}$$

such that the components of the particle four-momentum are expressed by the formulae

$$p_\nu = \frac{\partial S}{\partial x^\nu} = -m_0\, c\, g_{\mu\nu} u^\mu = -m_0\, c\, u_\nu \,, \quad \nu = \overline{1,4}\,. \tag{5.3.8}$$

3.1.2 The equations of gravitational field

We can deduce the equations of the gravitational field from a variational principle of the form

$$\delta S = \delta \int L\, d\omega = \delta \int L\, dx^1 dx^2 dx^3 dx^4 = 0\,, \tag{5.3.9}$$

but we have to take into account the fact that, if L is a scalar then this integral over the whole four-dimensional space is not covariant with respect to coordinate transformations. Under a coordinate transformation, the transformed volume element is multiplied by the transformation Jacobian, and we have to find a factor to compensate this multiplication by the transformation Jacobian. We note that, under a coordinate transformation, the metric tensor transforms according to the relation (5.1.29) which, in matrix notation, reads

$$\mathbf{g}' = \mathbf{J}\,\mathbf{g}\,\mathbf{J}^{\mathrm{T}}\,, \quad \mathbf{J} \equiv [J_{\mu\nu}] = \left[\frac{\partial x^\mu}{\partial x'^\nu}\right]. \tag{5.3.10}$$

The determinant of the matrix $\mathbf{J}$ is the Jacobian of the transformation ($J = \det \mathbf{J}$), and from (5.3.10) we obtain

$$\det \mathbf{g}' = J^2 \det \mathbf{g}\,. \tag{5.3.11}$$

Because in a Galilean reference frame $g \equiv \det \mathbf{g} = -1$, the sought factor has the form $\sqrt{-g}$. Hence, the variational principle that satisfies the principle of covariance, and can be used to derive the field equations, is

$$\delta \int L \sqrt{-g}\, d\omega = 0\,, \tag{5.3.12}$$

where L is a relativistic scalar.

The Lagrangian L has to be constructed from the field quantities, respectively from the components of the metric tensor, and if we assume that the field equations are partial differential equations of the second order, then $L = L(g_{\mu\nu}, \partial g_{\mu\nu}/\partial x^\rho)$. The simplest relativistic scalar which can be formed using the components of the metric tensor and their derivatives is the Ricci scalar (5.1.43), such that we may assume $L = R$. Consequently, we obtain the variational principle

$$\begin{aligned}\delta \int R\sqrt{-g}\, d\omega &= \int \left(\delta R\sqrt{-g} + R\,\delta\sqrt{-g}\right) d\omega \qquad (5.3.13)\\ &= \int \left[R_{\mu\nu}\,\delta(\sqrt{-g}\, g^{\mu\nu}) + \sqrt{-g}\, g^{\mu\nu}\,\delta R_{\mu\nu}\right] d\omega = 0\,.\end{aligned}$$

By observing that the variation of Ricci's tensor (5.1.42) is

$$\delta R_{\mu\nu} = -\left(\delta \left\{ {\lambda \atop \mu\ \nu} \right\}\right)_{;\lambda} + \left(\delta \left\{ {\lambda \atop \mu\ \lambda} \right\}\right)_{;\nu}, \qquad (5.3.14)$$

taking into account that

$$\delta\sqrt{-g} = -\frac{1}{2}\sqrt{-g}\, g^{\mu\nu}\,\delta g_{\mu\nu}\,,$$

and Ricci's theorem (Theorem 5.1.1), the variational principle (5.3.13) becomes

$$\begin{aligned}\int \Bigg\{ &\sqrt{-g}\left(R_{\mu\nu} - \frac{1}{2}g_{\mu\nu}R\right)\delta g^{\mu\nu} \\ &+ \frac{\partial}{\partial x^\lambda}\left[\sqrt{-g}\left(-g^{\mu\nu}\delta\left\{ {\lambda \atop \mu\ \nu} \right\} + g^{\mu\lambda}\delta\left\{ {\rho \atop \mu\ \rho} \right\}\right)\right]\Bigg\} d\omega = 0\,.\end{aligned}$$

The second term is a divergence leading to a surface integral which contains variations that, on the basis of the fundamental hypothesis of the variational principle, vanish on the boundary. Thus, we may write (5.3.13) in the form

$$\delta \int R\sqrt{-g}\, d\omega = \int \left[\sqrt{-g}\left(R_{\mu\nu} - \frac{1}{2}g_{\mu\nu}R\right)\right]\delta g^{\mu\nu}\, d\omega = 0\,, \qquad (5.3.15)$$

wherefrom it results the field equations in the absence of matter

$$R_{\mu\nu} - \frac{1}{2}g_{\mu\nu}R = 0\,, \quad \mu,\nu = \overline{1,4}\,, \qquad (5.3.16)$$

or, in terms of the components of Einstein's tensor (5.1.44),

$$G_{\mu\nu} = 0\,. \qquad (5.3.17)$$

The components of Einstein's tensor satisfy the relations $G^{\mu\nu}_{;\nu} = 0$, and also $g^{\mu\nu} G_{\mu\nu} = -R$, such that the field equations (5.3.16) are equivalent to the equations

$$R_{\mu\nu} = 0\,. \tag{5.3.18}$$

In the presence of matter, the variational principle for the field equations takes the form

$$\delta \int (R - 2\kappa\, L_m)\, \sqrt{-g}\, d\omega = 0\,, \tag{5.3.19}$$

where κ is *Einstein's gravitational constant*, and L_m is the Lagrangian density of the matter field or the total Lagrangian density also including the interactions (if there are different kinds of matter fields). In the case when L_m has the simple form

$$L_m = L_m(g_{\mu\nu}, \psi_a, \partial\psi_a/\partial x^\mu)\,, \tag{5.3.20}$$

where ψ_a are state functions for the fields, the second term in (5.3.19) becomes

$$\delta \int L_m\, \sqrt{-g}\, d\omega = -\frac{1}{2} \int T_{\mu\nu}\, \delta g^{\mu\nu}\, \sqrt{-g}\, d\omega\,, \tag{5.3.21}$$

where

$$T_{\mu\nu} = -\frac{2}{\sqrt{-g}} \left\{ \left[\frac{\partial(\sqrt{-g}\, L_m)}{\partial g^{\mu\nu}_{,\lambda}} \right]_{,\lambda} - \frac{\partial(\sqrt{-g}\, L_m)}{\partial g^{\mu\nu}} \right\}$$

is the *energy-momentum* tensor, and the comma indicates partial differentiation with respect to x^λ. The equations (5.3.16) take place if $\partial L_m/\partial\psi_a = 0$ and $g_{\mu\nu}$ and ψ_a are independently varied, because in such cases we have

$$\delta \int L_m\, \sqrt{-g}\, d\omega = \int \left\{ \left[\frac{\partial(\sqrt{-g}\, L_m)}{\partial g^{\mu\nu}} \right] \delta g^{\mu\nu} + \left[\frac{\partial(\sqrt{-g}\, L_m)}{\partial \psi_a} \right] \delta\psi_a \right\} d\omega\,.$$

Finally, from (5.3.15), (5.3.19), and (5.3.21) we obtain the field equations

$$R_{\mu\nu} - \frac{1}{2}\, g_{\mu\nu}\, R = \kappa\, T_{\mu\nu}\,, \quad \mu, \nu = \overline{1,4}\,, \tag{5.3.22}$$

or $G_{\mu\nu} = \kappa\, T_{\mu\nu}$.

Because the divergence of Einstein's tensor vanishes ($G^{\mu\nu}_{;\nu} = 0$) it follows that the divergence of the energy-momentum tensor also vanishes ($T^{\mu\nu}_{;\nu} = 0$), and this result represents the *theorem of energy and momentum conservation* for the gravitational field.

In the gravitational Riemann space, the Maxwell equations are obtained by applying Hamilton's principle to Lagrangian density

$$L = -\frac{1}{4\mu_0}\, F_{\mu\nu}\, F^{\mu\nu} + j^\mu\, A_\mu + L_e\,, \tag{5.3.23}$$

where L_e is the Lagrange density for charged particles. Now, the electromagnetic field tensor has the components

$$F_{\mu\nu} = A_{\mu;\nu} - A_{\nu;\mu}\,, \tag{5.3.24}$$

and the equations of the electromagnetic field have the same form as (5.2.38) and (5.2.40) if we replace the partial derivatives by the corresponding covariant derivatives

$$F^{\mu\nu}_{;\nu} = \mu_0\, j^\mu\,, \qquad \left(\epsilon^{\mu\nu\rho\sigma}\,\frac{F_{\mu\nu}}{\sqrt{-g}}\right)_{;\sigma} = 0\,. \tag{5.3.25}$$

3.2 Conservation laws in the Riemann space

3.2.1 The free particle in the Riemann space

The explicit form of the components of the metric tensor $g_{\mu\nu}(\mathbf{x})$, determined by Einstein's equations (5.3.22), depends on the intrinsic relation between the matter and fields, the relative motion of the matter and the fields existing in the reference frame, as well as the choice of the system of coordinates. In the general theory of relativity the presence of gravitational matter and field sources is equivalent to the curvature of the space-time, described by a Riemannian four-dimensional geometry with the first fundamental form (5.1.45). Thus, it has been said that the general theory of relativity makes the geometry part of physics. In any reference frame, the Riemann-Christoffel curvature tensor (5.1.40) indicates if the space-time is curved or flat. In the absence of gravitational matter and field sources, the metric tensor satisfies the equations (5.3.16) and the trajectory of a particle is a geodesic determined by the equations (5.3.6). Note that, in both cases of curved and flat space-time the equations of the geodesics have the same form.

There is a close relation between the group of rigid motions in a Riemannian n-dimensional space $\mathcal{V}_n$ and the first integrals of the motion along the geodesics, expressed by the following two theorems:

THEOREM 5.3.1 *If a Riemannian space $\mathcal{V}_n$ admits an m-parameter group of rigid motions, then the number of independent linear first integrals of the geodesic equations is equal to m, and reciprocally.*

THEOREM 5.3.2 *The maximum number of essential parameters of a group of rigid motions admitted by a Riemannian space $\mathcal{V}_n$ is $m = n(n+1)/2$, and it is attained only if the space has a constant curvature.*

Now, we will consider the motion of a free particle along a geodesic. Similarly to the case of Special Relativity (see Chap. 5 Sec. 1.2.3), for a Lagrangian density

$$L = L(s, x^\mu, dx^\mu/ds)\,, \tag{5.3.26}$$

we obtain the first integrals of the motion on a geodesic in the form

$$p_\mu \, \xi^\mu_{(\alpha)} = \text{const}, \quad \alpha = 1, 2, \ldots, m. \tag{5.3.27}$$

If the geodesic equations (5.3.6) are satisfied, then the generator of the conservation laws is

$$\frac{d}{dt}\left(m_0 c \, g_{\mu\nu} \, \frac{dx^\mu}{ds} \, \xi^\nu \right) = \frac{1}{2} \, m_0 c \, (\xi_{\mu;\nu} + \xi_{\nu;\mu}) \, \frac{dx^\mu}{ds} \frac{dx^\nu}{ds} \, . \tag{5.3.28}$$

Consequently, when the Killing equations (5.1.51) are satisfied, from (5.3.28) one obtains the constants of motion (independent linear first integrals), corresponding to the motion on the geodesic. For instance, in the case of the Schwartzschild metric

$$ds^2 = -\left(1 + \frac{\alpha}{4r}\right)^4 \left[(dx^1)^2 + (dx^2)^2 + (dx^3)^2\right] + \frac{\left(1 - \dfrac{\alpha}{4r}\right)^2}{\left(1 + \dfrac{\alpha}{4r}\right)^2} \, (dx^4)^2 \, , \tag{5.3.29}$$

where $r^2 = (x^1)^2 + (x^2)^2 + (x^3)^2$ and α is an integration constant, the Killing equations admit the solutions

$$\xi^\mu = \omega^\mu_\nu \, x^\nu \quad \mu = 1, 2, 3, \quad \xi^4 = k^4 \, ,$$

with ω^μ_ν and k^4 being constants. Then, by means of the generator (5.3.28) one obtains the conservation law of angular momentum

$$m_0 \, c \left(1 + \frac{\alpha}{4r}\right)^4 \left(x^i \, \frac{dx^j}{ds} - x^j \, \frac{dx^i}{ds} \right) = \text{const}, \quad i, j = 1, 2, 3, \tag{5.3.30}$$

and the conservation law of total mechanical energy

$$m_0 \, c \, \frac{\left(1 - \dfrac{\alpha}{4r}\right)^2}{\left(1 + \dfrac{\alpha}{4r}\right)^2} \, \frac{dt}{ds} = \text{const}. \tag{5.3.31}$$

In the case of conformal transformations (5.2.106)–(5.2.109) we may use the Lie derivative (5.2.115). If $\eta_\rho = 0$ $(\rho = \overline{1, 4})$ and $\eta_5 \neq 0$, then

$$(\mathsf{L}_{\boldsymbol{\xi}} \, g_{\mu\nu})_{;\lambda} = (\xi_{\mu;\nu} + \xi_{\nu;\mu})_{;\lambda} = 0 \, , \tag{5.3.32}$$

where $\xi_{\mu;\nu\lambda} = -\xi_\rho \, R^\rho_{\lambda\nu\mu}$. Thus, we obtain the first integrals

$$\frac{d}{ds}\left[\frac{d}{ds} \left(m_0 c \, g_{\mu\nu} \, \frac{dx^\mu}{ds} \, \xi^\nu \right) \right] = \frac{1}{2} \, m_0 \, c \, \frac{d}{ds} (2\eta_5) = 0. \tag{5.3.33}$$

An important quantity related to the conformal transformations is the *Weyl tensor* defined by the relation

$$C_{\rho\sigma\mu\nu} = R_{\rho\sigma\mu\nu} - \frac{1}{2}(g_{\rho\mu}R_{\nu\sigma} - g_{\rho\nu}R_{\mu\sigma} - g_{\sigma\mu}R_{\nu\rho} + g_{\sigma\nu}R_{\mu\rho}) - \frac{1}{6}(g_{\rho\nu}g_{\mu\sigma} - g_{\rho\mu}g_{\nu\sigma})R\,, \quad (5.3.34)$$

that has the property $C^{\rho}_{\mu\rho\sigma} = 0$. If the Weyl tensor vanishes at every point in the space, then the metric is *flat-conformal*, i.e., there exists a transformation which diagonalizes the metric tensor, the elements on the diagonal being functions of coordinates.

3.2.2 Conservation laws of the field

Let us consider the variation of the Lagrangian density $\sqrt{-g}\,R$, under an infinitesimal transformation of coordinates of the form (5.1.46). By applying the same method as in the case of the variation of Lagrangian action (5.3.13) we have

$$\delta(\sqrt{-g}\,R) = \delta(\sqrt{-g}\,g^{\mu\nu}R_{\mu\nu}) = -\sqrt{-g}\,G^{\mu\nu}\delta g_{\mu\nu} + \sqrt{-g}\,g^{\mu\nu}\delta R_{\mu\nu}\,. \quad (5.3.35)$$

Then, from the expression (5.3.14) of the variation $\delta R_{\mu\nu}$ and the equation (5.1.48) we may write

$$\delta g_{\mu\nu} = -\varepsilon\,(\xi_{\mu;\nu} + \xi_{\nu;\mu})\,,$$

$$\delta\left\{ \begin{matrix} \rho \\ \mu\ \nu \end{matrix} \right\} = \frac{1}{2}\,g^{\rho\lambda}\,(\delta g_{\mu\lambda;\nu} + \delta g_{\nu\lambda;\mu} - \delta g_{\mu\nu;\lambda})\,.$$

Therefore, the equation (5.3.35) becomes

$$\delta(\sqrt{-g}\,R) = \varepsilon\sqrt{-g}\left[2G^{\mu\nu}\xi_\mu - (g^{\rho\sigma}g^{\mu\nu} - g^{\rho\mu}g^{\nu\sigma})\,(\xi_{\rho;\sigma} + \xi_{\sigma;\rho})_{;\mu}\right]_{;\nu}\,. \quad (5.3.36)$$

Taking into account the relation

$$\xi_{\rho;\sigma\mu} - \xi_{\rho;\mu\sigma} = \xi_\lambda R^{\lambda}_{\rho\sigma\mu}\,,$$

we may also write

$$\delta(\sqrt{-g}\,R) = \varepsilon\sqrt{-g}\Big[2G^{\mu\nu}\xi_\mu - g^{\rho\sigma}g^{\mu\nu}R^{\lambda}_{\rho\sigma\mu}\xi_\lambda + g^{\rho\mu}g^{\nu\sigma}R^{\lambda}_{\rho\sigma\mu}\xi_\lambda - \frac{1}{2}\,(g^{\rho\sigma}g^{\mu\nu} - g^{\mu\sigma}g^{\nu\rho})\,\xi_{\rho;\mu\sigma}\Big]_{;\nu}\,.$$

Because the curvature tensor satisfies the relations

$$\begin{aligned} g^{\rho\lambda} g^{\mu\nu} R^{\sigma}_{\rho\lambda\mu} &= g^{\rho\lambda} g^{\mu\nu} g^{\sigma\kappa} R_{\kappa\rho\lambda\mu} = g^{\mu\nu} g^{\sigma\kappa} R^{\lambda}_{\kappa\mu\lambda} = g^{\mu\nu} R^{\sigma}_{\mu} = R^{\nu\sigma} , \\ g^{\rho\mu} g^{\nu\lambda} R^{\sigma}_{\rho\lambda\mu} &= g^{\rho\mu} g^{\nu\lambda} g^{\sigma\kappa} R_{\kappa\rho\lambda\mu} = -g^{\nu\lambda} g^{\sigma\kappa} R^{\mu}_{\kappa\lambda\mu} = -R^{\nu\sigma} , \end{aligned}$$

it results that

$$\delta(\sqrt{-g}\, R) = \varepsilon \sqrt{-g} \left[2G^{\mu\nu} \xi_{\mu} - 2R^{\mu\nu} \xi_{\mu} - \frac{1}{2} \left(g^{\rho\sigma} g^{\mu\nu} - g^{\mu\sigma} g^{\nu\rho} \right) \xi_{\rho;\mu\sigma} \right]_{;\nu} ,$$

and, by substituting the expression of Einstein's tensor (5.1.44), we finally obtain

$$\delta(\sqrt{-g}\, R) = \varepsilon \left\{ \sqrt{-g} \left[-R\, \xi^{\nu} - \frac{1}{2} \left(\xi^{\sigma;\nu} - \xi^{\nu;\sigma} \right)_{;\sigma} \right] \right\}_{;\nu} . \qquad (5.3.37)$$

Note that here, we are allowed to include $\sqrt{-g}$ in the covariant derivative with respect to x^{ν} on the basis that the rules of covariant differentiation for sums and products of tensors are the same as those for ordinary differentiation, and from Ricci's theorem (Theorem 5.1.1) we have

$$(\sqrt{-g})_{;\nu} = 0 .$$

The variation of the Lagrangian density with respect to an infinitesimal transformation of the form (5.1.46) can also be written as

$$\delta(\sqrt{-g}\, R) = -\varepsilon \frac{\partial}{\partial x^{\mu}} \left(\sqrt{-g}\, R\, \xi^{\mu} \right)$$

and, by equating with (5.3.37), we obtain the equation

$$-\frac{\partial}{\partial x^{\mu}} \left(\sqrt{-g}\, R\, \xi^{\mu} \right) = \left\{ \sqrt{-g} \left[-R\, \xi^{\nu} - \frac{1}{2} \left(\xi^{\sigma;\nu} - \xi^{\nu;\sigma} \right)_{;\sigma} \right] \right\}_{;\nu} ,$$

wherefrom

$$\left[\sqrt{-g} \left(\xi^{\sigma;\nu} - \xi^{\nu;\sigma} \right)_{;\sigma} \right]_{;\nu} \equiv 0 , \qquad (5.3.38)$$

because

$$-\frac{\partial}{\partial x^{\mu}} \left(\sqrt{-g}\, R\, \xi^{\mu} \right) = \left(\sqrt{-g}\, R\, \xi^{\mu} \right)_{;\mu} .$$

The equations (5.3.38) determine the *strong conservation laws* of the field. These conservation laws are expressed as the divergence of a tensor, which vanishes identically, independently of the fact that the field equations are satisfied or not. Now, the covariant generator of the conservation laws of the field, resulting from the symmetry properties of the Riemannian space, reads

$$\left[\sqrt{-g} \left(\xi^{\mu;\nu} - \xi^{\nu;\mu} \right) \right]_{;\mu\nu} = \frac{\partial}{\partial x^{\nu}} \left\{ \left[\sqrt{-g} \left(\xi^{\mu;\nu} - \xi^{\nu;\mu} \right) \right]_{;\mu} \right\} = 0 . \qquad (5.3.39)$$

For motions of the Riemannian space ($\xi^{\mu;\nu} + \xi^{\nu;\mu} = 0$), the generator of the conservation laws is

$$\frac{\partial}{\partial x^\nu}\left[\sqrt{-g}\,(\xi^{\mu;\nu})_{;\mu}\right] = 0\,. \tag{5.3.40}$$

These conservation laws can also be expressed in the integral form

$$\int \frac{\partial}{\partial x^\mu}\left[\sqrt{-g}\,(\xi^{\mu;\nu} - \xi^{\nu;\mu})\right] d\sigma_\nu = 0\,, \tag{5.3.41}$$

the integration being performed over a three-dimensional manifold in the space-time. For closed physical systems, (5.3.41) becomes

$$\begin{aligned}\int \frac{\partial}{\partial x^\mu}\left[\sqrt{-g}\,(\xi^{4;\mu} - \xi^{\mu;4})\right] d\sigma_4 &= \int \sum_{k=1}^{3} \frac{\partial}{\partial x^k}\left[\sqrt{-g}\left(\xi^{4;k} - \xi^{k;4}\right)\right] dV \\ &= \int \sum_{k=1}^{3} \sqrt{-g}\left(\xi^{4;k} - \xi^{k;4}\right) n_k\, dS\,,\end{aligned} \tag{5.3.42}$$

where n_k are the components of the exterior normal to the surface S, that represents the boundary of the volume V.

Thus, if the Riemannian space-time exhibits some symmetry properties, then the conserved physical quantities are given by the equations (5.3.42) successively eliminating the parameters which the Killing vectors ξ^μ depend on, as in the case of the theory of Special Relativity (see Chap. 5 Sec. 1.2.3).

Chapter 6

APPLICATIONS IN QUANTUM MECHANICS AND PHYSICS OF ELEMENTARY PARTICLES

1. Non-relativistic quantum mechanics

1.1 Invariance properties of quantum systems

1.1.1 Schrödinger's equation

The first form of Schrödinger's equation has been obtained as a result of the theoretical studies carried out to establish a differential equation able to model the energy levels of the electron in a hydrogen atom. Later on, in its general form, Schrödinger's equation reached the same central importance to quantum mechanics as Newton's laws of motion to classical mechanics.

According to Schrödinger's model, the energy levels of the electron are eigenvalues determining solutions of a differential equation, that are real, single-valued, bounded, and of class C^2. To obtain this differential equation Schrödinger started from the Hamilton-Jacobi equation for the hydrogen atom

$$\left(\frac{\partial S}{\partial x}\right)^2+\left(\frac{\partial S}{\partial y}\right)^2+\left(\frac{\partial S}{\partial z}\right)^2-2m\left(E+\frac{e^2}{r}\right)=0\,, \tag{6.1.1}$$

where S is the *action* associated to the electron, m, e, and E are respectively the electron mass, electric charge, and energy, while r is the distance between the electron and nucleus. Note that, the electrostatic energy represents the interaction between a proton, of charge $q_p = |e| > 0$, and an electron, of charge $q_e = -|e| < 0$, so that $q_p q_e = -e^2$. Schrödinger has replaced the action S in (6.1.1) by the function $S = k \ln \psi$, and considered that the left hand side of the Hamilton-Jacobi equation is a functional depending on the state function ψ and its derivatives, while the independent variables x, y, and z do not appear explicitly in the expression of the functional

$$\left(\frac{\partial \psi}{\partial x}\right)^2+\left(\frac{\partial \psi}{\partial y}\right)^2+\left(\frac{\partial \psi}{\partial z}\right)^2-\frac{2m}{k^2}\left(E+\frac{e^2}{r}\right)\psi^2=0\,. \tag{6.1.2}$$

Assuming that the variation of the left hand side in (6.1.2), integrated over the whole space, vanishes, it results the Euler-Lagrange equation

$$\Delta\psi + \frac{2m}{k^2}\left(E + \frac{e^2}{r}\right)\psi = 0\,. \tag{6.1.3}$$

Then, by setting $k = \hbar = h/2\pi$, where h is Planck's constant, one obtains the time-independent Schrödinger equation

$$-\frac{\hbar^2}{2m}\,\Delta\psi + V\,\psi = E\,\psi\,, \tag{6.1.4}$$

with $V = -e^2/r$ representing the *potential energy* of the electron in the electrostatic field generated by the nucleus. Schrödinger has shown that the formula for the energy levels of the electron in a hydrogen atom, derived from the eigenvalues E of (6.1.4), coincides with the the Balmer series formula in atomic spectroscopy.

The time-independent Schrödinger equation can be generalized by assuming that the potential energy V is a function $V(x, y, z)$ associated to a conservative force field $\mathbf{F} = -\nabla V$. Thus, the equation (6.1.4) describes the motion of a particle of mass m in the conservative force field $\mathbf{F}$. Furthermore, for a conservative system we may consider that the solutions of (6.1.4) depend on time as

$$\psi(\mathbf{r}, t) = u(\mathbf{r})\, e^{-i\frac{E}{\hbar}\,t}\,, \tag{6.1.5}$$

and satisfy the time-dependent Schrödinger equation

$$-\frac{\hbar^2}{2m}\,\Delta\psi + V\,\psi = i\hbar\,\frac{\partial\psi}{\partial t}\,. \tag{6.1.6}$$

Note that, we may also consider that the solutions of (6.1.4) have a time dependence of the form $\exp\left(-i\omega t\right)$, that is the time dependence of a plane wave, and using Planck's relation $E = \hbar\omega$ we obtain (6.1.5).

In a Lagrangian formulation (6.1.6) becomes the Euler-Lagrange equation of a variational problem, for the Lagrangian density

$$L = i\hbar\,\overline{\psi}\,\frac{\partial\psi}{\partial t} - \frac{\hbar^2}{2m}\,\nabla\overline{\psi}\cdot\nabla\psi - V(\mathbf{r}, t)\,\overline{\psi}\,\psi\,, \tag{6.1.7}$$

if we vary $\overline{\psi}$. Also, by varying ψ we get an equation that is the complex conjugate of (6.1.6)

$$-\frac{\hbar^2}{2m}\,\Delta\overline{\psi} + V\,\overline{\psi} = -i\hbar\,\frac{\partial\overline{\psi}}{\partial t}\,. \tag{6.1.8}$$

Then, the generalized momentum canonically conjugate to ψ is

$$\pi = \frac{\partial L}{\partial\dot{\psi}} = i\hbar\,\overline{\psi}\,, \tag{6.1.9}$$

where $\dot{\psi} = \partial\psi/\partial t$ and, because $\dot{\overline{\psi}}$ does not appear in (6.1.7), it follows that the generalized momentum canonically conjugate to $\overline{\psi}$ is $\widetilde{\pi} = 0$. By means of (6.1.9) we can eliminate $\overline{\psi}$ in (6.1.7), thus obtaining the Hamiltonian density

$$\mathcal{H} = \pi\,\dot{\psi} - L = -\frac{i\hbar}{2m}\,\nabla\pi\cdot\nabla\psi - \frac{i}{\hbar}\,V\,\pi\,\psi\,. \tag{6.1.10}$$

Consequently, the system of canonical equations reads

$$\begin{aligned} \dot{\psi} &= \quad [\mathcal{H}]^{\pi} = -\frac{i}{\hbar}\,V\,\psi + \frac{i\hbar}{2m}\,\Delta\psi\,, \\ \dot{\pi} &= -[\mathcal{H}]^{\psi} = \quad \frac{i}{\hbar}\,V\,\pi - \frac{i\hbar}{2m}\,\Delta\pi\,, \end{aligned} \tag{6.1.11}$$

where we have used the Euler-Lagrange derivative (4.1.94). It is easy to see that the first equation in (6.1.11), multiplied by $i\hbar$, is identical to (6.1.6), while the second equation, multiplied by $-i\hbar$, is identical to (6.1.8).

Generally, it is assumed that the *wave function* ψ is able to provide all the information we may obtain relative to a particle (microobject). Max Born has been the first to introduce the hypothesis that the expression

$$\rho(\mathbf{r},t)\,dxdydz = |\psi|^2\,dxdydz$$

represents the probability to find the particle inside the considered volume element, at the moment t, in the vicinity of the point $\mathbf{r}$. Thus, the wave function ψ unambiguously determines the probability with which one observer may obtain a given value for a certain physical quantity, as the result of an interaction (generally uncontrollable) between the particle and the device used to measure that physical quantity. Hence, to each physical quantity A we may associate an operator A, that admits a complete set of eigenfunctions φ_n defined by the eigenvalues λ_n ($\mathsf{A}\,\varphi_n = \lambda_n\,\varphi_n$). In terms of the eigenfunctions φ_n, the wave function ψ can be represented as a series

$$\psi = \sum_n c_n\,\varphi_n\,, \tag{6.1.12}$$

where the square modulus of the coefficients $|c_n|^2$ gives the probability to obtain, as a result of a measurement, the value λ_n for the physical quantity A.

It can be shown that the operators associated to the energy and momentum are the differential operators

$$\mathsf{E} = i\hbar\,\frac{\partial}{\partial t}\,,\quad \mathsf{p}_k = -i\hbar\,\frac{\partial}{\partial x_k}\,,\quad k = 1,2,3, \tag{6.1.13}$$

while the operators associated to the coordinates are the multiplicative operators

$$\mathsf{x}_k = x_k\,,\quad k = 1,2,3. \tag{6.1.14}$$

(For the spatial coordinates we will use the notations x, y, z or x_1, x_2, x_3 depending on the context.) If the classical expression of a physical quantity $A = A(p_k, x_k)$ is known, then the operator associated to A follows from the relation $\mathsf{A} = A(\mathsf{p}_k, \mathsf{x}_k)$, where we have to take into account certain rules regarding the composition of operators. Consequently, the operator of the potential energy is

$$\mathsf{V} = V(\mathsf{x}_k) = V(x_k)\,,$$

and applying it to the wave function ψ means the multiplication of ψ by $V(x_k)$. The operator of the kinetic energy is

$$\mathsf{T} = \frac{\mathsf{p}^2}{2m} = -\frac{\hbar^2}{2m}\,\nabla^2 = -\frac{\hbar^2}{2m}\,\Delta\,.$$

Because $H = T + V$, the operator associated to the Hamiltonian has to be of the form

$$\mathsf{H} = \mathsf{T} + \mathsf{V} = -\frac{\hbar^2}{2m}\,\Delta + V \tag{6.1.15}$$

such that, the time-independent Schrödinger equation (6.1.4) can be regarded as the eigenvalue equation for the Hamiltonian operator

$$\mathsf{H}\,\psi = E\,\psi\,. \tag{6.1.16}$$

1.1.2 The system of Natural Units ($\hbar = c = 1$)

Some physical constants are of universal importance and appear in many formulae and equations. Hence, it is convenient to choose units in which they become unity and dimensionless. When this is done physics appears more transparent, but we have to convert to more common units at the end of calculations. Accordingly, throughout this chapter, we choose *Natural Units* and set $\hbar = c = 1$, as is commonly done in elementary particle physics.

In the system of Natural Units, the actions and velocities become dimensionless quantities, and are measured in units of elementary quanta $\hbar$ and light velocity in vacuum c, respectively. It follows that any velocity is bounded by c, so that $v \leq 1$. Also, we use action (A), velocity (V), and energy (E) instead of the basic quantities of classical mechanics, length (L), mass (M), and time (T). The unit of energy is $[E] = 1\,\mathrm{GeV}$, which roughly corresponds to the rest mass of a proton

$$m_p c^2 \equiv m_p = 938.272\,\mathrm{MeV} \simeq 1\,\mathrm{GeV}\,.$$

We thus obtain a complete system of physical units which can be used to express any physical quantity. Now, the length, mass, and time are derived quantities measured in GeV or GeV^{-1}, according to the following table

Quantity	Dimensions	Units
Length	$[L] = [\hbar c/E]$	GeV^{-1}
Mass	$[M] = [E/c^2]$	GeV
Time	$[T] = [\hbar/E] = [L/c]$	GeV^{-1}

The conversion of length units from GeV^{-1} to Fermi (fm) follows from the values of c and $\hbar$

$$1 = \hbar\, c = 0.197327\,\mathrm{GeV\ fm} \simeq 0.2\,\mathrm{GeV\ fm}\,.$$

Thus, we find

$$1\,\mathrm{GeV}^{-1} = 0.2\,\mathrm{fm} = 0.2 \times 10^{-15}\,\mathrm{m}\,.$$

Also, the conversion formula for time units from GeV^{-1} to seconds can be derived from

$$1 = \hbar = 6.582119 \times 10^{-25}\,\mathrm{GeV\,s}\,,$$

such that

$$1\,\mathrm{GeV}^{-1} = 6.582119 \times 10^{-25}\,\mathrm{s}\,.$$

Finally, as $c = 1/\sqrt{\varepsilon_0\,\mu_0} = 1$, we may add the Heaviside-Lorentz units in which $\varepsilon_0 = \mu_0 = 1$.

1.1.3 The symmetry group of Schrödinger's equation

Like in the case of classical mechanics (see Chap. 4 Sec. 3.3) the maximal symmetry group of Schrödinger's equation is the symmetry group of the time-dependent equation (6.1.6) for a free particle (i.e., $V = 0$). To determine this group, denoted Sch_3, we use the general method shown in Chap. 3 Sec. 4.3.1. Hence, we consider the equation

$$-\frac{1}{2m}\,\Delta\, u = i\,\frac{\partial u}{\partial t}\,, \tag{6.1.17}$$

comprising the independent variables $\mathbf{q} \equiv (x, y, z, t)$ and the dependent variable $u = \psi(\mathbf{q})$. We will denote by $X = \mathbb{R}^4$ the space of independent variables and by U the one-dimensional space of the dependent variable. The solution of (6.1.17) will be identified with its graph, which is a one-dimensional manifold in the direct product space $X \times U$. Thus, the symmetry group of (6.1.17) is a continuous group of transformations acting on $X \times U$, such that the solutions of (6.1.17) are transformed between them.

According to the Theorem 3.4.5, in Cartesian coordinates, the infinitesimal generator of Sch_3 is

$$\mathsf{X} = \xi \frac{\partial}{\partial x} + \eta \frac{\partial}{\partial y} + \zeta \frac{\partial}{\partial z} + \tau \frac{\partial}{\partial t} + \varphi \frac{\partial}{\partial u}, \tag{6.1.18}$$

where the coefficients of the differential operators are functions of $\mathbf{q}$. Also, the first order prolongation of X has the form

$$\mathsf{pr}^{(1)}\mathsf{X} = \mathsf{X} + \varphi^x \frac{\partial}{\partial u_x} + \varphi^y \frac{\partial}{\partial u_y} + \varphi^z \frac{\partial}{\partial u_z} + \varphi^t \frac{\partial}{\partial u_t}, \tag{6.1.19}$$

with the coefficients

$$\begin{aligned}
\varphi^x &= \left(\frac{\partial}{\partial x} + u_x \frac{\partial}{\partial u}\right)\Lambda + u_{xx}\xi + u_{xy}\eta + u_{xz}\zeta + u_{xt}\tau\,,\\
\varphi^y &= \left(\frac{\partial}{\partial y} + u_y \frac{\partial}{\partial u}\right)\Lambda + u_{yx}\xi + u_{yy}\eta + u_{yz}\zeta + u_{yt}\tau\,,\\
\varphi^z &= \left(\frac{\partial}{\partial z} + u_z \frac{\partial}{\partial u}\right)\Lambda + u_{zx}\xi + u_{zy}\eta + u_{zz}\zeta + u_{zt}\tau\,,\\
\varphi^t &= \left(\frac{\partial}{\partial t} + u_t \frac{\partial}{\partial u}\right)\Lambda + u_{tx}\xi + u_{ty}\eta + u_{tz}\zeta + u_{tt}\tau\,,
\end{aligned} \tag{6.1.20}$$

where

$$\Lambda = \varphi - u_x\xi - u_y\eta - u_z\zeta - u_t\tau\,,$$

while the second order prolongation is

$$\mathsf{pr}^{(2)}\mathsf{X} = \mathsf{pr}^{(1)}\mathsf{X} + \varphi^{xx} \frac{\partial}{\partial u_{xx}} + \varphi^{yy} \frac{\partial}{\partial u_{yy}} + \varphi^{zz} \frac{\partial}{\partial u_{zz}} + \dots, \tag{6.1.21}$$

with the coefficients

$$\begin{aligned}
\varphi^{xx} &= \left(\frac{\partial}{\partial x} + u_x \frac{\partial}{\partial u}\right)^2\Lambda + u_{xxx}\xi + u_{xxy}\eta + u_{xxz}\zeta + u_{xxt}\tau\,,\\
\varphi^{yy} &= \left(\frac{\partial}{\partial y} + u_y \frac{\partial}{\partial u}\right)^2\Lambda + u_{yyx}\xi + u_{yyy}\eta + u_{yyz}\zeta + u_{yyt}\tau\,,\\
\varphi^{zz} &= \left(\frac{\partial}{\partial z} + u_z \frac{\partial}{\partial u}\right)^2\Lambda + u_{zzx}\xi + u_{zzy}\eta + u_{zzz}\zeta + u_{zzt}\tau\,.
\end{aligned} \tag{6.1.22}$$

Note that we do not need other coefficients in the second order prolongation of X because the associated derivatives vanish when applied to (6.1.17). Also, the equation (6.1.17) is a partial differential equation of the second order and therefore, we need only the second order prolongation of X to apply the condition (3.4.55). Thus, we obtain the equation

$$\mathrm{pr}^{(2)}\mathsf{X}\left[iu_t + \frac{1}{2m}(u_{xx}+u_{yy}+u_{zz})\right] = i\varphi^t + \frac{1}{2m}(\varphi^{xx}+\varphi^{yy}+\varphi^{zz}) = 0\,, \tag{6.1.23}$$

which has to be solved under the restriction

$$iu_t = -\frac{1}{2m}(u_{xx}+u_{yy}+u_{zz})\,.$$

By substituting the expressions (6.1.20) and (6.1.22) into (6.1.23), and assuming that the derivatives of the function u are linearly independent, we are led to the system of partial differential equations

$$\begin{aligned}
&\tau_x=\tau_y=\tau_z=\tau_u=0\,, && \xi_u=\eta_u=\zeta_u=0\,,\\
&\xi_x=\eta_y=\zeta_z=\frac{1}{2}\,\tau_t\,, && \eta_x+\xi_y=\zeta_y+\eta_z=\xi_z+\zeta_x=0\,,\\
&i\xi_t+\frac{1}{2m}\,\Delta\xi-\frac{1}{m}\,\varphi_{ux}=0\,, && i\eta_t+\frac{1}{2m}\,\Delta\eta-\frac{1}{m}\,\varphi_{uy}=0\,,\\
&i\zeta_t+\frac{1}{2m}\,\Delta\zeta-\frac{1}{m}\,\varphi_{uz}=0\,, && i\varphi_t+\frac{1}{2m}\,\Delta\varphi=0\,,
\end{aligned}$$

having the general solution

$$\begin{aligned}
\xi &= x(\delta-\alpha t)+\rho_2 z-\rho_3 y+v_1 t+\omega_1\,,\\
\eta &= y(\delta-\alpha t)+\rho_3 x-\rho_1 z+v_2 t+\omega_2\,,\\
\zeta &= z(\delta-\alpha t)+\rho_1 y-\rho_2 x+v_3 t+\omega_3\,,\\
\tau &= -\alpha t^2+2\delta t+\beta\,,\\
\varphi &= \left(\gamma+im\,\mathbf{v}\cdot\mathbf{r}+\tfrac{3}{2}\,\alpha t-\tfrac{1}{2}\,m\alpha\,\mathbf{r}^2\right)f+h\,,
\end{aligned} \tag{6.1.24}$$

where f and h are two arbitrary solutions of (6.1.17), $\mathbf{r}=(x,y,z)$, and α, β, γ, δ, $\boldsymbol{\rho}=(\rho_1,\rho_2,\rho_3)$, $\mathbf{v}=(v_1,v_2,v_3)$, $\boldsymbol{\omega}=(\omega_1,\omega_2,\omega_3)$ are scalar, respectively vector, arbitrary constants (the parameters of Sch_3). Consequently, the infinitesimal generators of Sch_3 are obtained by replacing the expressions of the coefficients (6.1.24) into (6.1.18) and successively set to zero all parameters excepting one. Thus, we have

$$\begin{aligned}
\omega_1 &= 1\ :\ \mathsf{X}_1=\partial_x\,,\\
\omega_2 &= 1\ :\ \mathsf{X}_2=\partial_y\,,\\
\omega_3 &= 1\ :\ \mathsf{X}_3=\partial_z\,,\\
\beta &= 1\ :\ \mathsf{X}_4=\partial_t\,,
\end{aligned} \tag{6.1.25}$$

$$\begin{aligned} \rho_1 &= 1 \; : \mathsf{X}_5 = y\,\partial_z - z\,\partial_y \,, \\ \rho_2 &= 1 \; : \mathsf{X}_6 = z\,\partial_x - x\,\partial_z \,, \\ \rho_3 &= 1 \; : \mathsf{X}_7 = x\,\partial_y - y\,\partial_x \,, \end{aligned} \tag{6.1.26}$$

$$\begin{aligned} v_1 &= 1 \; : \mathsf{X}_8 \;\; = t\,\partial_x + im\,xu\,\partial_u \,, \\ v_2 &= 1 \; : \mathsf{X}_9 \;\; = t\,\partial_y + im\,yu\,\partial_u \,, \\ v_3 &= 1 \; : \mathsf{X}_{10} = t\,\partial_z + im\,zu\,\partial_u \,, \end{aligned} \tag{6.1.27}$$

$$\begin{aligned} \delta &= 1 \; : \mathsf{X}_{11} = \mathbf{r}\cdot\nabla + 2t\,\partial_t \,, \\ \alpha &= 1 \; : \mathsf{X}_{12} = -t\,\mathbf{r}\cdot\nabla - t^2\,\partial_t + \left(\tfrac{3}{2}\,t - \tfrac{1}{2}\,im\,\mathbf{r}^2\right)u\,\partial_u \,, \\ \gamma &= 1 \; : \mathsf{X}_{13} = u\,\partial_u \,, \end{aligned} \tag{6.1.28}$$

$$h \neq 0 \; : \; \mathsf{X}_{14} = h(\mathbf{r},t)\,\partial_u \,. \tag{6.1.29}$$

It is evident that the first ten infinitesimal generators are related to the infinitesimal generators of the Galilei-Newton group (4.3.57). Also, using the usual notations in quantum mechanics, that is

$$\mathsf{H} = i\,\partial_t \,, \quad \mathbf{P} = -i\nabla \,, \quad \mathbf{J} = \mathbf{r}\times\mathbf{P} \,, \quad \mathbf{K} = it\nabla \,, \tag{6.1.30}$$

and redefining $\gamma = -3\delta/2 - \mu$, the infinitesimal generator (6.1.18) takes the form

$$\begin{aligned} i\,\mathsf{X} = \alpha &\left[\mathsf{A} - i\left(\frac{3}{2}\,t - \frac{1}{2}\,im\,\mathbf{r}^2\right)u\partial_u\right] + \delta\left[\mathsf{D} - \frac{3}{2}\,iu\partial_u\right] + \beta\mathsf{H} - \boldsymbol{\omega}\cdot\mathbf{P} \\ &+ \mathbf{v}\cdot\left[\mathbf{K} - m\mathbf{r}\,(1 + u\partial_u)\right] - \boldsymbol{\rho}\cdot\mathbf{J} - i\mu\,u\partial_u + ih\,\partial_u \,, \end{aligned} \tag{6.1.31}$$

where

$$\mathsf{A} = -i\left(t^2\partial_t + t\,\mathbf{r}\cdot\nabla\right), \quad \mathsf{D} = i\left(2t\partial_t + \mathbf{r}\cdot\nabla\right). \tag{6.1.32}$$

If for each infinitesimal generator X_i of the one-dimensional subgroups in Sch_3 we integrate the Lie equations

$$\frac{d\mathbf{r}}{d\lambda_i} = \mathsf{X}_i\mathbf{r} \,, \quad \frac{du}{d\lambda_i} = \mathsf{X}_i u \,, \quad \mathbf{r}(0) = \mathbf{r} \,, \quad u(0) = u \,, \tag{6.1.33}$$

where λ_i represents the parameter associated to the infinitesimal generator X_i $(i = \overline{1,14})$, then we can determine the finite transformations corresponding to each subgroup in Sch_3. Thus, we obtain the following transformations:

1 *expansions*

$$\alpha \; : \; (\mathbf{r},t,u) \to \left(\frac{\mathbf{r}}{1+\alpha t}, \frac{t}{1+\alpha t}, u\,(1+\alpha t)^{3/2}\exp\left[-\frac{im\mathbf{r}^2\,\alpha}{2(1+\alpha t)}\right]\right), \tag{6.1.34}$$

2 *dilatations*

$$\delta\ :\ (\mathbf{r},t,u)\to\left(\mathbf{r}e^{\delta},te^{2\delta},ue^{-3\delta/2}\right)\equiv\left(\mathbf{r}d,td^{2},ud^{-3/2}\right),\quad d=e^{\delta}\,,\tag{6.1.35}$$

3 *three-dimensional space translations*

$$\boldsymbol{\omega}\ :\ (\mathbf{r},t,u)\to(\mathbf{r}+\boldsymbol{\omega},t,u)\,,\tag{6.1.36}$$

4 *time translations*

$$\beta\ :\ (\mathbf{r},t,u)\to(\mathbf{r},t+\beta,u)\,,\tag{6.1.37}$$

5 *three-dimensional proper rotations*

$$\boldsymbol{\rho}\ :\ (\mathbf{r},t,u)\to(\mathsf{R}\,\mathbf{r},t,u)\,,\quad \mathsf{R}\in\mathsf{SO}(3)\,,\tag{6.1.38}$$

6 *pure Galilean transformations*

$$\mathbf{v}\ :\ (\mathbf{r},t,u)\to\left(\mathbf{r}+\mathbf{v}t,t,u\,\exp\left[im\,\mathbf{v}\cdot\mathbf{r}+\frac{i}{2}m\mathbf{v}^{2}t\right]\right).\tag{6.1.39}$$

We also mention the transformations

$$\mu\ :\ (\mathbf{r},t,u)\to(\mathbf{r},t,ue^{\mu})\,,\tag{6.1.40}$$

$$h\ :\ (\mathbf{r},t,u)\to(\mathbf{r},t,u+\lambda h)\,.\tag{6.1.41}$$

The invariance of the equation (6.1.17) with respect to the transformations (6.1.40) and (6.1.41) shows that the equation is linear, while the invariance with respect to the transformations (6.1.36) and (6.1.37) shows that the equation has constant coefficients. Actually, the Schrödinger group Sch_3 is a twelve-parameter Lie group comprising the space-time transformations that result by combining the transformations (6.1.34)–(6.1.39) in the form

$$(\mathbf{r},t)\to g(\mathbf{r},t)\equiv\left(d\,\frac{\mathsf{R}\mathbf{r}+\mathbf{v}t+\boldsymbol{\omega}}{1+\alpha(t+\beta)},\ d^{2}\,\frac{t+\beta}{1+\alpha(t+\beta)}\right),\tag{6.1.42}$$

such that the elements $g\in\mathsf{Sch}_3$ can be characterized by the symbol

$$g\equiv(d,\alpha,\beta,\boldsymbol{\omega},\mathbf{v},\mathsf{R})=(d)(\alpha)(\beta,\boldsymbol{\omega},\mathbf{v},\mathsf{R})\,.\tag{6.1.43}$$

One can define a projective representation of Sch_3 in the Hilbert space of square integrable functions, solutions of (6.1.17), with the inner product

$$< u_1, u_2 >= \int \overline{u_1(\mathbf{r},t)}\, u_2(\mathbf{r},t)\, dxdydz \,.$$

Thus, to each element $g \in \mathsf{Sch}_3$ we associate an operator $\mathsf{T}(g)$ that acts upon the wave functions according to the relation

$$u(\mathbf{r},t) \to (\mathsf{T}(g)u)(\mathbf{r},t) = f_g[g^{-1}(\mathbf{r},t)]\, u[g^{-1}(\mathbf{r},t)] \,. \tag{6.1.44}$$

Note that the explicit form of the functions f_g appears in the finite transformations (6.1.34), (6.1.35), and (6.1.39). The operators $\mathsf{T}(g)$ form a projective representation of Sch_3 if they satisfy the relation

$$\mathsf{T}(g_2)\, \mathsf{T}(g_1) = \omega(g_2, g_1)\, \mathsf{T}(g_2 g_1) \,, \tag{6.1.45}$$

where ω is a system of factors having the property

$$\omega(g_3 g_2, g_1)\, \omega(g_3, g_2) = \omega(g_3, g_2 g_1)\, \omega(g_2, g_1) \,,$$

that follows from the associativity of the composition law defined on Sch_3. Hence, the functions f_g have to satisfy the relation

$$f_{g_2}[g_1(\mathbf{r},t)]\, f_{g_1}(\mathbf{r},t) = \omega(g_2, g_1)\, f_{g_2 g_1}(\mathbf{r},t) \,. \tag{6.1.46}$$

From the expressions (6.1.34), (6.1.35), and (6.1.39), taking into account the symbol (6.1.43), we have

$$\begin{aligned} f_{(\alpha)} &= C\,(1+\alpha t)^{3/2} \exp\left[-\frac{im\mathbf{r}^2\,\alpha}{2(1+\alpha t)}\right], \\ f_{(d)} &= C\, d^{-3/2}\,, \\ f_{(\beta,\boldsymbol{\omega},\mathbf{v},\mathsf{R})} &= C\, \exp\left[im\left(\frac{1}{2}\,\mathbf{v}^2 t + \mathsf{R}\mathbf{r}\cdot\mathbf{v}\right)\right], \end{aligned} \tag{6.1.47}$$

where C is an arbitrary constant. If we choose $C = 1$, and combine the functions (6.1.47) according to (6.1.43) and (6.1.46) it results

$$\begin{aligned} & f_g(\mathbf{r},t) = d^{-3/2} s_g^{3/2} \exp\left[-\frac{im}{2s_g}\, h_g\right], \\ & s_g = 1 + \alpha(t+\beta)\,, \\ & h_g = \alpha\mathbf{r}^2 + 2\mathsf{R}\mathbf{r}\cdot[\alpha\boldsymbol{\omega} - (1+\alpha\beta)\mathbf{v}] + \alpha\boldsymbol{\omega}^2 - (1+\alpha\beta)t\mathbf{v}^2 + 2t\alpha\boldsymbol{\omega}\cdot\mathbf{v}\,. \end{aligned} \tag{6.1.48}$$

With this value of C, and the composition law on Sch_3, which can be obtained from (6.1.42), we can determine the system of factors $\omega(g_2, g_1)$ by successively applying two transformations on a point $(\mathbf{r}, t)$ and taking into account (6.1.46).

The transformation (6.1.42) can also be written in the form

$$
\begin{aligned}
t' &= \frac{d^2}{s_g}\,(t+\beta)\,,\\
x'_k &= \frac{d}{s_g}\,[R_{kj}x_j + v_k t + \omega_k]\,, \qquad k = 1,2,3\,;
\end{aligned}
$$

therefore, the Jacobian of the transformation of spatial coordinates is

$$
J \equiv \det\left[\frac{\partial(x'_k)}{\partial(x_j)}\right] = \frac{d^3}{s_g^3}\,\det\,[R_{kj}] = \frac{d^3}{s_g^3}\,. \tag{6.1.49}
$$

Consequently,

$$
\begin{aligned}
&< \mathsf{T}(g)u_1, \mathsf{T}(g)u_2 > \\
&\quad = \int |f_g[g^{-1}(\mathbf{r},t)]|^2\,\overline{u_1[g^{-1}(\mathbf{r},t)]}u_2[g^{-1}(\mathbf{r},t)]\,dx_1dx_2dx_3\\
&\quad = \int |f_g(\mathbf{r},t)|^2\,\overline{u_1(\mathbf{r},t)}u_2(\mathbf{r},t)\,dx'_1dx'_2dx'_3\\
&\quad = \int (d^{-3/2}s_g^{3/2})^2\,\overline{u_1(\mathbf{r},t)}u_2(\mathbf{r},t)\,\frac{d^3}{s_g^3}\,dx_1dx_2dx_3\\
&\quad = < u_1, u_2 >\,,
\end{aligned}
$$

where $(\mathbf{r}, t) \to g(\mathbf{r}, t) \equiv (\mathbf{r}', t')$, which proves that the representation $\mathsf{T}(g)$ is unitary. This representation is also irreducible, because its restriction to the Galilei-Newton group is irreducible.

The structure of the Lie algebra of Schrödinger's group is determined by the commutation relations satisfied by the infinitesimal generators. Hence, by using the expressions of the infinitesimal generators $\mathbf{J}$, $\mathbf{K}$, $\mathbf{P}$, H, A, and D, given in (6.1.30) and (6.1.32), we obtain the commutation relations

$$
\begin{aligned}
&[\mathsf{J}_l, \mathsf{J}_m] = i\,\epsilon_{klm}\,\mathsf{J}_k\,, \quad [\mathsf{K}_l, \mathsf{K}_m] = [\mathsf{P}_l, \mathsf{P}_m] = 0\,,\\
&[\mathsf{J}_l, \mathsf{P}_m] = i\,\epsilon_{klm}\,\mathsf{P}_k\,, \quad [\mathsf{K}_l, \mathsf{P}_m] = [\mathsf{P}_l, \mathsf{H}] = 0\,,\\
&[\mathsf{J}_l, \mathsf{K}_m] = i\,\epsilon_{klm}\,\mathsf{K}_k\,, \quad [\mathsf{K}_l, \mathsf{H}] = i\mathsf{P}_l\,, \quad [\mathsf{J}_l, \mathsf{H}] = 0\,,\\
&[\mathsf{A}, \mathsf{J}_l] = [\mathsf{A}, \mathsf{K}_l] = 0, \ [\mathsf{A}, \mathsf{P}_l] = -i\,\mathsf{K}_l, \ [\mathsf{A}, \mathsf{H}] = i\,\mathsf{D}, \ [\mathsf{A}, \mathsf{D}] = -2i\,\mathsf{A},\\
&[\mathsf{D}, \mathsf{J}_l] = 0\,, \quad [\mathsf{D}, \mathsf{K}_l] = 2i\,\mathsf{K}_l\,, \quad [\mathsf{D}, \mathsf{P}_l] = -i\,\mathsf{P}_l\,, \quad [\mathsf{D}, \mathsf{H}] = -2i\,\mathsf{H}\,.
\end{aligned} \tag{6.1.50}
$$

It is easy to observe that the Galilei-Newton group G, with the infinitesimal generators (4.3.57) rescaled as in (6.1.30), is a subgroup of Sch_3. The subgroup

$$
\mathsf{G}_1 = \{\,(1)(0)(0, \boldsymbol{\omega}, \mathbf{v}, \mathsf{R})\,|\,\boldsymbol{\omega} \in \mathbb{R}^3, \mathbf{v} \in \mathbb{R}^3, \mathsf{R} \in \mathsf{SO}(3)\,\} \lhd \mathsf{Sch}_3
$$

has the infinitesimal generators $\mathbf{J}$, $\mathbf{K}$, $\mathbf{P}$, and is an invariant subgroup of Schrödinger's group. Also, the subgroup G_1 is the maximal invariant subgroup of the Galilei-Newton group and is called the *isochronous Galilei-Newton group*. In terms of the subgroup of spatial translations $\mathsf{T} = \{(1)(0)(0, \boldsymbol{\omega}, \mathbf{0}, \mathsf{E})\}$, the subgroup of pure Galilean transformations $\Gamma = \{(1)(0)(0, \mathbf{0}, \mathbf{v}, \mathsf{E})\}$, and the subgroup of proper rotations $\mathsf{SO}(3) = \{(1)(0)(0, \mathbf{0}, \mathbf{0}, \mathsf{R})\}$, where $\mathsf{E} \in \mathsf{SO}(3)$ represents the identity element, the subgroup G_1 can be decomposed in the form

$$\mathsf{G}_1 = (\mathsf{T} \times \Gamma) \wedge \mathsf{SO}(3)\,. \tag{6.1.51}$$

The infinitesimal generators H, A, and D determine the subgroup

$$\mathsf{G}_2 = \{(d)(\alpha)(\beta, \mathbf{0}, \mathbf{0}, \mathsf{E})\} \subset \mathsf{Sch}_3\,.$$

By means of the structure constants from (6.1.50) we can find the adjoint representation of G_2 in the matrix form

$$\mathrm{ad}\,\mathsf{H} = \begin{bmatrix} 0 & 0 & 2i \\ 0 & 0 & 0 \\ 0 & -i & 0 \end{bmatrix}, \quad \mathrm{ad}\,\mathsf{A} = \begin{bmatrix} 0 & 0 & 0 \\ 0 & 0 & -2i \\ i & 0 & 0 \end{bmatrix}, \quad \mathrm{ad}\,\mathsf{D} = \begin{bmatrix} -2i & 0 & 0 \\ 0 & 2i & 0 \\ 0 & 0 & 0 \end{bmatrix},$$

and Cartan's tensor

$$g_{ij} = \mathrm{Tr}\,(\mathrm{ad}\,\mathsf{x}_i\,\mathrm{ad}\,\mathsf{x}_j)\,, \quad \mathsf{x}_i, \mathsf{x}_j \in \{\mathsf{H}, \mathsf{A}, \mathsf{D}\}\,.$$

According to Cartan's criterion, because $\det[g_{ij}] \neq 0$ the subgroup G_2 is a semisimple Lie group. We introduce now the infinitesimal generators $\mathsf{a}_+ = i\mathsf{A}$, $\mathsf{a}_- = i\mathsf{H}$, and $\mathsf{a}_3 = i\mathsf{D}$, which satisfy the commutation relations

$$[\mathsf{a}_+, \mathsf{a}_-] = -\mathsf{a}_3\,, \quad [\mathsf{a}_+, \mathsf{a}_3] = 2\mathsf{a}_+\,, \quad [\mathsf{a}_-, \mathsf{a}_3] = -2\mathsf{a}_-\,.$$

Thus, we prove that the subgroup G_2 is isomorphic to the group $\mathsf{SL}(2, \mathbb{R})$ and, finally, get the decomposition

$$\mathsf{Sch}_3 = \mathsf{G}_1 \wedge \mathsf{SL}(2, \mathbb{R})\,. \tag{6.1.52}$$

In the case of Schrödinger's equation in n variables $\mathbf{x} = (x_1, x_2, \ldots, x_n)$, with a general potential

$$-\frac{1}{2m}\,\Delta_n\,u + V(\mathbf{x}, t) = i\,\frac{\partial u}{\partial t}\,,$$

it can be shown that the maximal symmetry group is formed by the coordinate transformations

$$g(\mathbf{x}, t) = \left(\frac{\mathsf{R}\mathbf{x} + \mathbf{y}(t)}{d(t)}\,, \int \frac{1}{d(t)^2}\,dt\right) \tag{6.1.53}$$

and the associated functions

$$f(\mathbf{x},t) = d^{n/2} \times \exp\left\{-\frac{im}{2}\left[\frac{\dot{d}}{d}\mathbf{x}^2 + 2\,\mathsf{R}\mathbf{x}\cdot\left(\frac{\dot{d}}{d}\mathbf{y} - \dot{\mathbf{y}}\right) + \frac{\dot{d}}{d}\mathbf{y}^2 - \dot{\mathbf{y}}\cdot\mathbf{y} + l(t)\right]\right\},$$

where $d(t)$, $l(t)$, and $\mathbf{y}(t)$ are real functions, and $\mathsf{R} \in \mathsf{SO}(n)$. The functions $d(t), l(t)$, and $\mathbf{y}(t)$, as well as the integration constant in (6.1.53) are determined by the condition

$$V[g(\mathbf{x},t)] - d^2\,V(\mathbf{x},t) = \frac{m}{2}\,d\,\ddot{d}\,\mathbf{x}^2 + m\,\mathsf{R}\mathbf{x}\cdot(d\,\ddot{d}\,\mathbf{y} - d^2\,\ddot{\mathbf{y}}) + \frac{m}{2}\,(d\,\ddot{d}\,\mathbf{y}^2 - d^2\,\ddot{\mathbf{y}}\cdot\mathbf{y} + d^2\,\dot{l})\,. \quad (6.1.54)$$

The order of the symmetry group thus obtained is

$$N(n) = 2 + (n+1)(n+2)/2\,.$$

For instance, by applying this method in the particular case $n = 3$, we may find the symmetry group of Schrödinger's equation for the following potentials:

1 the motion of a free particle ($V = 0$)

$$(\mathbf{r},t) \to \left(\frac{\mathsf{R}\mathbf{r} + \mathbf{v}t + \boldsymbol{\omega}}{\gamma t + \delta},\ \frac{\alpha t + \beta}{\gamma t + \delta}\right), \quad (6.1.55)$$

that is an expression identical to (6.1.42) if we redefine the parameters,

2 the free fall of a particle of mass m ($V = -m\mathbf{g}\cdot\mathbf{r}$)

$$(\mathbf{r},t) \to \left(\frac{\mathsf{R}\mathbf{r} + \mathbf{v}t + \boldsymbol{\omega} + \frac{(\alpha t+\beta)^2}{2(\gamma t+\delta)}\,\mathbf{g} - \frac{1}{2}\,t^2\,\mathsf{R}\mathbf{g}}{\gamma t + \delta},\ \frac{\alpha t + \beta}{\gamma t + \delta}\right), \quad (6.1.56)$$

$\mathbf{g}$ being the standard gravitational acceleration,

3 the harmonic oscillator of mass m ($V = m\,\omega^2\,\mathbf{r}^2/2$)

$$(\mathbf{r},t) \to \left(\left[\frac{1 + \tan^2\omega t}{(\alpha\,\tan\omega t + \beta)^2 + (\gamma\,\tan\omega t + \delta)^2}\right]^{1/2}(\mathsf{R}\mathbf{r} + \mathbf{v}\,\sin\omega t + \mathbf{b}\,\cos\omega t),\ \frac{1}{\omega}\,\arctan\frac{\alpha\,\tan\omega t + \beta}{\gamma\,\tan\omega t + \delta}\right), \quad (6.1.57)$$

where ω represents the angular frequency of oscillations.

Note that, in contrast to the case of general transformations of the Newton-Galilei group, that preserve the time interval between two events as well as the spatial distance between two simultaneous events, the general transformation (6.1.42) does not preserve neither the time interval between two events

$$t_2 - t_1 \to t_2' - t_1' = \frac{d^2\,(t_2 - t_1)}{[1 + \alpha(t_2 + \beta)][1 + \alpha(t_1 + \beta)]}\,, \tag{6.1.58}$$

nor the spatial distance between two simultaneous events

$$|\mathbf{x}_2 - \mathbf{x}_1|_{t_2 = t_1 = t} \to |\mathbf{x}_2' - \mathbf{x}_1'| = \frac{d\,|\mathbf{x}_2 - \mathbf{x}_1|}{1 + \alpha(t + \beta)}\,, \tag{6.1.59}$$

which may arise problems related to the causality.

1.1.4 Local and global conservation laws

The transformations of Schrödinger's group Sch_3 preserve the action integral

$$S = \int \left[\frac{1}{2i}\,(u\dot{\overline{u}} - \overline{u}\dot{u}) - \frac{1}{2m}\,\nabla\,\overline{u}\cdot\nabla\,u \right] dx_1 dx_2 dx_3 dx_4\,, \tag{6.1.60}$$

where we have added the divergence type term

$$-\frac{1}{2i}\,\frac{\partial}{\partial t}\,(\overline{u}u)$$

to the Lagrangian density (6.1.7), and set $V = 0$. Thus, the local conservation laws follow from Noether's theorem (see Chap. 4 Sec. 1.4) and have the form (4.1.102) with the independent variables x_1, x_2, x_3, $x_4 = t$, and the dependent variables $u^1 = u$, $u^2 = \overline{u}$. From (6.1.31) and (6.1.34)–(6.1.39) we obtain the infinitesimal variations of the independent and dependent variables, corresponding to the group parameters, in the form

$$\begin{aligned} \delta t &= \beta\,, & \delta x_i &= 0\,, & \delta u &= \delta\overline{u} = 0\,,\\ \delta t &= 0\,, & \delta x_i &= \omega_i\,, & \delta u &= \delta\overline{u} = 0\,,\\ \delta t &= 0\,, & \delta x_i &= R_{ij}x_j\,, & \delta u &= \delta\overline{u} = 0\,, \end{aligned} \tag{6.1.61}$$

$$\delta t = 0\,,\quad \delta x_i = v_i\,t\,,\quad \delta u = im\,\mathbf{v}\cdot\mathbf{r}\,u\,,\quad \delta\overline{u} = -im\,\mathbf{v}\cdot\mathbf{r}\,\overline{u}\,, \tag{6.1.62}$$

$$\delta t = 2\varepsilon\,t\,,\quad \delta x_i = \varepsilon\,x_i\,,\quad \delta u = -\frac{3}{2}\,\varepsilon\,u\,,\quad \delta\overline{u} = -\frac{3}{2}\,\varepsilon\,\overline{u}\,, \tag{6.1.63}$$

where we have denoted by ε the parameter δ in (6.1.35) to avoid any confusion, and

$$\begin{aligned} \delta t = \alpha\,t^2\,,\quad \delta x_i = \alpha\,t\,x_i\,,\quad \delta u &= -\frac{\alpha}{2}\left(3t - im\,\mathbf{r}^2\right)u\,,\\ \delta\overline{u} &= -\frac{\alpha}{2}\left(3t + im\,\mathbf{r}^2\right)\overline{u}\,. \end{aligned} \tag{6.1.64}$$

Here, $R_{ij} = -R_{ji}$, $i, j = 1, 2, 3$, and $\mathbf{r} = (x_1, x_2, x_3)$. With these notations and $\delta f_i = 0$ in (4.1.102) the local conservation laws associated to Schrödinger's group are

$$\frac{\partial}{\partial t}\left(\frac{1}{2m}\nabla\overline{u}\cdot\nabla u\right) + \nabla\cdot\left[-\frac{1}{2m}\left(\dot{u}\nabla\overline{u} + \dot{\overline{u}}\nabla u\right)\right] = 0\,, \tag{6.1.65a}$$

$$\begin{aligned}\frac{\partial}{\partial t}\left[\frac{1}{2i}\left(\overline{u}u_k - u\overline{u}_k\right)\right] &+ \frac{\partial}{\partial x_j}\Bigg\{-\left[\frac{1}{2m}\nabla\overline{u}\cdot\nabla u + \frac{1}{2i}\left(\overline{u}\dot{u} - u\dot{\overline{u}}\right)\right]\delta_{kj} \\ &+ \frac{1}{2m}\left(\overline{u}_k u_j + u_k\overline{u}_j\right)\Bigg\} = 0\,, \quad k = 1, 2, 3,\end{aligned} \tag{6.1.65b}$$

$$\begin{aligned}&\frac{\partial}{\partial t}\left\{\frac{1}{2i}\left[\overline{u}\left(u_k x_j - u_j x_k\right) - u\left(\overline{u}_k x_j - \overline{u}_j x_k\right)\right]\right\} \\ &+\frac{\partial}{\partial x_l}\Bigg\{-\left[\frac{1}{2m}\nabla\overline{u}\cdot\nabla u + \frac{1}{2i}\left(\overline{u}\dot{u} - u\dot{\overline{u}}\right)\right]\left(\delta_{lk}x_j - \delta_{lj}x_k\right) \\ &-\frac{1}{2m}\left[\overline{u}_l\left(u_k x_j - u_j x_k\right) + u_l\left(\overline{u}_k x_j - \overline{u}_j x_k\right)\right]\Bigg\} = 0\,, \quad j, k = 1, 2, 3,\end{aligned} \tag{6.1.65c}$$

$$\begin{aligned}&\frac{\partial}{\partial t}\left[-\frac{1}{2i}\left(\overline{u}u_k - u\overline{u}_k\right)t + m\,\overline{u}ux_k\right] \\ &+\frac{\partial}{\partial x_l}\Bigg\{\left[\frac{1}{2m}\nabla\overline{u}\cdot\nabla u + \frac{1}{2i}\left(\overline{u}\dot{u} - u\dot{\overline{u}}\right)\right]\delta_{kl}\,t \\ &+\frac{1}{2i}\left(\overline{u}u_l - u\overline{u}_l\right)x_k + \frac{1}{2m}\left(\overline{u}_l u_k + u_l\overline{u}_k\right)t\Bigg\} = 0\,, \quad k = 1, 2, 3,\end{aligned} \tag{6.1.65d}$$

$$\begin{aligned}&\frac{\partial}{\partial t}\left[-\frac{1}{2i}\left(\overline{u}\nabla u - u\nabla\overline{u}\right)\cdot\mathbf{r} + \frac{1}{m}\nabla\overline{u}\cdot\nabla u\,t\right] \\ &+\nabla\cdot\Bigg\{\frac{1}{2i}\left(\overline{u}\dot{u} - u\dot{\overline{u}}\right)\mathbf{r} + \frac{1}{m}\left(\dot{u}\nabla\overline{u} + \dot{\overline{u}}\nabla u\right)t \\ &+\frac{1}{2m}\left[\nabla u\times\left(\mathbf{r}\times\nabla\overline{u}\right) - \nabla u\left(\mathbf{r}\cdot\nabla\overline{u}\right) - \frac{3}{2}\left(u\nabla\overline{u} + \overline{u}\nabla u\right)\right]\Bigg\} = 0\,,\end{aligned} \tag{6.1.65e}$$

$$\frac{\partial}{\partial t}\left[-\frac{1}{2i}\left(\overline{u}\nabla u - u\nabla\overline{u}\right)\cdot\mathbf{r}\,t + \frac{m}{2}\,\overline{u}u\,\mathbf{r}^2 + \frac{1}{2m}\,\nabla\overline{u}\cdot\nabla u\,t^2\right]$$
$$+\nabla\cdot\left\{\frac{1}{2i}\left(\overline{u}\dot{u} - u\dot{\overline{u}}\right)\mathbf{r}\,t + \frac{1}{4i}\left(\overline{u}\nabla u - u\nabla\overline{u}\right)\mathbf{r}^2 - \frac{1}{2m}\left[\nabla\overline{u}\left(\mathbf{r}\cdot\nabla u\right)t\right.\right.$$
$$\left.\left.-\nabla\overline{u}\times\left(\mathbf{r}\times\nabla u\right)t + \left(\dot{u}\nabla\overline{u} + \dot{\overline{u}}\nabla u\right)t^2 + \frac{3}{2}\left(u\nabla\overline{u} + \overline{u}\nabla u\right)t\right]\right\} = 0\,. \tag{6.1.65f}$$

Note that all these local conservation laws are of the form

$$\frac{\partial\rho}{\partial t} + \nabla\cdot\mathbf{j} = 0 \tag{6.1.66}$$

and, if the integrals of the divergence of $\mathbf{j}$ vanish, that is

$$\int_V \nabla\cdot\mathbf{j}\,dx_1dx_2dx_3 = \int_{\partial V}\mathbf{j}\cdot\mathbf{n}\,d\sigma\,,$$

where $\mathbf{n}$ is the exterior normal on the boundary of V, then we can obtain the *global conservation laws*

$$\frac{d}{dt}\int_V \rho\,dx_1dx_2dx_3 = 0\,. \tag{6.1.67}$$

The conditions that have to be imposed on the wave function to obtain global conservation laws from the local conservation laws (6.1.65) are more restrictive than the usual conditions of quantum mechanics. If these stronger conditions are satisfied then, integrating by parts the volume integrals and expressing the spatial derivatives of u and $\overline{u}$ in terms of the momentum operator $\mathbf{P} = -i\nabla$, we may obtain relations of the form

$$\frac{d}{dt}\int_V \overline{u}\,\mathsf{O}\,u\,dx_1dx_2dx_3 = 0\,, \tag{6.1.68}$$

where O is one of the infinitesimal generators of the Lie algebra associated to Schrödinger's group, that is $\mathsf{O}\in\{\mathbf{P},\mathbf{J},\mathbf{K},\mathsf{H},\mathsf{A},\mathsf{D}\}$.

The Lagrangian density in (6.1.60) is a homogeneous bilinear form in u, $\overline{u}$ and their first order derivatives. If A_1, A_2, and A_3 are three elements (not necessarily distinct) of the Lie algebra of Sch_3, then the operator

$$\mathsf{O} = \frac{1}{2}\left(\mathsf{A}_1\mathsf{A}_2\mathsf{A}_3 + \mathsf{A}_3\mathsf{A}_2\mathsf{A}_1\right)$$

also belongs to the Lie algebra and, according to Noether's theorem, we may obtain a local conservation law associated to this operator. Consequently, for a Lagrangian density of such a form there exists an infinite number of local conservation laws.

Schrödinger's group Sch_3 is the symmetry group of Schrödinger's equation even in the case of a nonzero potential of the form $V = \mathrm{const}/r^2$, and the associated conservation laws follow from (6.1.65) – (6.1.68) by adding the terms corresponding to $V \neq 0$. Such terms are obtained from the Lagrangian density (6.1.7) with $V \neq 0$. It can be shown that, in the case of a potential $V = \mathrm{const}/r^2$ the Hamilton-Jacobi equation

$$\frac{\partial S}{\partial t} + H = 0\,, \quad H = \frac{\mathbf{p}^2}{2m} + V\,, \quad \mathbf{p} = \nabla S \tag{6.1.69}$$

admits the group Sch_3 as symmetry group. Even if the equations (6.1.69) are not the result of a variational principle, assuming a Lagrangian of the form $L = m\dot{\mathbf{r}}^2/2 - V$, we may employ (4.3.2) to get the local conservation laws. Thus, we use the variations of the independent and dependent variables given by (6.1.61) – (6.1.64), and choose $\delta f = 0$ for the variations (6.1.61) and (6.1.63), $\delta f = m\,\mathbf{v}\cdot\mathbf{r}$ for (6.1.62), and $\delta f = m\alpha\,\mathbf{r}^2/2$ for (6.1.64). From these local conservation laws we obtain the following global conservation laws:

$$\begin{aligned}
&\frac{dH}{dt} = 0\,, \quad \frac{d\mathbf{p}}{dt} = \mathbf{0}\,, \quad \frac{d\mathbf{J}}{dt} = \mathbf{0}\,, \quad \mathbf{J} = \mathbf{r}\times\mathbf{p}\,,\\
&\frac{d\mathbf{G}}{dt} = \mathbf{0}\,, \quad \mathbf{G} = m\mathbf{r} - \mathbf{p}\,t\,, \quad \frac{dD}{dt} = 0\,, \quad D = 2H\,t - \mathbf{r}\cdot\mathbf{p}\,,\\
&\frac{dA}{dt} = 0\,, \quad A = Ht^2 - \mathbf{r}\cdot\mathbf{p}\,t + \frac{1}{2}\,m\mathbf{r}^2\,.
\end{aligned} \tag{6.1.70}$$

These twelve conservation laws are prime integrals for the motion equations. The first four prime integrals in (6.1.70) have also been obtained in Chap. 4 Sec. 3.1. We mention that the last two prime integrals in (6.1.70), associated to the motion of a free particle, or to an interaction of the form $1/r^2$, have been used by Jacobi but he did not considered them as conservation laws.

In conclusion we may say that, in the case of spherically symmetric potentials $V = V(r)$ Schrödinger's equation and the Hamilton-Jacobi equation are both invariant relative the Galilei-Newton group. In the absence of interactions ($V = 0$), Schrödinger's equation is invariant relative to Sch_3, while the Hamilton-Jacobi equation is invariant with respect to the maximal symmetry group described in Chap. 4 Sec. 3.3, that contains Sch_3 as a subgroup. Note that the Galilei-Newton group is a subgroup of Sch_3. In the particular case of a spherically symmetric potential $V = 1/r^2$, both equations are invariant with respect to Sch_3.

1.1.5 The notion of invariance in quantum mechanics

Let $\mathcal{E}$ be a space associated to a quantum system and G a group of transformations acting in $\mathcal{E}$ such that, to each point $\mathbf{x} \in \mathcal{E}$ it corresponds a point $\mathbf{x}' \in \mathcal{E}$

determined by the relation

$$g : \mathbf{x} \to \mathbf{x}' = g\,\mathbf{x}\,, \quad g \in G\,. \tag{6.1.71}$$

Let us now consider a functional space $\mathcal{H}$ defined on $\mathcal{E}$, and associate to each element $g \in G$ a linear operator $\mathsf{T}(g)$ acting in $\mathcal{H}$ according to the rule: if $\psi \in \mathcal{H}$ is a function of $\mathbf{x} \in \mathcal{E}$ then, by applying $\mathsf{T}(g)$ on ψ it results a function $\psi' \in \mathcal{H}$, such that

$$\psi'(\mathbf{x}') \equiv (\mathsf{T}(g)\psi)(\mathbf{x}') = \psi(\mathbf{x})\,; \tag{6.1.72}$$

that is, the value of the transformed function ψ' at the image point $\mathbf{x}'$ is the same as the value of the initial function ψ at the point $\mathbf{x}$. Note that the relation (6.1.72) can also be written in the form

$$(\mathsf{T}(g)\psi)(\mathbf{x}) = \psi(g^{-1}\mathbf{x})\,, \tag{6.1.73}$$

thus depending only on the initial point $\mathbf{x} \in \mathcal{E}$. Also, the set of operators $\{\mathsf{T}(g) | g \in G\}$, defined in this way, form a representation of G in $\mathcal{H}$, hence they satisfy the relations

$$\mathsf{T}(g_1 g_2) = \mathsf{T}(g_1)\,\mathsf{T}(g_2)\,, \quad \mathsf{T}(g^{-1}) = \mathsf{T}(g)^{-1}\,. \tag{6.1.74}$$

DEFINITION 6.1.1 *The function ψ is called invariant under the action of the operator $\mathsf{T}(g)$ (or with respect to the transformation $g \in G$) if $\mathsf{T}(g)\psi \equiv \psi$, that is*

$$(\mathsf{T}(g)\psi)\,(\mathbf{x}) \equiv \psi(\mathbf{x}) = \psi(g^{-1}\mathbf{x}) = \psi(g\mathbf{x})\,.$$

Let Ω be an operator acting in the space $\mathcal{H}$. Thus, to each function $\psi \in \mathcal{H}$ we associate, by means of Ω, a function $\varphi \in \mathcal{H}$, determined by the relation

$$\varphi(\mathbf{x}) = \Omega(\mathbf{x})\,\psi(\mathbf{x})\,. \tag{6.1.75}$$

Then,

$$\mathsf{T}(g)\,(\Omega(\mathbf{x})\,\psi(\mathbf{x})) = (\mathsf{T}(g)\varphi)\,(\mathbf{x}) = \varphi(g^{-1}\mathbf{x}) = \Omega(g^{-1}\mathbf{x})\,\psi(g^{-1}\mathbf{x})$$

and

$$\begin{aligned}
\mathsf{T}(g)\,(\Omega(\mathbf{x})\,\psi(\mathbf{x})) &= \mathsf{T}(g)\Omega(\mathbf{x})\mathsf{T}(g^{-1})\mathsf{T}(g)\,\psi(\mathbf{x}) \\
&= \left(\mathsf{T}(g)\Omega(\mathbf{x})\mathsf{T}(g)^{-1}\right)\psi(g^{-1}\mathbf{x}) \\
&= \Omega(g^{-1}\mathbf{x})\,\psi(g^{-1}\mathbf{x}) = \Omega(g^{-1}\mathbf{x})\,(\mathsf{T}(g)\psi)(\mathbf{x}) \\
&= \Omega(g^{-1}\mathbf{x})\,\psi'(\mathbf{x}) = \Omega'(\mathbf{x})\,\psi'(\mathbf{x})\,,
\end{aligned}$$

where we used the notations

$$\Omega'(\mathbf{x}) = \Omega(g^{-1}\mathbf{x}) = \mathsf{T}(g)\Omega(\mathbf{x})\mathsf{T}(g^{-1})\,, \tag{6.1.76}$$

such that $\Omega'(g\mathbf{x}) = \Omega(\mathbf{x})$. This means that the transformed operator Ω', at the point $\mathbf{x}' = g\mathbf{x}$, is the same as the initial operator Ω at the point $\mathbf{x}$.

DEFINITION 6.1.2 *If the operator Ω and the transformed operator Ω' are identical at a given point $\mathbf{x}$, that is*

$$\Omega'(\mathbf{x}) = \Omega(\mathbf{x}) , \tag{6.1.77}$$

then we say that the operator Ω is **invariant relative to the transformations** *$g \in G$.*

Consequently, $\Omega(g\mathbf{x}) = \Omega(\mathbf{x})$ or $\mathsf{T}(g)\Omega(\mathbf{x})\mathsf{T}(g)^{-1} = \Omega(\mathbf{x})$, and it follows that

$$[\mathsf{T}(g), \Omega(\mathbf{x})] \equiv \mathsf{T}(g)\Omega(\mathbf{x}) - \Omega(\mathbf{x})\mathsf{T}(g) = 0 . \tag{6.1.78}$$

Let

$$\Omega(\mathbf{r}, t)\, \psi(\mathbf{r}, t) = 0 \tag{6.1.79}$$

be a wave equation, where $\Omega(\mathbf{r}, t)$ is a differential operator, and $\mathbf{r} = (x, y, z)$. If $g \in G$ is a transformation, and $\{\mathsf{T}(g) \,|\, g \in G\}$ a representation of G in the Hilbert space of wave functions ψ, that is

$$g : (\mathbf{r}, t) \to (\mathbf{r}', t) = g(\mathbf{r}, t) , \quad \psi \to \psi' = \mathsf{T}(g)\psi , \tag{6.1.80}$$

then we say that the wave equation (6.1.79) is invariant with respect to the transformation g if $\psi'(\mathbf{r}', t)$ is also a solution of (6.1.79), hence

$$\Omega(\mathbf{r}, t) \left[(\mathsf{T}(g)\psi) \left(g(\mathbf{r}, t) \right) \right] = 0 . \tag{6.1.81}$$

If we assume that $\{\mathsf{T}(g) \,|\, g \in G\}$ is a projective representation of G, then the operators $\mathsf{T}(g)$ satisfy the relations (6.1.45), and their action upon the wave functions is defined by the equation (6.1.44) that, up to the factor f_g, is analogous to (6.1.73). When the space $\mathcal{E}$ is the coordinate space, then the group G is the *maximal symmetry group* of the wave equation (6.1.79). From (6.1.44) and (6.1.81) we have

$$\Omega(\mathbf{r}, t) \left[f_g(\mathbf{r}, t)\, \psi(\mathbf{r}, t) \right] = 0 , \quad \mathbf{r} \in \mathcal{E} , \tag{6.1.82}$$

such that the determination of G reduces to finding the solutions of this equation. If we use the Lie algebra associated to G then, the infinitesimal transformations have the form $\mathsf{T}(g) = 1 + i\,\varepsilon\,\mathsf{X}$, where X is the infinitesimal generator corresponding to the transformation g, and the invariance condition of the equation (6.1.79) reads

$$\Omega(\mathbf{r}, t)\, (1 + i\,\varepsilon\,\mathsf{X})\, \psi(\mathbf{r}, t) = i\varepsilon\, \Omega(\mathbf{r}, t)\, \mathsf{X}\, \psi(\mathbf{r}, t) = 0 , \tag{6.1.83}$$

for any function $\psi(\mathbf{r}, t)$ solution of (6.1.79). By postulating that the only operator that satisfies the equation (6.1.79) is the operator $\Omega(\mathbf{r}, t)$, the condition (6.1.83) becomes

$$[\Omega(\mathbf{r}, t), \mathsf{X}] = i\lambda(\mathbf{r}, t)\,\Omega(\mathbf{r}, t)\,, \tag{6.1.84}$$

that is independent of ψ, and where $\lambda(\mathbf{r}, t)$ is a function depending on the infinitesimal generator X. To solve the equation (6.1.84), we need the analytic form of the operators X and Ω. Note that $\lambda(\mathbf{r}, t)$ in the condition (6.1.84) is a multiplicative operator (i.e., $\lambda(\mathbf{r}, t)$ does not contain differential operators). If Ω is a differential operator of the second order, then it can be shown that the infinitesimal generator X, solution of (6.1.84), must be a linear differential operator of the first order

$$\mathsf{X}(\mathbf{r}, t) = \mathbf{a}(\mathbf{r}, t) \cdot (-i\nabla) + b(\mathbf{r}, t)\,(i\,\partial_t) + c(\mathbf{r}, t)\,. \tag{6.1.85}$$

In the case of Schrödinger's equation, the infinitesimal generator obtained by solving (6.1.84) is identical to the infinitesimal generator (6.1.31) restricted to its spatial part acting in X, that is the part comprising only the operators $\mathbf{P}$, $\mathbf{J}$, $\mathbf{K}$, H, A, and D.

The fundamental equation of non-relativistic quantum mechanics is the time-independent Schrödinger equation (6.1.16). Assuming a degeneracy of the eigenvalues, we introduce two indexes to label the eigenvalues and eigenfunctions, and write (6.1.16) in the form

$$\mathsf{H}\,\psi_{n\alpha} = E_n\,\psi_{n\alpha}\,. \tag{6.1.86}$$

The eigenfunctions $\psi_{n\alpha}$ form a complete orthonormal set. We denote by $\mathcal{H}$ the Hilbert space spanned by all the eigenfunctions $\psi_{n\alpha}$, and by $\mathcal{H}_n \subset \mathcal{H}$ the subspace spanned by $\psi_{n\alpha}$ for a fixed n.

Now, we can determine a number of essential physical properties of a quantum system, depending on the symmetry properties of this system, without explicitly solving the equation (6.1.86). Thus, if G is a symmetry group of the quantum system, then the Hamiltonian of this system is invariant relative to G, and from (6.1.78) we have

$$\mathsf{T}(g)\,\mathsf{H} = \mathsf{H}\,\mathsf{T}(g)\,, \quad \forall g \in G\,. \tag{6.1.87}$$

Also, the wave equation (6.1.86) has to be invariant with respect to G, hence

$$\mathsf{T}(g)\,\mathsf{H}\,\psi_{n\alpha} = E_n\,(\mathsf{T}(g)\,\psi_{n\alpha})\,,$$

such that

$$\mathsf{H}\,(\mathsf{T}(g)\,\psi_{n\alpha}) = E_n\,(\mathsf{T}(g)\,\psi_{n\alpha})\,. \tag{6.1.88}$$

Therefore, if $\psi_{n\alpha}$ is a solution of (6.1.86) corresponding to the eigenvalue E_n, then $\mathsf{T}(g)\,\psi_{n\alpha}$ is also a solution of (6.1.86), corresponding to the same eigenvalue E_n. Consequently, this new solution belongs to the subspace $\mathcal{H}_n$ and can be expressed as a series in terms of the eigenfunctions $\psi_{n\alpha}$ (fixed n)

$$\mathsf{T}(g)\,\psi_{n\alpha} = \sum_{\beta} A_{\beta\alpha}(g)\,\psi_{n\beta}\,. \tag{6.1.89}$$

Because the eigenfunctions form a complete orthonormal set, the coefficients in (6.1.89) form a unitary square matrix

$$\left[\mathbf{A}(g)^{\dagger}\,\mathbf{A}(g)\right]_{\alpha\gamma} = \left[\mathbf{A}(g)\,\mathbf{A}(g)^{\dagger}\right]_{\alpha\gamma} = \delta_{\alpha\gamma}\,, \quad \mathbf{A}(g) = [A_{\alpha\beta}(g)]$$

and, in particular, we have

$$\sum_{\beta} |A_{\beta\alpha}(g)|^2 = 1\,. \tag{6.1.90}$$

If $g_1, g_2 \in G$, then

$$\mathsf{T}(g_1)\,\psi_{n\alpha} = \sum_{\beta'} A_{\beta'\alpha}(g_1)\,\psi_{n\beta'}\,, \tag{6.1.91}$$

$$\mathsf{T}(g_2)\,\psi_{n\alpha} = \sum_{\beta''} A_{\beta''\alpha}(g_2)\,\psi_{n\beta''}\,, \tag{6.1.92}$$

and assuming that $g = g_1 g_2$ we obtain

$$\mathsf{T}(g_2)\,\mathsf{T}(g_1)\,\psi_{n\alpha} = \sum_{\beta} A_{\beta\alpha}(g)\,\psi_{n\beta} = \sum_{\beta'}\sum_{\beta''} A_{\beta''\beta'}(g_2)\,A_{\beta'\alpha}(g_1)\,\psi_{n\beta''}\,,$$

such that

$$A_{\beta\alpha}(g) = \sum_{\beta'} A_{\beta\beta'}(g_2)\,A_{\beta'\alpha}(g_1)\,,$$

or

$$\mathbf{A}(g) = \mathbf{A}(g_2 g_1) = \mathbf{A}(g_2)\mathbf{A}(g_1)\,.$$

We have thus shown that the set of matrices $\mathbf{A}(g)$, $g \in G$, defined in (6.1.89) form a unitary representation of G, associated to the eigenvalue E_n. The set of eigenfunctions $\psi_{n\alpha}$ (fixed n) form a basis for this representation of G.

The representation $\mathbf{A} = \{\mathbf{A}(g) | g \in G\}$, determined by means of the eigenfunctions associated to one energy level E_n, is necessarily irreducible. Otherwise, we may partition the set of eigenfunctions $\psi_{n\alpha}$, corresponding to the same eigenvalue E_n, into at least two subsets such that the functions in each subset transform among them under the action of the operators $\mathsf{T}(g)$, $\forall g \in G$, according to the relation (6.1.89). If we know the irreducible representations of G, then we can determine the degeneracy of the states of the studied physical system. Also, we may classify the energy states by means of these irreducible representations, and find the transformation properties of the wave functions of different energy states, without solving Schrödinger's equation.

If G is a symmetry group of the Hamiltonian H then, the condition (6.1.87) is fulfilled, and the operators H and $\mathsf{T}(g)$ admit a common system of eigenfunctions. Let ψ_i and ψ_f be two such eigenfunctions that satisfy the relations

$$\mathsf{T}(g)\,\psi_i = t_i\,\psi_i\,, \quad \mathsf{T}(g)\,\psi_f = t_f\,\psi_f\,.$$

The transition probability between the states i and f is given by the relation

$$P_{if} = |<\psi_f, \mathsf{H}\,\psi_i>|^2\,.$$

Now, we evaluate the inner product

$$\begin{aligned} 0 &= <\psi_f, [\mathsf{T}(g), \mathsf{H}]\psi_i> = <\psi_f, (\mathsf{T}(g)\mathsf{H} - \mathsf{H}\mathsf{T}(g))\psi_i> \\ &= <\psi_f, \mathsf{T}(g)\mathsf{H}\psi_i> - <\psi_f, \mathsf{H}\mathsf{T}(g)\psi_i> \\ &= <\mathsf{T}(g)^\dagger\psi_f, \mathsf{H}\psi_i> - <\psi_f, \mathsf{H}\mathsf{T}(g)\psi_i> = (\bar{t}_f - t_i)\,<\psi_f, \mathsf{H}\psi_i>\,. \end{aligned}$$

If $P_{if} \neq 0$, then it follows that $\bar{t}_f - t_i = 0$ and, assuming that the operators $\mathsf{T}(g)$ are Hermitian ($\mathsf{T}(g) = \mathsf{T}(g)^\dagger$), that is $\bar{t}_f = t_f$, we obtain the relation $t_f = t_i$, which represents the conservation of the eigenvalue (quantum number) t when the physical system undergoes a transition between the states determined by the eigenfunctions ψ_i and ψ_f. Thus, the operators $\mathsf{T}(g)$ will correspond to an observable, t being the value of this physical quantity, that is obtained as the result of a measurement.

1.1.6 The invariance under rotations in quantum mechanics

By definition, the wave function ψ, which is the solution of Schrödinger's equation, describes the state of the considered quantum system. The function ψ, also called *state vector*, belongs to a separable Hilbert space $\mathcal{H}$. Also, the function ψ is square integrable ($\|\psi\| = 1$), such that the description of the quantum system is not unique, because all the wave functions $\omega\,\psi$ with $|\omega| = 1$ are solutions of the same Schrödinger's equation, and $\|\omega\,\psi\| = 1$. The set $\Psi = \{\omega\,\psi\,|\,|\omega| = 1\}$ is called a *ray of vectors* in the Hilbert space.

DEFINITION 6.1.3 *The inner product of two rays Φ and Ψ, is defined by the relation*

$$\langle\Phi, \Psi\rangle = |<\varphi, \psi>|\,,$$

where $<\varphi, \psi>$ is the inner product of the wave functions $\varphi \in \Phi$ and $\psi \in \Psi$.

In the case of a quantum system, to each observable A (momentum, energy, position etc.) it corresponds a Hermitian operator A acting in the Hilbert space $\mathcal{H}$. Let $\{A, B, C, \ldots\}$ be a set of observables, $\{\mathsf{A}, \mathsf{B}, \mathsf{C}, \ldots\}$ the associated operators forming a commutative set (i.e., what is called set of *compatible variables*), and $\{a, b, c, \ldots\}$ the corresponding eigenvalues. Because the operators $\{\mathsf{A}, \mathsf{B}, \mathsf{C}, \ldots\}$ commute they share a common set of linearly independent eigenfunctions $u_{abc\ldots}$. The probability to obtain the values $\{a, b, c, \ldots\}$ as the result of a simultaneous measurement of observables $\{A, B, C, \ldots\}$ is given by a formula that represents a generalization of the coefficients in (6.1.12)

$$P_{abc\ldots} = |< u_{abc\ldots}, \Psi >|^2 \equiv \langle U_{abc\ldots}, \Psi \rangle^2 , \qquad (6.1.93)$$

where $U_{abc\ldots}$ is the ray corresponding to the eigenfunctions $u_{abc\ldots}$. The set of the eigenvalues and their probabilities form a statistical distribution. If we measure an observable A, then rotate the measuring device by a rotation R, and we follow the same procedure to measure the same observable (let us call it now A') the two statistical distributions of A and A' will be identical. In these two situations, the quantum system is described by the rays Ψ and Ψ', respectively, and the rotations define a one-to-one mapping between the rays

$$\Psi' = D(\mathsf{R})\,\Psi . \qquad (6.1.94)$$

Because the statistical distributions of the initial and rotated systems are identical, for more than one observable $\{A, B, C, \ldots\}$ we obtain

$$P_{abc\ldots} = P'_{abc\ldots} , \qquad (6.1.95)$$

that is

$$\langle U_{abc\ldots}, \Psi \rangle^2 = \left\langle U'_{abc\ldots}, \Psi' \right\rangle^2 \equiv \langle D(\mathsf{R})\, U_{abc\ldots}, D(\mathsf{R})\,\Psi \rangle^2 , \qquad (6.1.96)$$

for every *eigenray* $U_{abc\ldots}$ and every ray Ψ. But, any ray Φ can be regarded as the eigenray of a system of operators such that, replacing U by Φ, we may write

$$\langle \Phi, \Psi \rangle^2 = \left\langle \Phi', \Psi' \right\rangle^2 \equiv \langle D(\mathsf{R})\,\Phi, D(\mathsf{R})\,\Psi \rangle^2 . \qquad (6.1.97)$$

Consequently, the relation (6.1.95) holds if the mapping $D(\mathsf{R})$ is established in such a way that the equation (6.1.97) is satisfied $\forall\, \Psi, \Phi \subset \mathcal{H}$.

THEOREM 6.1.1 (WIGNER) *Let D be a mapping of rays in a Hilbert space, such that*

$$\langle D\,\Phi, D\,\Psi \rangle^2 = \langle \Phi, \Psi \rangle^2 , \quad \forall\, \Phi, \Psi \subset \mathcal{H} . \qquad (6.1.98)$$

Then, there exists an operator τ defined everywhere in $\mathcal{H}$, having the properties:

- *if $\psi \in \Psi$, then $\tau\,\psi \in D\,\Psi$, $\forall\, \psi \in \Psi$ and $\forall\, \Psi \subset \mathcal{H}$;*
- *the operator τ is unitary or antiunitary;*

- *the operator τ is determined up to a constant phase factor.*

The mapping $D(\mathsf{R})$ defined in (6.1.94) satisfies the equation (6.1.97), that is identical to (6.1.98). Therefore, there exists a correspondence between the rotation R and an operator $\tau(\mathsf{R})$ having the properties stated by Wigner's theorem. It also follows from (6.1.94) that $\psi' \equiv \tau(\mathsf{R})\psi \in \Psi'$ and, considering two successive rotations ($\mathsf{R}''\mathsf{R}' = \mathsf{R}$), we have

$$\tau(\mathsf{R}'')\,\tau(\mathsf{R}') = \omega(\mathsf{R}'', \mathsf{R}')\,\tau(\mathsf{R}''\mathsf{R}')\,, \quad |\omega(\mathsf{R}'', \mathsf{R}')| = 1\,. \tag{6.1.99}$$

Thus, the operator function $\tau(\mathsf{R})$ defines a projective representation of the rotation group in the space $\mathcal{H}$. Also, this function operator is nonsingular because, according to Wigner's theorem, the operator τ is unitary or antiunitary. It can be shown that, in general, the rotation group has only unitary projective representations in the space $\mathcal{H}$, and these representations can be reduced to the usual representations of $\mathsf{O}(3)$.

1.2 The angular momentum. The spin

1.2.1 The eigenvalues and eigenvectors of angular momentum

In classical mechanics the angular momentum of a particle is given by the formula

$$\mathbf{K} = \mathbf{r} \times \mathbf{p}\,, \tag{6.1.100}$$

where $\mathbf{p} = m\mathbf{v}$ is the linear momentum of the particle. In quantum mechanics, the quantities $\mathbf{r}$ and $\mathbf{p}$ are replaced by the operators from (6.1.13) and (6.1.14), such that to the angular momentum of a particle it corresponds the vector operator

$$\mathbf{L} = \mathbf{r} \times (-i\nabla)\,. \tag{6.1.101}$$

The components of $\mathbf{L}$ are the differential operators

$$\begin{aligned}
\mathsf{L}_1 &= -i\left(y\,\frac{\partial}{\partial z} - z\,\frac{\partial}{\partial y}\right),\\
\mathsf{L}_2 &= -i\left(z\,\frac{\partial}{\partial x} - x\,\frac{\partial}{\partial z}\right),\\
\mathsf{L}_3 &= -i\left(x\,\frac{\partial}{\partial y} - y\,\frac{\partial}{\partial x}\right),
\end{aligned} \tag{6.1.102}$$

that satisfy the commutation relations

$$[\mathsf{L}_j, \mathsf{L}_k] = i\,\epsilon_{jkl}\,\mathsf{L}_l\,, \quad j,k = 1,2,3, \tag{6.1.103}$$

and each of them commutes with the operator

$$\mathsf{L}^2 = \mathsf{L}_1^2 + \mathsf{L}_2^2 + \mathsf{L}_3^2\,. \tag{6.1.104}$$

The operators (6.1.102) can be replaced by the operators

$$\mathsf{L}_+ = \mathsf{L}_1 + i\,\mathsf{L}_2\,, \quad \mathsf{L}_- = \mathsf{L}_1 - i\,\mathsf{L}_2\,, \quad \mathsf{L}_z = \mathsf{L}_3\,, \tag{6.1.105}$$

where L_+ and L_- are one another Hermitian conjugate. The operators (6.1.105) satisfy the commutation relations

$$[\mathsf{L}_z, \mathsf{L}_\pm] = \pm \mathsf{L}_\pm\,, \quad [\mathsf{L}_+, \mathsf{L}_-] = 2\mathsf{L}_z\,,$$

and

$$\left[\mathsf{L}^2, \mathsf{L}_+\right] = \left[\mathsf{L}^2, \mathsf{L}_-\right] = \left[\mathsf{L}^2, \mathsf{L}_z\right] = 0\,, \quad \mathsf{L}^2 = \frac{1}{2}\left(\mathsf{L}_+\,\mathsf{L}_- + \mathsf{L}_-\,\mathsf{L}_+\right) + \mathsf{L}_z^2\,.$$

Note that the operators (6.1.102) are identical to the infinitesimal generators of the rotation group (3.1.7) in Cartesian coordinates (see (3.1.13)). Also, the operators (6.1.105) coincide to the infinitesimal generators (3.1.8). Consequently, each rotation operator is a function of the components of the angular momentum, and can be expressed in the form

$$\mathsf{R}(\alpha, \beta, \gamma) = e^{-i\alpha\,\mathsf{L}_3}\, e^{-i\beta\,\mathsf{L}_2}\, e^{-i\gamma\,\mathsf{L}_3}\,. \tag{6.1.106}$$

Because L^2 commutes with each component of $\mathbf{L}$, we can find a complete set of eigenvectors common to L^2 and one of the $\mathbf{L}$ components, for instance L_3 that appears in both forms (6.1.102) and (6.1.105). In what follows we will use Dirac's notation $|l, m\rangle$ for an eigenvector common to L^2 and L_3, that corresponds to the eigenvalues l and m (l is the eigenvalue of the angular momentum operator, while m is the projection of this observable onto the z axis). With this notation the equations (3.2.10) and (3.2.12) read

$$\begin{aligned} \mathsf{L}^2\,|l, m\rangle &= l(l+1)\,|l, m\rangle\,, \\ \mathsf{L}_3\,|l, m\rangle &= m\,|l, m\rangle\,. \end{aligned} \tag{6.1.107}$$

Therefore, the eigenvalues of L^2 have the form $l(l+1)$, $l = 0, 1, 2, \ldots$, and the eigenvectors $|l, m\rangle$ are the spherical harmonics $Y_{lm}(\theta, \varphi)$, defined in (3.2.11). For a given l, the set of eigenvectors $\{Y_{lm}\,|\,|m| \leq l\}$ span a Hilbert space $\mathcal{H}^{(l)}$ of square integrable functions over the sphere, of dimension $2l+1$, and invariant relative to rotations.

A wave function ψ is invariant with respect to rotations if and only if the application of the angular momentum operator $\mathbf{L}$ to this function gives a zero result ($\mathbf{L}\,\psi = \mathbf{0}$). Note that actually, it is sufficient to have $\mathsf{L}^2\psi = 0$. In particular, this is the case of all wave functions that do not explicitly depend on the coordinates x, y, and z, but only on the modulus $r = |\mathbf{r}|$. Schrödinger's equation (6.1.16) is invariant with respect to rotations if $[\mathbf{L}, \mathsf{H}] = 0$ and, in this case, the angular momentum is conserved (see Chap. 6 Sec. 1.1.5). At the same time, if **L** and H commute, then L^2 and H also commute, and this property simplifies the problem of finding the eigenfunctions of H. It is sufficient to look for the eigenfunctions of H in the set of eigenfunctions that are common to L^2 and L_3. By means of the method of separation of variables, applied to Schrödinger's equation with a spherically symmetric potential $V = V(r)$, we can show that the energy eigenvalues are independent of m. Thus, to each eigenvalue E_{nl} associated to a given value l, it corresponds at least one set of $2l+1$ eigenfunctions, spanning a subspace $\mathcal{H}^{(l)} \subset \mathcal{H}$ that is invariant relative to rotations.

1.2.2 The spin and the total angular momentum

The hypothesis of the electron spin has been introduced in quantum mechanics because the model based on Schrödinger's equation was unable to explain the fine structure of spectral lines produced by complex atoms. Experimental studies of these atoms in magnetic fields has revealed that, each electron in motion on its orbit exhibits a magnetic momentum collinear to the electron angular momentum **l**

$$\boldsymbol{\mu} = -\frac{|e|}{2m_e}\,\mathbf{l}\,, \tag{6.1.108}$$

where m_e is the mass of the electron, and e is the electric charge of the electron. This magnetic moment has been associated to an intrinsic angular momentum (called the electron *spin*) of magnitude $1/2$. Later on, the experimental results shown that the nucleons (protons and neutrons) also possess a spin of magnitude $1/2$, of which existence can be directly demonstrated by measuring the magnetic moment of the nucleon.

If s is the spin operator of a particle of spin $1/2$ then, in analogy to the angular momentum, s^2 will have the eigenvalue $s(s+1) = 3/4$, and each component of s can take one of the values $\pm 1/2$. Therefore, the components of s are operators acting in a two-dimensional space, with a basis formed by the two eigenvectors that are common to s^2 and s_3. In addition to the commutation relations (6.1.103), that are characteristic to angular momentum operators, the components of s satisfy the relations

$$\mathsf{s}_1^2 = \mathsf{s}_2^2 = \mathsf{s}_3^2 = \frac{1}{4}\,, \quad \mathsf{s}_+^2 = \mathsf{s}_-^2 = 0\,.$$

Because

$$\mathsf{s}_+^2 = (\mathsf{s}_1 + i\,\mathsf{s}_2)^2 = \mathsf{s}_1^2 - \mathsf{s}_2^2 + i\,(\mathsf{s}_1\,\mathsf{s}_2 + \mathsf{s}_2\,\mathsf{s}_1)\,,$$

it follows that

$$\mathsf{s}_1\,\mathsf{s}_2 + \mathsf{s}_2\,\mathsf{s}_1 = 0\,,$$

hence the operators s_1, s_2, and s_3, anti-commute each another. All these conditions are satisfied if we choose the operator $\mathbf{s}$ in the form

$$\mathbf{s} = \frac{1}{2}\boldsymbol{\sigma}, \tag{6.1.109}$$

where $\boldsymbol{\sigma} \equiv (\boldsymbol{\sigma}_1, \boldsymbol{\sigma}_2, \boldsymbol{\sigma}_3)$ represents the *Pauli matrices*

$$\boldsymbol{\sigma}_1 = \begin{bmatrix} 0 & 1 \\ 1 & 0 \end{bmatrix}, \quad \boldsymbol{\sigma}_2 = \begin{bmatrix} 0 & -i \\ i & 0 \end{bmatrix}, \quad \boldsymbol{\sigma}_3 = \begin{bmatrix} 1 & 0 \\ 0 & -1 \end{bmatrix}. \tag{6.1.110}$$

Pauli's matrices satisfy the relations

$$\begin{aligned} &\boldsymbol{\sigma}_1^2 = \boldsymbol{\sigma}_2^2 = \boldsymbol{\sigma}_3^2 = \mathbf{1}, \\ &\boldsymbol{\sigma}_j \boldsymbol{\sigma}_k = -\boldsymbol{\sigma}_k \boldsymbol{\sigma}_j = i\,\varepsilon_{jkl}\,\boldsymbol{\sigma}_l, \quad j,k,=1,2,3, \\ &\boldsymbol{\sigma}_1 \boldsymbol{\sigma}_2 \boldsymbol{\sigma}_3 = i\,\mathbf{1}, \\ &\mathrm{Tr}\,\boldsymbol{\sigma}_1 = \mathrm{Tr}\,\boldsymbol{\sigma}_2 = \mathrm{Tr}\,\boldsymbol{\sigma}_3 = 0, \\ &\det \boldsymbol{\sigma}_1 = \det \boldsymbol{\sigma}_2 = \det \boldsymbol{\sigma}_3 = -1, \end{aligned} \tag{6.1.111}$$

and the vectors from the two-dimensional space of the spin operator $\mathbf{s}$ are called *spinors*.

The variables which define the state of a particle of spin $1/2$ are of two kinds: *orbit variables* $(\mathbf{r}, \mathbf{p})$ and *spin variables* $(\mathbf{s})$. Because the orbit and spin variables commute each another, the space of the particle states is the direct product $\mathcal{H}^{(0)} \times \mathcal{H}^{(s)}$, where $\mathcal{H}^{(0)}$ is the space of states of a zero spin particle, while $\mathcal{H}^{(s)}$ is the two-dimensional space of spinors. The matrices (6.1.110) are identical to the matrices of the infinitesimal generators of the $\mathsf{SU}(2)$ group, such that the space $\mathcal{H}^{(s)}$ will be a space invariant with respect to the transformations in $\mathsf{SU}(2)$. Also, the total angular momentum $\mathbf{j}$ of a particle is

$$\mathbf{j} = \mathbf{l} + \mathbf{s}, \tag{6.1.112}$$

where the operator $\mathbf{J}$, associated to the observable $\mathbf{j}$, satisfies the commutation relations (6.1.103).

We have shown in Chap. 6 Sec. 1.1.5 that, if a Hamiltonian is invariant relative to a symmetry transformation group G, then the eigenfunctions associated to an energy level form a basis for a representation of G. Thus, the invariant subspaces in the Hilbert space $\mathcal{H}$ of wave functions, solutions of (6.1.86), are associated to the energy levels E_n and we obtain a decomposition of $\mathcal{H}$ into a direct sum of invariant subspaces

$$\mathcal{H} = \sum_n{}^{\oplus} \mathcal{H}^{(n)}. \tag{6.1.113}$$

In the case of the invariance of the Hamiltonian relative to the rotation group $\mathsf{SO}(3)$, the solutions of the wave equation (6.1.86), in spherical coordinates, are

$$\psi_{nlm}(r,\theta,\varphi)=R_{nl}(r)\,Y_{lm}(\theta,\varphi)\,, \tag{6.1.114}$$

where l and m determine the representation and an element of the representation, respectively. For a fixed n, the subspace $\mathcal{H}^{(n)}$ decomposes into rotationally invariant subspaces

$$\mathcal{H}^{(n)}=\sum_{l}{}^{\oplus}\mathcal{H}^{(nl)}\,; \tag{6.1.115}$$

the invariant subspace $\mathcal{H}^{(nl)}$ being of dimension $2l+1$, and having the basis $\{Y_{lm}\,|\,|m|\leq l\}$. Each eigenvector in the basis represents a state of angular momentum l, with the projection onto the z axis $l_z=m$.

The vector addition of two angular momenta is performed according to a relation of the form (3.3.47), adapted to the case of the rotation group

$$Y_{l_1m_1}(\theta,\varphi)\,Y_{l_2m_2}(\theta,\varphi)=\sum_{l=|l_1-l_2|}^{l_1+l_2}C^{\,l_1\;l_2\;l}_{\,m_1\,m_2\,m_1+m_2}\,Y_{l\,m_1+m_2}(\theta,\varphi)\,. \tag{6.1.116}$$

In the same manner, the vector addition of two spins $\mathbf{s}_1$ and $\mathbf{s}_2$, which are associated, respectively, to the $\mathsf{SU}(2)$ representations of weights s_1 and s_2, is given by the formula

$$|s_1,s_{1z}\rangle\;|s_2,s_{2z}\rangle=\sum_{s=|s_1-s_2|}^{s_1+s_2}C^{\,s_1\;s_2\;s}_{\,s_{1z}\,s_{2z}\,s_z}\;|s,s_z\rangle\,, \tag{6.1.117}$$

where $s_z=s_{1z}+s_{2z}$. Because the irreducible representations of $\mathsf{SO}(3)$ are locally isomorphic to the even representations of $\mathsf{SU}(2)$, the vector addition of an angular momentum state $|l,m\rangle$ and a spin state $|s,s_z\rangle$ can be performed by means of the Clebsch-Gordan coefficients for the $\mathsf{SU}(2)$ group

$$|l,m\rangle\;|s,s_z\rangle=\sum_{j=|l-s|}^{l+s}C^{\,l\;s\;J}_{\,m\,s_z\,J_z}\;|j,j_z\rangle\,, \tag{6.1.118}$$

where j is the value of the total angular momentum (note that the eigenvalues of the total angular momentum operator $\mathbf{J}$ are $j(j+1)$), while $j_z=m+s_z$ is the eigenvalue of J_z.

1.2.3 Irreducible tensor operators. The Wigner-Eckart theorem

Let $\{\mathbf{e}_1,\mathbf{e}_2,\mathbf{e}_3\}$ be a basis in the geometric space $\mathcal{G}$, and let $\{\mathsf{A}_{\kappa q}\,|\,|q|\leq\kappa\}$ be a set of operators acting in a Hilbert space $\mathcal{H}$ defined on $\mathcal{G}$. Also, let

$\{\mathsf{A}'_{\kappa q} \,|\, |q| \leq \kappa\}$ be the set of transformed operators corresponding to a rotation of the basis, defined by the rotation operator R.

DEFINITION 6.1.4 *If the relation*

$$\mathsf{A}'_{\kappa q} = \sum_{p=-\kappa}^{\kappa} \tau^{\kappa}_{pq}(\mathsf{R})\, \mathsf{A}_{\kappa p} \tag{6.1.119}$$

takes place, and if there exists an irreducible representation $\mathsf{T}(\mathsf{R})$ *of the rotation group, such that*

$$\mathsf{A}'_{\kappa q} = \mathsf{T}(\mathsf{R})\, \mathsf{A}_{\kappa q}\, \mathsf{T}(\mathsf{R})^{\dagger}\,, \tag{6.1.120}$$

then we say that the set of operators $\{\mathsf{A}_{\kappa q} \,|\, |q| \leq \kappa\}$ *forms an* **irreducible tensor operator of weight** κ.

THEOREM 6.1.2 *If the set of operators* $\{\mathsf{A}_{\kappa q} \,|\, |q| \leq \kappa\}$ *forms an irreducible tensor operator, then*

$$\begin{aligned} [\mathsf{H}_3,\, \mathsf{A}_{\kappa q}] &= q\, \mathsf{A}_{\kappa q}\,, \\ [\mathsf{H}_+, \mathsf{A}_{\kappa q}] &= \alpha^{\kappa}_{q+1}\, \mathsf{A}_{\kappa, q+1}\,, \\ [\mathsf{H}_-, \mathsf{A}_{\kappa q}] &= \alpha^{\kappa}_{q}\, \mathsf{A}_{\kappa, q-1}\,, \end{aligned} \tag{6.1.121}$$

where

$$\alpha^{\kappa}_{q} = \sqrt{(\kappa + q)(\kappa - q + 1)}\,,$$

while H_+, H_-, *and* H_3 *are the Hermitian infinitesimal generators of the irreducible representation* $\mathsf{T}(\mathsf{R})$, *that transforms the tensor operator under rotations.*

Reciprocally, if the operators $\{\mathsf{A}_{\kappa q} \,|\, |q| \leq \kappa\}$ satisfy the conditions (6.1.121), where H_+, H_-, and H_3 are the Hermitian infinitesimal generators of an irreducible representation $\mathsf{T}(\mathsf{R})$ of the rotation group, then the set of operators $\{\mathsf{A}_{\kappa q} \,|\, |q| \leq \kappa\}$forms an irreducible tensor operator of weight κ, that transforms under rotations by means of the operator $\mathsf{T}(\mathsf{R})$.

On the basis of the definition of an invariant operator, any scalar can be regarded as an irreducible tensor operator of weight zero. Indeed, because $\tau^{0}_{pq} = \delta_{pq}$, from (6.1.119) and (6.1.120) we obtain $\mathsf{A} = \mathsf{T}(\mathsf{R})\, \mathsf{A}\, \mathsf{T}(\mathsf{R})^{\dagger}$. Furthermore, a vector operator is an irreducible tensor operator of weight one. Thus, the operators

$$\mathsf{A}_{1,1} = -\frac{1}{\sqrt{2}}\mathsf{H}_+\,, \quad \mathsf{A}_{1,0} = \mathsf{H}_3\,, \quad \mathsf{A}_{1,-1} = \frac{1}{\sqrt{2}}\mathsf{H}_-$$

satisfy the conditions (6.1.121) and therefore, the infinitesimal generators H_1, H_2, and H_3 form an irreducible vector operator. Also, the spin operators $\boldsymbol{\sigma}_1$, $\boldsymbol{\sigma}_2$, and $\boldsymbol{\sigma}_3$ are the components of an irreducible vector operator. The spherical harmonics $\{Y_{lm} \mid |m| \leq l\}$ transform under rotations according to the relation

$$Y'_{lm}(\theta', \varphi') = \sum_{m'=-l}^{l} D^{l}_{m'm}(\mathsf{R})\, Y_{lm'}(\theta, \varphi)\,,$$

that enables us to show that the set $\{Y_{lm} \mid |m| \leq l\}$ forms an irreducible tensor operator of weight l.

Most of the calculations in quantum mechanics require the evaluation of matrix elements $\langle j_2, m_2 | \mathsf{T}_{jm} | j_1, m_1 \rangle$ of an irreducible tensor operator T_{jm} between two states $|j_1, m_1\rangle$ and $|j_2, m_2\rangle$. The transformations of angular momentum states $|l, l_z\rangle$ are defined by the operators $\mathsf{D}(l)$ in a representation of weight l of $\mathsf{SO}(3)$[1], while the transformations of spin states $|s, s_z\rangle$ are defined by the operators $\mathsf{D}(s)$ in a representation of weight s of $\mathsf{SU}(2)$. Because the irreducible representations of $\mathsf{SO}(3)$ are locally isomorphic to the even representations of $\mathsf{SU}(2)$, we may use only $\mathsf{SU}(2)$ representations and thus, the transformations of total angular momentum states $|j, m\rangle$, where $m = l_z + s_z$, can be defined by the operators $\mathsf{D}(j)$ in a representation of weight j of $\mathsf{SU}(2)$. Consequently, the transformations of the states $|j_1, m_1 >$ and $|j_2, m_2\rangle$, and the tensor operator T_{jm} are defined by the $\mathsf{SU}(2)$ representation operators $\mathsf{D}(j_1)$, $\mathsf{D}(j_2)$, and $\mathsf{D}(j)$, respectively, such that the transformations of the matrix element $\langle j_2, m_2 | \mathsf{T}_{jm} | j_1, m_1 \rangle$ will be defined by the direct product of $\mathsf{SU}(2)$ representations

$$\mathsf{D}(j_1) \otimes \mathsf{D}(j) \otimes \mathsf{D}(j_2) = \sum_{j'} \mathsf{D}(j')\,.$$

The analytic form of the matrix element is

$$\langle j_2, m_2 | \mathsf{T}_{jm} | j_1, m_1 \rangle = \int \overline{\psi_{j_2 m_2}}\, \mathsf{T}_{jm}\, \psi_{j_1 m_1}\, dx dy dz\,,$$

but the action of the tensor operator T_{jm} onto the wave function $\psi_{j_1 m_1}$ can be expressed in the form

$$\mathsf{T}_{jm}\, \psi_{j_1 m_1} = \sum_{j''} C^{j_1\, j\, j''}_{m_1\, m\, m+m_1}\, \psi_{j'', m+m_1}\,,$$

such that, in spherical coordinates, we may write

$$\langle j_2, m_2 | \mathsf{T}_{jm} | j_1, m_1 \rangle = \sum_{j''} C^{j_1\, j\, j''}_{m_1\, m\, m+m_1} \int r^2 dr \int \overline{\psi_{j_2 m_2}}\, \psi_{j'', m+m_1}\, d\omega\,.$$

[1] In this chapter, to avoid any confusion with standard operators in quantum mechanics, the representations of $\mathsf{SO}(3)$ and $\mathsf{SU}(2)$ will be denoted by $\mathsf{D}(l)$ and $\mathsf{D}(j)$ instead of T^l and T^j, respectively.

The orthogonality of the states $|j_2, m_2\rangle \equiv \psi_{j_2 m_2}$ and $|j'', m+m_1\rangle \equiv \psi_{j'', m+m_1}$ implies that the integral over the sphere vanishes unless $j' = j_2$ therefore,

$$\langle j_2, m_2 | \mathsf{T}_{jm} | j_1, m_1 \rangle = C^{j_1\ j\ j_2}_{m_1\, m\, m_2} \int f(r)\, r^2 dr\,.$$

The remaining integral, denoted by

$$\langle j_2 \| \mathsf{T}_j \| j_1 \rangle \equiv \int f(r)\, r^2 dr\,,$$

is called *reduced matrix element* and it depends only on the total angular momenta j_1, j, and j_2, while the dependence of the matrix element of T_{jm} on m_1, m, and m_2 is contained in the Clebsch-Gordan coefficient.

THEOREM 6.1.3 (WIGNER-ECKART) *The matrix element of the component m of an irreducible tensor operator T_{jm}, between the basis vectors $|j_1, m_1\rangle$ and $|j_2, m_2\rangle$ of two $\mathsf{SU}(2)$ irreducible representations, is equal to the product of the corresponding Clebsch-Gordan coefficient and a quantity that is independent of m_1, m, and m_2*

$$\langle j_2, m_2 | \mathsf{T}_{jm} | j_1, m_1 \rangle = C^{j_1\ j\ j_2}_{m_1\, m\, m_2} \langle j_2 \| \mathsf{T}_j \| j_1 \rangle\,, \quad m_2 = m + m_1\,. \quad (6.1.122)$$

1.2.4 SU(2) statistical weights

One aspect of theoretical interpretation of the experimental results in the elementary particle physics, in the case of multiple production interactions, consists in finding the ratios of probabilities of different final states, that have the same multiplicity. These probabilities are determined by some quantities called statistical weights that can be theoretically calculated under the hypothesis of interaction invariance relative to a transformation group G. Thus, we say that the interaction is invariant relative to the group G if:

1 the particles involved in the interaction are described by wave functions that belong to the basis of an irreducible representation of G;

2 the interaction Lagrangian is invariant with respect to the transformations in G.

If we denote by $\mathsf{D}(\delta)$ an irreducible representation of G, then the final state of the interaction is described by a wave function $|f\rangle = |\delta, \rho\rangle$, where δ represents the quantum numbers that characterize the representation D, while ρ represents the quantum numbers required to completely determine the considered state inside the representation space of D. The multiple production of particles being an interaction in which take part many particles, it is reasonable to consider that some aspects of the phenomenon exhibit a statistical character. Consequently, we will introduce a second hypothesis (the *statistical hypothesis*) by which, we assume that the probabilities of the final states determined by the same quantum numbers δ and ρ are equal.

Let us consider a multiple production interaction that is invariant relative to the $\mathsf{SU}(2)$ group. Thus, we assume the conservation of the total angular momentum (spin) $\mathbf{j}$ and its projection m onto a certain axis. The statistical weight associated to a final state, characterized by $\mathbf{j}$ and m, will be equal to the number of independent combinations leading to this final state. If the final state consists of n particles of spin $\mathbf{j}_k$, $k = \overline{1,n}$, then the statistical weight of a final state of total spin $\mathbf{j}$ is equal to the multiplicity of the irreducible representation $\mathsf{D}(j)$ of $\mathsf{SU}(2)$, in the decomposition of the direct product of the irreducible representations $\mathsf{D}(j_k)$, into irreducible representations. Because to the direct product of representations it corresponds the product of the respective characters, we may write

$$\prod_{k=1}^{n} \chi^{j_k} = \sum_{j} g(j; j_1, j_2, \ldots, j_n)\, \chi^{j} \,, \tag{6.1.123}$$

where χ^{j_k} is the character of the irreducible representation $\mathsf{D}(j_k)$. The multiplicities g are obtained by means of the orthogonality relation satisfied by the characters

$$g(j; j_1, j_2, \ldots, j_n) = \frac{1}{\pi} \int_{-\pi}^{\pi} \sin^2\alpha \, \frac{\sin{(2j+1)\alpha}}{\sin\alpha} \prod_{k=1}^{n} \frac{\sin{(2j_k+1)\alpha}}{\sin\alpha} \, d\alpha \,, \tag{6.1.124}$$

where α is the class parameter of $\mathsf{SU}(2)$. Also, the integration element and the expressions of characters have been derived in (3.3.72) and (3.3.73). When the final state consists of n particles of spin $1/2$, then the relation (6.1.124) becomes

$$g'(j; n) = \frac{2j+1}{n+1} \begin{pmatrix} n+1 \\ \frac{n}{2} - j \end{pmatrix} , \tag{6.1.125}$$

where $\begin{pmatrix} n \\ k \end{pmatrix}$ represents the binomial coefficient.

The character of an irreducible representation $\mathsf{D}(j)$ can also be expressed as a Gegenbauer polynomial $P_{2j}^{(1)}(\cos\alpha)$. If we denote the character of the fundamental representation of $\mathsf{SU}(2)$ by $w = 2\cos\alpha$, then we have

$$\chi^{j}(\alpha) = \sum_{k=0}^{2j} (-1)^{j-\frac{k}{2}} \begin{pmatrix} j + \frac{k}{2} \\ k \end{pmatrix} w^{k} \,, \tag{6.1.126}$$

and from the integral (6.1.124) we obtain

$$g(j; j_1, j_2, \ldots, j_n) = \sum_{k_1, k_2, \ldots, k_n} a^{j_1}_{k_1} a^{j_2}_{k_2} \ldots a^{j_n}_{k_n} g'(j; k_1 + k_2 + \ldots + k_n), \tag{6.1.127}$$

where a^j_k is the coefficient of w^k in (6.1.126) and the multiplicities g' are defined in (6.1.125). From (6.1.127) it follows the statistical weight of a system comprising p particles of spin $1/2$, r particles of spin 1, and s particles of spin $3/2$, in a state of total spin j

$$g''(j; p, r, s) = (2j+1) \sum_{i,k} \frac{(-1)^{r-i}(-2)^{s-k}}{p+s+2(i+k)+1} \binom{r}{i} \binom{s}{k} \times \binom{p+s+2(i+k)+1}{\frac{p+s}{2}+i+k-j}. \tag{6.1.128}$$

The statistical weights thus obtained can be used when the projections m_k, $k = \overline{1,n}$, of the spins (or total angular momenta) $\mathbf{j}_k$ are not known. If these projections are known, then the final state of a multiple production process is described by the state vector

$$|f\rangle = |j_1, m_1, j_2, m_2, \ldots, j_n, m_n\rangle, \tag{6.1.129}$$

that is also equal to the direct product of the state vectors $|j_k, m_k\rangle$, $k = \overline{1,n}$. To obtain a final state defined by the quantum numbers j and m we have to couple the state vectors $|j_k, m_k\rangle$, associated to the particles in the final state, according to the scheme: couple the spins j_1 and j_2 to a total spin j_{12}, then couple j_{12} and j_3 to a total spin j_{123}, and so on. We denote this coupling scheme by

$$|j_{12}, j_{123}, \ldots; j, m\rangle \equiv |j_1\, j_2\, (j_{12})\, j_3\, (j_{123})\, \ldots\rangle.$$

Also, in terms of the $\mathsf{SU}(2)$ Clebsch-Gordan coefficients we have

$$|j^2, j^3, \ldots; j, m\rangle = C^{j_1\, j_2\, j^2}_{m_1\, m_2\, m^2} \ldots C^{j^{n-1}\, j_n\, j}_{m^{n-1}\, m_n\, m} |j_1, m_1\rangle \ldots |j_n, m_n\rangle, \tag{6.1.130}$$

where $j^k = j_{12\ldots k}$, $m^k = m_1 + \ldots + m_k$, and $m = m_1 + \ldots + m_n$. If the initial state of the interaction is of the form $|i\rangle = |j', m'\rangle$, and we follow the coupling scheme (6.1.130) then, the probability to observe the final state (6.1.129) is given by the relation

$$P \propto |\langle f|\mathsf{T}|i\rangle|^2 = |\langle j_1, m_1, j_2, m_2, \ldots, j_n, m_n|\, \mathsf{T}\, |j', m'\rangle|^2,$$

T being the transition operator. Then, by inverting the relation (6.1.130) we obtain

$$P \propto |\langle j^2, j^3, \dots; j, m|\mathsf{T}|j', m'\rangle|^2 \left(C^{j_1\ j_2\ j^2}_{m_1\ m_2\ m^2} \cdots C^{j^{n-1}\ j_n\ j}_{m^{n-1}\ m_n\ m} \right)^2$$

and, on the basis of the hypothesis that the interaction is invariant relative to the $\mathsf{SU}(2)$ group, it follows that $P = 0$ if $j \neq j'$ or $m \neq m'$.

The relation (6.1.130) is not the only coupling scheme leading to a final state $|j, m\rangle$ but, according to the statistical hypothesis, all the coupling schemes leading to the same final state $|j, m\rangle$ are equally probable. Consequently, the probability to observe a final state with given j and m is proportional to the statistical weight

$$P^j_{m_1\ m_2\ \dots\ m_n} = \sum_\alpha \left(C^{j_1\ j_2\ j^2}_{m_1\ m_2\ m^2} \cdots C^{j^{n-1}\ j_n\ j}_{m^{n-1}\ m_n\ m} \right)^2, \qquad (6.1.131)$$

where α stands for the set of quantum numbers $\{j^2, j^3, \dots, j^{n-1}\}$. Also, in terms of the state vectors (6.1.129), this statistical weight takes the form

$$P^j_{m_1\ m_2\ \dots\ m_n} = \left[\sum_\alpha |\langle j_1, m_1, j_2, m_2, \dots, j_n, m_n|\alpha; j, m\rangle|^2 \right]$$
$$\times \left[\sum_{m_i} \sum_\alpha |\langle j_1, m_1, j_2, m_2, \dots, j_n, m_n|\alpha; j, m\rangle|^2 \right]^{-1}, \qquad (6.1.132)$$

the second factor being introduced for normalization. Therefore, to calculate the statistical weight associated to a system of n particles, in a state $|j, m\rangle$, we have to sum the squares of the projections of the state vector (6.1.129) onto all the vectors $|\alpha; j, m\rangle$ (with fixed j and m). The state vectors (6.1.129) span a representation space for the $\mathsf{SU}(2)$ group. In terms of the Eulerian parametrization, an $\mathsf{SU}(2)$ transformation (rotation in the spin space) will change these state vectors according to the relation

$$|j_1, m_1, j_2, m_2, \dots, j_n, m_n\rangle = \sum_{m_i'} D^{j_1}_{m_1 m_1'}(\varphi, \theta, \psi) \dots D^{j_n}_{m_n m_n'}(\varphi, \theta, \psi)$$
$$\times |j_1, m_1', j_2, m_2', \dots, j_n, m_n'\rangle,$$

where the matrix $\mathbf{D}^{j_k}(\varphi, \theta, \psi)$ belongs to the $\mathsf{SU}(2)$ irreducible representation $\mathsf{D}(j_k)$ (see Chap. 3 Sec. 3.2.2), while the matrices having the elements

$$D_{m_1 \dots m_n, m_1' \dots m_n'}(\varphi, \theta, \psi) = D^{j_1}_{m_1 m_1'}(\varphi, \theta, \psi) \dots D^{j_n}_{m_n m_n'}(\varphi, \theta, \psi) \qquad (6.1.133)$$

form a representation of $\mathsf{SU}(2)$, that is the direct product of irreducible representations $\mathsf{D}(j_1) \otimes \mathsf{D}(j_2) \otimes \dots \otimes \mathsf{D}(j_n)$. This representation can be decomposed into a direct sum of irreducible representations by a suitable unitary transformation. The decomposition is equivalent to the determination of all the invariant subspaces in the space spanned by the state vectors (6.1.129). In particular, to the irreducible representation $\mathsf{D}(j)$ it corresponds a subspace of dimension $2j+1$, which is invariant relative to the transformations in $\mathsf{D}(j)$. The vectors $|\alpha; j, m\rangle$ necessarily belong to such an invariant subspace and, because we may have more than one invariant subspaces of dimension $2j+1$ (i.e., the multiplicity of the irreducible representation $\mathsf{D}(j)$, in the decomposition of the direct product of irreducible representations, can be greater than one), we will denote by $\Gamma(j)$ the direct sum of all these invariant subspaces. The projection of an arbitrary vector in the space spanned by the state vectors $|j_1, m_1, j_2, m_2, \ldots, j_n, m_n\rangle$ onto the invariant subspace $\Gamma(j)$ is obtained by means of the projection operator (see Chap. 3 Sec. 3.3.5)

$$\mathsf{P}^{\Gamma(j)} = \frac{2j+1}{16\pi^2} \int_0^{4\pi} d\varphi \int_0^{\pi} \sin\theta \, d\theta \int_0^{2\pi} d\psi \, \overline{\chi^j(\varphi, \theta, \psi)} \times D_{m_1 \ldots m_n, m_1' \ldots m_n'}(\varphi, \theta, \psi) \,. \tag{6.1.134}$$

Here,

$$\int_{\mathsf{SU}(2)} d\mathsf{U} = \frac{1}{16\pi^2} \int_0^{4\pi} d\varphi \int_0^{\pi} \sin\theta \, d\theta \int_0^{2\pi} d\psi$$

is the invariant integral over $\mathsf{SU}(2)$ in the Eulerian parametrization, defined in (2.2.39). Because any projection operator is Hermitian and idempotent, the square of the projection of the state vector $|j_1, m_1, j_2, m_2, \ldots, j_n, m_n\rangle$ onto $\Gamma(j)$ takes the simple form

$$\begin{aligned} &\left| \mathsf{P}^{\Gamma(j)} |j_1, m_1, j_2, m_2, \ldots, j_n, m_n\rangle \right|^2 \\ &\quad = \langle j_1, m_1, j_2, m_2, \ldots, j_n, m_n | \mathsf{P}^{\Gamma(j)} | j_1, m_1, j_2, m_2, \ldots, j_n, m_n\rangle \,, \end{aligned} \tag{6.1.135}$$

such that the statistical weights (6.1.132) are given by the diagonal elements of the matrix associated to the projector operator $\mathsf{P}^{\Gamma(j)}$ in the space spanned by the state vectors $|j_1, m_1, j_2, m_2, \ldots, j_n, m_n\rangle$, that is

$$P^j_{m_1 \, m_2 \ldots m_n} = \frac{2j+1}{16\pi^2} \int_0^{4\pi} d\varphi \int_0^{\pi} \sin\theta \, d\theta \int_0^{2\pi} d\psi \, \overline{\chi^j(\varphi, \theta, \psi)} \times \prod_{k=1}^{n} D^{j_k}_{m_k m_k}(\varphi, \theta, \psi) \,. \tag{6.1.136}$$

By substituting into (6.1.136) the expressions of the matrix elements D^j_{mm} and the character χ^j,

$$\begin{aligned} D^j_{mm}(\varphi,\theta,\psi) &= \left(\frac{1+\cos\theta}{2}\right)^m P^{0,2m}_{j-m}(\cos\theta)\, e^{-im(\varphi+\psi)}\,, \\ \chi^j(\varphi,\theta,\psi) &= \sum_{m=-j}^{j} D^j_{mm}(\varphi,\theta,\psi)\,, \end{aligned} \tag{6.1.137}$$

where the Jacobi polynomial is given by the formula

$$P^{0,2m}_{j-m}(w) = 2^{m-j}\sum_{\nu=0}^{j-m}\binom{j-m}{\nu}\binom{j+m}{j-m-\nu}(w-1)^{j-m-\nu}(w+1)^{\nu}\,,$$

we finally obtain the expression of the statistical weights in terms of an integral of a product of Jacobi's polynomials

$$\begin{aligned} P^j_{m_1\,m_2\,\ldots\,m_n} &= \frac{2j+1}{16\pi^2}\int_0^{4\pi} d\varphi \int_0^{\pi} \sin\theta\, d\theta \int_0^{2\pi} d\psi \left(\frac{1+\cos\theta}{2}\right)^{m_1+m_2+\ldots+m_n} \\ &\quad\times \prod_{k=1}^{n} P^{0,2m_k}_{j_k-m_k}(\cos\theta)\, e^{-i(m_1+m_2+\ldots+m_n)(\varphi+\psi)} \\ &\quad\times \sum_{m'=-j}^{j}\left(\frac{1+\cos\theta}{2}\right)^{m'} P^{0,2m'}_{j-m'}(\cos\theta)\, e^{im'(\varphi+\psi)} \\ &= \frac{2j+1}{2^{2m+1}}\int_{-1}^{1}(1+w)^{2m} P^{0,2m}_{j-m}(w)\prod_{k=1}^{n} P^{0,2m_k}_{j_k-m_k}(w)\, dw\,. \end{aligned} \tag{6.1.138}$$

Note that, in the last equation $m = m_1 + m_2 + \ldots + m_n$ is fixed, and we used the function substitution $w = \cos\theta$. We have also used the general formula

$$\frac{1}{2p\pi}\int_0^{2p\pi} e^{iq\alpha}\, d\alpha = \delta_{q0}\,, \quad p, q \in \mathbb{Z}\,,$$

to calculate the integrals over φ and ψ.

In (6.1.138) we have assumed that $m \geq 0$ which is not a restriction because, on the basis of (6.1.131) and the properties of Clebsch-Gordan coefficients (3.3.53), it is easy to prove that

$$P^j_{m_1\,m_2\,\ldots\,m_n} = P^j_{-m_1\,-m_2\,\ldots\,-m_n}\,.$$

Also, the statistical weights do not depend on the order of the particles (i.e., the order of the state vectors $|j_k, m_k\rangle$). Thus, if we do not label explicitly particles of the same kind (that differ only by the quantum number m_k, but have the same spin), then we have to sum over all different permutations of the configuration $\{m_1, m_2, \ldots, m_n\}$, that is to multiply the statistical weight (6.1.138) by the number of permutations of this configuration. We will denote the obtained statistical weight by ${}^*P^j_{m_1\, m_2\, \ldots\, m_n}$. Therefore, the probability associated, for instance, to a final state comprising n particles of identical spin 1, and among them n_+ have $m = +1$, n_0 have $m = 0$, and n_- have $m = -1$, is proportional to the statistical weight

$$
{}^*P^j_{n_+\, n_0\, n_-} = \frac{n!}{n_+!\, n_0!\, n_-!}\, P^j_{\underbrace{+1\,\ldots\,+1}_{n_+}\,\underbrace{0\,\ldots\,0}_{n_0}\,\underbrace{-1\,\ldots\,-1}_{n_-}} \cdot
$$

Moreover, if we sum the statistical weights ${}^*P^j_{m_1\, m_2\, \ldots\, m_n}$ over all configurations with $m = m_1 + m_2 + \ldots + m_n =$ const, then we obtain

$$
\sum_{m_1, m_2, \ldots, m_n} {}^*P^j_{m_1\, m_2\, \ldots\, m_n} = g(j; j_1, j_2, \ldots, j_n)\,,
$$

where the multiplicities $g(j; j_1, j_2, \ldots, j_n)$ are given by (6.1.124). Note that, the multiplicities $g(j; j_1, j_2, \ldots, j_n)$ also represent the number of independent spin functions which can be formed by coupling the spins $\mathbf{j}_1, \mathbf{j}_2, \ldots, \mathbf{j}_n$ to a final state of given j and m.

2. Internal symmetries of elementary particles

2.1 The isospin and the $\mathsf{SU}(2)$ group

2.1.1 Classification of elementary particles

Experimental studies of nuclear forces shown that the interactions between protons and neutrons do not depend on the electric charge of the particles, and these interactions are invariant with respect to the substitution $p \rightleftarrows n$. This suggests that, as far as strong interactions are concerned, the proton and neutron may be regarded as two states of the same particle. W. Heisenberg generalized this result by introducing the *hypothesis of isospin invariance* of strong interactions between mesons and baryons. Thus, it is assumed that the mesons and baryons are grouped into *multiplets*, which elements have different electric charges, but close masses. These multiplets correspond to the irreducible representations of $\mathsf{SU}(2)$, regarded as a rotation group in a three-dimensional space (the isospin space). The infinitesimal generators of the $\mathsf{SU}(2)$ group are now denoted by T_i, $i = 1, 2, 3$. They satisfy the commutation relations of the type (2.2.46), that are

$$
[\mathsf{T}_j, \mathsf{T}_k] = i\, \epsilon_{jkl}\, \mathsf{T}_l\,, \quad j, k = 1, 2, 3,
$$

and are called *isospin operators*. For each irreducible representation of $\mathsf{SU}(2)$ (i.e., for each isospin multiplet), the Casimir operator $\mathsf{T}^2 = \mathsf{T}_1^2 + \mathsf{T}_2^2 + \mathsf{T}_3^2$ is a multiple of the identity element $\mathsf{E} \in \mathsf{SU}(2)$ (i.e., $\mathsf{T}^2 = T(T+1)\,\mathsf{E}$), and the eigenvalue (quantum number) T is called the *isospin* of the corresponding multiplet.

The particles in an isospin multiplet differ by their electric charge and are considered as states of a single particle, determined by the eigenvalues of the operator T_3. It is also assumed that the infinitesimal generators T_i commute with the operators of angular momentum J_i and other operators associated to different physical quantities, for instance the *baryon number* B, the *strangeness* S etc. Hence, according to Schur's lemma (see Lemma 1.1.2), all the particles in an isospin multiplet have the same spin, baryon number, strangeness etc. Note that, the relation between the electric charge Q of an isospin state defined by T_3, its baryon number and strangeness is given by the *Gell-Mann – Nishijima formula*

$$Q = T_3 + \frac{B+S}{2}\,, \tag{6.2.1}$$

where the quantity $Y = B + S$ is called *hypercharge.*

In this formulation, the nucleons (p, n) form an *isospin doublet* with $T = 1/2$, and the wave functions of the nucleons are the components of a covariant iso-spinor. Thus, we obtain

$$|p\rangle = \begin{bmatrix} 1 \\ 0 \end{bmatrix} \otimes \psi(\mathbf{r})\,, \quad |n\rangle = \begin{bmatrix} 0 \\ 1 \end{bmatrix} \otimes \psi(\mathbf{r})\,, \tag{6.2.2}$$

where $|p\rangle$ and $|n\rangle$ are the wave functions of the proton and neutron, respectively, while $\psi(\mathbf{r})$ is the solution of the time-independent Schrödinger equation. The wave functions $|p\rangle$ and $|n\rangle$, having $T_3 = \pm 1/2$, form the basis of the representation space for the fundamental representation $\mathsf{D}(1/2)$ of $\mathsf{SU}(2)$. Also, the baryon Λ ($T = T_3 = 0$) is associated to a singlet, that corresponds to the space of the one-dimensional representation of $\mathsf{D}(0)$ of $\mathsf{SU}(2)$, while the π mesons, which have three electric charge states ($T = 1$, $T_3 = -1, 0, 1$) form the basis of the space for the regular representation $\mathsf{D}(1)$ of $\mathsf{SU}(2)$. The infinitesimal generators of the fundamental representation $\mathsf{D}(1)$ are associated to the matrices

$$\mathbf{T}_1 = \begin{bmatrix} 0 & 0 & 0 \\ 0 & 0 & -i \\ 0 & i & 0 \end{bmatrix}, \quad \mathbf{T}_2 = \begin{bmatrix} 0 & 0 & i \\ 0 & 0 & 0 \\ -i & 0 & 0 \end{bmatrix}, \quad \mathbf{T}_3 = \begin{bmatrix} 0 & -i & 0 \\ i & 0 & 0 \\ 0 & 0 & 0 \end{bmatrix}. \tag{6.2.3}$$

Now, if we denote by $\boldsymbol{\pi} = (\pi_1, \pi_2, \pi_3)$ the components of a vector in the space of operators T_1, T_2, and T_3, then we may write

$$\begin{bmatrix} \pi_1 \\ \pi_2 \\ \pi_3 \end{bmatrix} = \pi_1 \begin{bmatrix} 1 \\ 0 \\ 0 \end{bmatrix} + \pi_2 \begin{bmatrix} 0 \\ 1 \\ 0 \end{bmatrix} + \pi_3 \begin{bmatrix} 0 \\ 0 \\ 1 \end{bmatrix}.$$

Because the $\mathsf{SU}(2)$ group is of rank one, we may diagonalize one of the matrices in (6.2.3), for instance $\mathbf{T}_3$. The eigenvalues of $\mathbf{T}_3$ are $1, 0, -1$, and the unitary matrix that diagonalizes $\mathbf{T}_3$ is

$$\mathbf{U} = \frac{1}{\sqrt{2}} \begin{bmatrix} 1 & -i & 0 \\ 0 & 0 & \sqrt{2} \\ 1 & i & 0 \end{bmatrix}.$$

Thus, by means of the transformation $\mathbf{T}'_k = \mathbf{U}\,\mathbf{T}_k\,\mathbf{U}^\dagger$ we obtain the new matrices

$$\mathbf{T}'_1 = \frac{1}{\sqrt{2}} \begin{bmatrix} 0 & -1 & 0 \\ -1 & 0 & 1 \\ 0 & 1 & 0 \end{bmatrix}, \; \mathbf{T}'_2 = \frac{1}{\sqrt{2}} \begin{bmatrix} 0 & i & 0 \\ -i & 0 & -i \\ 0 & i & 0 \end{bmatrix}, \; \mathbf{T}'_3 = \begin{bmatrix} 1 & 0 & 0 \\ 0 & 0 & 0 \\ 0 & 0 & -1 \end{bmatrix}, \tag{6.2.4}$$

such that, in the transformed basis, the components of the vector $\boldsymbol{\pi}$ correspond to the electric charges of the triplet states

$$\begin{bmatrix} \pi^+ \\ \pi_0 \\ \pi^- \end{bmatrix} = \mathbf{U} \begin{bmatrix} \pi_1 \\ \pi_2 \\ \pi_3 \end{bmatrix},$$

respectively

$$\pi^+ = \frac{\pi_1 - i\pi_2}{\sqrt{2}}, \quad \pi^0 = \pi_3\,, \quad \pi^- = \frac{\pi_1 + i\pi_2}{\sqrt{2}}.$$

The triplet of baryons Σ^+, Σ_0, and Σ^-, is also associated to the regular representation $\mathsf{D}(1)$, while the four baryon resonances Δ are associated to the four-dimensional irreducible representation $\mathsf{D}(3/2)$. Therefore, we may group into multiplets all the known elementary particles. The isospin $\mathsf{SU}(2)$ group provides us two quantum numbers. The projection of the isospin onto the z axis (T_3) is an additive quantum number which specifies the states inside a multiplet. Hence, T_3 has to determine the electric charge of the multiplet members. At the same time, the Casimir operator T^2 commutes with all the infinitesimal generators, and thus it is a scalar of $\mathsf{SU}(2)$. The eigenvalue T is conserved in an interaction, but is not an additive quantum number, and determines only the multiplet. Consequently, the state vectors of particles are indexed by two quantum numbers, and have the form $|T, T_3\rangle$.

In the same manner as in the case of $\mathsf{SU}(3)$ (see Chap. 2 Sec. 3.1.1), the unimodular unitary transformations of $\mathsf{SU}(2)$ can be defined in a two-dimensional linear complex space $\mathbb{C}^2$. Moreover, if $\{\xi_\alpha | \alpha = 1, 2, \ldots, 2T+1\}$ is the covariant basis in the representation space of $\mathsf{D}(T)$, then the contravariant basis

$\{\xi^\alpha|\alpha = 1, 2, \ldots, 2T+1\}$ is defined by the relation $\xi^\alpha \xi_\alpha =$ invariant. In the case of unitary transformations of $\mathsf{SU}(2)$ we have

$$\boldsymbol{\xi} \rightarrow \boldsymbol{\xi}' = \mathbf{U}\,\boldsymbol{\xi}\,, \quad \boldsymbol{\xi}^\dagger \rightarrow \boldsymbol{\xi}'^\dagger = \boldsymbol{\xi}^\dagger\,\mathbf{U}^\dagger = \boldsymbol{\xi}^\dagger\,\mathbf{U}^{-1}\,,$$

and it follows that $\boldsymbol{\xi}^\dagger \cdot \boldsymbol{\xi} = \overline{\xi}_\alpha \xi_\alpha =$ invariant. Hence, in the case of unitary transformations, a contravariant representation is the complex conjugate of the corresponding covariant representation ($\xi^\alpha = \overline{\xi}_\alpha$). The covariant and contravariant representations of $\mathsf{SU}(2)$ are equivalent, and the basis vectors transform according to the relations

$$\begin{aligned} \xi_\alpha &\rightarrow \xi'_\alpha = (1 + \boldsymbol{\omega}\cdot\boldsymbol{\tau})\,\xi_\alpha\,, \\ \xi^\alpha &\rightarrow \xi'^\alpha = (1 - \boldsymbol{\omega}\cdot\overline{\boldsymbol{\tau}})\,\xi^\alpha\,, \end{aligned}$$

where $\boldsymbol{\tau}$ are identical to Pauli's matrices (6.1.110), while ω_1, ω_2, and ω_3 represent the group parameters.

Let

$$\boldsymbol{\xi} = \begin{bmatrix} p \\ n \end{bmatrix}$$

be a vector in the space of the fundamental representation of $\mathsf{SU}(2)$. We identify the operation $\boldsymbol{\xi} \rightarrow \overline{\boldsymbol{\xi}}$ as the operation of *charge conjugation* (or the transformation *particle* $\rightarrow$ *antiparticle*) applied to the particles p and n

$$\overline{\boldsymbol{\xi}} = \mathsf{C}\,\boldsymbol{\xi} \tag{6.2.5}$$

and defined as

$$\boldsymbol{\xi} = \begin{bmatrix} p \\ n \end{bmatrix}, \begin{matrix} T_3 = & 1/2 \\ T_3 = & -\,1/2 \end{matrix}, \quad \overline{\boldsymbol{\xi}} = \mathsf{C}\,\boldsymbol{\xi} = \begin{bmatrix} \overline{p} \\ \overline{n} \end{bmatrix}, \begin{matrix} T_3 = & -\,1/2 \\ T_3 = & 1/2 \end{matrix}.$$

For π mesons the conjugation of charge reads

$$\mathsf{C}\begin{bmatrix} \pi^+ \\ \pi_0 \\ \pi^- \end{bmatrix} = \begin{bmatrix} \pi^- \\ \pi_0 \\ \pi^+ \end{bmatrix}.$$

The system nucleon-antinucleon is described by the four-component tensor $M_\alpha^\beta = \xi_\alpha \xi^\beta$, $\alpha, \beta = 1, 2$, that can be decomposed into the sum

$$M_\alpha^\beta = \frac{1}{2}\,\mathsf{Tr}\,\mathbf{M}\,\delta_{\alpha\beta} + \left(\xi_\alpha \xi^\beta - \frac{1}{2}\,\mathsf{Tr}\,\mathbf{M}\,\delta_{\alpha\beta}\right) = \frac{1}{2}\,\mathsf{Tr}\,\mathbf{M}\,\delta_{\alpha\beta} + M'^\beta_\alpha\,,$$

where $\mathsf{Tr}\,\mathbf{M} = \xi_\alpha \xi^\alpha$, comprising a multiple of the identity and a tensor of zero trace. This decomposition corresponds to the decomposition of the direct product of irreducible representations into a direct sum of irreducible representations

$$\mathsf{D}(1/2) \otimes \mathsf{D}(1/2) = \mathsf{D}(0) \oplus \mathsf{D}(1)\,.$$

The normalized wave function associated to the representation $\mathsf{D}(0)$ is

$$\varphi_0^0 = \frac{\overline{p}p + \overline{n}n}{\sqrt{2}}\,,$$

while for $\mathsf{D}(1)$ we have the irreducible tensor

$$\mathbf{M}' = \left[M'^{\beta}_{\alpha} \right] = \begin{bmatrix} (\overline{p}p - \overline{n}n)/2 & \overline{n}p \\ \overline{p}n & -(\overline{p}p - \overline{n}n)/2 \end{bmatrix} \tag{6.2.6}$$

or, as a three-dimensional iso-vector

$$\boldsymbol{\varphi} = \begin{bmatrix} \overline{n}p \\ (\overline{p}p - \overline{n}n)/\sqrt{2} \\ \overline{p}n \end{bmatrix}. \tag{6.2.7}$$

Note that the π mesons are also associated to the representation $\mathsf{D}(1)$, such that the components of $\boldsymbol{\varphi}$ have the same isospin quantum numbers as the components π^+, π^0, and π^-. At the same time, the scalar φ_0^0 has the same isospin quantum numbers as the meson η^0. Therefore, we may infer that from the point of view of isospin properties, the wave functions of the π mesons follow the same transformation rules as the bound states $\overline{N}N$. Thus, we may assume that the meson η $(T = 0)$ and the mesons π $(T = 1)$ are bound states of the nucleon-antinucleon system. Finally, the multiplet π can be described by an iso-vector of components π_1, π_2, and π_3, or an iso-vector of components π^+, π^0, and π^-, or a matrix of zero trace

$$\boldsymbol{\pi} = \left[\pi^{\beta}_{\alpha} \right] = \begin{bmatrix} \pi^0/\sqrt{2} & \pi^+ \\ \pi^- & -\pi^0/\sqrt{2} \end{bmatrix}. \tag{6.2.8}$$

The matrix in (6.2.8) can also be written in terms of π_1, π_2, and π_3, in the form

$$\boldsymbol{\pi} = \left[\pi^{\beta}_{\alpha} \right] = \frac{1}{\sqrt{2}} \begin{bmatrix} \pi_3 & \pi_1 - i\pi_2 \\ \pi_1 + i\pi_2 & -\pi_3 \end{bmatrix} = \frac{1}{\sqrt{2}} \boldsymbol{\tau} \cdot \boldsymbol{\pi}\,, \quad \boldsymbol{\pi} = \begin{bmatrix} \pi_1 \\ \pi_2 \\ \pi_3 \end{bmatrix}. \tag{6.2.9}$$

A similar representation can be used for the triplet Σ. The baryon resonances Δ $(T = 3/2)$ form a multiplet represented by a covariant spinor of the third rank, having the components

$$\Delta_{111} = \Delta^{++}, \quad \Delta_{112} = \Delta_{121} = \Delta_{211} = \frac{1}{\sqrt{3}}\,\Delta^{+},$$

$$\Delta_{122} = \Delta_{212} = \Delta_{221} = \frac{1}{\sqrt{3}}\,\Delta^{0}, \quad \Delta_{222} = \Delta^{-}.$$

2.1.2 The Lagrangian of strong interactions

In the preceding section we have assumed that the wave functions of mesons and baryons transform in the same way as the basis vectors in the representation spaces of the irreducible representations of the isospin $\mathsf{SU}(2)$ group. The isospin invariance of strong interactions means that the interaction Lagrangian is invariant with respect to all transformations of the isospin group. Therefore, the interaction Lagrangian (L), as well the interaction Hamiltonian (H), are scalars (irreducible tensor operators of weight zero). Moreover, according to Wigner-Eckart theorem (see Theorem 6.1.3), the transitions between states of different isospin are forbidden because, the matrix elements of the Hamiltonian operator vanish unless the isospins of the initial and final states are identical:

$$\langle T'', T_3' | \mathsf{H} | T', T_3'' \rangle = C^{T''\,0\,T'}_{T_3''\,0\,T_3'} \langle T'' \| \mathsf{H} \| T' \rangle \propto \delta_{T''\,T'}\, \delta_{T_3''\,T_3'} \langle T'' \| \mathsf{H} \| T' \rangle\,. \tag{6.2.10}$$

The Lagrangian of strong interactions has to depend on the wave functions (that are iso-scalars, iso-spinors, or iso-tensors) of all the particles involved in the interaction. Also, the Lagrangian has to be an iso-scalar such that we have to consider a linear combination of all the iso-scalars that can be formed from the wave functions associated to the particles in the initial and final states. For instance, the interaction Lagrangian for the coupling of a meson π with a nucleon-antinucleon pair is a linear combination of products of the form $\xi_\alpha \xi^\beta \pi^\gamma_\delta$, where the nucleon, the antinucleon, and the meson are respectively represented by the iso-spinors ξ_α, ξ^β, and the irreducible tensor π^γ_δ. To form invariant expressions we have to sum these products with respect to each pair of upper and lower indices. As $\pi^\gamma_\gamma = 0$, we can construct only one invariant that is

$$\xi_\alpha \xi^\beta \pi^\alpha_\beta = \frac{1}{\sqrt{2}}\, N^\beta\, N_\alpha (\boldsymbol{\tau})^\alpha_\beta \cdot \boldsymbol{\pi} = \frac{1}{\sqrt{2}}\, (\overline{\mathbf{N}}\, \boldsymbol{\tau}\, \mathbf{N}) \cdot \boldsymbol{\pi}\,,$$

where we have denoted the nucleon and antinucleon states by

$$\mathbf{N} = \begin{bmatrix} p \\ n \end{bmatrix}, \quad \overline{\mathbf{N}} = [\overline{p}\; \overline{n}]\,. \tag{6.2.11}$$

In terms of these notations and the matrix (6.2.8), the interaction Lagrangian for the $\pi N \overline{N}$ coupling reads

$$\mathsf{L}_{\pi NN} = i\, g_{\pi NN}\, (\overline{\mathbf{N}}\, \boldsymbol{\tau}\, \mathbf{N}) \cdot \boldsymbol{\pi}$$

$$= i\, g_{\pi NN} \left[\sqrt{2}(\overline{p}\, n\, \pi^{+} + \overline{n}\, p\, \pi^{-}) + (\overline{p}\, p - \overline{n}\, n)\pi^{0} \right] . \quad (6.2.12)$$

Here, $g_{\pi NN}$ is the coupling constant, which is a measure of the intensity of interaction forces. Note that, as a direct consequence of the charge independence of strong interactions, we have only one coupling constant in (6.2.12).

If we consider the coupling $\pi\Sigma\Sigma$, then we can form two invariants

$$a = \overline{\Sigma}^{\beta}_{\alpha}\, \Sigma^{\gamma}_{\beta}\, \pi^{\alpha}_{\gamma} = \mathsf{Tr}\,(\overline{\mathbf{\Sigma}}\, \mathbf{\Sigma}\, \boldsymbol{\pi})\,, \quad b = \overline{\Sigma}^{\beta}_{\alpha}\, \pi^{\gamma}_{\beta}\, \Sigma^{\alpha}_{\gamma} = \mathsf{Tr}\,(\overline{\mathbf{\Sigma}}\, \boldsymbol{\pi}\, \mathbf{\Sigma})\,, \quad (6.2.13)$$

which differ only by sign. Hence, the Lagrangian for this coupling is

$$\mathsf{L}_{\pi\Sigma\Sigma} = i\, \sqrt{2}\, g_{\pi\Sigma\Sigma}\, \overline{\Sigma}^{\beta}_{\alpha}\, \Sigma^{\alpha}_{\gamma}\, \pi^{\gamma}_{\beta} = i\, g_{\pi\Sigma\Sigma}\, (\overline{\mathbf{\Sigma}} \times \mathbf{\Sigma}) \cdot \boldsymbol{\pi}\,,$$

where the last term has been obtained by substituting $[\pi^{\gamma}_{\beta}] = \boldsymbol{\tau} \cdot \boldsymbol{\pi}/\sqrt{2}$ (and similar expressions for $\overline{\Sigma}$ and Σ), and also taking into account the relation satisfied by Pauli's matrices

$$\boldsymbol{\tau}_j \boldsymbol{\tau}_k = i\, \epsilon_{jkl}\, \boldsymbol{\tau}_l + \delta_{jk}\, \mathbf{E}\,,$$

$\mathbf{E}$ being the identity matrix.

2.1.3 The decay of resonances

The invariance of the Lagrangian (and the Hamiltonian) relative to the isospin group implies the invariance of all the matrix elements associated to different transitions representing decay processes, scattering, and multiple production in strong interactions. As a consequence, relations between the amplitudes of these processes can be derived.

Let us consider the case of the decay of $f^0(600)$ meson ($T = 0$) into two π mesons. The π mesons in the final state have different four-momenta ($\mathbf{q}$) and we will denote them by $\pi(\mathbf{q}_1)$ and $\pi(\mathbf{q}_2)$. According to the hypothesis of isospin invariance the matrix elements corresponding to this decay are scalars of the form

$$\begin{aligned} M(f^0 \to 2\pi) &= g\, f^0\, \overline{\pi}^{\alpha}_{\beta}(\mathbf{q}_1)\, \overline{\pi}^{\beta}_{\alpha}(\mathbf{q}_2) \\ &= g\, f^0 \left[\overline{\pi^{+}}(\mathbf{q}_1)\, \overline{\pi^{-}}(\mathbf{q}_2) + \overline{\pi^{-}}(\mathbf{q}_1)\, \overline{\pi^{+}}(\mathbf{q}_2) + \overline{\pi^{0}}(\mathbf{q}_1)\, \overline{\pi^{0}}(\mathbf{q}_2) \right] . \end{aligned}$$

It follows that the probability of the decay $f^0 \to \pi^{+}\pi^{-}$ is two times larger than the probability of the decay $f^0 \to 2\pi^0$.

In the case of the decay of mesons with $T = 1$, in addition to the isospin invariance, we have to take into account the invariance with respect to charge conjugation, defined as

$$\mathsf{C}\, \psi^{\alpha}_{\beta}\, \mathsf{C}^{-1} = \pm \psi^{\beta}_{\alpha}\,, \quad (6.2.14)$$

where the sign has to be the same for all the particles in a multiplet. It can be shown that for the π mesons we have to choose the plus sign, while for ρ mesons, which form the triplet $\{\rho^+, \rho^0, \rho^-\}$ of spin $s = 1$, the corresponding sign is minus, such that

$$\mathsf{C}\,\pi^\alpha_\beta\,\mathsf{C}^{-1} = \pi^\beta_\alpha\,, \quad \mathsf{C}\,\rho^\alpha_\beta\,\mathsf{C}^{-1} = -\rho^\beta_\alpha\,.$$

Thus, the invariants which can be formed for the decay $\rho \to 2\pi$, are of the form (6.2.13), and the matrix element of the decay is

$$\begin{aligned}
M(\rho \to 2\pi) &= g\,\rho^\alpha_\beta\,\overline{\pi}^\gamma_\alpha(\mathbf{q}_1)\,\overline{\pi}^\beta_\gamma(\mathbf{q}_2) = -i\,\frac{g}{\sqrt{2}}(\overline{\boldsymbol{\pi}} \times \overline{\boldsymbol{\pi}}) \cdot \boldsymbol{\rho} \\
&= \frac{g}{\sqrt{2}}\,\rho^+\left[\overline{\pi^+}(\mathbf{q}_1)\,\overline{\pi^0}(\mathbf{q}_2) - \overline{\pi^+}(\mathbf{q}_2)\,\overline{\pi^0}(\mathbf{q}_1)\right] \\
&+ \frac{g}{\sqrt{2}}\,\rho^0\left[\overline{\pi^+}(\mathbf{q}_1)\,\overline{\pi^-}(\mathbf{q}_2) - \overline{\pi^+}(\mathbf{q}_2)\,\overline{\pi^-}(\mathbf{q}_1)\right] \\
&+ \frac{g}{\sqrt{2}}\,\rho^-\left[\overline{\pi^-}(\mathbf{q}_1)\,\overline{\pi^0}(\mathbf{q}_2) - \overline{\pi^-}(\mathbf{q}_2)\,\overline{\pi^0}(\mathbf{q}_1)\right].
\end{aligned}$$

Therefore, the decay probabilities (w) satisfy the relations

$$w(\rho^+ \to \pi^+\pi^0) = w(\rho^0 \to \pi^+\pi^-) = w(\rho^- \to \pi^-\pi^0)\,,\ w(\rho^0 \to 2\pi^0) = 0\,.$$

The decay $\rho^0 \to 2\pi^0$ is forbidden because there is no term of the form $\rho^0\overline{\pi}^0\overline{\pi}^0$ in the expression of the matrix element.

Another example is the decay of the baryon resonance $\Delta(1232)$ with $T = 3/2$, into a meson π and a nucleon N. The decay matrix element is

$$M(\Delta \to N\pi) = g\,\overline{N}^\alpha\,\overline{\pi}^{\beta\gamma}\,\Delta_{\alpha\beta\gamma}\,,$$

where $\pi^{\beta\gamma} = \epsilon^{\gamma\alpha}\,\pi^\beta_\alpha$ $(\alpha, \beta, \gamma = 1, 2)$, and $\epsilon^{21} = -\epsilon^{12} = 1$, $\epsilon^{11} = \epsilon^{22} = 0$. Replacing the components of the iso-spinors by the corresponding particles, we obtain the expression

$$\begin{aligned}
M(\Delta \to N\pi) &= g\,\overline{\pi^+}\,\overline{p}\,\Delta^{++} + \frac{g}{\sqrt{3}}\left(\sqrt{2}\,\overline{\pi^0}\,\overline{p} + \overline{\pi^+}\,\overline{n}\right)\Delta^+ \\
&+ \frac{g}{\sqrt{3}}\left(\sqrt{2}\,\overline{\pi^0}\,\overline{n} + \overline{\pi^-}\,\overline{p}\right)\Delta^0 + g\,\overline{\pi^-}\,\overline{n}\,\Delta^-
\end{aligned}$$

such that, the decay probabilities satisfy the relations

$$\begin{aligned}
&w(\Delta^{++} \to p\pi^+) = 3\,w(\Delta^+ \to n\pi^+) = \frac{3}{2}\,w(\Delta^+ \to p\pi^0) \\
&= 3\,w(\Delta^0 \to p\pi^-) = \frac{3}{2}\,w(\Delta^0 \to n\pi^0) = w(\Delta^- \to n\pi^-)\,.
\end{aligned}$$

2.1.4 Strong interactions

Let us consider a strong interaction of the form

$$a + b \to c + d\,. \tag{6.2.15}$$

The initial and final states are described by the vectors $|i\rangle$ and $|f\rangle$, respectively, while the *interaction* (the transition from the state i to the state f) is represented by the matrix element $\langle f|\mathsf{S}|i\rangle$, where S is the operator that symbolizes the interaction. Let $\mathbf{Q}_i = \mathbf{q}_a + \mathbf{q}_b$ and $\mathbf{Q}_f = \mathbf{q}_c + \mathbf{q}_d$ be, respectively, the four-momenta of the initial and final states of the interaction (6.2.15), where $\mathbf{q}_a$ represents the four-momentum of particle a, $\mathbf{q}_b$ represents the four-momentum of particle b etc. We may emphasize the four-momentum conservation by writing the matrix element associated to the interaction (6.2.15) in the form

$$\langle f|\mathsf{S}|i\rangle = \delta_{fi} + i\,(2\pi)^4\,\delta(\mathbf{Q}_f - \mathbf{Q}_i)\,N\,\langle f|\mathsf{T}|i\rangle\,, \tag{6.2.16}$$

where δ_{fi} is the Kronecker symbol, $\delta(\mathbf{Q}_f - \mathbf{Q}_i)$ represents Dirac's delta function, N is a normalisation factor, and T is the *transition operator*. By definition, the *scattering amplitude* is given by the matrix element of the transition operator

$$A = \langle f|\mathsf{T}|i\rangle\,, \tag{6.2.17}$$

while the *cross-section* of the interaction is proportional to the transition probability per unit of time, that is

$$\sigma \propto |A|^2\,. \tag{6.2.18}$$

In the case of the interaction π – nucleon, there are six interactions of elastic scattering type ($a + b \to a + b$) and two interactions of charge exchange type ($\pi^- p \to \pi^0 n$ and $\pi^+ n \to \pi^0 p$). Note that, according to the principle of micro-reversibility

$$\sigma(\pi^0 n \to \pi^- p) = \sigma(\pi^- p \to \pi^0 n)\,, \quad \sigma(\pi^0 p \to \pi^+ n) = \sigma(\pi^+ n \to \pi^0 p)\,.$$

The π mesons are associated to the irreducible representation $\mathsf{D}(1)$ of the isospin $\mathsf{SU}(2)$ group, while the nucleons are associated to the irreducible representation $\mathsf{D}(1/2)$. Thus, the total isospin of the initial and final states can be calculated on the basis of the decomposition of the direct product

$$\mathsf{D}(1) \otimes \mathsf{D}(1/2) = \mathsf{D}(1/2) \oplus \mathsf{D}(3/2)\,. \tag{6.2.19}$$

Hence, we may obtain only two independent scattering amplitudes that correspond to a total isospin $T = 1/2$ and $T = 3/2$, respectively. This means that we can associate the system πN to the irreducible representations $\mathsf{D}(1/2)$ and $\mathsf{D}(3/2)$, respectively.

The cross-sections of the charge exchange interactions are equal, and this represents a particular case of isospin invariance called *charge symmetry*. In our case, the charge symmetry is expressed in the form

$$\sigma(\pi^- p \to \pi^0 n) = \sigma(\pi^+ n \to \pi^0 p)\,; \tag{6.2.20}$$

that is, the cross-section of the interaction remains unchanged if we change the sign of the third component (T_3) of the isospin associated to each particle involved in the interaction, at the same time leaving the isospin T unchanged. The general proof of this symmetry property exhibited by strong interactions is based on the Wigner-Eckart theorem (see Theorem 6.1.3), and the first property of Clebsch-Gordan coefficients shown in (3.3.53).

We also have

$$\begin{aligned}
\sigma(\pi^+ p \to \pi^+ p) &= \sigma(\pi^- n \to \pi^- n)\,, \\
\sigma(\pi^0\, p \to \pi^0\, p) &= \sigma(\pi^0\, n \to \pi^0\, n)\,, \\
\sigma(\pi^- p \to \pi^- p) &= \sigma(\pi^+ n \to \pi^+ n)\,,
\end{aligned}$$

such that, the relations between the cross-sections of $\pi^+ p \to \pi^+ p$, $\pi^0 p \to \pi^0 p$, $\pi^- p \to \pi^- p$, and $\pi^- p \to \pi^0 n$ interactions, determine the relations satisfied by the cross-sections of all eight π – nucleon interactions. Because interactions of the type $\pi^0 N \to \pi^0 N$ cannot be measured experimentally we will consider only the relations between the cross-sections of $\pi^+ p \to \pi^+ p$, $\pi^- p \to \pi^- p$, and $\pi^- p \to \pi^0 n$ interactions. Firstly, by means of Clebsch-Gordan coefficients we express the initial and final states of these interactions in terms of pure isospin states $|T, T_3\rangle$

$$|\pi^+, p\rangle \equiv |\pi^+\rangle\,|\,p\,\rangle = C^{1\ 1/2\ 3/2}_{1\ 1/2\ 3/2}\left|\frac{3}{2}, \frac{3}{2}\right\rangle = \left|\frac{3}{2}, \frac{3}{2}\right\rangle,$$

$$\begin{aligned}
|\pi^-, p\rangle \equiv |\pi^-\rangle\,|\,p\,\rangle &= C^{\ \,1\ \ 1/2\ \ \ 1/2}_{-1\ 1/2\ -1/2}\left|\frac{1}{2}, -\frac{1}{2}\right\rangle + C^{\ \,1\ \ 1/2\ \ \ 3/2}_{-1\ 1/2\ -1/2}\left|\frac{3}{2}, -\frac{1}{2}\right\rangle \\
&= -\sqrt{\frac{2}{3}}\left|\frac{1}{2}, -\frac{1}{2}\right\rangle + \sqrt{\frac{1}{3}}\left|\frac{3}{2}, -\frac{1}{2}\right\rangle,
\end{aligned}$$

$$\begin{aligned}
|\pi^0, n\rangle \equiv |\pi^0\rangle\,|\,n\,\rangle &= C^{0\ \ \ 1/2\ \ \ 1/2}_{0\ -1/2\ -1/2}\left|\frac{1}{2}, -\frac{1}{2}\right\rangle + C^{1\ \ \ 1/2\ \ \ 3/2}_{0\ -1/2\ -1/2}\left|\frac{3}{2}, -\frac{1}{2}\right\rangle \\
&= \sqrt{\frac{1}{3}}\left|\frac{1}{2}, -\frac{1}{2}\right\rangle + \sqrt{\frac{2}{3}}\left|\frac{3}{2}, -\frac{1}{2}\right\rangle.
\end{aligned}$$

The amplitudes of the considered interactions are given by (6.2.17) and, assuming that the transition operator is an iso-scalar, and taking into account the orthogonality of the wave functions, we obtain

$$\begin{aligned} A_+ &= \langle \pi^+, p|\mathsf{T}|\pi^+, p\rangle = A_{3/2}\,, \\ A_- &= \langle \pi^-, p|\mathsf{T}|\pi^-, p\rangle = \frac{2}{3}\,A_{1/2} + \frac{1}{3}\,A_{3/2}\,, \\ A_0 &= \langle \pi^0, n|\mathsf{T}|\pi^-, p\rangle = -\frac{\sqrt{2}}{3}\,A_{1/2} + \frac{\sqrt{2}}{3}\,A_{3/2}\,, \end{aligned} \tag{6.2.21}$$

where the matrix elements

$$A_{1/2} = \langle 1/2, T_3|\mathsf{T}|1/2, T_3\rangle, \quad A_{3/2} = \langle 3/2, T_3|\mathsf{T}|3/2, T_3\rangle,$$

are independent of T_3. It is evident that these amplitudes (in general, complex numbers) satisfy the relation $A_+ - A_- = \sqrt{2}\,A_0$, while the corresponding cross-sections satisfy the triangular relation

$$|\sqrt{\sigma_+} - \sqrt{\sigma_-}| \le \sqrt{2\,\sigma_0} \le \sqrt{\sigma_+} + \sqrt{\sigma_-}\,.$$

On the basis of empirical data we may assume that $|A_{3/2}| \gg |A_{1/2}|$, which leads to the following ratios for the partial cross-sections

$$\sigma_+ : \sigma_- : \sigma_0 = 9 : 1 : 2\,, \tag{6.2.22}$$

while the total cross-sections, that are the sums of individual processes (partial cross-sections), satisfy the relation

$$\frac{\sigma(\pi^+ + p)}{\sigma(\pi^- + p)} = 3\,. \tag{6.2.23}$$

We may also obtain relations between the cross-sections of pure isospin transitions, because

$$\begin{aligned} |A_+|^2 &= |A_{3/2}|^2\,, \\ |A_-|^2 &= \frac{4}{9}\,|A_{1/2}|^2 + \frac{1}{9}\,|A_{3/2}|^2 + \frac{4}{9}\,\mathsf{Re}\left(\overline{A_{1/2}}\,A_{3/2}\right), \\ |A_0|^2 &= \frac{2}{9}\,|A_{1/2}|^2 + \frac{2}{9}\,|A_{3/2}|^2 - \frac{4}{9}\,\mathsf{Re}\left(\overline{A_{1/2}}\,A_{3/2}\right). \end{aligned}$$

Hence,

$$\sigma_{1/2} = \frac{1}{2}\,[3(\sigma_- + \sigma_0) - \sigma_+]\,, \quad \sigma_{3/2} = \sigma_+\,.$$

The same results are obtained if we use the method of invariants. Thus, the scattering amplitude has to be an invariant relative to the isospin group, comprising two iso-tensors (π_α^β and $\pi_{\alpha'}^{\beta'}$) and two iso-spinors (N_γ and $N^{\gamma'}$).

There are only three invariants that can be constructed from these elements

$$\begin{aligned} a_1 &= \pi^{\beta}_{\alpha}\, N^{\alpha}\, \pi^{\gamma}_{\beta}\, N_{\gamma} = \pi_i\, \pi_j\, (\boldsymbol{\tau}_i\, \boldsymbol{\tau}_j)^{\gamma}_{\alpha}\, N^{\alpha}\, N_{\gamma}\,, \\ a_2 &= \pi^{\beta}_{\alpha}\, N^{\gamma}\, \pi^{\alpha}_{\beta}\, N_{\gamma} = \pi_i\, \pi_j\, (\boldsymbol{\tau}_i\, \boldsymbol{\tau}_j)^{\alpha}_{\alpha}\, N^{\gamma}\, N_{\gamma}\,, \\ a_3 &= \pi^{\beta}_{\alpha}\, N^{\gamma}\, \pi^{\alpha}_{\gamma}\, N_{\beta} = \pi_i\, \pi_j\, (\boldsymbol{\tau}_j\, \boldsymbol{\tau}_i)^{\beta}_{\gamma}\, N^{\gamma}\, N_{\beta}\,. \end{aligned}$$

Here, the last expression in each line has been obtained by substituting the form (6.2.9) for the π mesons. Only two of these invariants are independent because, from $\boldsymbol{\tau}_i\, \boldsymbol{\tau}_j + \boldsymbol{\tau}_j\, \boldsymbol{\tau}_i = \mathbf{0}$ it follows that $a_3 = -a_1$. Also, the product $\boldsymbol{\tau}_i\, \boldsymbol{\tau}_j$ can be written in the form

$$\boldsymbol{\tau}_i\, \boldsymbol{\tau}_j = \delta_{ij}\, \mathbf{E} + \frac{1}{2}\, [\boldsymbol{\tau}_i, \boldsymbol{\tau}_j]\,,$$

where $\mathbf{E}$ is the identity matrix in the isospin space. Consequently, the invariant scattering amplitude of the interaction π – nucleon is a linear combination of two invariants, having the form

$$\pi_i\, \pi_j\, (\mathbf{T}_{ij})^{\beta}_{\alpha}\, N^{\alpha}\, N_{\beta}\,, \tag{6.2.24}$$

where the matrix

$$\mathbf{T}_{ij} = A\, \delta_{ij}\, \mathbf{E} + \frac{1}{2}\, B\, [\boldsymbol{\tau}_i, \boldsymbol{\tau}_j]$$

is a linear combination of a term proportional to the identity matrix, and a term containing a matrix of trace zero. Note that, the iso-scalar term

$$(\boldsymbol{\pi} \cdot \boldsymbol{\pi})(\overline{\mathbf{N}}\, \mathbf{E}\, \mathbf{N}) = (\boldsymbol{\pi} \cdot \boldsymbol{\pi})(\overline{\mathbf{N}}\, \mathbf{N})$$

is associated to the scattering amplitude A, while the scattering amplitude B is associated to the term

$$\pi_k\, \pi_l\, i\, \epsilon_{klm}\, (\overline{\mathbf{N}}\, \boldsymbol{\tau}_m\, \mathbf{N}) = i(\boldsymbol{\pi} \times \boldsymbol{\pi}) \cdot (\overline{\mathbf{N}}\, \boldsymbol{\tau}\, \mathbf{N})\,,$$

which has a total isospin $T = 0$, but comprises the iso-vector operator $(\overline{\mathbf{N}}\, \boldsymbol{\tau}\, \mathbf{N})$. A direct comparison between the four-particle couplings defined in (6.2.24), and the initial and final states of different π – nucleon interactions, shows that the scattering amplitudes A and B can be expressed in terms of the scattering amplitudes (6.2.21) as

$$\begin{aligned} A &= \frac{1}{3}\, A_{1/2} + \frac{2}{3}\, A_{3/2}\,, \\ B &= \frac{1}{3}\, A_{1/2} - \frac{1}{3}\, A_{3/2}\,. \end{aligned}$$

Generally, the cross-sections of multiple production processes

$$a + b \to c_1 + c_2 + \ldots + c_n \tag{6.2.25}$$

are proportional to statistical weights of the form (6.1.138), in which we have to replace the spin states by the isospin states. Also, in the case of final states comprising sets of identical particles, for instance a final state of isospin T, consisting of n particles, of which n_+, n_0, and n_- particles are, respectively π^+, π^0, and π^- mesons, we may use the formula

$$^*P^T_{n_+\, n_0\, n_-} = \frac{n!}{n_+!\, n_0!\, n_-!}\, P^T_{\underbrace{+1\ldots+1}_{n_+}\,\underbrace{0\ldots0}_{n_0}\,\underbrace{-1\ldots-1}_{n_-}}\,.$$

These statistical weights allow us to derive ratios for the cross-sections of different interactions of the type (6.2.25), which have the same number of particles in the final state.

2.2 The unitary spin and the $\mathsf{SU}(3)$ group

2.2.1 Classification of elementary particles

In general, if the elementary particles are associated to certain irreducible representations of a group G, and the interaction Hamiltonian (or equally well, the interaction Lagrangian) is an irreducible tensor operator relative to G, then we say that the particles and their interactions are described by a symmetry model generated by G. We also say that the interactions between the elementary particles exhibit the symmetry generated by G if they are invariant relative to G, that is the interaction Hamiltonian is the simplest irreducible tensor (i.e., a scalar) relative to G. Actually, this is called an *exact symmetry* because, when the interaction Hamiltonian is an irreducible tensor of higher weight relative to G, we have a model of *broken symmetry* generated by G. In both cases, we assume that the strongly interacting particles (*hadrons*) are associated to basis vectors in the representation space of irreducible representations of G.

The symmetry generated by the $\mathsf{SU}(3)$ group is called *unitary symmetry*, and this model is intended to describe the strong interactions. Thus, the *leptons* (i.e., electrons, positrons, ν and μ mesons etc.) are not included in the $\mathsf{SU}(3)$ classification of elementary particles, and their wave functions are scalars relative to $\mathsf{SU}(3)$. The lowest irreducible representations of $\mathsf{SU}(3)$ are $\mathsf{D}(1,0)$ and $\mathsf{D}(0,1)$, and we identify the generators in the Cartan subalgebra of $\mathsf{SU}(3)$ by $\sqrt{3}\,\mathsf{H}_1 = \mathsf{T}_3$ (the third component of the isospin) and $2\,\mathsf{H}_2 = \mathsf{Y}$ (the hypercharge). The particles associated to the representation $\mathsf{D}(1,0)$ are called *quarks*, while the particles associated to the representation $\mathsf{D}(0,1)$ are called *antiquarks*. A model based on these particles, which have fractional electric charge and hypercharge (see Chap. 2 Sec. 3.1.4), will be discussed in Chap. 6 Sec. 3.2.1.

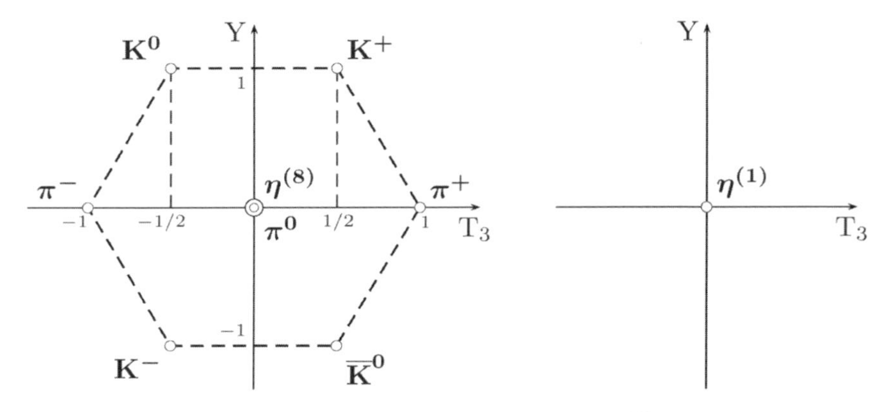

Figure 6.1. The nonet of pseudoscalar mesons ($J^{PC} = 0^{-+}$).

In 1961, M. Gell-Mann and Y. Ne'eman independently proposed a scheme for classifying the strongly interacting particles, called the *Eightfold Way*[2]. The particles and resonances associated to an irreducible representation of $\mathsf{SU}(3)$ are grouped into a *multiplet*, and this classification of elementary particles into multiplets is based on two hypotheses:

- all the particles in a multiplet have the same spin, because the operators in $\mathsf{SU}(3)$ commute with the operators in the Lorentz group;
- the location of a particle inside a multiplet is determined by its quantum numbers T_3, and Y translated into weights, that is $m_1 = T_3/\sqrt{3}$ and $m_2 = Y/2$.

Note that, in the limit of *exact* $\mathsf{SU}(3)$ *symmetry* all the particles in a multiplet have the same mass.

The number of elementary particles is quite huge so that we will show only some basic multiplets (a full list of elementary particles and their properties can be found at **http://pdg.lbl.gov**). Thus, there are three main *nonets* of mesons associated to the irreducible representations in the decomposition of the direct product

$$\mathsf{D}(1,0) \otimes \mathsf{D}(0,1) = \mathsf{D}(0,0) \oplus \mathsf{D}(1,1) \ : \qquad (6.2.26)$$

the pseudoscalar mesons (Fig. 6.1), the vector mesons (Fig. 6.2), and the tensor mesons (Fig. 6.3). In the caption of these figures J, P, and C represent the quantum numbers of spin, parity and charge conjugation, respectively. It is assumed that the octet state $\eta^{(8)}$ mixes with the singlet state $\eta^{(1)}$ to produce the pseudoscalar mesons η and $\eta'(958)$. Also, the states $\phi^{(8)}$ and $\phi^{(1)}$ produce the

[2] After Buddha's Eightfold Path to Enlightenment and bliss.

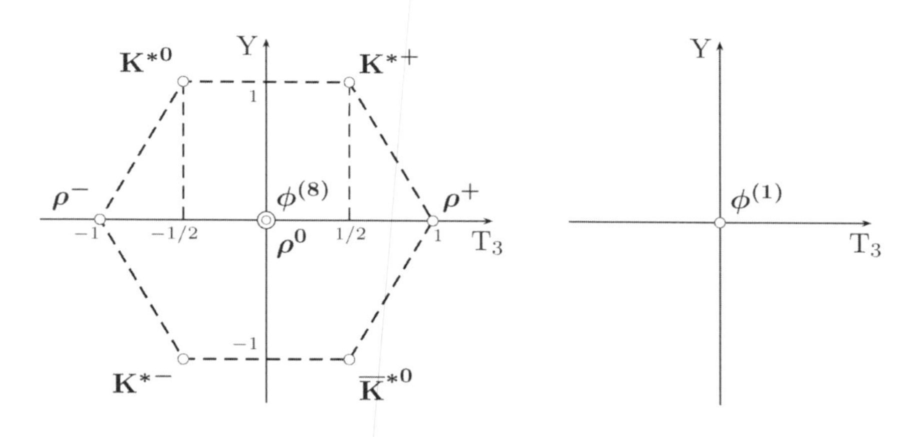

Figure 6.2. The nonet of vector mesons ($J^{PC} = 1^{--}$).

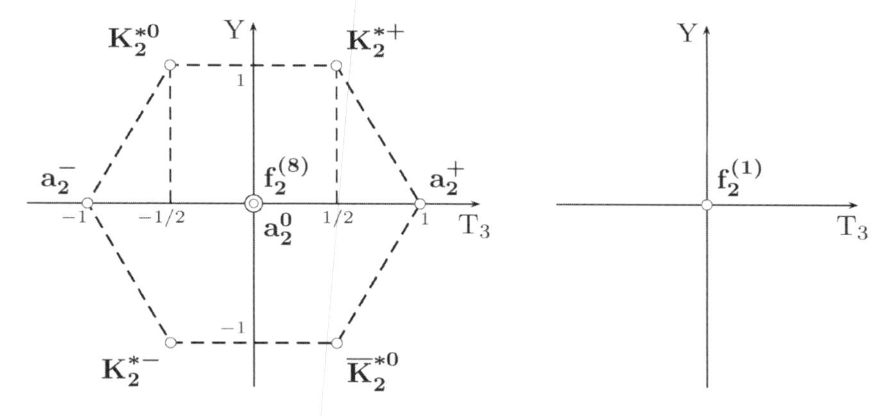

vector mesons $\omega(782)$ and $\phi(1020)$, while the states $f_2^{(8)}$ and $f_2^{(1)}$ produce the tensor mesons $f_2(1270)$ and $f_2'(1525)$.

The baryons and baryon resonances are classified respectively, into octets and decuplets, according to the decomposition

$$\mathsf{D}(1,0)\otimes\mathsf{D}(1,0)\otimes\mathsf{D}(1,0) = \mathsf{D}(0,0)\oplus\mathsf{D}(1,1)\oplus\mathsf{D}(1,1)\oplus\mathsf{D}(3,0)\,. \quad (6.2.27)$$

Thus, the basic baryons of spin $J = 1/2$ form an octet (Fig. 6.4), while the basic baryon resonances of spin $3/2$ are grouped into a decuplet (Fig. 6.5). There are other baryons of higher spin, but all of them are classified into octets and singlets. It is worth-while to add here that, among the first successes of this classification scheme there were the prediction of the existence of the pseudoscalar meson η^0 to complete the octet in Fig. 6.1, and the baryon resonance Ω^- from the decuplet in Fig. 6.5. The experimental discovery Ω^- in 1963, shown how deeply the

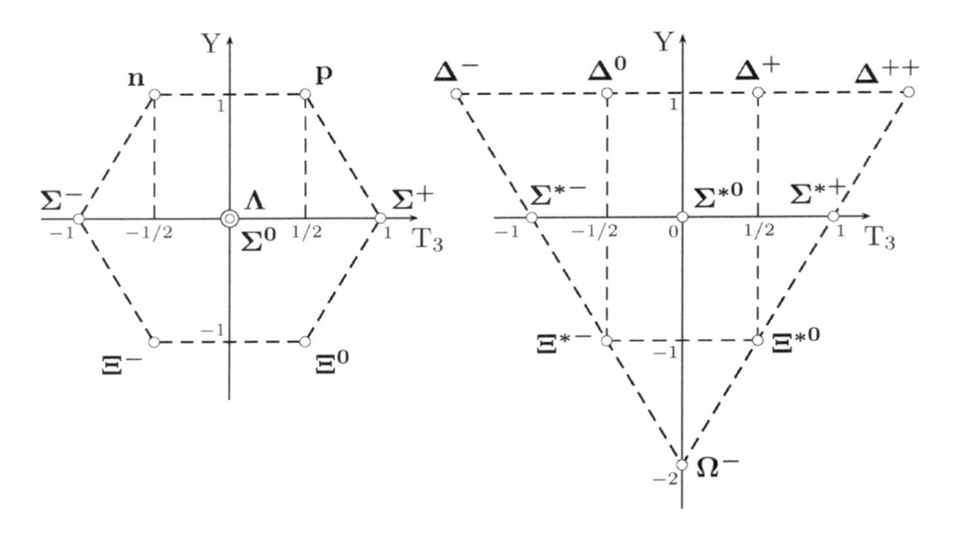

Figure 6.4. The octet of basic baryons $J^P = 1/2^+$.

Figure 6.5. The decuplet of basic baryon resonances $J^P = 3/2^+$.

unitary symmetry is rooted in the world of hadrons, and its importance in modelling the strong interactions.

2.2.2 Mass formulae

The basis vectors in a representation space of the $\mathsf{SU}(3)$ irreducible representations are labelled according to the decomposition (see Chap. 3 Sec. 3.2.3)

$$\mathsf{SU}(3) \supset \mathsf{SU}(2)_T \otimes \mathsf{U}(1)_Y ,$$

where $\mathsf{SU}(2)_T$ is the isospin group, and the infinitesimal generator of $\mathsf{U}(1)_Y$ is the hypercharge operator. Thus, a basis vector in the representation space is denoted by $|\mu, \nu\rangle$, where $\mu \equiv \{p, q\}$ specifies the representation, and $\nu \equiv \{T, T_3, Y\}$ determines the position of the state vector inside the corresponding multiplet. Actually, the particles grouped into a multiplet have different masses, which contradicts the hypothesis of invariance relative to the $\mathsf{SU}(3)$ group. To accommodate this mass difference in the $\mathsf{SU}(3)$ symmetry model it is considered that the $\mathsf{SU}(3)$ symmetry is broken due to a *medium-strong interaction*. Consequently, the interaction Hamiltonian is written as a sum of two operators

$$\mathsf{H}_{\mathrm{s}} = \mathsf{H}_{\mathrm{exact}} + \mathsf{H}_{\mathrm{m}} ,$$

where $\mathsf{H}_{\mathrm{exact}}$ acts as a scalar relative to the $\mathsf{SU}(3)$ group, while H_{m} is the Hamiltonian of medium-strong interactions responsible for the mass differences inside a multiplet. Because the operator H_{m} does not break the $\mathsf{SU}(2)_T$ isospin symmetry inside an $\mathsf{SU}(3)$ multiplet, it commutes with all the infinitesimal generators of the $\mathsf{SU}(2)_T$ group, $\mathsf{E}_{\pm 1}$ and $\mathsf{T}_3 = \sqrt{3}\,\mathsf{H}_1$. Also, the hypercharge is conserved in strong and medium-strong interactions such that H_m commutes with the hypercharge operator $\mathsf{Y} = 2\,\mathsf{H}_2$. According to the commutation relations satisfied by the infinitesimal generators of $\mathsf{SU}(3)$, it follows that H_m must have the same behaviour as the component $T = T_3 = Y = 0$ of an irreducible tensor operator, relative to the transformations in the $\mathsf{SU}(3)$ representation $\mathsf{D}(1,1)$. Only the irreducible representations $\mathsf{D}(p,q)$ with $p = q$ contain weights corresponding to $T = T_3 = Y = 0$ hence, we may introduce a mass operator

$$\mathsf{M} = \mathsf{M}_{000}^{(0,0)} + \mathsf{M}_{000}^{(1,1)} + \mathsf{M}_{000}^{(2,2)} + \dots .$$

Thus, taking into account that the elementary particles are classified only into multiplets associated to the irreducible representations $\mathsf{D}(0,0) = 1$, $\mathsf{D}(1,1) = 8$, and $\mathsf{D}(3,0) = 10$, and also that $\mathsf{M}_{000}^{(0,0)}$ is an $\mathsf{SU}(3)$ scalar, the evaluation of the matrix elements of the mass operator in a given representation leads to the *Gell-Mann-Okubo mass formula*

$$m_\nu = m_0 + m_1 Y + m_2 \left[T(T+1) - \frac{1}{4} Y^2 \right] . \tag{6.2.28}$$

Here, m_ν is the mass of the state $|\mu, \nu\rangle$, while the coefficients m_0, m_1, and m_2 depend on μ only, and are thus constant for a given multiplet. The mass formula (6.2.28) holds in the case of particles having a half-integer spin (*fermions*). For particles of integer spin (*bosons*) we have to replace m_ν by m_ν^2.

If we apply the Gell-Mann-Okubo mass formula to the baryon octet (Fig. 6.4), and eliminate the coefficients m_0, m_1, and m_2, we obtain a mass relation satisfied by the iso-multiplet masses

$$2\,(m_\mathrm{N} + m_\Xi) - 3\,(m_\Lambda + m_\Sigma) = 0 . \tag{6.2.29}$$

Furthermore, for the decuplet of baryon resonances (Fig. 6.5) we have

$$m_\Omega - m_{\Xi^*} = m_{\Xi^*} - m_{\Sigma^*} = m_{\Sigma^*} - m_\Delta . \tag{6.2.30}$$

In the case of pseudoscalar mesons (Fig. 6.1), we are led to a relation of the form

$$3\,m_{\eta^{(8)}}^2 + m_\pi^2 = 4\,m_K^2 , \tag{6.2.31}$$

and it follows that $m_{\eta^{(8)}} \approx 566\,\mathrm{MeV}$. The pseudoscalar meson η has a mass $m_\eta = 548\,\mathrm{MeV}$, but there is another candidate having the same quantum numbers $T = T_3 = Y = 0$, that is the meson η' with a mass $m_{\eta'} = 958\,\mathrm{MeV}$. The solution of this problem is to assume that the mesons η and η' are, in fact, a mixture of states $\eta^{(8)}$ and $\eta^{(1)}$

$$\begin{aligned} \eta &= \eta^{(8)}\,\cos\theta - \eta^{(1)}\,\sin\theta , \\ \eta' &= \eta^{(8)}\,\sin\theta + \eta^{(1)}\,\cos\theta , \end{aligned}$$

the *mixing angle* (also called *Cabibbo's angle*) being determined by the experimental data for η and η' ($\theta \approx -10.1°$). The same method can be applied in the case of vector and tensor mesons.

In 1972, I. Lucač and L. Toth derived a very interesting set of mass formulae, based on the elliptical coordinates (3.1.36) for the $\mathsf{SU}(3)$ parametrization. Thus, by means of the relations (2.3.59) and (3.1.33) the infinitesimal generators of $\mathsf{SU}(3)$ can be represented as a matrix

$$[\mathsf{B}^i_j] = \begin{bmatrix} \mathsf{Q} & \mathsf{T}_+ & \mathsf{V}_+ \\ \mathsf{T}_- & \mathsf{Y}-\mathsf{Q} & \mathsf{U}_+ \\ \mathsf{V}_- & \mathsf{U}_- & -\mathsf{Y} \end{bmatrix}, \tag{6.2.32}$$

where $\mathsf{Q} = \mathsf{T}_3 + \mathsf{Y}/2$ is the operator of the electric charge, and

$$\mathsf{T}_3 = \sqrt{3}\,\mathsf{H}_1\,, \quad \mathsf{Y} = 2\,\mathsf{H}_2\,, \quad \mathsf{V}_3 = \frac{1}{2}\,\mathsf{T}_3 + \frac{3}{4}\,\mathsf{Y}\,, \quad \mathsf{U}_3 = -\frac{1}{2}\,\mathsf{T}_3 + \frac{3}{4}\,\mathsf{Y}\,.$$

The breaking of the $\mathsf{SU}(3)$ symmetry is introduced by adding to the strong interaction Hamiltonian $\mathsf{H}_{\text{exact}}$ a term of the form

$$\mathsf{H}_{\text{b}} = u^i_j\,\mathsf{T}^j_i\,, \tag{6.2.33}$$

where u^i_j is an irreducible tensor represented by an arbitrary unitary matrix of dimensions 3×3, and zero trace. Without affecting the generality of the method, we may consider that this matrix is diagonalized, its diagonal elements satisfy the relations

$$u^1_1 \geq u^2_2 \geq u^3_3\,, \quad u^1_1 + u^2_2 + u^3_3 = 0\,,$$

and can be parametrized in the form

$$u^1_1 = -q_0\,e_1\,, \quad u^2_2 = -q_0\,e_2\,, \quad u^3_3 = -q_0\,e_3\,,$$

where q_0 is an arbitrary constant, while e_k are defined in (3.1.38). Hence, the matrix $\mathbf{u}$ is characterized by two independent parameters q_0 and k^2, that satisfy the conditions

$$0 < q_0 = \frac{1}{3}\,(u^1_1 - u^3_3) < \infty\,, \quad 0 \leq k^2 = \frac{u^1_1 - u^2_2}{u^1_1 - u^3_3} \leq 1\,.$$

We thus obtain

$$\begin{aligned} [u^i_j] &= q_0\,k'^2 \begin{bmatrix} 1 & 0 & 0 \\ 0 & 1 & 0 \\ 0 & 0 & -2 \end{bmatrix} + q_0\,k^2 \begin{bmatrix} 2 & 0 & 0 \\ 0 & -1 & 0 \\ 0 & 0 & -1 \end{bmatrix} \\ &= 2\sqrt{3}\,q_0\,k'^2\,\mathbf{F}_8 + 3\,q_0\,k^2 \left(\mathbf{F}_3 + \frac{1}{\sqrt{3}}\,\mathbf{F}_8 \right) \end{aligned}$$

$$= 6\, q_0\, k'^2\, \mathbf{H}_2 + 3\, q_0\, k^2\, (\sqrt{3}\, \mathbf{H}_1 + \mathbf{H}_2)$$
$$= 3\, q_0\, k'^2\, \mathbf{Y} + 3\, q_0\, k^2 \left(\mathbf{T}_3 + \frac{1}{2}\, \mathbf{Y} \right)$$
$$= 3\, q_0\, (k'^2\, \mathbf{Y} + k^2\, \mathbf{Q})\,, \tag{6.2.34}$$

where $k'^2 = 1 - k^2$, and the relations between different bases of the $\mathsf{SU}(3)$ algebra are given in (2.3.13) and (2.3.49). Note that the tensor u^i_j comprises two terms: one breaking the exact $\mathsf{SU}(3)$ symmetry ($k = 0$), and the other breaking the isospin invariance ($k' = 0$). Consequently, for $k' \neq 0$ and $k \neq 0$, this model introduces simultaneously the mass differences due to the hypercharge and electric charge.

The irreducible tensor operator T^j_i can be expressed as a series in terms of the infinitesimal generators

$$\mathsf{T}^j_i = \alpha\, \boldsymbol{\delta}^j_i + \beta\, \mathsf{B}^j_i + \gamma\, \mathsf{B}^j_l\, \mathsf{B}^l_i + \dots\,.$$

Limiting this series to the quadratic term, and identifying the diagonal infinitesimal generators in the elliptic parametrization of $\mathsf{SU}(3)$,

$$\mathsf{C}^{(2)}\,, \quad \mathsf{C}^{(3)}\,, \quad \mathsf{B}^1_1\,, \quad \mathsf{B}^3_3\,, \quad \mathsf{C}_{\mathsf{UVT}} = e_1\, \mathsf{C}^{(2)}_{\mathsf{U}} + e_2\, \mathsf{C}^{(2)}_{\mathsf{V}} + e_3\, \mathsf{C}^{(2)}_{\mathsf{T}}\,,$$

defined in Chap. 3 Sec. 2.2.3, one may write the operator (6.2.33) in the form

$$\begin{aligned} u^i_j\, \mathsf{T}^j_i = &- q_0\, \beta\, (e_1\, \mathsf{B}^1_1 + e_2\, \mathsf{B}^2_2 + e_3\, \mathsf{B}^3_3) - \frac{3}{2}\, q_0\, \gamma\, e_2\, \mathsf{C}^{(2)} \\ &+ q_0\, \gamma \left[k'^2\, \mathsf{C}^{(2)}_{\mathsf{T}} - k^2\, \mathsf{C}^{(2)}_{\mathsf{U}} - \frac{1}{4}\, k'^2\, (\mathsf{B}^3_3)^2 + \frac{1}{4}\, k^2\, (\mathsf{B}^1_1)^2 \right]. \end{aligned}$$

The eigenvalues of the operator $k'^2\, \mathsf{C}^{(2)}_{\mathsf{T}} - k^2\, \mathsf{C}^{(2)}_{\mathsf{U}}$, that is equivalent to the operator $\mathsf{C}_{\mathsf{UVT}}$, are

$$\begin{aligned} \lambda^{u(\tau)}_{\mathsf{QY}}(k^2) = &\ \varepsilon^{u(\tau)}_{m_1 m_2 m_3}(k^2) + \frac{1}{4}\, (k'^2\, Y^2 - k^2\, Q^2) - \frac{1}{6}\, (p - q)(k'^2\, Y + k^2\, Q) \\ &+ (1 - 2k^2) \left[\frac{1}{9}\, (p^2 + pq + q^2 + 3p + 3q) - \frac{2}{3}\, (Q^2 - QY + Y^2) \right], \end{aligned}$$

where

$$Q = 2 \left(m_1 - \frac{m}{3} \right), \quad Y = -2 \left(m_3 - \frac{m}{3} \right), \quad m = m_1 + m_2 + m_3 = \frac{p - q}{2}\,.$$

In the particular cases $k^2 = 0$ and $k^2 = 1$ these eigenvalues become

$$\lambda^{u(\tau)}_{\mathsf{QY}}(0) = T(T + 1)\,, \quad \lambda^{u(\tau)}_{\mathsf{QY}}(1) = -U(U + 1)\,.$$

This method of symmetry breaking leads to the mass formula

$$m_\nu = a_0 + a_1\,(k'^2\,Y + k^2\,Q) + a_2\left[4\,\lambda^{u(\tau)}_{\mathsf{QY}}(k^2) - k'^2\,Y^2 + k^2\,Q^2\right], \quad (6.2.35)$$

where a_0, a_1, and a_2 are constants, and $k'^2 = 1 - k^2$. Applying this mass formula to the baryon octet (Fig. 6.4), and eliminating the parameters a_0, a_1, a_2, and k^2, one obtains four linearly independent mass relations

$$\begin{aligned}
&m_p + m_n + m_{\Sigma^+} + m_{\Sigma^-} + m_{\Xi^0} + m_{\Xi^-} = 3(m_\Lambda + m_{\Sigma^0})\,,\\
&m_p + m_{\Sigma^-} + m_{\Xi^0} = m_n + m_{\Sigma^+} + m_{\Xi^-}\,,\\
&m_p^2 + m_{\Sigma^-}^2 + m_{\Xi^0}^2 = m_n^2 + m_{\Sigma^+}^2 + m_{\Xi^-}^2\,,\\
&2[(m_p + m_{\Xi^-})^2 + (m_n + m_{\Xi^0})^2 + (m_{\Sigma^+} + m_{\Sigma^-})^2]\\
&\qquad\qquad = 3(m_\Lambda + m_{\Sigma^0})^2 + 8m_{\Sigma^0}^2\,.
\end{aligned}$$

For the decuplet of baryon resonances (Fig. 6.5) the mass formula reads

$$m_\nu = a_0 + a_1\,(k'^2\,Y + k^2\,Q)\,,$$

and it holds for all irreducible representations of the type $\mathsf{D}(p,0)$ or $\mathsf{D}(0,q)$. By eliminating the parameters a_0, a_1, and k^2, it results seven linearly independent mass relations

$$\begin{aligned}
m_{\Sigma^{*0}} &= (m_{\Sigma^{*+}} + m_{\Sigma^{*-}})/2 = (m_{\Xi^{*0}} + m_{\Delta^0})/2 = (m_{\Xi^{*-}} + m_{\Delta^+})/2\\
&= (m_{\Delta^{++}} + m_{\Delta^-} + m_{\Omega^-})/3 = (m_{\Delta^-} + m_{\Xi^{*0}} + m_{\Sigma^{*+}})/3\\
&= (m_{\Delta^{++}} + m_{\Sigma^{*-}} + m_{\Xi^{*-}})/3 = (m_{\Omega^-} + m_{\Delta^0} + m_{\Delta^+})/3\,.
\end{aligned}$$

In the case of mesons, the particle and antiparticle masses are equal and we have to set $a_1 = 0$ in the mass formula (6.2.35). Also, we have to use the squares of the meson masses. Note that the parameter k^2/k'^2 characterizes the magnitude ratio of electromagnetic and strong interactions. Thus, the setting $k^2 = 0$ means that we neglect the electromagnetic interactions.

2.2.3 Strong interactions

The consequences of the $\mathsf{SU}(3)$ symmetry of strong interactions follow from the classification of elementary particles into $\mathsf{SU}(3)$ multiplets, assuming a scalar behaviour of the interaction Hamiltonian relative to the $\mathsf{SU}(3)$ group, and applying the Wigner-Eckart theorem (see Theorem 6.1.3) to the matrix elements of the transition operator. Actually, this is a generalization of the technique used in the case of isospin symmetry.

For instance, a strong interaction of the form

$$a + b \to c + d \quad (6.2.36)$$

is characterized by the transition amplitude (see Chap. 6 Sec. 2.1.4)

$$A = \langle f|\,\mathsf{T}\,|i\rangle \equiv \langle c,d|\,\mathsf{T}\,|a,b\rangle, \tag{6.2.37}$$

where $|a,b\rangle = |a\rangle|b\rangle \equiv |\mu_a,\nu_a\rangle|\mu_b,\nu_b\rangle$ is the product of the $\mathsf{SU}(3)$ states associated to the particles a and b, and similarly for the particles c and d. We use the orthogonality relations (3.3.62) to invert (3.3.58), that is

$$\begin{aligned} |a,b\rangle &= |\mu_a,\nu_a\rangle|\mu_b,\nu_b\rangle = \sum_{\mu,\gamma,\nu} \begin{pmatrix} \mu_a & \mu_b & \mu_\gamma \\ \nu_a & \nu_b & \nu \end{pmatrix} |\mu_\gamma,\nu\rangle, \\ |c,d\rangle &= |\mu_c,\nu_c\rangle|\mu_d,\nu_d\rangle = \sum_{\mu',\gamma',\nu'} \begin{pmatrix} \mu_c & \mu_d & \mu'_{\gamma'} \\ \nu_c & \nu_d & \nu' \end{pmatrix} |\mu'_{\gamma'},\nu'\rangle. \end{aligned} \tag{6.2.38}$$

In the limit of exact $\mathsf{SU}(3)$ symmetry, the transition operator T is a scalar such that, according to the Wigner-Eckart theorem, we have

$$\langle \mu'_{\gamma'},\nu'|\,\mathsf{T}\,|\mu_\gamma,\nu\rangle = \mathcal{A}(\mu_\gamma \to \mu'_{\gamma'})\,\delta_{\mu\mu'}\,\delta_{\nu\nu'}\,, \tag{6.2.39}$$

where $\mathcal{A}(\mu_\gamma \to \mu'_{\gamma'})$ represents the $\mathsf{SU}(3)$ *invariant transition amplitudes*. Consequently,

$$A = \sum_{\mu,\nu,\gamma,\gamma'} \begin{pmatrix} \mu_c & \mu_d & \mu_{\gamma'} \\ \nu_c & \nu_d & \nu \end{pmatrix} \begin{pmatrix} \mu_a & \mu_b & \mu_\gamma \\ \nu_a & \nu_b & \nu \end{pmatrix} \mathcal{A}(\mu_\gamma \to \mu_{\gamma'})\,, \tag{6.2.40}$$

which leads to a set of linear relations between the *interaction amplitudes* and *invariant amplitudes*. By eliminating the invariant amplitudes we obtain relations between the interaction amplitudes which, in turn, can be converted into relations between the cross-sections of different interactions, relations that can be checked against experimental data. To model the breaking of $\mathsf{SU}(3)$ symmetry, we assume that the transition operator comprises two terms $\mathsf{T} = \mathsf{T}_{\text{exact}} + \mathsf{T}_{\text{m}}$, where $\mathsf{T}_{\text{exact}} = \mathsf{T}^{(0,0)}_{000}$ is an $\mathsf{SU}(3)$ scalar, while $\mathsf{T}_{\text{m}} = \mathsf{T}^{(1,1)}_{000}$ is the component $T = T_3 = Y = 0$ of an irreducible tensor operator, relative to the transformations in the $\mathsf{SU}(3)$ representation $\mathsf{D}(1,1)$. Again, by eliminating the invariant amplitudes from the matrix elements of the operators $\mathsf{T}^{(0,0)}_{000}$ and $\mathsf{T}^{(1,1)}_{000}$, we obtain relations between interaction amplitudes as well as between cross-sections.

Similarly to the case of isospin symmetry, the baryons, antibaryons, and pseudoscalar mesons can be represented respectively, by the matrices

$$\mathbf{B} = \begin{bmatrix} \frac{1}{\sqrt{2}}\,\Sigma^0 + \frac{1}{\sqrt{6}}\,\Lambda & \Sigma^+ & p \\ \Sigma^- & -\frac{1}{\sqrt{2}}\,\Sigma^0 + \frac{1}{\sqrt{6}}\,\Lambda & n \\ \Xi^- & \Xi^0 & -\frac{2}{\sqrt{6}}\,\Lambda \end{bmatrix}, \tag{6.2.41}$$

$$\overline{\mathbf{B}} = \begin{bmatrix} \frac{1}{\sqrt{2}}\,\overline{\Sigma^0} + \frac{1}{\sqrt{6}}\,\overline{\Lambda} & \overline{\Sigma^-} & \overline{\Xi^-} \\ \overline{\Sigma^+} & -\frac{1}{\sqrt{2}}\,\overline{\Sigma^0} + \frac{1}{\sqrt{6}}\,\overline{\Lambda} & \overline{\Xi^0} \\ \overline{p} & \overline{n} & -\frac{2}{\sqrt{6}}\,\overline{\Lambda} \end{bmatrix}, \tag{6.2.42}$$

$$\mathbf{P} = \begin{bmatrix} \frac{1}{\sqrt{2}}\,\pi^0 + \frac{1}{\sqrt{6}}\,\eta & \pi^+ & K^+ \\ \pi^- & -\frac{1}{\sqrt{2}}\,\pi^0 + \frac{1}{\sqrt{6}}\,\eta & K^0 \\ K^- & \overline{K^0} & -\frac{2}{\sqrt{6}}\,\eta \end{bmatrix}, \tag{6.2.43}$$

the most general Lagrangian density of the antibaryon-baryon-meson coupling being

$$L = g_1\,\mathsf{Tr}\,(\overline{\mathbf{B}}\,\mathbf{B}\,\mathbf{P}) + g_2\,\mathsf{Tr}\,(\overline{\mathbf{B}}\,\mathbf{P}\,\mathbf{B}),$$

or

$$L = \sqrt{2}\,g_s[\mathsf{Tr}(\overline{\mathbf{B}}\,\mathbf{P}\,\mathbf{B}) + \mathsf{Tr}(\overline{\mathbf{B}}\,\mathbf{B}\,\mathbf{P})] + \sqrt{2}\,g_a[\mathsf{Tr}(\overline{\mathbf{B}}\,\mathbf{P}\,\mathbf{B}) - \mathsf{Tr}(\overline{\mathbf{B}}\,\mathbf{B}\,\mathbf{P})]. \tag{6.2.44}$$

Here, g_s and g_a are the constants of the symmetric and antisymmetric coupling, respectively, while $g_1 = \sqrt{2}(g_s - g_a)$ and $g_2 = \sqrt{2}(g_s + g_a)$. By substituting the matrices (6.2.41) – (6.2.43) into (6.2.44) we obtain the Lagrangian density

$$\begin{aligned} L =\; & (g_s + g_a)\,(\overline{\mathbf{N}}\,\boldsymbol{\tau}\,\mathbf{N}) \cdot \boldsymbol{\pi} + (\sqrt{3}\,g_a - \frac{1}{\sqrt{3}}\,g_s)\overline{\mathbf{N}}\,\mathbf{N}\,\eta \\ & - (\sqrt{3}\,g_a + \frac{1}{\sqrt{3}}\,g_s)(\overline{\mathbf{N}}\,\mathbf{K}\,\Lambda + \overline{\Lambda}\,\overline{\mathbf{K}}\,\mathbf{N}) \\ & + (g_s - g_a)[\overline{\boldsymbol{\Sigma}} \cdot (\overline{\mathbf{K}}\,\boldsymbol{\tau}\,\mathbf{N}) + (\overline{\mathbf{N}}\,\boldsymbol{\tau}\,\mathbf{K}) \cdot \boldsymbol{\Sigma}] \\ & + 2\,i\,g_a\,(\overline{\boldsymbol{\Sigma}} \times \boldsymbol{\Sigma}) \cdot \boldsymbol{\pi} + \frac{2}{\sqrt{3}}\,g_s\,(\overline{\boldsymbol{\Sigma}} \cdot \boldsymbol{\Sigma})\eta + \frac{2}{\sqrt{3}}\,g_s\,(\overline{\Lambda}\,\boldsymbol{\Sigma} + \overline{\boldsymbol{\Sigma}}\,\Lambda) \cdot \boldsymbol{\pi} \\ & - \frac{2}{\sqrt{3}}\,g_s\,\overline{\Lambda}\,\Lambda\,\eta + (g_a - g_s)\,(\overline{\boldsymbol{\Xi}}\,\boldsymbol{\tau}\,\boldsymbol{\Xi}) \cdot \boldsymbol{\pi} - (\sqrt{3}\,g_a + \frac{1}{\sqrt{3}}\,g_s)\overline{\boldsymbol{\Xi}}\,\boldsymbol{\Xi}\,\eta \\ & + \frac{2}{\sqrt{3}}\,g_s\,(\overline{\boldsymbol{\Xi}}\,\mathbf{K}_c\,\Lambda + \overline{\Lambda}\,\overline{\mathbf{K}}_c\,\boldsymbol{\Xi}) - (g_s + g_a)\,[\overline{\boldsymbol{\Sigma}} \cdot (\overline{\mathbf{K}}_c\,\boldsymbol{\tau}\,\boldsymbol{\Xi}) + (\overline{\boldsymbol{\Xi}}\,\boldsymbol{\tau}\,\mathbf{K}_c) \cdot \boldsymbol{\Sigma}]. \end{aligned} \tag{6.2.45}$$

Here we used the notations of the isospin symmetry formulation. Thus, $\boldsymbol{\Sigma}$ and $\boldsymbol{\pi}$ are iso-vectors, Λ and η are iso-scalars, and

$$\mathbf{N} = \begin{bmatrix} p \\ n \end{bmatrix}, \quad \mathbf{K} = \begin{bmatrix} K^+ \\ K^0 \end{bmatrix}, \quad \mathbf{K}_c = \begin{bmatrix} \overline{K^0} \\ -K^- \end{bmatrix}, \quad \boldsymbol{\Xi} = \begin{bmatrix} -\Xi^0 \\ \Xi^- \end{bmatrix},$$

are iso-spinors. Note that, all the coupling constants of the $\overline{\mathbf{B}}\mathbf{B}\mathbf{P}$ system are expressed in terms of two constants g_s and g_a. The coupling constant $g_{\overline{N}N\pi} = g_s + g_a$ has a value $g^2/(4\pi) \approx 14$, therefore it is reasonable to introduce a parameter α (the only constant in the Lagrangian density) defined as $\alpha = g_a/(g_s + g_a)$. In the same manner as in the case of isospin symmetry, by means of the Lagrangian density (6.2.45), we may derive relations between different interaction amplitudes.

2.2.4 SU(3) statistical weights

Similarly to the case of isospin symmetry, we will consider a system of particles, each of them being associated to an irreducible representation $\mathsf{D}(\mu_k)$ of $\mathsf{SU}(3)$, where the symbol μ_k stands for the pair $\{p_k, q_k\} \in \mathbb{N}^2$. The statistical weight of such a system of particles, corresponding to a state of unitary spin belonging to the space of an irreducible representation $\mathsf{D}(\mu)$, is equal to the number of independent unitary spin functions, that can be constructed from the product of the n unitary spin functions associated to the particles in the system. In other words, this statistical weight is equal to the multiplicity of the irreducible representation $\mathsf{D}(\mu)$, in the decomposition of the direct product of irreducible representations

$$\mathsf{D}(\mu_1) \otimes \mathsf{D}(\mu_2) \otimes \ldots \otimes \mathsf{D}(\mu_n) = \sum_{\mu}^{\oplus} g(\mu; \mu_1, \mu_2, \ldots, \mu_n)\, \mathsf{D}(\mu_\gamma)\,. \quad (6.2.46)$$

Note that we used here the label μ not μ_γ because we only have to count the representations of the same dimension, irrespective their particular symmetry properties. Hence, the evaluation of the unitary statistical weights reduces to the calculation of the multiplicities $g(\mu; \mu_1, \mu_2, \ldots, \mu_n)$, which are directly related to the characters of irreducible representations. Thus, in terms of the class parameters, the character of fundamental representation of $\mathsf{SU}(3)$ is given by (3.3.73)

$$\chi(\boldsymbol{\alpha}) = \sum_{k=1}^{3} e^{i\alpha_k} = \sum_{k=1}^{3} z_k\,, \qquad -\pi \leq \alpha_k \leq \pi\,, \quad (6.2.47)$$

where $\alpha_1 + \alpha_2 + \alpha_3 = 0$, or $\overline{z_3} = z_1 z_2$. If $\mu = \{p, q\}$ determines the irreducible representation $\mathsf{D}(\mu) \equiv \mathsf{D}(p, q)$ of $\mathsf{SU}(3)$, having the character $\chi_q^p(\boldsymbol{\alpha})$, and the irreducible representations $\mathsf{D}(\mu_k)$ have the characters $\chi_{q_k}^{p_k}(\boldsymbol{\alpha})$, then, according to (3.3.39), we may write

$$g(\mu; \mu_1, \mu_2, \ldots, \mu_n) = \int \overline{\chi_q^p(\boldsymbol{\alpha})} \prod_{k=1}^{n} \chi_{q_k}^{p_k}(\boldsymbol{\alpha})\, dV\,, \quad (6.2.48)$$

with

$$dV = \frac{1}{(2\pi)^3\, 3!}\, \Delta(\overline{z})\, \Delta(z)\, d\alpha_1 d\alpha_2\,, \quad \Delta(z) = \prod_{k<j} (z_k - z_j)\,.$$

The character of any irreducible representation $\mathsf{D}(\mu)$ of $\mathsf{SU}(3)$ can be expressed as a polynomial in terms of the character of the fundamental representation (6.2.47), and its complex conjugate, that corresponds to the contragredient fundamental representation. Also, note that $\overline{z_k} = 1/z_k$. Therefore, the integrals in (6.2.48) become sums of integrals of the form

$$\int_{-\pi}^{\pi} \int_{-\pi}^{\pi} z_k^m\, z_j^n\, d\alpha_k\, d\alpha_j = (2\pi)^2\, \delta_{m0}\, \delta_{n0}\,.$$

For instance, in the case of a system consisting of $N = n+m+m'$ particles, such that n particles are classified in the octet associated to the irreducible representation $8 = \mathsf{D}(1,1)$, while the remaining m and m' particles are classified in the decuplets associated to the irreducible representations $10 = \mathsf{D}(3,0)$ and $\overline{10} = \mathsf{D}(0,3)$, respectively, we have to evaluate integrals of the form

$$J(n,m,m') = \int \left(\overline{\chi_1^1}\right)^n \left(\chi_0^3\right)^m \left(\chi_3^0\right)^{m'} dV\,,$$

where we have omitted the arguments of characters. Taking into account the expression of the characters

$$\begin{aligned}
\chi_1^1 &= \chi\,\overline{\chi} - 1\,,\\
\chi_0^3 &= \overline{\chi_3^0} = \chi^3 - 2\,\overline{\chi} + 1\,,\\
\chi_2^2 &= \left(\chi_1^1\right)^2 - \chi_0^3 - \chi_3^0 - 2\,\chi_1^1 - 1\,,
\end{aligned}$$

one can derive the relations

$$\begin{aligned}
g(0,0;n,m,m') &= J(n,m,m')\,,\\
g(1,1;n,m,m') &= J(n+1,m,m')\,,\\
g(3,0;n,m,m') &= J(n,m,m'+1)\,,\\
g(0,3;n,m,m') &= J(n,m+1,m')\,,\\
g(2,2;n,m,m') &= J(n+2,m,m') - g(0,0;n,m,m') - 2g(1,1;n,m,m')\\
&\quad - g(3,0;n,m,m') - g(0,3;n,m,m')\,.
\end{aligned}$$

If we know the irreducible representation $\mu = \{p,q\}$ associated to each particle in the system, as well as the quantum numbers $\nu = \{T, T_3, Y\}$ of each particle, then we can calculate the statistical weights by means of the $\mathsf{SU}(3)$

parametrization (2.3.61) and the technique described in Chap. 6 Sec. 1.2.4. Thus, we denote the state of the n-particle system by

$$|f\rangle = |\mu_1, \nu_1, \mu_2, \nu_2, \ldots, \mu_n, \nu_n\rangle ,$$

and seek the projection of $|f\rangle$ (the direct product of the states $|\mu_k, \nu_k\rangle$) onto the subspace $\mathcal{R}_\mu$ that transforms according to the irreducible representation $\mathsf{D}(\mu)$. The method of evaluation of this projection is based on the projection operator defined in (3.3.74), that is

$$\mathsf{P}^\mu = N_\mu \int \overline{\chi^\mu(g)}\, \mathcal{D}_{\nu_1\nu_2\ldots\nu_n\, \nu_1'\nu_2'\ldots\nu_n'}(g)\, dg\,, \tag{6.2.49}$$

which projects a vector in the space spanned by the vectors $|f\rangle$ onto the invariant subspace $\mathcal{R}_\mu$. Here, $N_\mu = (p+1)(q+1)(p+q+2)/2$ is the dimension of the irreducible representation $\mathsf{D}(\mu)$, $\chi^\mu(g)$ is the character of irreducible representation $\mathsf{D}(\mu)$, and the integration element has been defined in (2.3.65). The last term of the integrand in (6.2.49) represents the matrix elements of the transformation operator in the space spanned by the state vectors $|f\rangle$

$$\mathcal{D}_{\nu_1\nu_2\ldots\nu_n\, \nu_1'\nu_2'\ldots\nu_n'}(g) = \prod_{k=1}^{n} \mathcal{D}^{\mu_k}_{\nu_k\nu_k'}(g)\,, \tag{6.2.50}$$

where the matrix elements $\mathcal{D}^{\mu_k}_{\nu_k\nu_k'}(g)$ have been defined in (3.3.35). The projection operator P^μ is Hermitian and idempotent, that is

$$|\,\mathsf{P}^\mu|f\rangle\,|^2 = \langle f\,|\,\mathsf{P}^\mu\,|\,f\rangle\,,$$

such that the statistical weight will be given by the formula

$$P^\mu_{\nu_1\nu_2\ldots\nu_n} \equiv |\,\mathsf{P}^\mu\,|f\rangle|^2 = N_\mu \int \overline{\chi^\mu(g)} \prod_{k=1}^{n} \mathcal{D}^{\mu_k}_{\nu_k\nu_k}(g)\, dg\,. \tag{6.2.51}$$

In terms of the statistical weights, the transition probability for an interaction of the form

$$a + b \to c_1 + c_2 + \ldots c_n\,,$$

reads

$$w = \sum_{\mu,\gamma} \begin{pmatrix} \mu_a & \mu_b & \mu_\gamma \\ \nu_a & \nu_b & \nu \end{pmatrix}^2 P^\mu_{\nu_1\nu_2\ldots\nu_n}\,. \tag{6.2.52}$$

When the considered final state contains identical particles one has to multiply the corresponding statistical weights by the number of permutations of the identical particles, such that the probability associated to the final state is proportional to

$$^{*}P^{\mu}_{\nu_1\nu_2\ldots\nu_n} = \prod_{k=1}^{n} \frac{[n(\mu_k)]!}{\prod_{j=1}^{N_{\mu_k}} [n(\mu_k,\nu_j)]!} P^{\mu}_{\nu_1\nu_2\ldots\nu_n} , \tag{6.2.53}$$

where $n(\mu_k, \nu_j)$ represents the number of particles that have the same quantum number ν_j and are associated to the irreducible representation $\mathsf{D}(\mu_k)$,

$$n(\mu_k) = \sum_{j=1}^{N_{\mu_k}} n(\mu_k, \nu_j) ,$$

and $N_{\mu_k} = (p_k+1)(q_k+1)(p_k+q_k+2)/2$. Also, the sum of statistical weights (6.2.53) over all configurations $\{\nu_1, \nu_2 \ldots, \nu_n\}$ of $|f\rangle$ gives the multiplicity of the irreducible representation $\mathsf{D}(\mu)$

$$\sum_{\nu_1,\nu_2\ldots,\nu_n} {}^{*}P^{\mu}_{\nu_1\nu_2\ldots\nu_n} = g(\mu; \mu_1, \mu_2, \ldots, \mu_n) .$$

3. Relativistic quantum mechanics

3.1 Basic equations. Symmetry groups

3.1.1 Dirac's equation

In 1928, P. A. M. Dirac introduced the basic wave equation of relativistic quantum mechanics, able to describe particles that have a half-integer spin (fermions). Dirac started from the generic form of wave equation in non-relativistic quantum mechanics

$$\mathsf{H}\,\Psi = i\,\frac{\partial \Psi}{\partial t} , \tag{6.3.1}$$

that contains a first order time derivative. In such cases, to preserve the Lorentz invariance of the equation the space coordinate derivatives have to be of the first order as well. On the other hand, the eigenvalues of the Hamiltonian operator represent the total energy and, in the theory of Special Relativity, the total energy E and the momentum $\mathbf{p}$ of a particle of mass m satisfy the relation

$$E^2 = \mathbf{p}^2 + m^2 \tag{6.3.2}$$

where we used $c = 1$. Hence, one may try to find the square root of (6.3.2) as a linear form

$$E = \boldsymbol{\alpha} \cdot \mathbf{p} + m\,\beta , \tag{6.3.3}$$

where the quantities $\boldsymbol{\alpha} = \{\alpha_1, \alpha_2, \alpha_3\}$ and β satisfy the relations

$$\begin{aligned} &\alpha_i\,\alpha_j + \alpha_j\,\alpha_i = 2\,\delta_{ij}\,, \\ &\alpha_i\,\beta + \beta\,\alpha_i = 0\,, \qquad i,j = 1,2,3, \\ &\beta^2 = 1\,. \end{aligned} \tag{6.3.4}$$

It is evident that the quantities $\boldsymbol{\alpha} = \{\alpha_1, \alpha_2, \alpha_3\}$ and β cannot be numbers. Dirac proved that these parameters are matrices, the lowest order being 4×4, while the wave function Ψ in (6.3.1) is a spinor with four components. It follows that the Hamiltonian corresponding to (6.3.3) has the form

$$\mathsf{H} = \boldsymbol{\alpha}\cdot\mathbf{p} + m\,\beta = -i\,\boldsymbol{\alpha}\cdot\nabla + m\,\beta\,, \tag{6.3.5}$$

such that the wave equation (Dirac's equation) which describe the particle motion is

$$(-i\,\boldsymbol{\alpha}\cdot\nabla + m\,\beta)\,\Psi = i\,\frac{\partial\Psi}{\partial t}\,. \tag{6.3.6}$$

Dirac's equation (6.3.6) can be derived from the Lagrangian density

$$L = i\,\overline{\psi_\rho}\,\frac{\partial\psi_\rho}{\partial t} + i\,\overline{\psi_\rho}(\boldsymbol{\alpha})_{\rho\sigma}\cdot\nabla\psi_\sigma - m\,\overline{\psi_\rho}(\beta)_{\rho\sigma}\psi_\sigma\,, \tag{6.3.7}$$

where $\psi_\rho(\mathbf{r},t)$ are the components of the spinor Ψ, and the summation is from 1 to 4 with respect to the repeated indices. From (6.3.7) we obtain the Euler-Lagrange derivative

$$[L]^{\overline{\psi_\rho}} = \frac{\partial L}{\partial\overline{\psi_\rho}} = i\,\frac{\partial\psi_\rho}{\partial t} + i\,(\boldsymbol{\alpha})_{\rho\sigma}\cdot\nabla\psi_\sigma - m\,(\beta)_{\rho\sigma}\psi_\sigma\,, \tag{6.3.8}$$

such that the Euler-Lagrange equations $[L]^{\overline{\psi_\rho}} = 0$ take the form

$$-i\,(\boldsymbol{\alpha})_{\rho\sigma}\cdot\nabla\psi_\sigma + m\,(\beta)_{\rho\sigma}\psi_\sigma = i\,\frac{\partial\psi_\rho}{\partial t}\,, \qquad \rho = \overline{1,4}, \tag{6.3.9}$$

that is identical to (6.3.6) written in terms of the components ψ_ρ of the spinor Ψ. The generalized momentum canonically conjugate to ψ_ρ is

$$\pi_\rho(\mathbf{r},t) = \frac{\partial L}{\partial\dot{\psi}_\rho} = i\,\overline{\psi_\rho}\,, \tag{6.3.10}$$

while $\widetilde{\pi}_\rho = 0$, like in the case of Schrödinger equation. Therefore, the Hamiltonian density that corresponds to the Lagrangian density (6.3.7) is

$$\begin{aligned} \mathcal{H} = \pi_\rho\,\frac{\partial\psi_\rho}{\partial t} - L &= -\pi_\rho\left[(\boldsymbol{\alpha})_{\rho\sigma}\cdot\nabla\psi_\sigma + i\,m\,(\beta)_{\rho\sigma}\psi_\sigma\right] \\ &= -i\,\overline{\psi_\rho}\left[(\boldsymbol{\alpha})_{\rho\sigma}\cdot\nabla\psi_\sigma + i\,m\,(\beta)_{\rho\sigma}\psi_\sigma\right]. \end{aligned} \tag{6.3.11}$$

Consequently, the Hamiltonian of a free particle reads

$$H = \int \mathcal{H}\, dV = -i \int \left(\Psi^\dagger \boldsymbol{\alpha} \cdot \nabla\Psi + im\, \Psi^\dagger \beta\, \Psi \right) dV$$
$$= -i \int \left[\frac{1}{2} \left(\Psi^\dagger \boldsymbol{\alpha} \cdot \nabla\Psi - \nabla\Psi^\dagger \cdot \boldsymbol{\alpha}\Psi \right) + im\, \Psi^\dagger \beta\, \Psi \right] dV\,, \quad (6.3.12)$$

with $\Psi^\dagger \equiv (\overline{\psi_1},\, \overline{\psi_2},\, \overline{\psi_3},\, \overline{\psi_4})$.

By means of Pauli's matrices (6.1.110) we can obtain a solution of (6.3.4) expressed as Hermitian block matrices

$$\alpha_1 = \begin{bmatrix} \mathbf{0} & \boldsymbol{\sigma}_1 \\ \boldsymbol{\sigma}_1 & \mathbf{0} \end{bmatrix},\ \alpha_2 = \begin{bmatrix} \mathbf{0} & \boldsymbol{\sigma}_2 \\ \boldsymbol{\sigma}_2 & \mathbf{0} \end{bmatrix},\ \alpha_3 = \begin{bmatrix} \mathbf{0} & \boldsymbol{\sigma}_3 \\ \boldsymbol{\sigma}_3 & \mathbf{0} \end{bmatrix},\ \beta = \begin{bmatrix} \mathbf{e} & \mathbf{0} \\ \mathbf{0} & -\mathbf{e} \end{bmatrix}. \quad (6.3.13)$$

However, this solution is not unique because we may apply an arbitrary linear transformation to the matrices (6.3.13) to obtain a new solution of (6.3.4). Now, we may introduce the matrices

$$\gamma^k = \beta\, \alpha_k = -\alpha_k\, \beta\,, \quad k = 1, 2, 3, \quad \gamma^4 = \beta\,, \quad (6.3.14)$$

that satisfy the anti-commutation relations

$$\{\gamma^\mu, \gamma^\nu\} \equiv \gamma^\mu\gamma^\nu + \gamma^\nu\gamma^\mu = 2\, g^{\mu\nu}\,, \quad (6.3.15)$$

with $g^{\mu\nu} = g_{\mu\nu}$ being the metric tensor defined in (2.4.4).

Thus, we may write Dirac's equation (6.3.9) in a form that reveals the properties of this equation with respect to the proper Lorentz transformations

$$[i\, (\gamma^\mu)_{\rho\sigma}\, \partial_\mu - m\, \delta_{\rho\sigma}]\, \psi_\sigma = 0\,, \quad \rho = \overline{1, 4}\,, \quad (6.3.16)$$

where $(x^\mu) = (x^1, x^2, x^3, x^4 = t)$, $(\partial_\mu) = (\nabla, \partial_4)$, and

$$\gamma^1 = \begin{bmatrix} \mathbf{0} & \boldsymbol{\sigma}_1 \\ -\boldsymbol{\sigma}_1 & \mathbf{0} \end{bmatrix},\ \gamma^2 = \begin{bmatrix} \mathbf{0} & \boldsymbol{\sigma}_2 \\ -\boldsymbol{\sigma}_2 & \mathbf{0} \end{bmatrix},\ \gamma^3 = \begin{bmatrix} \mathbf{0} & \boldsymbol{\sigma}_3 \\ -\boldsymbol{\sigma}_3 & \mathbf{0} \end{bmatrix},\ \gamma^4 = \begin{bmatrix} \mathbf{e} & \mathbf{0} \\ \mathbf{0} & -\mathbf{e} \end{bmatrix}. \quad (6.3.17)$$

With $(\mathsf{p}_\mu) = -i(\partial_\mu)$, we may also write Dirac's equation (6.3.16) as

$$(\gamma^\mu \mathsf{p}_\mu + m)\, \Psi = 0\,. \quad (6.3.18)$$

Note that $\gamma^\mu \mathsf{p}_\mu$ is not an invariant relative to the Lorentz transformations since $\{\gamma^\mu\}$ is just a set of numerical matrices which do not change with Lorentz transformations. It can be shown that, if we define an adjoint Dirac spinor $\widetilde{\Psi} = \Psi^\dagger\, \gamma^4$, then $\widetilde{\Psi}\, \Psi$ is a Lorentz scalar, and $(\widetilde{\Psi}\, \gamma^\mu\, \Psi)$ is a four-vector. The Hermitian conjugate of the matrices (6.3.14) is given by

$$(\gamma^\mu)^\dagger = g^{\mu\nu}\, \gamma^\nu = g_{\mu\nu}\, \gamma^\nu\,, \quad \mu = \overline{1, 4}\,, \quad (6.3.19)$$

leading to the equation satisfied by the components of the adjoint Dirac spinor

$$\left[i\,(\gamma^{\mu\mathrm{T}})_{\rho\sigma}\,\partial_\mu + m\,\delta_{\rho\sigma}\right]\widetilde{\psi}_\sigma = 0\,, \quad \rho = \overline{1,4}\,. \tag{6.3.20}$$

In the derivation of (6.3.20) we used the relations

$$[(\gamma^\mu \mathsf{p}_\mu + m)\,\Psi]^\dagger = (\gamma^\mu \mathsf{p}_\mu \Psi)^\dagger + m\,\Psi^\dagger = (\mathsf{p}_\mu \Psi)^\dagger\,(\gamma^\mu)^\dagger + m\,\Psi^\dagger\,,$$

$$(\mathsf{p}_\mu \Psi)^\dagger = \overline{\mathsf{p}_\mu}\,\Psi^\dagger = i\,\partial_\mu\,\Psi^\dagger = -\mathsf{p}_\mu\,\Psi^\dagger\,,$$

and

$$(\gamma^\mu)^\dagger\,\gamma^4 = \left(\gamma^4\gamma^\mu\right)^\dagger = \gamma^4\gamma^\mu\,.$$

Now, taking into account that $\gamma^4 = \beta$ and $x^4 = t$, we may write the Lagrangian density (6.3.7) in the form

$$\begin{aligned} L &= i\,\Psi^\dagger\,\partial_t\Psi + i\,\Psi^\dagger\,\boldsymbol{\alpha}\cdot\nabla\,\Psi - m\,\Psi^\dagger\,\beta\,\Psi \\ &= i\,\Psi^\dagger\,\beta\beta\,\partial_t\Psi + i\,\Psi^\dagger\,\beta\beta\,\boldsymbol{\alpha}\cdot\nabla\,\Psi - m\,\Psi^\dagger\,\beta\,\Psi \\ &= i\,\widetilde{\Psi}\,\gamma^4\,\partial_4\Psi + i\,\widetilde{\Psi}\,\beta\,\boldsymbol{\alpha}\cdot\nabla\,\Psi - m\,\widetilde{\Psi}\,\Psi \\ &= i\,\widetilde{\Psi}\,\gamma^\mu\,\partial_\mu\Psi - m\,\widetilde{\Psi}\,\Psi \\ &= -\widetilde{\Psi}\,\gamma^\mu\,\mathsf{p}_\mu\Psi - m\,\widetilde{\Psi}\,\Psi\,. \end{aligned} \tag{6.3.21}$$

Therefore, both the Lagrangian density (6.3.7) and the Hamiltonian density (6.3.11) are Lorentz scalars, while the Dirac equation (6.3.16) as well its adjoined form (6.3.20) are covariant relative to Lorentz transformations.

If the considered particle has an electric charge q, and is moving in an electromagnetic field, defined by the scalar potential φ and the vector potential $\mathbf{A}$ then, according to (5.2.14), we have to replace in (6.3.5) the momentum $\mathbf{p}$ by $\mathbf{p} - q\mathbf{A}$, and add the electrostatic potential energy $q\varphi$. Consequently, Dirac's equation becomes

$$[\boldsymbol{\alpha}\cdot(\mathbf{p} - q\,\mathbf{A}) + m\,\beta + q\,\varphi\,\mathbf{I}]\,\Psi = i\,\frac{\partial\Psi}{\partial t}\,, \tag{6.3.22}$$

where $\mathbf{I}$ represents the 4×4 identity matrix. A detailed analysis of this equation brings out a magnetic momentum ($\mathbf{m}$) which depends on the particle spin ($\mathbf{s}$) by a relation of the form $\mathbf{m} = (q/m)\mathbf{s}$. Experimental data shown that this relation is satisfied only for electrons, positrons, and $\mu^\pm$ mesons, hence Dirac's equation gives exact results only for these particles.

3.1.2 The Klein-Gordon equation

The Klein-Gordon equation is a relativistic version of the Schrödinger equation, that describes particles of integer spin (bosons). The derivation of this equation starts from the expression of relativistic energy-momentum relation (6.3.2), which we use to form a classical Hamiltonian

$$H = \sqrt{\mathbf{p}^2 + m^2}\,.$$

Now, we may replace the Hamiltonian and classical momentum by quantum mechanical operators, which yields the free particle wave equation

$$i\,\partial_t\,\varphi(\mathbf{x}) = \left[\sqrt{(-i\nabla)^2 + m^2}\right]\varphi(\mathbf{x})\,,$$

where $\mathbf{x} \equiv (x^\mu) = (x^1, x^2, x^3, x^4 = t)$. This equation, does not satisfy the main condition required by Special Relativity, that is to be invariant with respect to proper Lorentz transformations, and the square root introduces ambiguities. Both these problems are corrected by taking the square of the energy-momentum relation, which leads to the equation

$$E^2\varphi(\mathbf{x}) = (i\,\partial_t)^2\varphi(\mathbf{x}) = \left[(-i\nabla)^2 + m^2\right]\varphi(\mathbf{x})\,,$$

or

$$(\partial_t^2 - \Delta + m^2)\varphi(\mathbf{x}) = 0\,.$$

This is the Klein-Gordon equation, generally written in the form

$$\left(\Box - m^2\right)\varphi(\mathbf{x}) = 0\,, \tag{6.3.23}$$

where the operator

$$\Box = \Delta - \partial_t^2$$

is called D'Alembertian. We may also use the form

$$\left(\partial_\mu\partial^\mu + m^2\right)\varphi(\mathbf{x}) = 0\,, \tag{6.3.24}$$

in terms of the operator $\partial_\mu\partial^\mu = g^{\mu\nu}\partial_\mu\partial_\mu = -\Box$, with $g^{\mu\nu} = g_{\mu\nu}$ being the metric tensor defined in (2.4.4). The D'Alembertian is a Lorentz invariant operator and thus, if a wave function $\varphi(\mathbf{x})$ satisfies the Klein-Gordon equation in one reference frame, then the same form of equation is satisfied by the transformed wave function $\varphi'(\mathbf{x}')$ provided that $\varphi(\mathbf{x})$ is a scalar function relative to proper Lorentz transformations, that is $\varphi(\mathbf{x}) \to \varphi'(\mathbf{x}') = \varphi(\mathbf{x})$. Therefore, the Klein-Gordon equation is Lorentz invariant.

The Klein-Gordon equation (6.3.23) describes scalar particles ($\varphi(-\mathbf{x}) = \varphi(\mathbf{x})$) of mass m and spin $s = 0$. It has been proved that the same equation describes the pseudoscalar particles ($\varphi(-\mathbf{x}) = -\varphi(\mathbf{x})$) of spin $s = 0$, as well as the vector ($s = 1$) and tensor ($s = 2$) particles. The latter are represented respectively, by vector and tensor wave functions, such that each component of these functions satisfies the Klein-Gordon equation. Generally, it is admitted that the Klein-Gordon equation describes the motion of particles that have an integer spin (bosons).

The solutions of Dirac's equation are also solutions of Klein-Gordon equation, but the reciprocal is not true. Thus, if we expand the expression

$$(i\partial_t + \boldsymbol{\alpha} \cdot \mathbf{p} + m\beta)\,(i\partial_t - \boldsymbol{\alpha} \cdot \mathbf{p} - m\beta)\,\Psi = 0\,,$$

we obtain

$$-\partial_t^2 \Psi - m(\beta\boldsymbol{\alpha} + \boldsymbol{\alpha}\beta) \cdot \mathbf{p}\Psi - (\boldsymbol{\alpha} \cdot \mathbf{p})^2 \Psi - m^2 \Psi = 0\,.$$

The second term vanishes by virtue of anti-commutation relations (6.3.4), and we also have

$$\begin{aligned}(\boldsymbol{\alpha} \cdot \mathbf{p})^2 &= \sum_{j,k} (\alpha_j p_j)\,(\alpha_k p_k) = \frac{1}{2}\Big(\sum_{j,k} \alpha_j \alpha_k p_j p_k + \sum_{j,k} \alpha_k \alpha_j p_k p_j \Big) \\ &= \frac{1}{2} \sum_{j,k} (\alpha_j \alpha_k + \alpha_k \alpha_j)\, p_j p_k = \sum_{j,k} \delta_{jk}\, p_j p_k = \mathbf{p}^2 = -\Delta\,.\end{aligned}$$

It follows that the following equation holds

$$-\partial_t^2 \Psi + \Delta\Psi - m^2 \Psi = 0$$

which is identical to the Klein-Gordon equation (6.3.23). In conclusion, if a four-component spinor is a solution of Dirac's equation (6.3.6), then each of its components satisfies the Klein-Gordon equation (6.3.23).

3.1.3 Symmetry groups

To determine the maximal symmetry groups corresponding to Dirac's and Klein-Gordon equations, one may apply the method described in Chap. 6 Sec. 1.1.3. Equally well, one may consider the expression (6.1.85) for the infinitesimal generator of the symmetry group and solve the equation (6.1.84). Thus, by substituting $\Omega(\mathbf{r}, t) = \Box - m^2$ in (6.1.84) we obtain

$$[\,\Box, \mathsf{X}\,] = i\,\lambda(\mathbf{r}, t)\,(\Box - m^2)\,. \tag{6.3.25}$$

Now, there are two distinct cases: $m \neq 0$ and $m = 0$. For $m \neq 0$, it can be shown that the maximal symmetry group of the Klein-Gordon equation (6.3.23) is the Poincaré group (see Chap. 2 Sec. 4.1.2, and Chap. 5 Sec. 2.2.1). If $m = 0$, then the maximal symmetry group of Klein-Gordon equation is the conformal group (see Chap. 5 Sec. 2.2.2). This result is consistent with the invariance of Maxwell equations with respect to the conformal group, taking into account that the wave equations (5.2.86), in their source-free form (i.e., $\mathbf{j} = \mathbf{0}$ and $\rho = 0$), are nothing else but the system of Klein-Gordon equations that describe a photon (vector particle). If we consider the infinitesimal generators of the conformal group according to its subgroup structure

1 the homogeneous Lorentz transformations

$$\mathsf{M}_{\mu\nu} = x_\mu \partial_\nu - x_\nu \partial_\mu \,,$$

2 the translations in $\mathcal{M}_4$

$$\mathsf{P}_\mu = \partial_\mu \,,$$

3 the Weyl-Zeemann dilatations

$$\mathsf{D} = x^\mu \partial_\mu \,,$$

4 the special conformal transformations

$$\mathsf{C}_\mu = 2x_\mu x^\nu \partial_\nu - \mathbf{x}^2 \partial_\mu \,,$$

where $\mu, \nu = \overline{1,4}$ then, the general form of the infinitesimal generator of the conformal group is

$$\mathsf{Z} = \frac{1}{2}\,\omega^{\mu\nu}\,\mathsf{M}_{\mu\nu} + \rho^\mu\,\mathsf{P}_\mu + \delta\,\mathsf{D} + \tau^\mu\,\mathsf{C}_\mu \,, \tag{6.3.26}$$

where $\omega^{\mu\nu}$, ρ^μ, δ, and τ^μ are real constants. Consequently, the condition (6.1.84) becomes

$$[\,\square, \mathsf{Z}\,] = -i\,(2\,\delta + 4\,\tau^\mu x_\mu)\,\square \,. \tag{6.3.27}$$

This result shows that, for $m = 0$, the Klein-Gordon equation is invariant relative to the transformations of the conformal group, as the D'Alembertian satisfies a condition of the form (6.1.84) with $\lambda(\mathbf{r}, t) = -2\,\delta - 4\,\tau^\mu x_\mu$, even if the *form* of the D'Alembertian operator is not preserved by the conformal group because, following such a transformation,

$$\square \to \left(1 + 2\,\tau^\mu x_\mu + \boldsymbol{\tau}^2 \mathbf{x}^2\right) \square \,. \tag{6.3.28}$$

In the case of Dirac's equation we have $\Omega(\mathbf{r}, t) = \mathsf{H} - i\partial_t$. When the Hamiltonian H does not explicitly depend on time variable, then it can play the rolê of an infinitesimal generator for the time translations. Thus, we obtain two kinds of representations of the algebra associated to the maximal symmetry group of Dirac's equation, depending on how we consider the infinitesimal generator of time translations as being the operator $i\partial_t$ (*representations of the first kind*), or the Hamiltonian H (*representations of the second kind*). The representations of the first kind are obtained by solving (6.1.84), that is

$$[\,\mathsf{H} - i\,\partial_t, \mathsf{X}_\mathrm{I}\,] = i\,\lambda(\mathbf{r}, t)\,(\mathsf{H} - i\,\partial_t) \,. \tag{6.3.29}$$

If we choose the operator H as the infinitesimal generator of time translations, then we observe that, replacing $i\partial_t$ by H, the right side in (6.3.30) vanishes, hence the representations of the second kind satisfy the condition

$$[\,\mathsf{H} - i\,\partial_t, \mathsf{X}_{\mathrm{II}}\,] = 0\,. \tag{6.3.30}$$

Thus, for $m \neq 0$, we have

$$[\,\mathsf{H} - i\,\partial_t, \mathsf{L}\,] = 0\,, \tag{6.3.31}$$

where L is the general infinitesimal generator of the Poincaré group. If $m = 0$, the maximal symmetry group of Dirac's equation is the conformal group, and the general infinitesimal generator (6.3.26) of this group satisfies the conditions

$$\begin{aligned} [\,\mathsf{H}_0 - i\,\partial_t, \mathsf{Z}_{\mathrm{I}}\,] &= i\,\left(\delta + 2\,\tau^4\,t\right)(\mathsf{H}_0 - i\,\partial_t)\,, \\ [\,\mathsf{H}_0 - i\,\partial_t, \mathsf{Z}_{\mathrm{II}}\,] &= 0\,, \end{aligned} \tag{6.3.32}$$

where

$$\mathsf{H}_0 = \mathsf{H}\Big|_{m=0}\,.$$

We may conclude that the maximal symmetry group of both Dirac and Klein-Gordon equations, for massless particles, is the conformal group, while for particles of mass $m \neq 0$ it becomes the restriction of the conformal group to the Poincaré group.

3.1.4 The conservation of electric charge

Noether's theorem shows the connection between the conservation laws of physical quantities and the symmetry groups of motion equations. Thus, it has been pointed out (see Chap. 5 Sec. 2.1.5) that the equations satisfied by the free electromagnetic field are invariant relative to the gauge transformations of the second kind

$$A_\mu \to A_\mu - \partial_\mu \varepsilon(\mathbf{x})\,, \quad \mu = \overline{1,4}\,, \tag{6.3.33}$$

where $(A_\mu) = (-\mathbf{A}, \varphi)$, and $\varepsilon(\mathbf{x})$ is a scalar function that satisfies the equation $\Box\,\varepsilon(\mathbf{x}) = 0$ (i.e., the necessary condition to preserve the Lorentz gauge $\partial_\mu A^\mu = 0$). If we study the motion of a relativistic charged particle in an electromagnetic field, defined by the four-potential (A_μ), then we have to use (6.3.22). Multiplying (6.3.22) to the left by β, substituting $x^4 = t$, and $\mathbf{p}$ by $-i\nabla$, we have

$$\left[\boldsymbol{\gamma} \cdot (-i\nabla - q\mathbf{A}) + \boldsymbol{\gamma}^4\,(-i\,\partial_4 + q\varphi) + m\,\mathbf{I}\right]\Psi = 0\,,$$

where $\boldsymbol{\gamma} = (\gamma^1, \gamma^2, \gamma^3)$, or

$$\left[i\,\boldsymbol{\gamma} \cdot (\nabla - iq\,\mathbf{A}) + i\,\boldsymbol{\gamma}^4\,(\partial_4 + iq\,\varphi) - m\,\mathbf{I}\right]\Psi = 0\,,$$

leading to a motion equation of the form (6.3.16), that is

$$[i\,(\gamma^{\mu})_{\rho\,\sigma}\,(\partial_{\mu} + iq\,A_{\mu}) - m\,\delta_{\rho\,\sigma}]\,\psi_{\sigma} = 0\,, \quad \rho = \overline{1,4}\,. \tag{6.3.34}$$

Therefore, the motion equations of a relativistic charged particle in an electromagnetic field are obtained by replacing the operator ∂_{μ} in (6.3.16) and (6.3.20) by $\partial_{\mu} + iqA_{\mu}$ and $\partial_{\mu} - iqA_{\mu}$, respectively. It also results that the motion equation (6.3.34) can be derived from the Lagrangian density

$$\begin{aligned} L' &= i\,\widetilde{\Psi}\,\gamma^{\mu}\,(\partial_{\mu} + iq\,A_{\mu})\,\Psi - m\,\widetilde{\Psi}\,\Psi\,, \\ &= i\,\widetilde{\Psi}\,\gamma^{\mu}\,\partial_{\mu}\,\Psi - m\,\widetilde{\Psi}\,\Psi - q\,\widetilde{\Psi}\,\gamma^{\mu}\,\Psi\,A_{\mu}\,, \end{aligned} \tag{6.3.35}$$

obtained from (6.3.21) by means of the substitution $\partial_{\mu} \rightarrow \partial_{\mu} + iqA_{\mu}$. We may now infer that $L' = L_0 + L_{\text{int}}$, where L_0 is the Lagrangian density (6.3.7) of a free particle, and $L_{\text{int}} = -q\,\widetilde{\Psi}\,\gamma^{\mu}\,\Psi\,A_{\mu}$, that is of the form $j^{\mu}A_{\mu}$ and represents the interaction between the moving charge and the electromagnetic field (see (5.2.35)). Note that the total Lagrangian density for our problem has to contain the Lagrangian density (5.2.28) of the free electromagnetic field

$$L_e = -\frac{1}{4}\,F_{\mu\nu}F^{\mu\nu} = -\frac{1}{4}\,(\partial_{\mu}A_{\nu} - \partial_{\nu}A_{\mu})\,(\partial^{\mu}A^{\nu} - \partial^{\nu}A^{\mu})\,, \tag{6.3.36}$$

such that we have

$$L = L_0 + L_e + L_{\text{int}}\,. \tag{6.3.37}$$

The fact that we added the Lagrangian density L_e does not affect the motion equation (6.3.34) because, in the derivation of this equation we consider the variations of $\widetilde{\Psi}$, but the term L_e in (6.3.37) has its rôle in relation to Noether's theorem.

The gauge transformation (6.3.33) will change the differential operators in (6.3.34) and its adjoint counterpart, according to the relations

$$\begin{aligned} (\partial_{\mu} + iq\,A_{\mu})\,\Psi &\rightarrow [\partial_{\mu} + iq\,A_{\mu} - iq\,\partial_{\mu}\varepsilon(\mathbf{x})]\,\Psi\,, \\ (\partial_{\mu} - iq\,A_{\mu})\,\overline{\Psi} &\rightarrow [\partial_{\mu} - iq\,A_{\mu} + iq\,\partial_{\mu}\varepsilon(\mathbf{x})]\,\overline{\Psi}\,. \end{aligned} \tag{6.3.38}$$

Hence, the Lagrangian density (6.3.36) of the free electromagnetic field will remain unchanged, while L' will contain terms of the form $(iq\,\partial_{\mu}\varepsilon)\Psi$ and $(iq\,\partial_{\mu}\varepsilon)\widetilde{\Psi}$. The latter expression appears when we consider the real form of the total Lagrangian density $L \rightarrow L +$ Hermitian conjugate. To compensate such terms, we have to assume that the gauge transformations of the second kind (6.3.33) are accompanied by the transformations

$$\psi_{\rho} \rightarrow \exp{[iq\,\varepsilon(\mathbf{x})]}\,\psi_{\rho}\,, \quad \overline{\psi}_{\rho} \rightarrow \exp{[-iq\,\varepsilon(\mathbf{x})]}\,\overline{\psi}_{\rho}\,, \tag{6.3.39}$$

called *gauge transformations of the first kind*. Combining the two kinds of gauge transformations, (6.3.33) and (6.3.39), we obtain

$$\begin{aligned}(\partial_\mu + iq\,A_\mu)\,\Psi &\to [(\partial_\mu + iq\,A_\mu)\,\Psi]\exp\left[iq\,\varepsilon(\mathbf{x})\right],\\(\partial_\mu - iq\,A_\mu)\,\overline{\Psi} &\to \left[(\partial_\mu - iq\,A_\mu)\,\overline{\Psi}\right]\exp\left[-iq\,\varepsilon(\mathbf{x})\right].\end{aligned} \tag{6.3.40}$$

Because the Lagrangian density L' is a bilinear function of Ψ and $\widetilde{\Psi}$ (i.e., it does not contain terms of the form $\Psi\,\Psi$ or $\overline{\Psi}\,\overline{\Psi}$), the exponentials from (6.3.40) vanish and therefore, the total Lagrangian density L is invariant relative to the combined gauge transformations (6.3.33) and (6.3.39).

The combined gauge transformations form an Abelian group, and we can determine its infinitesimal generators by considering the transformations

$$\begin{aligned}A_\mu &\to A_\mu - \partial_\mu\,\varepsilon\,,\\\psi_\rho &\to \psi_\rho + iq\,\varepsilon\,\psi_\rho\,,\\\overline{\psi}_\rho &\to \overline{\psi}_\rho - iq\,\varepsilon\,\overline{\psi}_\rho\,,\end{aligned} \tag{6.3.41}$$

where $\mu, \rho = \overline{1,4}$, and $\varepsilon = \varepsilon(\mathbf{x})$ is an infinitesimal variation. Also, assuming that $\delta x_\mu = 0$, the variations of the field functions have the form

$$\begin{aligned}\delta A_\mu &= -\partial_\mu\,\varepsilon\,,\\\delta\psi_\rho &= \quad iq\,\varepsilon\,\psi_\rho\,,\\\delta\overline{\psi}_\rho &= -iq\,\varepsilon\,\overline{\psi}_\rho\,.\end{aligned} \tag{6.3.42}$$

Consequently, from Noether's theorem in Special Relativity we obtain the continuity equation

$$\partial_\mu\left[iq\,\frac{\partial L}{\partial(\partial_\mu\,\psi_\rho)}\,\psi_\rho\,\varepsilon - iq\,\frac{\partial L}{\partial(\partial_\mu\,\overline{\psi}_\rho)}\,\overline{\psi}_\rho\,\varepsilon - \frac{\partial L}{\partial(\partial_\mu\,A_\nu)}\,\partial_\nu\,\varepsilon\right] = 0\,, \tag{6.3.43}$$

which follows directly from (5.2.50), if we set $\delta f^\mu = 0$.

Because $\partial_\mu A_\nu$ appears only in the Lagrangian density of the free electromagnetic field, and from (5.2.32) we have $\partial L/\partial(\partial_\mu A_\nu) = -F^{\mu\nu}$, the last term in (6.3.43) reads

$$\partial_\mu\left(F^{\mu\nu}\,\partial_\nu\,\varepsilon\right) = \partial_\mu\left[\partial_\nu\left(F^{\mu\nu}\varepsilon\right) - \varepsilon\,\partial_\nu\,F^{\mu\nu}\right]. \tag{6.3.44}$$

The tensor $F^{\mu\nu}$ is skew-symmetric therefore $\partial_\mu\,\partial_\nu\,(F^{\mu\nu}\,\varepsilon) = 0$. Also, in our case of a constant metric tensor, and units $\hbar = c = 1$, the field equations (5.2.38) become

$$\partial_\nu\,F^{\mu\nu} = j^\mu\,, \quad \mu = \overline{1,4}.$$

Note that the equations (5.2.38) have been obtained from a Lagrangian density comprising L_{int} and L_e, so that j^μ represents the current generated by the motion of the particle. Finally, by using the relation $\partial_\mu \partial_\nu F^{\mu\nu} = 0$, we obtain

$$\partial_\mu (\varepsilon \, \partial_\nu F^{\mu\nu}) = (\partial_\mu \varepsilon) \, j^\mu \, .$$

Substituting this expression into (6.3.43) leads to the continuity equation

$$\partial_\mu \left[iq \frac{\partial L}{\partial(\partial_\mu \psi_\rho)} \psi_\rho \, \varepsilon - iq \frac{\partial L}{\partial(\partial_\mu \overline{\psi}_\rho)} \overline{\psi}_\rho \, \varepsilon \right] - j^\mu \, \partial_\mu \varepsilon = 0$$

or

$$\left\{ j^\mu - iq \left[\frac{\partial L}{\partial(\partial_\mu \psi_\rho)} \psi_\rho - \frac{\partial L}{\partial(\partial_\mu \overline{\psi}_\rho)} \overline{\psi}_\rho \right] \right\} \partial_\mu \varepsilon$$
$$- \left\{ \partial_\mu \, iq \left[\frac{\partial L}{\partial(\partial_\mu \psi_\rho)} \psi_\rho - \frac{\partial L}{\partial(\partial_\mu \overline{\psi}_\rho)} \overline{\psi}_\rho \right] \right\} \varepsilon = 0 \, .$$

Because, the quantities ε and $\partial_\mu \varepsilon$ are arbitrary, it results that

$$j^\mu = iq \left[\frac{\partial L}{\partial(\partial_\mu \psi_\rho)} \psi_\rho - \frac{\partial L}{\partial(\partial_\mu \overline{\psi}_\rho)} \overline{\psi}_\rho \right] , \tag{6.3.45}$$

and

$$\partial_\mu \, j^\mu = 0 \, . \tag{6.3.46}$$

The equation (6.3.45) determines the expression of the current density from the Lagrangian density, and this current density satisfies the continuity equation (6.3.46). Also, from (6.3.45) it follows that $j^\mu = 0$ for a real particle field ($\overline{\Psi} = \Psi$). Consequently, $L_{\text{int}} = 0$, and this means that a real particle field cannot be coupled with the electromagnetic field. On the other hand, the gauge transformations of the first kind (6.3.39) can be defined only for a complex field such that, the gauge invariance leads to the conclusion that only a complex field can be associated to an electric charge.

Generally, the current density is $(j^\mu) = (\mathbf{j}, \rho)$, so that we may find the density of the electric charge associated to the particle field from the relation

$$\rho = j^4 = iq \left[\frac{\partial L}{\partial(\partial_4 \psi_\rho)} \psi_\rho - \frac{\partial L}{\partial(\partial_4 \overline{\psi}_\rho)} \overline{\psi}_\rho \right] = -2q \, \mathsf{Im} \left[\frac{\partial L}{\partial(\partial_4 \psi_\rho)} \psi_\rho \right] ,$$

the total electric charge being

$$Q = \int \rho \, dx^1 dx^2 dx^3 \, .$$

As a consequence of the Lagrangian invariance with respect to the combined gauge transformations the current density (j^μ) satisfies the continuity equation (6.3.46), which also express the conservation of the total charge density ($dQ/dt = 0$).

In the case of a relativistic electron, of charge $e = -|e|$, we have a fermion field coupled with an electromagnetic field and the interaction Lagrangian is

$$L_{\text{int}} = |e|\, \widetilde{\Psi}\, \gamma^\mu\, \Psi\, A_\mu\,. \tag{6.3.47}$$

Hence, the current density is given by the formula

$$j^\mu = |e|\, \widetilde{\Psi}\, \gamma^\mu\, \Psi\,,$$

while the electric charge density and the total electric charge are, respectively,

$$\rho = j^4 = |e|\, \widetilde{\Psi}\, \gamma^4\, \Psi\,, \quad Q = |e| \int \widetilde{\Psi}\, \gamma^4\, \Psi\, dx^1 dx^2 dx^3\,.$$

3.2 Elementary particle interactions

3.2.1 The quark model

The *Eightfold Way* model turned out to be highly fruitful and its success in predicting new elementary particles pushed on the investigations to a deeper understanding of the rôle of SU(3) group in the world of elementary particles. Thus, a basic theorem from the general theory of SU(3) group (*Cartan's theorem*), states that all the irreducible representations of SU(3) can be obtained by decomposing direct products comprising the two fundamental representations D(1, 0) and D(0, 1). Accordingly, it is reasonable to assume that the elementary particles are composed of *quarks*[3] and *antiquarks* (*Gell-Mann's hypothesis*, 1964), two kinds of fundamental particles associated to the wave functions that span the representation spaces for D(1, 0) and D(0, 1).

The quantum numbers of quarks and antiquarks are determined by the weight diagrams shown in Figs. 2.6 (B) and (C), rescaled by

$$T_3 = \sqrt{3}\, m_1\,, \quad Y = 2\, m_2$$

(see Chap. 6 Sec. 2.2.1). Thus, we obtain the diagrams displayed in Fig. 6.6, where the quark and antiquark triplets have been denoted $\{u(\text{up}),\, d(\text{down}),\, s(\text{strange})\}$ and $\{\overline{u}, \overline{d}, \overline{s}\}$, respectively [4]. It follows that the quarks u and d form an isospin doublet, with $T = 1/2$ and $T_3 = \pm 1/2$, while s belongs to

[3] In 1964 Gell-Mann introduced the concept of quarks, adopting the name from a passage in James Joyce's novel *Finnegans Wake*. The American physicist George Zweig developed a similar theory independently that same year and called his fundamental particles *aces*.

[4] The original symbols for the quark and antiquark triplets have been respectively, $\{p, n, \lambda\}$ and $\{\overline{p}, \overline{n}, \overline{\lambda}\}$.

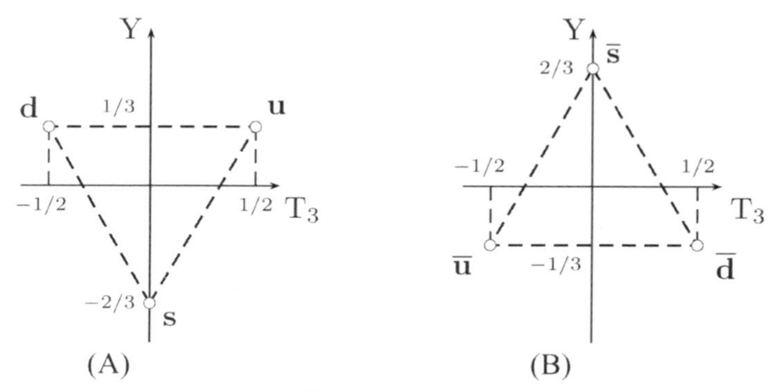

Figure 6.6. The triplets of quarks $J^P = 1/2^+$ (A) and antiquarks $J^P = 1/2^-$ (B).

an isospin singlet $T = T_3 = 0$. The electric charge of each quark is given by the Gell-Mann – Nishijima formula $Q = T_3 + Y/2$, such that $Q_u = 2/3e$, $Q_d = -1/3e$, and $Q_s = -1/3e$, where e represents the electric charge of an electron. In addition, each quark has a baryon number $B = 1/3$ and a spin-parity $J^P = 1/2^+$. From the definition of hypercharge ($Y = B + S$) the strangeness number of each quark is $S_u = S_d = 0$, $S_s = -1$. The antiquarks have opposite T_3, Y, B, S, Q, and $J^P = 1/2^-$.

According to the $\mathsf{SU}(3)$ classification of elementary particles, and the decomposition of the direct products of fundamental representations

$$\mathsf{D}(1,0) \otimes \mathsf{D}(0,1) = \mathsf{D}(0,0) \oplus \mathsf{D}(1,1)\,,$$

$$\mathsf{D}(1,0) \otimes \mathsf{D}(1,0) \otimes \mathsf{D}(1,0) = \mathsf{D}(0,0) \oplus \mathsf{D}(1,1) \oplus \mathsf{D}(1,1) \oplus \mathsf{D}(3,0)\,,$$

and

$$\mathsf{D}(0,1) \otimes \mathsf{D}(0,1) \otimes \mathsf{D}(0,1) = \mathsf{D}(0,0) \oplus \mathsf{D}(1,1) \oplus \mathsf{D}(1,1) \oplus \mathsf{D}(0,3)\,,$$

the quark model postulates that two types of particles exist in nature, comprising either a pair quark-antiquark ($q\,\overline{q}$), or three quarks (qqq), or three antiquarks ($\overline{q}\,\overline{q}\,\overline{q}$). A combination $q\,\overline{q}$ has a baryon number $B = 0$ and an integer spin, which are the characteristics of a meson, while a combination qqq has a baryon number $B = 1$ and a half-integer spin, thus constituting a baryon. Also, three antiquarks produce an antibaryon.

The quark model proposed by Murray Gell-Mann and George Zweig in 1964, was based on the hypothesis that only three types (or *flavours*) of quarks exist. This model was capable to explain all hadronic states observed before 1974. However, a difficulty arises when we try to build the wave function of a baryon resonance like Δ^{++} or Ω^-. For instance

$$|\Delta^{++}\rangle = |\,u\uparrow u\uparrow u\uparrow\,\rangle\,, \tag{6.3.48}$$

where the symbol "$\uparrow$" denotes a spin state $|J = 1/2, J_z = 1/2\rangle$. Hence, we have a situation where three identical quarks are all in the same quantum state. The quarks are fermions, and such a situation is in contradiction with the Fermi-Dirac statistics. Moreover, the baryon resonance Δ^{++} is also a fermion so that its wave function has to be antisymmetric under quark interchange, while the wave function (6.3.48) is totally symmetric. To correct the quark model a new quantum number, called *colour*, has been introduced. Thus, each type of quark can take on three possible colour values, conventionally called red, green, and blue. Antiquarks have the colours antired, antigreen, and antiblue, also known as cyan, magenta, and yellow. The colour quantum number allow us to construct a totally antisymmetric wave function for the baryon resonance Δ^{++}

$$|\Delta^{++}\rangle = \sum_{r,g,b} \epsilon^{rgb} \, | \, u_r \uparrow u_g \uparrow u_b \uparrow \rangle \, . \tag{6.3.49}$$

The three colours associated to each quark can be thought of as being associated to the fundamental representation of an $\mathsf{SU}(3)$ group, while the anti-colours are associated to the contragredient fundamental representation. This colour symmetry group is denoted by $\mathsf{SU}(3)_c$ to distinguish it from the initial $\mathsf{SU}(3)$ group, now called *flavour symmetry group*. The new quark model thus created was based on the $\mathsf{SU}(3) \times \mathsf{SU}(3)_c$ symmetry and the underlying theory of strong interactions due to the colours of quarks was called *quantum chromodynamics* (or simply QCD). Note that, actually, the symmetry group of non-relativistic quantum chromodynamics is

$$\mathsf{SU}(2)_{\text{spin}} \times \mathsf{SU}(3)_{\text{flavour}} \times \mathsf{SU}(3)_c \subset \mathsf{SU}(6) \times \mathsf{SU}(3)_c \, .$$

Later, in 1974, the experimental discovery of the *J/ψ* particle required a new quark "c" (and a new quantum number *charm*), thus leading to the $\mathsf{SU}(4)$ model. Actually, the $\mathsf{SU}(4)$ model has been introduced by S. L. Glashow, J. Iliopoulos, and L. Maiani in 1970, to describe the weak interactions. In this theoretical model the currents are constructed out of four quark fields and interact with a charged massive vector boson.

The Weinberg-Salam model of weak and electromagnetic interactions has been extended to hadrons assuming that we have four quartets of fundamental elementary particles namely, the leptons (e^-, μ^-, ν_e, ν_μ) and three coloured quartets of quarks (and the corresponding quartets of antiparticles), all being treated identically. This is the "Standard Model" of fundamental particles and interactions, based on the $\mathsf{SU}(3) \times \mathsf{SU}(2) \times \mathsf{U}(1)$ internal symmetry group.

In 1977 the discovery of the Υ (upsilon) particle necessitated another quark "b" (*bottom*) with a new quantum number attribute, and, in 1995, the sixth quark "t" (*top*) has been discovered at Fermilab. In total, there are six quarks characterized by the following additive quantum numbers:

	d	u	s	c	b	t
Electric charge	$-\frac{1}{3}$	$+\frac{2}{3}$	$-\frac{1}{3}$	$+\frac{2}{3}$	$-\frac{1}{3}$	$+\frac{2}{3}$
Isospin z-component	$-\frac{1}{2}$	$+\frac{1}{2}$	0	0	0	0
Strangeness	0	0	-1	0	0	0
Charm	0	0	0	$+1$	0	0
Bottomness	0	0	0	0	-1	0
Topness	0	0	0	0	0	$+1$

and six antiquarks. Each quark has spin $1/2$ and baryon number $1/3$, while the antiquarks have spin $1/2$ and baryon number $-1/3$. Also, by convention, each quark is assigned positive parity, so that each antiquark has negative parity. Now, the Standard Model includes the theory of strong interactions (QCD) and the unified theory of electromagnetic and weak interactions (electroweak interactions). The underlying theory of the Standard Model is complicated and beyond the scope of this book. Therefore, we suggest the interested reader to consult the excellent site **http://pdg.lbl.gov**.

3.2.2 Electromagnetic interactions

In Chap. 6 Sec. 2.2.2 we discussed the breaking of $\mathsf{SU}(3)$ symmetry due to a medium-strong interaction, which preserves the isospin symmetry. In the case of electromagnetic interactions, that depend on the electric charge of the involved particles, even the isospin symmetry is no longer an exact symmetry. Consequently, we may write the interaction Lagrangian for elementary particles in the form

$$\mathsf{L} = \mathsf{L}_{\mathrm{s}} + \mathsf{L}_{\mathrm{em}} + \mathsf{L}_{\mathrm{w}} , \tag{6.3.50}$$

where L_{s} is the Lagrangian of strong interactions, characterized by a coupling constant $g^2/4\pi\hbar c \approx 10$, L_{em} represents the Lagrangian of electromagnetic interactions, with the coupling constant $\alpha \equiv e^2/4\pi\hbar c \approx 1/137$, and L_{w} is the Lagrangian of weak interactions, with a coupling constant $G_{\mathrm{F}}(m_p c)^2/\hbar^3 \approx 10^{-5}$, where G_{F} represents *Fermi's constant* and m_p is the proton mass. In (6.3.50) only the Lagrangian of strong interactions L_{s} is invariant relative to the isospin group, therefore the isospin is conserved in such interactions, while in the case of electromagnetic or weak interactions the isospin is not conserved. The most important consequence of isospin symmetry breaking is the mass difference exhibited by the members of an isospin multiplet. If we neglect the weak interactions then, the symmetry group of the interaction Lagrangian

$\mathsf{L} = \mathsf{L}_\mathrm{s} + \mathsf{L}_\mathrm{em}$ is a subgroup of the isospin $\mathsf{SU}(2)$ group, and the transformations of this subgroup preserve the electric charge as well the isospin component T_3. In fact, such transformations are nothing else but rotations about the third coordinate axis in the isospin space. Hence, a preferred direction is set in the isospin space, in the same way as in the case of a magnetic field that defines a preferred direction in the real three-dimensional space (usually, the z-axis).

The electromagnetic interactions are described by Lagrangian densities of the type (6.3.47), associating a current to each particle. Thus, the currents associated to the proton and neutron are respectively, $\widetilde{\mathbf{p}}\,\gamma^\mu\,\mathbf{p}$ and $\widetilde{\mathbf{n}}\,\gamma^\mu\,\mathbf{n}$, where $\mathbf{p}$ and $\mathbf{n}$ are Dirac spinors. But the proton and the neutron are also members of an iso-doublet and, to point out the isospin properties of the currents, we introduce the linear combinations

$$\frac{1}{2}\left(\widetilde{\mathbf{p}}\,\gamma^\mu\,\mathbf{p} + \widetilde{\mathbf{n}}\,\gamma^\mu\,\mathbf{n}\right) = \frac{1}{2}\,\overline{\mathbf{N}}\,\gamma^\mu\,\mathbf{N}\,,$$

that is an iso-scalar, and

$$\frac{1}{2}\left(\widetilde{\mathbf{p}}\,\gamma^\mu\,\mathbf{p} - \widetilde{\mathbf{n}}\,\gamma^\mu\,\mathbf{n}\right) = \frac{1}{2}\,\overline{\mathbf{N}}\,\gamma^\mu\,\boldsymbol{\tau}_3\,\mathbf{N}\,,$$

which represents the third component of the iso-vector $\overline{\mathbf{N}}\,\gamma^\mu\,\boldsymbol{\tau}\,\mathbf{N}/2$. The iso-spinors $\mathbf{N}$ and $\overline{\mathbf{N}}$ have been defined in (6.2.11), but now we have to consider that their components are Dirac spinors, the adjoint spinors being associated to the antinucleons $\overline{p}$ and $\overline{n}$. Accordingly, the general form of the electromagnetic current associated to the nucleon is

$$j^\mu_\mathrm{em} = F_\mathrm{S}\,\frac{1}{2}\,\overline{\mathbf{N}}\,\gamma^\mu\,\mathbf{N} + F_\mathrm{V}\,\frac{1}{2}\,\overline{\mathbf{N}}\,\gamma^\mu\,\boldsymbol{\tau}_3\,\mathbf{N}\,, \tag{6.3.51}$$

where F_S and F_V depend on the four-momentum transfer between nucleons, and are called *form factors*. Because the electric charge is conserved in electromagnetic interactions, it follows that the electromagnetic current satisfies the continuity equation $\partial_\mu j^\mu_\mathrm{em} = 0$, and this means that T_3 is also conserved. The Lagrangian density for the interaction of nucleons with the electromagnetic field (i.e., the $NN\gamma$ coupling) is

$$L_{NN\gamma} = i\,g_{NN\gamma}\,j^\mu_\mathrm{em}\,A_\mu\,, \tag{6.3.52}$$

where A_μ represents the four-potential of the electromagnetic field. In fact, the four-potential (A_μ) is the wave function of the photon (denoted by γ), that satisfies the Klein-Gordon equation $\Box\,A_\mu = 0$.

We may consider the electromagnetic interactions in the $\mathsf{SU}(3)$ model, assuming that the Lagrangian of strong interactions L_s is invariant relative to the $\mathsf{SU}(3)$ transformations, while L_em induces the symmetry breaking (again, we neglect the weak interactions). In this case, the first problem we have to solve is to find the behaviour of the electromagnetic current under the $\mathsf{SU}(3)$ transformations. Thus, the Gell-Mann – Nishijima formula $\mathsf{Q} = \mathsf{T}_3 + \mathsf{Y}/2$ shows that the operator of the electric charge comprises two infinitesimal generators of the $\mathsf{SU}(3)$ group namely,

$$\mathsf{Q} = \sqrt{3}\,\mathsf{H}_1 + \mathsf{H}_2 = \mathsf{F}_3 + \frac{1}{\sqrt{3}}\,\mathsf{F}_8 = \mathsf{X}_1^1\,.$$

The electromagnetic current $j_{\rm em}^{\mu}(\mathbf{r}, t)$ is a conserved current ($\partial_\mu j_{\rm em}^{\mu} = 0$), which satisfies the relation

$$Q = \int j_{\rm em}^{4}(\mathbf{r}, t)\, dx dy dz\,,$$

such that we may infer that the operator $\mathsf{j}_{\rm em}$ has the same behaviour as the Q component of an octet operator. The electromagnetic interactions also preserve the U spin, defined by the decomposition

$$\mathsf{SU}(3) \supset \mathsf{U}(1)_{\mathsf{Q}} \times \mathsf{SU}(2)_{\mathsf{U}}$$

(see (2.3.59)), because the infinitesimal generators of the $\mathsf{SU}(2)_{\mathsf{U}}$ subgroup ($\mathsf{U}_\pm = \sqrt{6}\,\mathsf{E}_{\pm 3}$ and $\mathsf{U}_3 = \mathsf{Y} - \mathsf{Q}/2$) commute with the operator Q. Note that, when we use this decomposition of the $\mathsf{SU}(3)$ group, the electromagnetic current is considered as being scalar relative to $\mathsf{SU}(2)_{\mathsf{U}}$.

By means of the $\mathsf{SU}(3)$ model of electromagnetic interactions we can obtain relations between the magnetic momenta of elementary particles, electromagnetic mass differences, and cross-sections of meson photoproduction. We may also analyse the radiative decays of mesons and baryon resonances.

3.2.3 Weak interactions

Before 1971, the Lagrangian density of weak interactions, introduced in (6.3.50), was considered to be a product of two *weak currents*, analogous to the electromagnetic currents

$$L_{\rm w} = -\frac{G_{\rm F}}{\sqrt{2}}\,\overline{J}_\lambda\, J^\lambda + \text{ Hermitian conjugate}\,, \qquad (6.3.53)$$

where $G_{\rm F}/\sqrt{2}$ represents Fermi's coupling constant. The weak current J^λ comprises two currents; one for the hadrons denoted by h^λ (the *hadronic current*), the other for the leptons ($\mu^\pm$, $e^\pm$, ν_μ and ν_e) denoted by l^λ (the *leptonic current*). The leptonic current has a simple expression in terms of the Dirac spinors associated to the leptons while, the hadronic current depends on the quark wave functions. Both currents are of the V-A (vector minus axial vector) form, for instance

$$l^\lambda = \widetilde{\boldsymbol{\mu}}\,\gamma^\lambda\left(\mathbf{1} + \gamma^5\right)\boldsymbol{\nu}_\mu + \widetilde{\boldsymbol{e}}\,\gamma^\lambda\left(\mathbf{1} + \gamma^5\right)\boldsymbol{\nu}_e\,, \qquad (6.3.54)$$

with $\gamma^5 = \gamma^1\gamma^2\gamma^3\gamma^4$.

Despite its phenomenological character, the model based on (6.3.53) described correctly all experimental results that were inside its natural domain. Thus, the interaction current-current simultaneously satisfies all the theoretical selection rules to restrict the β couplings. This interaction is symmetric, incorporates the physical principle of universality, implies the two-component representation of neutrinos, the conservation of total lepton numbers, and preserves the invariance relative to the combined operation charge conjugation - space inversion, and time-reversal. By means of this theory one can calculate the decay probability of the muon, as well the asymmetry in the direction $\pi \to \mu \to e$. In the case of beta decay one obtains a good agreement with the experimental data, and the model leads to the same results as the experimental measurements of the spin polarisation of the emitted electron.

For instance, in the case of semi-leptonic decays with $\Delta S = 0$

$$n \to p + e^- + \overline{\nu}_e\,, \quad \pi^+ \to \pi^0 + e^+ + \nu_e\,, \quad \pi \to l + \nu_l\,, \quad l = \mu, e\,,$$

the Gell-Mann – Nishijima formula implies $\Delta Q = \Delta T_3$, for the hadronic current. Because $|\Delta Q| = 1$, it results that $|\Delta T_3| = 1$ and, accordingly $\Delta T = 1, 2, \ldots$. A confirmation of the selection rule $\Delta T = 1$ is observed in the interactions neutrino-nucleon at high energies

$$\nu + n \to \mu^- + \Delta^+\,, \quad \nu + p \to \mu^- + \Delta^{++}\,,$$

having the cross-section ratios

$$\frac{\sigma(\nu + n \to \mu^- + \Delta^+)}{\sigma(\nu + p \to \mu^- + \Delta^{++})} = \begin{cases} 1/3 & \text{if} \quad \Delta T = 1\,, \\ 3 & \text{if} \quad \Delta T = 2\,. \end{cases}$$

The electromagnetic current iso-vector with $|\Delta T_3| = 0$, and the two vector currents with $|\Delta T_3| = \pm 1$, constitute the components of an iso-vector current $\overline{\mathbf{N}}\,\gamma^\lambda\,\boldsymbol{\tau}\,\mathbf{N}/2$. The electromagnetic current having $|\Delta T_3| = 0$ is conserved, hence, the isospin invariance implies the conservation of the iso-vector current $\overline{\mathbf{N}}\,\gamma^\lambda\,\boldsymbol{\tau}\,\mathbf{N}/2$. Actually, this is a general result of group theory which states that quantities of the form $\overline{\Psi}\,\mathsf{O}\,\Psi$, where O are the infinitesimal generators of a group representation in the space spanned by the wave functions $\{\Psi\}$, are transformed like vectors, relative to the regular representation of the group, and are called *group currents*. Thus, the quantity $\overline{\mathbf{N}}\,\gamma^\lambda\,\boldsymbol{\tau}\,\mathbf{N}/2$ is a current vector in the Minkowski space, and also a current in the isospin space that is, it transforms like an iso-vector, relative to the regular representation of $\mathsf{SU}(2)$. Also, according to Noether's theorem, the isospin invariance leads to the conservation of these currents.

To introduce the hypercharge ($Y = B + S$) in this model one has to consider the behaviour of the current J^λ under the $\mathsf{SU}(3)$ transformations. The leptonic current remains unchanged, but the hadronic current will depend on the quark wave functions in a chiral $\mathsf{SU}(3) \times \mathsf{SU}(3)$ symmetry model of strong interactions.

The discovery of new elementary particles, as well the investigations in the direction of unifying weak and electromagnetic interactions, resulted in the replacement of the current-current theory by gauge theories, which will be discussed in the next section.

3.2.4 The electroweak theory

The idea of unifying the four basic forces (gravitational, weak, electromagnetic, and strong), so that a single set of equations can be used to predict all their characteristics, is very old and, at present, it is not know whether such a theory can be developed. However, the most successful attempt in this direction is the *electroweak* theory proposed during the late 1960s by S. Weinberg, A. Salam, and S. Glashow. The electroweak theory is a unified theory of electromagnetic and weak interactions, based on the $\mathsf{SU}(2)\times\mathsf{U}(1)$ symmetry. It regards the weak force and the electromagnetic force as different manifestations of a more fundamental force (electroweak), similarly to electricity and magnetism that appear as different aspects of the electromagnetic force. The outstanding success of the electroweak theory was the experimental discovery of the particles $W^{\pm}$ (1982) and Z^0 (1983), predicted theoretically. Here, we will outline the formulation of electroweak theory that, in essence, is a gauge theory. Most of theoretical methods presented here follow the excellent lectures given by J. Iliopoulos, at CERN, in 1975-1976.

The gauge invariance of electromagnetic interactions is a fundamental principle in mathematical modelling of two experimental results, namely: the conservation of electric charge (see Chap. 6 Sec. 3.1.4), and the fact that the rest mass of the photon is zero, expressed by the absence of a term of the form $A_\mu A^\mu$ in the Lagrangian density (6.3.36). In Chap. 6 Sec. 3.1.4, we shown that the interaction between a charged particle and an electromagnetic field is described by the Lagrangian density

$$L(\widetilde{\Psi}, \Psi, A) = \widetilde{\Psi}\,(i\gamma^\mu\,\partial_\mu - m)\,\Psi - \frac{1}{4}\,F_{\mu\nu}F^{\mu\nu} - q\,\widetilde{\Psi}\,\gamma^\mu\,\Psi\,A_\mu\,, \tag{6.3.55}$$

where all the quantities $\widetilde{\Psi}$, Ψ, $F_{\mu\nu}$, and A_μ are functions of $\mathbf{x} = (x^1, x^2, x^3, x^4)$. The first two terms in (6.3.55) represents the free fields, while the last term represents the interaction of fields. This last term is the Lagrangian density for the interaction of the current $j^\mu = -q\,\widetilde{\Psi}\,\gamma^\mu\,\Psi$ and the electromagnetic field A_μ. The Lagrangian density (6.3.55), as well the field equations derived from it, are invariant relative to the gauge transformations

$$\Psi \to \exp\left[iq\,\varepsilon(\mathbf{x})\right]\Psi\,, \quad \widetilde{\Psi} \to \exp\left[-iq\,\varepsilon(\mathbf{x})\right]\widetilde{\Psi}\,, \quad A_\mu \to A_\mu - \partial_\mu\varepsilon(\mathbf{x})\,, \tag{6.3.56}$$

where q is the electric charge of the particle, and $\varepsilon(\mathbf{x})$ is a scalar function that satisfies the Klein-Gordon equation $\Box\,\varepsilon(\mathbf{x}) = 0$. Consequently, by means of Noether's theorem, this invariance property leads to the conservation of the current

$$\partial_\mu j^\mu = 0\,. \tag{6.3.57}$$

The gauge invariance can be introduced in a more systematic way, enabling the generalization to higher symmetries. Firstly, we have to mention that the Lagrangian density used in field theories has the general form $L(\varphi^i, \partial_\mu \varphi^i)$, where $\varphi^i(\mathbf{x})$ are the fields (scalar, vector, tensor, spinor etc.) associated to the particles in the considered physical system. The independent variables are the space-time coordinates (x^μ), while the dependent variables are the fields (wave functions) $\varphi^i(\mathbf{x})$. Assuming that the motion equations (i.e., the Euler-Lagrange equations) are partial differential equations of at most second degree, the Lagrangian density will contain only the first order derivatives of the fields. This means that we are in a similar situation to that of time independent Lagrangians in classical mechanics. Therefore, the symmetry properties of the Lagrangian density are completely determined by the symmetry properties of the fields and their first derivatives.

We will now show that the Lagrangian density (6.3.55) can be obtained starting with the first term, and imposing the gauge invariance. The Lagrangian density of a free fermion field $\Psi(\mathbf{x})$ has the form (6.3.21), that is

$$L_0 = \widetilde{\Psi}(\mathbf{x})\,(i\gamma^\mu\,\partial_\mu - m)\,\Psi(\mathbf{x})\,, \tag{6.3.58}$$

and this expression is invariant under the phase transformation

$$\Psi(\mathbf{x}) \to \exp\,(i\,\varepsilon)\,\Psi(\mathbf{x})\,, \quad \partial_\mu\,\Psi(\mathbf{x}) \to \exp\,(i\,\varepsilon)\,\partial_\mu\,\Psi(\mathbf{x})\,, \tag{6.3.59}$$

where ε is an arbitrary, constant phase. Note that, the second equation in (6.3.59) follows from the first one, and shows that, under a phase transformation, the derivative of the field transforms like the field itself. The group of phase transformations is the one-parameter, unitary, Abelian group $\mathsf{U}(1)$ and, from Noether's theorem, the invariance of the Lagrangian density with respect to this group implies the existence of a conserved current

$$j^\mu = -\widetilde{\Psi}(\mathbf{x})\,\gamma^\mu\,\Psi(\mathbf{x})\,, \quad \partial_\mu\,j^\mu = 0\,, \tag{6.3.60}$$

which follows from (5.2.50), if we substitute $\delta\,\psi_\rho = i\,\varepsilon\,\psi_\rho$, $\delta\,\overline{\psi}_\rho = -i\,\varepsilon\,\overline{\psi}_\rho$, and $\delta\,f^\mu = 0$.

The next step is to assume that the phase ε is a function of $\mathbf{x}$, such that the transformation of the fermion field becomes

$$\Psi(\mathbf{x}) \to \exp\,[i\,\varepsilon(\mathbf{x})]\,\Psi(\mathbf{x})\,. \tag{6.3.61}$$

The group of transformations (6.3.61) is still an Abelian group, like in the previous case, but now the derivative of the field no longer transforms like the field itself because,

$$\partial_\mu \Psi(\mathbf{x}) \to \exp\left[i\,\varepsilon(\mathbf{x})\right] \partial_\mu \Psi(\mathbf{x}) + i\, \exp\left[i\,\varepsilon(\mathbf{x})\right] \Psi(\mathbf{x})\, \partial_\mu\, \varepsilon(\mathbf{x})\,. \qquad (6.3.62)$$

Consequently, the Lagrangian density (6.3.58) is no longer invariant under the transformations (6.3.61). Note that (6.3.61) defines a gauge transformation of the second kind.

To restore the invariance of the Lagrangian density we may use a standard method from differential geometry, that consists of replacing the usual derivative of the field by a covariant derivative which has the property that it transforms under (6.3.61) like the field itself

$$D_\mu \Psi(\mathbf{x}) \to \exp\left[i\,\varepsilon(\mathbf{x})\right] D_\mu \Psi(\mathbf{x})\,. \qquad (6.3.63)$$

We have already seen the definitions of the covariant derivative of a contravariant (see (5.1.35)) and a covariant (see (5.1.33)) vector, in relativity. There, the trick was to introduce the affine connection expressed by Christoffel's symbols of the second kind. Generally, the affine connection speaks of no particular coordinate system but is expressed in some coordinate system. Here, we are looking for the covariant derivative of a spinor and, the methods from differential geometry show that we have to introduce an affine connection related to a gauge field $A_\mu(\mathbf{x})$, which transforms like

$$A_\mu(\mathbf{x}) \to A_\mu(\mathbf{x}) - \frac{1}{q}\,\partial_\mu\, \varepsilon(\mathbf{x})\,. \qquad (6.3.64)$$

where q is a constant. We then define the covariant derivative by

$$D_\mu = \partial_\mu + iq\, A_\mu\,, \qquad (6.3.65)$$

and it is easy to verify that

$$D_\mu \Psi(\mathbf{x}) \to \left[\partial_\mu + iq\, A_\mu - i\,\partial_\mu\, \varepsilon(\mathbf{x})\right] \left[e^{i\,\varepsilon(\mathbf{x})}\, \Psi(\mathbf{x})\right] = e^{i\,\varepsilon(\mathbf{x})}\, D_\mu \Psi(\mathbf{x})\,.$$

The invariance with respect to gauge transformations is thus restored if we replace the operator ∂_μ by D_μ everywhere in L_0, that is

$$\begin{aligned} L_0 \to L_1 &= \widetilde{\Psi}(\mathbf{x})\,(i\gamma^\mu\, D_\mu - m)\,\Psi(\mathbf{x}) \\ &= \widetilde{\Psi}(\mathbf{x})\,(i\gamma^\mu\, \partial_\mu - m)\,\Psi(\mathbf{x}) - q\,\widetilde{\Psi}\,\gamma^\mu\,\Psi\, A_\mu\,. \end{aligned} \qquad (6.3.66)$$

Therefore, the Lagrangian density (6.3.66) is invariant under gauge transformations (6.3.61) and (6.3.64), if L_0 was invariant under (6.3.59).

Finally, if we want to interpret the gauge field $A_\mu(\mathbf{x})$ as the field representing the photon, then we must add to L_1 a term corresponding to its kinetic energy (i.e., the Lagrangian density of the free electromagnetic field). This term has to be by itself gauge invariant, so that we are led to the Lagrangian density (6.3.56), which is nothing else but the Lagrangian density of quantum electrodynamics, obtained by just imposing the invariance under gauge transformations. Note that, the Lagrangian density (6.3.56) does not contain a term proportional to $A_\mu A^\mu$, since such a term is not invariant under the gauge transformation (6.3.64). Therefore, we may say that gauge invariance forces the photon to be massless.

The group of the gauge transformations (6.3.61) is the unitary Abelian group $U(1)$, but the described method can easily be generalized to cases of higher symmetries. We thus consider n fields $\varphi^i(\mathbf{x})$, $i = \overline{1,n}$, and the corresponding Lagrangian density $L_0(\varphi^i, \partial_\mu \varphi^i)$, that is assumed to be invariant under the transformations of a m-parameter group G_0. The infinitesimal transformations of G_0 are defined as

$$\varphi^i(\mathbf{x}) \to \varphi^i(\mathbf{x}) + \varepsilon^\alpha \, (\mathsf{T}_\alpha)^i_j \, \varphi^j(\mathbf{x}) \,, \quad i,j = \overline{1,n}, \quad \alpha = \overline{1,m}, \tag{6.3.67}$$

where ε^α are small parameters, independent of $\mathbf{x}$ (these transformations are called *global transformations*), and T_α are the matrices of the infinitesimal generators of G_0 in the space spanned by $\varphi^i(\mathbf{x})$. The group G_0 can be any global, non-Abelian group ($\mathsf{SU}(2)$, $\mathsf{SU}(3)$ etc.). In this case, the derivatives of the fields transform like the fields themselves

$$\partial_\mu \varphi^i(\mathbf{x}) \to \partial_\mu \varphi^i(\mathbf{x}) + \varepsilon^\alpha(\mathbf{x}) \, (\mathsf{T}_\alpha)^i_j \, \partial_\mu \varphi^j(\mathbf{x}) \,. \tag{6.3.68}$$

Also, by means of Noether's theorem, we obtain m conserved currents.

Let G be the local version of G_0 that is, an m-parameter group of transformations acting on the fields φ^i according to the relation

$$\varphi^i(\mathbf{x}) \to \varphi^i(\mathbf{x}) + \varepsilon^\alpha(\mathbf{x}) \, (\mathsf{T}_\alpha)^i_j \, \varphi^j(\mathbf{x}) \,, \quad i,j = \overline{1,n}, \quad \alpha = \overline{1,m}, \tag{6.3.69}$$

where T_α are the matrices of the infinitesimal generators of G in the space spanned by $\varphi^i(\mathbf{x})$, and $\varepsilon^\alpha(\mathbf{x})$ are the infinitesimal parameters of G. Note that, the elements of G are finite transformations represented by the operators $\exp\left(\varepsilon^\alpha(\mathbf{x}) \, \mathsf{T}_\alpha\right)$, and (6.3.69) are local infinitesimal transformations corresponding to small parameters $\varepsilon^\alpha(\mathbf{x})$. Now, the derivatives of the fields transform as

$$\partial_\mu \varphi^i(\mathbf{x}) \to \partial_\mu \varphi^i(\mathbf{x}) + \varepsilon^\alpha(\mathbf{x}) \, (\mathsf{T}_\alpha)^i_j \, \partial_\mu \varphi^j(\mathbf{x}) + (\mathsf{T}_\alpha)^i_j \, \varphi^j(\mathbf{x}) \, \partial_\mu \, \varepsilon^\alpha(\mathbf{x}) \tag{6.3.70}$$

and it is the last term which breaks the invariance of L_0. The rule for restoring the invariance is the same as in the previous case. Thus, we introduce m gauge fields $W_\mu^\alpha(\mathbf{x})$ that transform like

$$W_\mu^\alpha(\mathbf{x}) \to W_\mu^\alpha(\mathbf{x}) + c_{\beta\gamma}^\alpha\, W_\mu^\gamma(\mathbf{x})\, \varepsilon^\beta(\mathbf{x}) + \frac{1}{q}\, \partial_\mu\, \varepsilon^\alpha(\mathbf{x})\,, \tag{6.3.71}$$

where q is a constant, and $c_{\beta\gamma}^\alpha$ are the structure constants of G, defined by

$$[\mathsf{T}_\beta, \mathsf{T}_\gamma] = c_{\beta\gamma}^\alpha\, \mathsf{T}_\alpha\,.$$

In terms of the gauge fields, the covariant derivative reads

$$D_\mu\, \varphi^i(\mathbf{x}) = \partial_\mu\, \varphi^i(\mathbf{x}) - q\, (\mathsf{T}_\alpha)^i_j\, W_\mu^\alpha\, \varphi^j(\mathbf{x})\,. \tag{6.3.72}$$

We may use the finite transformations of G to write (6.3.69) and (6.3.72) in the matrix form

$$\boldsymbol{\varphi} \to \exp[\varepsilon^\alpha(\mathbf{x})\, \mathsf{T}_\alpha]\, \boldsymbol{\varphi}\,, \quad D_\mu\, \boldsymbol{\varphi} = \partial_\mu\, \boldsymbol{\varphi} - q\, W_\mu^\alpha\, \mathsf{T}_\alpha\, \boldsymbol{\varphi}\,,$$

and verify that the covariant derivative transforms like the fields, that is

$$D_\mu\, \boldsymbol{\varphi} \to \exp[\varepsilon^\alpha(\mathbf{x})\, \mathsf{T}_\alpha]\, D_\mu\, \boldsymbol{\varphi}$$

or, in terms of infinitesimal transformations,

$$D_\mu\, \varphi^i(\mathbf{x}) \to D_\mu\, \varphi^i(\mathbf{x}) + \varepsilon^\alpha(\mathbf{x})\, (\mathsf{T}_\alpha)^i_j\, D_\mu\, \varphi^j(\mathbf{x})\,. \tag{6.3.73}$$

If we replace everywhere in L_0 the derivatives ∂_μ by D_μ, then we obtain a Lagrangian density $L_1(\varphi^i, D_\mu\, \varphi^i)$. Like in the case of $\mathsf{U}(1)$ symmetry, L_1 is invariant under (6.3.69) and (6.3.71), if $L_0(\varphi^i, \partial_\mu\, \varphi^i)$ was invariant relative to the global infinitesimal transformations (6.3.67).

We may also add the kinetic energy term for the gauge fields in the form

$$L_1 \to L = L_1 - \frac{\kappa}{4}\, G_{\mu\nu}^\alpha G_\alpha^{\mu\nu}\,, \tag{6.3.74}$$

where κ is a constant depending on the system of units, and

$$G_{\mu\nu}^\alpha = \partial_\mu\, W_\nu^\alpha(\mathbf{x}) - \partial_\nu\, W_\mu^\alpha(\mathbf{x}) - g\, c_{\beta\gamma}^\alpha\, W_\mu^\beta\, W_\nu^\gamma\,. \tag{6.3.75}$$

Note that for Abelian groups $c_{\beta\gamma}^\alpha = 0$ which simplifies the expressions (6.3.71) and (6.3.75).

The Lagrangian density L_1 does not contain terms of the form $W_\mu\, W^\mu$, even if it contains higher order products, so that the particles associated to gauge fields are massless ($\Box\, W_\mu^\alpha = 0$). The problem is that no massless vector bosons, other than the photon, are known, and it seems that the only useful symmetry models are those based on Abelian groups. However, if we assume that the gauge transformations are accompanied by a translation of fields, then the particles associated to gauge fields become *real particles* having nonzero mass. All these can be achieved by introducing the mechanism of spontaneous symmetry breaking.

A simple example of spontaneous breaking of global symmetry is the case of a complex scalar field $\varphi(\mathbf{x})$, whose equation of motion is determined by the Lagrangian density

$$L_1 = (\partial_\mu \varphi)(\partial^\mu \overline{\varphi}) - \rho^2 \varphi \overline{\varphi} - \lambda (\varphi \overline{\varphi})^2 , \tag{6.3.76}$$

where ρ and λ are two complex numbers. This expression is invariant with respect to the global transformations of $\mathrm{U}(1)$ (in fact, phase transformations)

$$\varphi(\mathbf{x}) \to e^{i\varepsilon} \varphi(\mathbf{x}) . \tag{6.3.77}$$

From (6.3.76) we may derive the Hamiltonian density of the system

$$\mathcal{H}_1 = (\partial_\mu \varphi)(\partial^\mu \overline{\varphi}) + \rho^2 \varphi \overline{\varphi} + \lambda (\varphi \overline{\varphi})^2 , \tag{6.3.78}$$

and infer that the potential function is

$$V(\varphi) = \rho^2 \varphi \overline{\varphi} + \lambda (\varphi \overline{\varphi})^2 . \tag{6.3.79}$$

The ground state of the physical system corresponds to $\varphi = \mathrm{const.}$, when $V(\varphi)$ is minimum. Note that, the existence of a minimum for $V(\varphi)$ requires $\lambda > 0$, otherwise $V(\varphi) \to -\infty$ for $|\varphi| \to \infty$. Therefore, the minimum of $V(\varphi)$ is given by the equation

$$\frac{\partial V}{\partial \varphi} = \left(\rho^2 + 2\lambda |\varphi|^2\right) \overline{\varphi} = 0 ,$$

which has two solutions

$$\overline{\varphi_0} = 0 , \quad |\varphi_0|^2 = -\frac{\rho^2}{2\lambda} .$$

When $\rho^2 \geq 0$, the second solution has to be left out because $|\varphi_0|^2 \geq 0$, such that we have only one minimum of $V(\varphi)$ for $\varphi_0 = 0$ (see Fig. 6.7 (A)), which is invariant under (6.3.77). If $\rho^2 < 0$, the solution $\varphi_0 = 0$ gives a local maximum of $V(\varphi)$ and has to be excluded therefore, we get a solution of the form $|\varphi_0|^2 = |\rho^2|/2\lambda \equiv \alpha^2$ (see Fig. 6.7 (B)). In this case, the magnitude of φ_0 is fixed, but its direction is arbitrary such that, the physical system has an infinite number of ground states laying on the circle of radius α in the complex plane φ (i.e., the ground state becomes degenerated). Starting from any point of this circle we obtain all the other points by transformations of the form (6.3.77), hence all the points on the circle are equivalent. Consequently, we may say that $\rho = 0$ is a critical value such that, for $\rho^2 \geq 0$ we have a symmetric stable solution, while for $\rho^2 < 0$ a *spontaneous symmetry breaking* occurs.

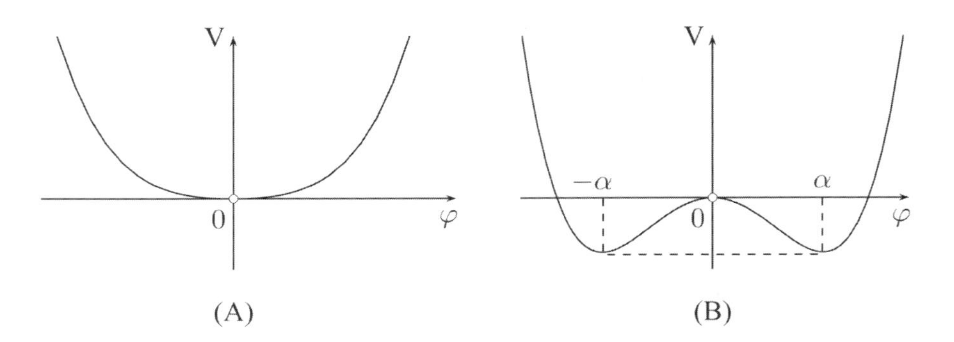

Let us choose a ground state on the circle of radius α, also lying on the real axis in the complex plane φ, and apply a translation of the form

$$\varphi(\mathbf{x}) = \frac{1}{\sqrt{2}}\left[\beta + \varphi_1(\mathbf{x}) + i\,\varphi_2(\mathbf{x})\right] , \tag{6.3.80}$$

where β is a constant, while $\varphi_1(\mathbf{x})$ and $\varphi_2(\mathbf{x})$ are real fields. Substituting into (6.3.76) we obtain the Lagrangian density

$$L_2 = \frac{1}{2}(\partial_\mu\varphi_1)(\partial^\mu\varphi_1) + \frac{1}{2}(\partial_\mu\varphi_2)(\partial^\mu\varphi_2) - \frac{1}{2}(2\lambda\beta^2)\,\varphi_1^2 + \text{coupling terms} . \tag{6.3.81}$$

This expression contains a term proportional to φ_1^2, hence the associated particle has a mass $m_1^2 = 2\lambda\beta^2$. Note that L_1 does not contain terms proportional to φ_2^2, such that for the second field we obtain a mass $m_2^2 = 0$. Consequently, the physical system consists of two interacting scalar particles. The appearance of the massless particle represents a particular case of a general theorem due to Goldstone, stating that to each generator of a spontaneously broken symmetry there corresponds a zero-mass particle, called the *Goldstone particle*.

Now, we consider the same model of (6.3.76), but assuming the invariance under local transformations of $\mathsf{U}(1)$

$$\varphi(\mathbf{x}) \to \exp\left[i\,\varepsilon(\mathbf{x})\right]\varphi(\mathbf{x}) , \quad A_\mu \to A_\mu - \frac{1}{q}\,\partial_\mu\,\varepsilon(\mathbf{x}) . \tag{6.3.82}$$

The corresponding form of the Lagrangian density results by substituting D_μ for ∂_μ and adding the Lagrangian density of the free electromagnetic field

$$L_1 = |(\partial_\mu + iq\,A_\mu)\,\varphi|^2 - \rho^2\,\varphi\overline{\varphi} - \lambda\,(\varphi\overline{\varphi})^2 - \frac{1}{4}\,F_{\mu\nu}F^{\mu\nu} . \tag{6.3.83}$$

As in the case of (6.3.76) we find that for $\lambda > 0$ and $\rho^2 < 0$ there appears a spontaneous breaking of the gauge symmetry. Then, substituting (6.3.80) into

(6.3.83) we obtain the Lagrangian density

$$L_2 = \frac{1}{2}(\partial_\mu \varphi_1)(\partial^\mu \varphi_1) + \frac{1}{2}(\partial_\mu \varphi_2)(\partial^\mu \varphi_2) - \frac{1}{2}(2\lambda\beta^2)\varphi_1^2$$
$$- \frac{1}{4} F_{\mu\nu}F^{\mu\nu} + \frac{q^2\beta^2}{2} A_\mu A^\mu - q\beta A_\mu \partial^\mu \varphi_2$$
$$+ \text{coupling terms}. \quad (6.3.84)$$

Now it seems that the photon has acquired a non-zero mass expressed by the coefficient of $A_\mu A^\mu$. The reason is that the Lagrangian density (6.3.84) contains fields which do not correspond to physical particles. If we consider a more general transformation

$$\begin{aligned} \varphi(\mathbf{x}) &= \frac{1}{\sqrt{2}}[\beta + \chi(\mathbf{x})]\exp[i\zeta(\mathbf{x})/\beta], \\ A_\mu(\mathbf{x}) &= B_\mu - \frac{1}{q\beta}\partial\zeta(\mathbf{x}) \end{aligned} \quad (6.3.85)$$

then, replacing (6.3.85) into (6.3.83) we have

$$L_3 = \frac{1}{2}(\partial_\mu \chi)(\partial^\mu \chi) - \frac{1}{2}(2\lambda\beta^2)\chi^2 - \frac{\lambda}{4}\chi^4 - \frac{1}{4}B_{\mu\nu}B^{\mu\nu}$$
$$+ \frac{q^2\beta^2}{2} B_\mu B^\mu + \frac{q^2}{2} B_\mu B^\mu (2\beta\chi + \chi^2) + \ldots \quad (6.3.86)$$

where $B_{\mu\nu} = \partial_\mu B_\nu - \partial_\nu B_\mu$. The Lagrangian density L_3 describes a vector massive particle (B_μ) and a scalar massive particle (χ), while the field ζ vanished. Even if L_3 is not invariant under any gauge transformations, it clearly shows the particle spectrum of the theory. Thus, in a spontaneously broken gauge symmetry the gauge vector bosons become massive while the massless Goldstone particles disappear. In conclusion, we started with a gauge invariant model, which contains gauge vector bosons of zero-mass, to obtain a model containing only massive particles. In this way, the weak interactions can be described by an interaction Lagrangian density

$$L_{\mathrm{w}} = g\, J^\mu(\mathbf{x})\, W_\mu(\mathbf{x}) + \text{ Hermitian conjugate}, \quad (6.3.87)$$

where J^μ is a sum of two currents (leptonic and hadronic), while W_μ is the wave function of a charged vector boson associated to the gauge fields. If we add the terms corresponding to the electromagnetic interactions, containing the photon wave function A_μ, then the same model is capable to describe the electromagnetic interactions.

The formulation presented here is a simplified version of the model of electroweak interactions, and shows only the main ideas leading to the actual model. Thus, it has been shown that, the model of electroweak interactions requires four particles: two vector bosons $W^{\pm}$ with electric charge for weak interactions, one neutral vector boson Z^0 for neutral currents, and the photon associated to the four-potential A_μ. To each gauge field it corresponds an infinitesimal generator of the symmetry group therefore, the simplest group that satisfies all the requirements is the $\mathsf{SU}(2) \times \mathsf{U}(1)$ group, leading to the model of electroweak interactions introduced by S. Weinberg, A. Salam, and S. Glashow. Now, the model of electroweak interactions is a component of the Standard Model that also includes the theory of strong interactions (QCD).

The masses of vector bosons, predicted by the model of electroweak interactions, are $m_W = (82 \pm 2.4)\,\mathrm{GeV}$, and $m_Z = 75\,\mathrm{Gev}$. In December 1982, the mass of the vector boson has been determined experimentally, its value being of $m_W = (81 \pm 5)\,\mathrm{GeV}$, in excellent agreement with the theoretical prediction. In June 1983, the existence of the third vector boson Z^0 has been confirmed experimentally. Recent experimental values of the masses of vector bosons are $m_W = (80.423 \pm 0.039)\,\mathrm{GeV}$, and $m_Z = (91.1876 \pm 0.0021)\,\mathrm{GeV}$. All these experimental data confirm the validity of the unified model of weak and electromagnetic interactions, and we may infer that there are only three fundamental forces in Nature: strong (nuclear), electroweak, and gravitational.

However, the recent discovery of exotic quark systems (i.e., quark systems formed by more than a pair quark-antiquark, three quarks, or three antiquarks) suggests that the Nature is much more complicated. Thus, in July 2003 nuclear physicists in Japan, Russia and the US have discovered the *pentaquark* (a particle comprising two up quarks, two down quarks and a strange antiquark) having a mass of 1.54 GeV, in agreement with theoretical predictions made by D. Diakonov of the Petersburg Nuclear Physics Institute in 1997. Also, in November 2003 a new sub-atomic particle called X(3872), an exotic meson containing two quarks and two antiquarks, has been discovered by the Belle collaboration at the KEK laboratory in Japan, and confirmed by the CDF collaboration at Fermilab in the US. Consequently, it seems that the current models of strong interactions need to be revised.

REFERENCES

Altmann, S. L. (1991). *Band theory of solids: an introduction from the point of view of symmetry.* Clarendon Press, Oxford.

Anderson, R. L. and Davison, S. M. (1974). A generalization of Lie's "counting" theorem for second order ordinary differential equations. *J. Math. Analysis and Appl.*, 48:301.

Arnold, V. (1976). *Méthodes mathématiques de la mécanique classique.* Mir, Moscow.

Baird, G. E. and Biedenharn, L. C. (1963). On the representations of the semisimple Lie groups, II. *J. Math. Phys.*, 4:1449.

Baird, G. E. and Biedenharn, L. C. (1964a). On the representations of the semisimple Lie groups, III. *J. Math. Phys.*, 5:1723.

Baird, G. E. and Biedenharn, L. C. (1964b). On the representations of the semisimple Lie groups, IV. *J. Math. Phys.*, 5:1730.

Baird, G. E. and Biedenharn, L. C. (1965). On the representations of the semisimple Lie groups, V. *J. Math. Phys.*, 6:1847.

Bateman, H. (1953). *Higher Transcendental Functions.* McGraw-Hill, New York.

Baumann, G. (2000). *Symmetry analysis of differential equations with Mathematica.* Springer / Telos, New York. System requirements for accompanying computer disc: Windows 9x, NT, Me, XP; Mac; UNIX.

Baumslag, B. and Chandler, B. (1968). *Theory and problems of group theory.* McGraw-Hill, New York.

Beju, I., Soós, E., and Teodorescu, P. P. (1976). *Tehnici de calcul vectorial cu aplicaţii (Vector calculus with applications).* Editura Tehnică, Bucharest.

Beju, I., Soós, E., and Teodorescu, P. P. (1983a). *Euclidean tensor calculus with applications.* Abacus Press, Tunbridge Wells, Kent. Also, published in Romanian by Editura Tehnică, Bucharest, 1977.

Beju, I., Soós, E., and Teodorescu, P. P. (1983b). *Spinor and non-Euclidean tensor calculus with applications.* Abacus Press, Tunbridge Wells, Kent. Also, published in Romanian by Editura Tehnică, Bucharest, 1979.

Berestetskii, V. B., Lifshitz, E. M., and Pitaevskii, L. P. (1972). *Théorie quantique relativiste,* volume I. Mir, Moscow.

Beşliu, C., Mihul, A., and Nicorovici, N. (1973). Quark model and strange particle interactions at high energies. *Rev. Roum. Phys.*, 18:1139.

Biedenharn, L. C. (1963). On the representations of the semisimple Lie groups, I. *J. Math. Phys.*, 4:436.

Boerner, H. (1970). *Representations of groups: with special considerations for the needs of modern physics*. North-Holland, Amsterdam, 2nd rev. edition.

Bogoliubov, N. N., Logunov, A. A., and Todorov, I. T. (1969). *Osnovi aksiomaticeskogo podhoda v kvantovoi teorii polea (Basics of axiomatic approach in quantum field theory)*. Nauka, Moscow.

Boyer, C. P. (1974). The maximal "kinematical" invariance group for an arbitrary potential. *Acta Physica Helvetica*, 47:589.

Boyer, C. P. and Peñafiel, M. N. (1976). Conformal symmetry of the Hamilton-Jacobi equation and quantization. *Nuovo Cimento B*, 31:195.

Brillouin, L. (1953). *Wave propagation in periodic structures: electric filters and crystal lattices*. Dover, New York, 2nd edition.

Carmeli, M. (1977). *Group theory and general relativity: representations of the Lorentz group and their applications to the gravitational field*. McGraw-Hill, London.

Cartan, E. (1966). *The theory of spinors*. Hermann, Paris.

Cerulus, F. (1961a). Closed formulae for the statistical weights. *Nuovo Cimento*, 19:528.

Cerulus, F. (1961b). Statistical theory of multiple meson production with angular momentum conservation. *Nuovo Cimento*, 22:958.

Chen, J.-Q., Ping, J., and Wang, F. (2002). *Group Representation Theory for Physicists*. World Scientific, Singapore, 2nd edition.

Creangă, I. and Luchian, T. (1963). *Introducere în calculul tensorial (Introduction to tensor calculus)*. Editura Didactică şi Pedagogică, Bucharest.

Dass, T. (1966). Conservation laws and symmetries. *Phys. Rev.*, 150:1251.

Davis, W. R. (1970). *Classical fields, particles and the theory of relativity*. Gordon and Breach, New York.

Diaz, J. M. (2000). Yet another Lie Mathematica package for Lie symmetries. MathSource, http://library.wolfram.com/infocenter/MathSource/4231/.

Dirac, P. A. M. (1958). *The principles of quantum mechanics*. Clarendon Press, Oxford.

Dragomir, P. and Dragomir, A. (1981). *Structuri algebrice (Algebraic structures)*. Editura Facla, Timişoara, Romania.

Dragoş, L. (1976). *Principiile mecanicii analitice (The principles of analytical mechanics)*. Editura Tehnică, Bucharest.

Eddington, A. S. (1921). *Espace, temps et gravitation*. Hermann, Paris.

Edmonds, A. E. (1957). *Angular momentum in quantum mechanics*. Princeton University Press, Princeton.

Einstein, A. (1905). Zur Elektrodynamik der bewegter Körper. *Annalen der Physik*, 17:891.

Einstein, A. (1916). Die Grundlage der allgemeinen Relativitätstheorie. *Annalen der Physik*, 49:769.

Einstein, A. (1950). *The Meaning of Relativity*. Princeton University Press, Princeton, New Jersey, 3rd rev. and enl. edition.

Eliezer, C., Leach, P., and Prince, G. (1982). Group theory, O.D.E.'s and classical mechanics. Technical report, La Trobe University, Bundoora, Victoria, Australia. A collection of papers produced between 1978 and 1981.

Emmerson, J. M. (1972). *Symmetry principles in particles physics*. Clarendon Press, Oxford.

Fliess, M. (1981). Fonctionnelles causales et indéterminées non-commutatives. *Bull. Soc. Math. de France*, 109:3.

Fulton, T., Röhrlich, F., and Witten, L. (1962). Conformal invariance in physics. *Rev. Mod. Phys.*, 34:442.

Gantmacher, F. P. (1960). *The theory of matrices*. Chelsea, New York.

Gantmacher, F. P. (1970). *Lectures in analytical mechanics*. Mir, Moscow.

Glashow, S. L. (1961). Partial symmetry of weak interactions. *Nuclear Physics*, 22:579.

Glashow, S. L., Iliopoulos, J., and Maiani, L. (1970). Weak interactions with lepton-hadron symmetry. *Physical Review D*, 2:1285.

Goldstein, H. (1956). *Classical mechanics*. Addison Wesley, Cambridge.

Greiner, W. and Müller, B. (1994). *Quantum mechanics: symmetries*. Springer, Berlin, 2nd edition.

Hagen, C. R. (1972). Scale and conformal transformations in Galilean-covariant field theory. *Phys. Rev. D*, 5:377.

Hagiwara *et al.*, K. (2002). Review of Particle Physics. *Physical Review D*, 66:010001+.

Hamel, G. (1949). *Theoretische Mechanik*. Springer, Berlin.

Hammermesh, M. (1962). *Group theory and its applications to physical problems*. Pergamon Press, London.

Havas, P. (1978). The conservation laws of non-relativistic classical and quantum mechanics for a system of interacting particles. *Acta Physica Helvetica*, 51:393.

Heine, V. (1993). *Group theory in quantum mechanics: an introduction to its present usage*. Dover, New York.

Hermann, R. (1975). Sophus Lie's 1880 transformation group paper. In *Lie groups: history, frontiers & applications*. Math. Sci. Press, Brookline, Mass.

Hill, E. L. (1951). Hamilton's principle and the conservation theorems of mathematical physics. *Rev. Mod. Phys.*, 23:253.

Holland, D. F. (1969). Finite transfromations of SU3. *J. Math. Phys.*, 10:531.

Hydon, P. E. (2000). Symmetries and first integrals of ordinary difference equations. *Proc. R. Soc. Lond. A*, 456:2835.

Ianuş, S. (1983). *Geometria diferenţială cu aplicatţii în teoria relativităţii (Differential geometry and its applications to the theory of relativity)*. Editura Academiei, Bucharest.

Ibragimov, N. H. (1985). *Transformation groups applied to mathematical physics*. Reidel, Boston.

Iliopoulos, J. (1976). An introduction to gauge theories. Preprint CERN 76-11, CERN, Geneva.

Ionescu-Pallas, N. (1969). *Introducere în mecanica teoretică modernă (Introduction to modern theoretical mechanics)*. Editura Academiei, Bucharest.

Ionescu-Pallas, N. (1980). *Relativitate generală şi cosmologie (General relativity and cosmology)*. Editura Ştiinţifică şi Enciclopedică, Bucharest.

Jing-Li Fu and Li-Qun Chen (2004). Form invariance, Noether symmetry and Lie symmetry of Hamiltonian systems in phase space. *Mech. Res. Comm.*, 31:9.

Jones, H. (1975). *The theory of Brillouin zones and electronic states in crystals*. North-Holland, Amsterdam, 2nd edition.

Jones, H. F. (1994). *Groups, representations and physics*. Institute of Physics, Bristol.

Kahan, T. (1960). *Théorie des groupes en physique classique et quantique*. Dunod, Paris.

Kästner, S. (1960). *Vektoren, Tensoren, Spinoren*. Akademie-Verlag, Berlin.

Kentaro, Y. (1965). *The theory of Lie derivatives and its applications*. North Holland, Amsterdam. Also P. Noordhaff Ltd., Groeningen.

Lagrange, J. L. (1788). *Mécanique analytique*. Paris.

Lánczos, K. (1970). *Space through the ages*. Academic Press, New York.

Landau, L. D. and Lifshitz, E. M. (1975). *The Classical Theory of Fields*. Pergamon Press, Oxford; Sydney, 4th rev. English edition. Translated from the Russian by Morton Hammermesh.

Landau, L. D. and Lifshitz, E. M. (1977). *Quantum mechanics : non-relativistic theory*. Pergamon Press, Oxford; New York, 3rd rev. and enl. edition. Translated from the Russian by J.B. Sykes and J.S. Bell.

Landau, L. D. and Lifshitz, E. M. (1999). *Mechanics*. Butterworth-Heinemann, Oxford, 3rd edition. Translated from the Russian by J.B. Sykes and J.S. Bell.

Lang, S. (1972). *Differential manifolds*. Addison-Wesley, Reading, Mass.

Lefebvre, J.-Y. and Metzger, P. (1974). Quelques exemples de groupes d'invariance d'équations aux dérivées partielles. *C. R. Acad. Sci., Sér. A*, 279:165.

Levi-Civita, T. (1924). *Questioni di meccanica classica e relativista*. Zanichelli, Bologna.

Levi-Civita, T. (1926). *The absolute differential calculus*. Hafner, New York.

Lévy-Leblond, J.-M. (1971). Galilean group and Galilean invariance. In *Group theory and its applications*, volume II. Academic Press, New York.

Lie, S. (1874). Zur Theorie des Integrabilitätsfaktors. *Christ. Forh., Aar*, page 242.

Lie, S. (1912). *Vorlesungen über Differentialgleichungen mit bekannten infinitesimalen Transformationen*. Teubner, Leipzig.

Lie, S. and Engel, F. (1888-1893). *Theorie der Transformationsgruppen*, volume I-III. Teubner, Leipzig. Reprinted by Chelsea, New York, 1970.

Lifshitz, E. M. and Pitaevskii, L. P. (1973). *Théorie quantique relativiste*, volume II. Mir, Moscow.

Littlewood, D. E. (1958). *The theory of group characters*. Clarendon Press, Oxford.

Lomont, J. S. (1959). *Applications of finite groups*. Academic Press, New York.

Lorentz, H. A. (1916). *The theory of electrons*. Teubner, Leipzig.

Lorentz, H. A., Einstein, A., Minkowski, H., and Weyl, H. (1952). *The principle of relativity*. Dover, New York. Collection of original memoirs of the special and general theory of relativity.

Lucač, I. (1972). The mass formulae in SU(3) - symmetry with elliptical basis. Preprint P2-6690, Joint Institute for Nuclear Research, Dubna, Russia.

Lucač, I. and Toth, L. (1972). On the complete set of observables on the sphere in the three-dimensional complex space. Preprint P2-6689, Joint Institute for Nuclear Research, Dubna, Russia.

Lyubarskii, G. Y. (1960). *The applications of group theory in physics*. Pergamon Press, London.

Mathews, J. and Walker, R. L. (1970). *Mathematical methods of physics*. Addison-Wesley, Redwood City, California, 2nd edition.

McConnell, J. (1965). Introduction to the group theory of elementary particles. *Comm. of the Dublin Inst. Adv. Studies*, Ser. A, 16.

McIsaac, P. R. (1975). Symmetry-induced modal characteristics of uniform waveguides, I, II. *IEEE Trans. Microwave Theory Tech.*, MTT-23:421, 429.

McWeeny, R. (1963). *Symmetry: an introduction to group theory and its applications*. Pergamon Press, Oxford.

Meijer, P. H. E. and Bauer, E. (1962). *Group theory: the application to quantum mechanics*. North-Holland, Amsterdam.

Messiah, A. (1970). *Quantum mechanics*. North-Holland, Amsterdam.

Metzger, P. (1974). Quelques autres exemples de groupes d'invariance d'équations aux dérivées partielles. *C. R. Acad. Sci., Sér. A*, 279:193.

Miller, Jr, W. (1972). *Symmetry groups and their applications*. Academic Press, New York.

Miller, Jr, W. (1977). *Symmetry and separation of variables*, volume 4 of *Encyclopedia of Mathematics and its Applications*. Addison-Wesley, Reading, Mass.

Minkowski, H. (1909). Raum und Zeit. *Phys. Zeitschr.*, 10:104.

Møller, C. (1972). *The theory of relativity*. Clarendon Press, Oxford, 2nd edition.

Morales, A. (1968). Higher symmetries. In *Proc. of the 1968 CERN School of Physics, El Escorial*. Preprint CERN 68-23.

Murnaghan, F. D. (1962). *The unitary and rotation groups*. Spartan Books, Washington, New Jersey.

Naimark, M. A. and Stern, A. I. (1982). *Theory of Group Representations*. Springer, New York.

Nguyen Van Hieu (1967). *Lektsii po teoriia unitarnoi simetrii elementarnykh chastits (Lessons on the theory of unitary symmetry of elementary particles)*. Atomizdat, Moscow.

Nicorovici, N. (1972a). Statistical weights of many-particle systems in unitary-spin space. *Nuclear Physics B*, 36:485.

Nicorovici, N. (1972b). SU(2) statistical weights for many-particle systems. *Rev. Roum. Phys.*, 17:1217.

Nicorovici, N. (1972c). Unitary statistical weights. *Nuclear Physics B*, 44:321.

Nicorovici-Porumbaru, N. (1983). Grupul de invarianţă al ecuaţiei Schrödinger pentru particula libera (The invariance group of the Schrödinger equation for a free particle). *Studii şi Cercetări de Fizică*, 35:3.

Niederer, U. (1972). The maximal kinematical invariance group of the free Schrödinger equation. *Acta Physica Helvetica*, 45:802.

Niederer, U. (1973). The maximal kinematical invariance group of the harmonic oscillator. *Acta Physica Helvetica*, 46:192.

Niederer, U. (1974). The maximal kinematical invariance group of Schrödinger equations with arbitrary potentials. *Acta Physica Helvetica*, 47:167.

Niederer, U. (1978). Schrödinger invariant generalized heat equation. *Acta Physica Helvetica*, 51:220.

Nyayapathi, V., Swamy, V. J., and Samuel, M. A. (1979). *Group theory made easy for scientists and engineers*. J. Wiley & Sons, New York.

Olver, P. J. (1979). *How to find the symmetry group of differential equations*, page 200. Volume 762 of Sattinger, 1979.

Olver, P. J. (1986). *Applications of Lie groups to differential equations*. Springer, New York.

Ougarov, V. (1974). *Théorie de la relativité restreinte*. Mir, Moscow.

Ovsiannikov, L. V. (1982). *Group analysis of differential equations*. Academic Press, New York. Russian edition published as "Gruppovoi analiz differentsial'nykh uravnenii" (Nauka, Moscow, 1978).

Özer, T. (2003). The solution of Navier equations of classical elasticity using Lie symmetry groups. *Mech. Res. Comm.*, 30:193.

Palmieri, C. and Vitale, B. (1970). On the inversion of Noether's theorem in the lagrangian formalism. *Nuovo Cimento A*, 66:299.

Partensky, A. (1971). On the eigenvalues of the invariant operators of the unitary unimodular group SU(n). Preprint LYCEN-7125, Institute de Physique Nucléaire, Université Claude Bernard de Lyon.

Perrin, M. (1975). *Représentations de Schrödinger et de Heisenberg des algébres d'invariance en mécanique quantique*. Thése, Université de Paris, VII.

Plăcinţeanu, I. (1958). *Mecanica vectorială şi analitică (Vector and analytic mechanics)*. Editura Tehnică, Bucharest.

Planck, M. (1906). Das Prinzip der Relativität und die Grundgleichungen der Mechanik. *Verband Deutscher Phys. Ges.*, 5:136.

Racah, G. (1965). *Group theory and spectroscopy*. Springer, Berlin.

Röhrlich, F. (1965). *Classical charged particles*. Addison-Wesley, Reading, Mass.

Roman, P. (1961). *Theory of elementary particles*. North-Holland, Amsterdam.

Rose, M. E. (1995). *Elementary theory of angular momentum*. Dover, New York.

Rosen, J. (1971). Conservation laws gauge-invariant lagrangians in classical mechanics. *American Journal of Physics*, 39:502.

Rosen, J. (1972). Noether's theorem in classical field theory. *Annals of Physics*, 69:349.

Sakoda, K. (2001). *Optical properties of photonic crystals*. Springer, Berlin.

Salam, A. (1968). Weak and electromagnetic interactions. In Svartholm, N., editor, *Elementary Particle Physics*, Proc. Nobel Symposium No 8, page 367. Almqvist and Wiksell, Stockholm.

Sattinger, D. H. (1979). *Group theoretic methods in bifurcation theory*, volume 762 of *Lecture Notes in Mathematics*. Springer, New York.

Schouten, J. A. (1961). *Tensor analysis for physicists*. Oxford University Press, New York.

Schröder, U. E. (1968). Noether's theorem and the conservation laws in classical field theory. *Fortschritte der Physik*, 16:357.

Sharp, R. T. and Kolman, B. (1977). *Group theoretical methods in physics*. Academic Press, New York.

Shokin, Y. I. (1983). *The method of differential approximation*. Springer Series in Computational Physics. Springer, Berlin.

Soper, D. E. (1976). *Classical Field Theory*. J. Wiley & Sons, New York.

Stendel, H. (1967). Zur Umkehrung des Noetherschen Satzes. *Annalen der Physik*, 20:110.

Stephani, H. (1990). *Differential equations: their solution using symmetries*. Cambridge University Press, New York.

Streitwolf, H.-W. (1971). *Group theory in solid-state physics*. Macdonald, London.

Synge, J. L. (1960). Classical dynamics. In *Handbuch der Physik*, volume III/1. Springer, Berlin.

Taylor, E. F. and Wheeler, J. A. (1966). *Space-time physics*. W. H. Freeman and Comp., San-Francisco, 2nd edition.

Teleman, C. and Teleman, M. (1973). *Elemente de teoria grupurilor cu aplicaţii în topologie şi fizică (Elements of group theory with applications in topology and physics)*. Editura Ş tiinţifică, Bucharest.

Teodorescu, P. P. (1984-2002). *Sisteme mecanice: modele clasice (Mechanical systems: classical models)*, volume I-IV. Editura Tehnică, Bucharest.

Teodorescu, P. P. and Ille, V. (1976-1980). *Teoria ealsticităţii şi introducere în mecanica solidelor deformabile (Theory of elasticity and introduction to mechanics of deformable bodies)*, volume I-III. Editura Dacia, Cluj-Napoca, Romania.

Teodorescu, P. P. and Nicorovici-Porumbaru, N. (1984). On some applications of symmetries in mechanics. Preprint nr. 74, The Seminar of Geometry and Topology, Faculty of Natural Sciences, University of Timişoara, Romania.

Teodorescu, P. P. and Nicorovici-Porumbaru, N. (1985). *Aplicaţii ale teoriei grupurilor în mecanică şi fizică (Applications of the theory of groups in mechanics and physics)*. Editura Tehnică, Bucharest.

Teodorescu, P. P. and Toma, I. (1989). An application of LEM to the postcritical study of a Bernoulli - Euler bar. *Rev. Roum. de Math. Pures et Appl.*, 34:591.

Teodorescu, P. P. and Toma, I. (2000). Non-linear damped pendulum treated by linear equivalence. *Mech. Res. Comm.*, 27:373.

Toma, I. (1980). On polynomial differential equations. *Bull. Math. Soc. Sci. Math. de la Roumanie*, 24(72):417.

Toma, I. (1995). *The linear equivalence method and its applications*. Flores, Bucharest. (In Romanian).

Trautmann, A. (1965). *Foundations and current problems of general relativity*. Prentice-Hall, Englewood Cliffs, New Jersey.

van der Waerden, B. L. (1974). *Group theory and quantum mechanics*. Springer, Berlin. Translation of "Die gruppentheoretische Methode in der Quantenmechanik".

Vilenkin, N. J. (1968). *Special functions and the theory of group representations*, volume 22 of *Translations of Mathematical Monographs*. American Mathematical Society, Providence, Rhode Island.

von Laue, M. (1922). *La theorié de la relativité: le principe de la relativité de la transformation de Lorentz*, volume I. Gauthier-Villars, Paris.

von Laue, M. (1926). *La theorié de la relativité: la relativité générale et la théorie de la gravitation d'Einstein*, volume II. Gauthier-Villars, Paris.

Vrănceanu, G. (1960-1965). *Leçons de géométrie différentielle*, volume I-III. Editions de l'Académie, Bucharest, Gauthier-Villars, Paris.

Vrănceanu, G. and Mihăileanu, N. (1968). *Introducere în teoria relativităţii (Introduction to the theory of relativity*. Editura Tehnică, Bucharest.

Vrănceanu, G. and Teleman, K. (1967). *Geometrie euclidiană, geometrii neeuclidiene, teoria relativităţii (The Euclidean geometry, non-Euclidean geometries, the theory of relativity)*. Editura Tehnică, Bucharest.

Weinberg, S. (1967). A model of lepton. *Phys. Rev. Lett.*, 19:1264.

Weinberg, S. (1972). *Gravitation and cosmology: principles and applications of the general theory of relativity*. John Wiley & Sons, New York.

Weyl, H. (1925). Theorie der Darstellung kontinuierlicher halb-einfacher Gruppen durch lineare Transformationen, I. *Math. Zeitschr.*, 23:271.

Weyl, H. (1926). Theorie der Darstellung kontinuierlicher halb-einfacher Gruppen durch lineare Transformationen, II, III, Nachtrag. *Math. Zeitschr.*, 24:328, 377, 789.

Weyl, H. (1950). *The theory of groups and quantum mechanics*. Dover, New York.

Weyl, H. (1952a). *Space, Time and Matter*. Dover, New York.

Weyl, H. (1952b). *Symmetry*. Princeton University Press, Princeton, New Jersey.

Weyl, H. (1964). *The classical groups: their invariants and representations*. Princeton University Press, Princeton, New Jersey, 2nd edition.

Wigner, E. (1959). *Group theory and its applications to the quantum mechanics and atomic spectra*. Academic Press, New York.

Wolfram, S. (1996). *The Mathematica Book*. Wolfram Media/Cambridge University Press, Champaign, IL, 3rd edition.

Yóurgrau, W. and Mandelstam, S. (1955). *Variational principles in dynamics and quantum mechanics*. Sir Isaac Pitman & Sons, London.

INDEX

Abelian
  group, 9, 12–13, 15, 23, 32, 52
  monoid, 8
  semigroup, 8
Ado's theorem, 48
Automorphism
  inner, 12, 35
  of group, 12–13, 15, 242, 282

Bernoulli's theorem, 265
Bianchi's identities, 287
Binary
  operation, 5
  relation, 15
Bracket
  Lagrange, 209
  Poisson, 46, 209

$C_{3v} \simeq S_3$, 33
Canonical
  coordinates, 207
    variations of, 220
  equations, 208, 210, 337
  transformations, 215, 217
    canonical matrix, 219
    complete, 215, 217,
    generating function, 218
    infinitesimal, 219
    Lagrange brackets, 218
    Poisson brackets, 218
    Symplectic group, 219, 238–239
    Symplectic matrix, 219
    univalent, 219
Cardinality, 9, 16, 18
Cartesian product, 5, 15, 267–268, 271
Casimir operator, 138
Cayley's model, 280
Cayley's theorem, 19–20
Central functions, 150
Centre, 15, 36, 118
Character of a representation, 31, 150
Christoffel symbols
  of the first kind, 205, 286, 325
  of the second kind, 205, 285–286, 325, 416
Class, 13
  conjugacy, 13
  equivalence, 16–17
Clebsch-Gordan
  coefficients, 160
    SO(3), 163
    SU(2), 159, 161, 163, 362, 365, 380
    SU(3), 165, 167
  series, 160
Commutator, 46
Composition law
  external, 7
  internal, 4
Conformal group, 321
  invariance of physical laws, 324
  linear representations, 322
  subgroups, 321
Conservation laws, 245
  angular momentum, 247
  centre of mass motion, 250
  conservation of electric charge, 317
  energy, 248
  general theory of relativity, 328
  momentum, 246
  rigid motions, 297
    general forms, 297, 299
    generator, 298
    Killing vectors, 299
    Noether's theorem, 298
  Special Relativity, 295
    angular four-momentum, 296
    centre of mass motion, 297
    four-momentum, 295–296
    general forms, 300
    Lorentz invariance, 295
Conserved quantities
  canonical transformations, 234

m-parameters Lie group, 235
one-parameter Lie group, 234
Contravariant
tensor, 284
vector, 283
Coordinates
canonical, 207
cyclic, 234
generalized, 202
normal, 265, 268–269
Coset, 16–17
Covariant
derivative, 285–286, 299, 416, 418
tensor, 284
vector, 284
Curvature, 280
Cyclic group, 14, 17

Derivative
Euler-Lagrange, 223
Lie, 288
Dirac's equation, 397
adjoint spinor, 398
charged particles, 399
Dirac's matrices, 398
Hamiltonian density, 397
Lagrangian density, 397, 399
symmetry group, 401
conformal group, 403
maximal symmetry group, 403
Poincaré's group, 403
representations of the first kind, 402
representations of the second kind, 403
Direct product of groups, 241
Direct product of representations, 30
Direct sum of representations, 25

Eightfold Way = unitary symmetry, 384
Einstein's equations, 328
Einstein's tensor, 287, 327
Electroweak theory, 414, 421
gauge transformations, 414–415
affine connection, 416
covariant derivative, 416
field derivative, 415–416
Lagrangian density, 415, 417
m-parameter groups, 417
covariant derivative, 418
gauge fields, 418
global transformations, 417
Lagrangian density, 418
local transformations, 417
spontaneous symmetry breaking, 419
gauge symmetry, 420
gauge vector bosons, 421
Goldstone particles, 420
Lagrangian density, 419, 421
vector bosons, 422
Element
conjugate, 13
idempotent, 6
identity, 6
neutral, 6
null, 6
order of an, 18
regular, 9
symmetrizable, 7
Endomorphism, 11
Epimorphism = surjective homomorphism, 11
Equation of conservation, 245
classical mechanics, 245
Equivalence relation, 15–16
Euclidean geometry, 279
Euclidean group, 240
n-dimensional, 240
faithful representation, 240
semidirect product, 242
rigid motions, 280
three-dimensional, 244
commutation relations, 244
infinitesimal generators, 244
semidirect product, 244
two-dimensional, 242
commutation relations, 243
faithful representation, 243
infinitesimal generators, 243
semidirect product, 243
Euler's equations = Lagrange's equations, 203

Factor group, 17
Factor set, 16
Faithful representation, 20
First integrals, 209–211

Galilei's group, 249
centre of mass motion, 250
infinitesimal transformations, 249
Galilei-Newton group, 256
infinitesimal generators, 257–258
invariants, 258
Lagrangian form, 258
Lie algebra, 257
matrix representation, 256
particle intrinsical energy, 259
subgroups, 257
Gauge transformations, 206, 315, 403
conservation of charge density, 407
current density, 406
group of, 405
invariant Lagrangian, 405
Noether's theorem, 405
of the first kind, 405
of the second kind, 404
second kind, 403
Gaunt's coefficients, 139, 163
General theory of relativity, 324
conformal transformations, 330

Weyl's tensor, 331
conservation laws, 330
Einstein's equations, 329
energy and momentum conservation, 328
field conservation laws, 331
integral form, 333
strong conservation laws, 332
geodesic equations, 325
geodesics, 324
gravitational field, 326
field equations, 327–328
Lagrangian, 327
Maxwell's equations, 328
rigid motions, 329
Schwartzschild metric, 330
angular momentum, 330
Killing equations, 330
total energy, 330
Generalized
accelerations, 206
coordinates, 202
momenta, 207
velocities, 202
Geometry
Cayley's model, 280
rigid motions, 281
the Absolute, 280
elliptical, 280
Euclidean, 279
hyperbolic, 280
metric of, 279
metric tensor, 283
Minkowskian, 290
non-Euclidean, 279
parabolic, 279
Riemannian, 279
GL(n, C), xiii, 108
elements of, 108
infinitesimal generators, 108
differential operators, 108
matrix form, 108
Lie algebra, 108
Lie subgroups, 109
GL(n, R), xiv, 109
Gravitational field, 326
field equations, 327–328
Lagrangian, 327
Group
Abelian, 9, 12–13, 15, 23, 32, 52
automorphism, 12–13, 15, 242, 282
commutative, 9
cyclic, 14, 17
definition, 8
direct product, 241
Euclidean, 240
factor, 17
finite, 9, 18
Lie, 43
of transformations, 10, 133
order of a, 9
permutation, 18
rigid motions, 288
rotation, 67
semidirect product, 242
semisimple, 17
simple, 17
symmetric, 18
Symplectic, 235
topological, 41
compact, 41–42, 52
noncompact, 41
Groupoid, 8

Hamilton's equations = canonical equations, 208
Hamilton's principle, 202
Hamilton-Jacobi equation, 212, 295, 304, 335
conformal group O(5, 2), 263
subgroup structure, 263
conformal group O(n+2, 2), 261–262
subgroup structure, 262
differential operators, 260
maximal symmetry group, 259–260
n-dimensional space, 260
three-dimensional space, 259
Hamiltonian, 207
Homeomorphism, 110
Homomorphism, 11
bijective, 11
canonical, 17
injective, 11
surjective, 11
Hurwitz integral = invariant integral, 42

Identical representation, 20
Identity representation, 20
Inner product, 52
Integrals of motion = first integrals, 209
Invariant integral, 41
Invariant subspace, 21
Inversion, 63
Irreducible representation, 22
Isomorphism, 12
Isoscalar factors, 165

Killing's equations, 288
Killing's vectors, 288
Klein-Gordon equation, 400
symmetry group, 401
conformal group, 401
maximal symmetry group, 403
Poincaré's group, 401

Lagrange brackets, 209
Lagrange's equations, 203, 205
Lagrange's theorem, 17
Lagrangian action, 202
Lagrangian, 202

gravitational field, 327
Maxwell's equations, 306, 316
two-particle interaction, 251
vibrating systems, 270
Legendre's transform, 207
Lie algebra, 50
dimension of, 51
Casimir operator, 138
M(n), 107
rank, 52
semisimple, 51
simple, 51
Lie derivative, 288
Lie differential, 288
Lie group, 43
dimension of, 43
infinitesimal generators, 46
infinitesimal transformations, 48
linear transformations, 48
m-parameters, 234
conserved quantities, 235
one-parameter, 230
canonical transformations, 231, 233–234
conserved quantities, 234
parameter group, 48
parameters, 43
rank, 52
structure constants, 46
Lie subalgebra, 51
Cartan, 51
commutative, 51
Lie theorems, 47
Lorentz gauge, 315
conservation of electric charge, 317
Lorentz group, 113–114, 116, 305
classes, 116
infinitesimal generators, 121
inhomogeneous Lorentz group, 116
invariant integral, 119
Lorentz transformations, 112, 114
homogeneous, 116
improper, 116
improper, non-orthochronous, 117
improper, orthochronous, 117
inhomogeneous, 116
proper, 116
proper, non-orthochronous, 117
proper, orthochronous, 116
special, 113
parametrization, 118
subgroup of proper rotations, 117

M(n) Lie algebra, xiii, 107
Maschke's theorem, 52
Matrices, 61
Matrix representation, 21
Maxwell's equations, 302, 328
conformal group, 321
electromagnetic field tensor, 305
four-vector potential, 303
Lagrangian density, 306, 316
Metric tensor, 283
infinitesimal transformations, 287
Lie derivative, 288
Lie differential, 288
Minkowskian geometry, 290
Minkowskian space, 111, 290
Monoid, 8–9
Monomorphism = injective homomorphism, 11

Natural Units, 338
Noether's theorem
direct, 225, 227, 245
reciprocal, 229, 253
SO(3), 254
time translation, 255
translations, 253
Special Relativity, 309–310
Non-Euclidean geometry, 279

O(2, 1), 282
infinitesimal generators, 282
matrix representation, 282
O(3), 69
conjugacy classes, 70
functions defined on, 74
invariant integral, 75
parametrization, 74
subgroups = point groups, 70
O(n), xiv, 110
Order
of a group = cardinality, 9
of an element, 18
Ordinary differential equations, 176
linear equivalence method, 177
exponential mapping, 178
Lie derivative, 179
linear equivalent partial differential equation, 179
linear equivalent system of differential equations, 180
nonlinear damped pendulum, 183
system of first order, 176
invariance, 176
reduction, 177

Partial differential equations, 186
dependent variables, 186
independent variables, 186
infinitesimal generators, 187
k-th prolongation, 187, 189
invariance, 187
method of separation of variables, 189
Cartesian coordinates, 189
Helmholtz equation, 189
Lie algebra, 192
polar coordinates, 190

separable system of coordinates, 194–195
symmetry group, 190
symmetry operators of the second order, 193
similarity (invariant) solutions, 196
axially-symmetric problem, 199
infinitesimal generators, 197
Navier equation, 196
solution graph, 186
symmetry group, 187–188
Partition, 15–17
Pauli spin matrices, 51, 77, 87, 104, 361, 398
Permutation, 18
group = symmetric group, 18
Phase space, 207
Poincaré's group, 116, 318
infinitesimal generators, 318–319
Killing vectors, 318
structure constants, 320
subgroups, 320
Point groups, 70
Point transformations, 213
canonical, 216
Symplectic group, 219
invariance of kinetic energy, 214
invariance of Lagrange's equations, 214
Poisson brackets, 46, 209
Poisson-Jacobi identity, 211
Projection operators, 173
Pseudo-sphere, 280
tractrix, 280

Quantum mechanics
angular momentum operator, 358
commutation relations, 359
eigenvalues, 359
eigenvectors, 359
rotation group, 359
spherical harmonics, 359
coordinate operators, 337
Dirac's equation, 397
energy operator, 337
Hamiltonian operator, 338
eigenfunctions, 361
invariance under rotations, 356
Wigner's theorem, 358
invariance, 351
invariant function, 352
invariant operator, 353
wave equation, 353
irreducible tensor operator, 363
spherical harmonics, 364
spin operators, 363
Klein-Gordon equation, 400
momentum operator, 337
Natural Units, 338
Schrödinger's equation, 335
spin operators, 360
spin, 360
Pauli's matrices, 361
spinors, 361
total angular momentum, 362
wave function, 337
Quark model, 407
antibaryons, 408
baryons, 408
charm, 409
SU(4) symmetry, 409
colour, 409
chromodynamics, 409
electroweak theory, 409
exotic quark systems, 422
mesons, 408
Standard Model, 409
Quaternions, 78
Quotient group = factor group, 17
Quotient set = factor set, 16

Relation
binary, 15
congruence, 17
equivalence, 15
Representation, 20
adjoint, 24–25
character of, 31
dimension of, 20
equivalent, 22
faithful, 20
fully reducible, 28
identical, 20
identity, 20
irreducible, 22
matrix form, 21
multiplicity of a, 28
operators, 20
reducible, 22
regular, 29
left, 133
right, 133
space, 20
triangular, 28
unitary, 52
Ricci's scalar, 287, 327
Ricci's symbol, 49
Ricci's tensor, 287
Ricci's theorem, 286
Riemann-Christoffel tensor, 286
Riemannian geometry, 279
Riemannian space, 283
Bianchi's identities, 287
curvature tensor, 286
Einstein's tensor, 287
group of rigid motions, 288
isometries, 288
Killing's equations, 288
Killing's vectors, 288
Ricci's scalar, 287

Ricci's tensor, 287
Rigid motions
Cayley's model, 281
conservation laws, 297, 299
Euclidean group, 280
O(2, 1), 282
infinitesimal generators, 282
matrix representation, 282
Poincaré's group, 318
proper Lorentz transformations, 281
Rotation, 62
active, 68
group = SO(3), 67
improper, 63
operator, 63
passive, 68
proper, 63

$S_3 \simeq C_{3v}$, 33
centre, 36
character table, 39
conjugacy classes, 35
cosets, 35
factor group, 36
internal composition law, 34
regular representation, 36, 272
representations, 37
subgroup structure, 34
vibrating systems, 269–270
Scalar product = inner product, 52
Sch_3, 339
conservation laws, 348
global, 351
local, 349
elements, 343
Galilei-Newton group, 345
infinitesimal generators, 341
isochronous Galilei-Newton group, 346
Lie algebra, 345
projective representations, 344
proper rotations, 346
pure Galilean transformations, 346
SL(2, R), 346
subgroups, 342
Schrödinger's equation, 335
canonical equations, 337
Hamiltonian density, 337
Lagrangian density, 336
Lagrangian formulation, 336
symmetry group, 339
global conservation laws, 351
local conservation laws, 349
n-dimensional, 346
three-dimensional = Sch_3, 339
wave function, 337
Schur's lemmas, 23
Schur-Auerbach theorem, 52
Schwartzschild metric, 330
Semidirect product of groups, 242
Semigroup, 8
SL(n, C), xiv, 109
SL(n, R), xiv, 109
Small oscillations, 264
Bernoulli's theorem, 265
decomposition of representations, 266
degeneracy, 267
distinct frequencies, 267
Lagrangian invariance, 266
normal coordinates, 265, 268–269
normal mode, 264
resonant frequencies, 264
secular equation, 264
similarity transformation, 269
Sylvester's theorem, 264
symmetry considerations, 265
vibrating systems, 269
SO(3), 67, 138
conjugacy classes, 67
functions defined on, 74
infinitesimal generators, 75
differential operators, 124
invariant integral, 75
parametrization, 70
Eulerian, 70, 152
exponential (canonical), 71
representations, 88, 123
characters, 154
Clebsch-Gordan coefficients, 163
left regular, 138
unitary matrices, 153
spherical harmonics, 138, 151
stationary subgroup, 138
SO(n), xiv, 110, 126
infinitesimal generators, 127
differential operators, 127
invariant integral, 126
parametrization, 171
representations, 127
rotations in E_n, 126
Special Relativity, 289
action integral, 291
conformal group, 324
conservation laws, 295
angular four-momentum, 296
centre of mass motion, 297
four-momentum, 295–296
general forms, 300
integrals of motion, 301
Killing vectors, 300
mass-energy relation, 302
energy-momentum formula, 292
equations of motion, 306
field equations, 307–308
four-momentum, 294
gauge invariance, 315

gauge transformations, 315
generalized velocity, 293
Hamilton's principle, 293
Hamilton-Jacobi equation, 295
Hamiltonian, 292
Lagrangian, 291
Lorentz gauge, 315
conservation of electric charge, 317
Maxwell's equations, 302
electromagnetic field tensor, 305
four-vector potential, 303
generalized momentum, 304
Hamilton's principle, 304
Hamilton-Jacobi equation, 304
Hamiltonian, 304
Lagrangian action, 303, 306
Lagrangian density, 306, 316
Lorentz invariants, 305
Noether's theorem, 309–310
Minkowskian space, 290
rotations in M_4, 312
symmetry transformations, 309
translations in M_4, 310
Speiser's method, 167
Speiser's theorem, 168
SU(2) symmetry model, 366, 371
π-nucleon interaction, 379
charge conjugation, 374
cross-section, 379
decay of resonances, 377
Gell-Mann – Nishijima formula, 372
isospin multiplets, 372
baryon triplet, 373
mesons triplet, 372
nucleons doublet, 372
isospin, 371
Lagrangian of strong interactions, 376
multiple production, 382
nucleon-antinucleon system, 374–375
scattering amplitudes, 379
SU(2) statistical weights, 366
identical particles, 371
multiplicities, 366
projection operator, 369
rotation matrices, 368
SU(2), 76, 373
elements of, 87
functions defined on, 82
scalar product, 86
infinitesimal generators, 86
Cartan basis, 104
differential operators, 128–129
invariant integral, 83
Cayley-Klein parametrization, 84
Eulerian parametrization, 86
exponential parametrization, 85
Lie algebra, 130, 140
Cartan basis, 130
Cartan subalgebra, 140
Casimir operator, 141
parametrization, 76
Cayley-Klein, 76
Eulerian, 81
exponential, 78
representations, 88, 127, 373
characters, 156
Clebsch-Gordan coefficients, 159, 161, 163, 362, 365, 380
decomposition of direct product, 160
harmonic functions, 141
matrix elements, 156
regular, 154
unitary irreducible, 141, 154
Wigner's 3-j symbol, 163
SU(3) symmetry model = unitary symmetry, 383
SU(3), 89
canonical decompositions, 105
elements of, 93, 106
infinitesimal generators, 90
Cartan basis, 94
differential operators, 130, 132
Gell-Mann basis, 92
matrices of, 102
Okubo basis, 92
invariant integral, 107
Lie algebra, 90
Cartan basis, 94
Cartan subalgebra, 93
Casimir operator, 143
elliptical parametrization, 145
Gell-Mann basis, 92
invariant operators, 143
Okubo basis, 92
root diagrams, 93
structure constants, 93
parametrization, 104
elliptical, 132
representations, 89
basis labels, 156
Clebsch-Gordan coefficients, 165, 167
direct product, 164
elliptical coordinates, 148
fundamental, 90, 101
irreducible, 100
isoscalar factors, 165
matrix elements, 157, 166
regular, 101
Speiser's method, 167
Speiser's theorem, 168
subgroup structure, 104
weight diagrams, 96
dominant weight, 100
equivalent weights, 100
Weyl reflections, 100

SU(n), xiv, 109–110
  canonical decomposition, 111
  class parameters, 172
    character of fundamental representation, 173
  elements of, 110
Subgroup, 13
  index of, 16
  invariant, 14–15
Sylvester's theorem, 264
Symmetric group, 18, 133
  of degree 3, 33
Symmetry group
  partial differential equations, 187–188
  symmetry transformations, 224
Symmetry transformations, 223–224, 245
  divergence transformations, 223
  equation of conservation, 225
  equations of motion, 221
  gauge transformations, 223
  Hessian, 227
  infinitesimal transformation, 221
  integral of motion, 228
  Jacobian, 221
  Lagrangian variation, 228
  Noether's theorem, 225, 227
  reciprocal of Noether's theorem, 229
  rotations, 246
  Special Relativity, 309
  symmetry group, 224
  time translations, 248
  translations, 245
Symplectic group, 235
  canonical transformations, 238–239
  elements, 235, 237
  representations, 237
  Symplectic matrices, 237

Tensor
  contravariant, 284
  covariant, 284
  curvature, 286
  Einstein's, 287
  electromagnetic field, 305
  metric, 283–284
    Ricci's theorem, 286
  Ricci's, 287
  Riemann-Christoffel, 286
  Weyl's, 331
Topological group, 41
  central functions, 150
  compact, 41–42
  direct product of representations, 159
    Clebsch-Gordan coefficients, 160
    Clebsch-Gordan series, 160
  functions defined on a, 74
  noncompact, 41
  projection operators, 173
  representation character, 150
Topological space, 40
Transformation group, 10, 133
  effective, 133
  invariant operators, 137
  regular representation, 133
    left, 133
    right, 133
  representations, 135
    of class 1, 136
  spherical functions, 136
  stationary subgroup, 134
  transitive, 134
  translations, 133
    left, 133
    right, 133
  zonal harmonics, 136
Transformation, 9
  group, 10
  inverse of a, 10
Two-particle interaction
  Galilean invariance, 252
  invariance of Lagrangian, 251
  Lagrangian, 251

U(n), xiv, 109–110
  class parameters, 172
    character of fundamental representation, 173
    invariant integral, 173
  parametrization, 170
Unitary representation, 52
Unitary symmetry, 383
  broken symmetry, 386
    cross-sections, 391
    decuplet mass formula, 390
    elliptical coordinates, 388
    Gell-Mann-Okubo mass formula, 387
    interaction amplitudes, 391
    octet mass formula, 390
  electromagnetic interactions, 410, 412
    $NN\gamma$ coupling, 411
    Lagrangian density, 411
  Gell-Mann – Nishijima formula, 412
  multiplets, 384
    baryon octet, 385
    decuplet of baryon resonances, 385
    meson nonets, 384
  quark model, 407
  strong interactions, 390
    cross-sections, 391
    Lagrangian density, 392
  SU(3) statistical weights, 393, 395
    identical particles, 396
    multiplicities, 393
    projection operator, 395
    transition probability, 395
  weak interactions, 412
    conserved currents, 413
    hadronic current, 412

Lagrangian density, 412
leptonic current, 412
semi-leptonic decays, 413

Vector
contravariant, 283
covariant, 284
Vibrating systems, 269
broken symmetry, 274
eigenvalues, 276–277
representations, 275
Lagrangian, 270
rotations, 271
$S_3 \simeq C_{3v}$, 269–270
eigenvalues, 273
normal mode, 274
regular representation, 272
similarity transformation, 273
small oscillations, 269

Weyl's tensor, 331
Wigner's 3-j symbol, 163
Wigner's theorem, 358
Wigner-Eckart theorem, 365

Fundamental Theories of Physics

Series Editor: Alwyn van der Merwe, University of Denver, USA

1. M. Sachs: *General Relativity and Matter.* A Spinor Field Theory from Fermis to Light-Years. With a Foreword by C. Kilmister. 1982 ISBN 90-277-1381-2
2. G.H. Duffey: *A Development of Quantum Mechanics.* Based on Symmetry Considerations. 1985 ISBN 90-277-1587-4
3. S. Diner, D. Fargue, G. Lochak and F. Selleri (eds.): *The Wave-Particle Dualism.* A Tribute to Louis de Broglie on his 90th Birthday. 1984 ISBN 90-277-1664-1
4. E. Prugovečki: *Stochastic Quantum Mechanics and Quantum Spacetime.* A Consistent Unification of Relativity and Quantum Theory based on Stochastic Spaces. 1984; 2nd printing 1986 ISBN 90-277-1617-X
5. D. Hestenes and G. Sobczyk: *Clifford Algebra to Geometric Calculus.* A Unified Language for Mathematics and Physics. 1984 ISBN 90-277-1673-0; Pb (1987) 90-277-2561-6
6. P. Exner: *Open Quantum Systems and Feynman Integrals.* 1985 ISBN 90-277-1678-1
7. L. Mayants: *The Enigma of Probability and Physics.* 1984 ISBN 90-277-1674-9
8. E. Tocaci: *Relativistic Mechanics, Time and Inertia.* Translated from Romanian. Edited and with a Foreword by C.W. Kilmister. 1985 ISBN 90-277-1769-9
9. B. Bertotti, F. de Felice and A. Pascolini (eds.): *General Relativity and Gravitation.* Proceedings of the 10th International Conference (Padova, Italy, 1983). 1984 ISBN 90-277-1819-9
10. G. Tarozzi and A. van der Merwe (eds.): *Open Questions in Quantum Physics.* 1985 ISBN 90-277-1853-9
11. J.V. Narlikar and T. Padmanabhan: *Gravity, Gauge Theories and Quantum Cosmology.* 1986 ISBN 90-277-1948-9
12. G.S. Asanov: *Finsler Geometry, Relativity and Gauge Theories.* 1985 ISBN 90-277-1960-8
13. K. Namsrai: *Nonlocal Quantum Field Theory and Stochastic Quantum Mechanics.* 1986 ISBN 90-277-2001-0
14. C. Ray Smith and W.T. Grandy, Jr. (eds.): *Maximum-Entropy and Bayesian Methods in Inverse Problems.* Proceedings of the 1st and 2nd International Workshop (Laramie, Wyoming, USA). 1985 ISBN 90-277-2074-6
15. D. Hestenes: *New Foundations for Classical Mechanics.* 1986 ISBN 90-277-2090-8; Pb (1987) 90-277-2526-8
16. S.J. Prokhovnik: *Light in Einstein's Universe.* The Role of Energy in Cosmology and Relativity. 1985 ISBN 90-277-2093-2
17. Y.S. Kim and M.E. Noz: *Theory and Applications of the Poincaré Group.* 1986 ISBN 90-277-2141-6
18. M. Sachs: *Quantum Mechanics from General Relativity.* An Approximation for a Theory of Inertia. 1986 ISBN 90-277-2247-1
19. W.T. Grandy, Jr.: *Foundations of Statistical Mechanics.* Vol. I: *Equilibrium Theory.* 1987 ISBN 90-277-2489-X
20. H.-H von Borzeszkowski and H.-J. Treder: *The Meaning of Quantum Gravity.* 1988 ISBN 90-277-2518-7
21. C. Ray Smith and G.J. Erickson (eds.): *Maximum-Entropy and Bayesian Spectral Analysis and Estimation Problems.* Proceedings of the 3rd International Workshop (Laramie, Wyoming, USA, 1983). 1987 ISBN 90-277-2579-9
22. A.O. Barut and A. van der Merwe (eds.): *Selected Scientific Papers of Alfred Landé.* [*1888-1975*]. 1988 ISBN 90-277-2594-2

Fundamental Theories of Physics

23. W.T. Grandy, Jr.: *Foundations of Statistical Mechanics.* Vol. II: *Nonequilibrium Phenomena.* 1988 ISBN 90-277-2649-3
24. E.I. Bitsakis and C.A. Nicolaides (eds.): *The Concept of Probability.* Proceedings of the Delphi Conference (Delphi, Greece, 1987). 1989 ISBN 90-277-2679-5
25. A. van der Merwe, F. Selleri and G. Tarozzi (eds.): *Microphysical Reality and Quantum Formalism, Vol. 1.* Proceedings of the International Conference (Urbino, Italy, 1985). 1988 ISBN 90-277-2683-3
26. A. van der Merwe, F. Selleri and G. Tarozzi (eds.): *Microphysical Reality and Quantum Formalism, Vol. 2.* Proceedings of the International Conference (Urbino, Italy, 1985). 1988 ISBN 90-277-2684-1
27. I.D. Novikov and V.P. Frolov: *Physics of Black Holes.* 1989 ISBN 90-277-2685-X
28. G. Tarozzi and A. van der Merwe (eds.): *The Nature of Quantum Paradoxes.* Italian Studies in the Foundations and Philosophy of Modern Physics. 1988 ISBN 90-277-2703-1
29. B.R. Iyer, N. Mukunda and C.V. Vishveshwara (eds.): *Gravitation, Gauge Theories and the Early Universe.* 1989 ISBN 90-277-2710-4
30. H. Mark and L. Wood (eds.): *Energy in Physics, War and Peace.* A Festschrift celebrating Edward Teller's 80th Birthday. 1988 ISBN 90-277-2775-9
31. G.J. Erickson and C.R. Smith (eds.): *Maximum-Entropy and Bayesian Methods in Science and Engineering.* Vol. I: *Foundations.* 1988 ISBN 90-277-2793-7
32. G.J. Erickson and C.R. Smith (eds.): *Maximum-Entropy and Bayesian Methods in Science and Engineering.* Vol. II: *Applications.* 1988 ISBN 90-277-2794-5
33. M.E. Noz and Y.S. Kim (eds.): *Special Relativity and Quantum Theory.* A Collection of Papers on the Poincaré Group. 1988 ISBN 90-277-2799-6
34. I.Yu. Kobzarev and Yu.I. Manin: *Elementary Particles. Mathematics, Physics and Philosophy.* 1989 ISBN 0-7923-0098-X
35. F. Selleri: *Quantum Paradoxes and Physical Reality.* 1990 ISBN 0-7923-0253-2
36. J. Skilling (ed.): *Maximum-Entropy and Bayesian Methods.* Proceedings of the 8th International Workshop (Cambridge, UK, 1988). 1989 ISBN 0-7923-0224-9
37. M. Kafatos (ed.): *Bell's Theorem, Quantum Theory and Conceptions of the Universe.* 1989 ISBN 0-7923-0496-9
38. Yu.A. Izyumov and V.N. Syromyatnikov: *Phase Transitions and Crystal Symmetry.* 1990 ISBN 0-7923-0542-6
39. P.F. Fougère (ed.): *Maximum-Entropy and Bayesian Methods.* Proceedings of the 9th International Workshop (Dartmouth, Massachusetts, USA, 1989). 1990 ISBN 0-7923-0928-6
40. L. de Broglie: *Heisenberg's Uncertainties and the Probabilistic Interpretation of Wave Mechanics.* With Critical Notes of the Author. 1990 ISBN 0-7923-0929-4
41. W.T. Grandy, Jr.: *Relativistic Quantum Mechanics of Leptons and Fields.* 1991 ISBN 0-7923-1049-7
42. Yu.L. Klimontovich: *Turbulent Motion and the Structure of Chaos.* A New Approach to the Statistical Theory of Open Systems. 1991 ISBN 0-7923-1114-0
43. W.T. Grandy, Jr. and L.H. Schick (eds.): *Maximum-Entropy and Bayesian Methods.* Proceedings of the 10th International Workshop (Laramie, Wyoming, USA, 1990). 1991 ISBN 0-7923-1140-X
44. P. Pták and S. Pulmannová: *Orthomodular Structures as Quantum Logics.* Intrinsic Properties, State Space and Probabilistic Topics. 1991 ISBN 0-7923-1207-4
45. D. Hestenes and A. Weingartshofer (eds.): *The Electron.* New Theory and Experiment. 1991 ISBN 0-7923-1356-9

Fundamental Theories of Physics

46. P.P.J.M. Schram: *Kinetic Theory of Gases and Plasmas.* 1991 ISBN 0-7923-1392-5
47. A. Micali, R. Boudet and J. Helmstetter (eds.): *Clifford Algebras and their Applications in Mathematical Physics.* 1992 ISBN 0-7923-1623-1
48. E. Prugovečki: *Quantum Geometry.* A Framework for Quantum General Relativity. 1992 ISBN 0-7923-1640-1
49. M.H. Mac Gregor: *The Enigmatic Electron.* 1992 ISBN 0-7923-1982-6
50. C.R. Smith, G.J. Erickson and P.O. Neudorfer (eds.): *Maximum Entropy and Bayesian Methods.* Proceedings of the 11th International Workshop (Seattle, 1991). 1993 ISBN 0-7923-2031-X
51. D.J. Hoekzema: *The Quantum Labyrinth.* 1993 ISBN 0-7923-2066-2
52. Z. Oziewicz, B. Jancewicz and A. Borowiec (eds.): *Spinors, Twistors, Clifford Algebras and Quantum Deformations.* Proceedings of the Second Max Born Symposium (Wrocław, Poland, 1992). 1993 ISBN 0-7923-2251-7
53. A. Mohammad-Djafari and G. Demoment (eds.): *Maximum Entropy and Bayesian Methods.* Proceedings of the 12th International Workshop (Paris, France, 1992). 1993 ISBN 0-7923-2280-0
54. M. Riesz: *Clifford Numbers and Spinors* with Riesz' Private Lectures to E. Folke Bolinder and a Historical Review by Pertti Lounesto. E.F. Bolinder and P. Lounesto (eds.). 1993 ISBN 0-7923-2299-1
55. F. Brackx, R. Delanghe and H. Serras (eds.): *Clifford Algebras and their Applications in Mathematical Physics.* Proceedings of the Third Conference (Deinze, 1993) 1993 ISBN 0-7923-2347-5
56. J.R. Fanchi: *Parametrized Relativistic Quantum Theory.* 1993 ISBN 0-7923-2376-9
57. A. Peres: *Quantum Theory: Concepts and Methods.* 1993 ISBN 0-7923-2549-4
58. P.L. Antonelli, R.S. Ingarden and M. Matsumoto: *The Theory of Sprays and Finsler Spaces with Applications in Physics and Biology.* 1993 ISBN 0-7923-2577-X
59. R. Miron and M. Anastasiei: *The Geometry of Lagrange Spaces: Theory and Applications.* 1994 ISBN 0-7923-2591-5
60. G. Adomian: *Solving Frontier Problems of Physics: The Decomposition Method.* 1994 ISBN 0-7923-2644-X
61. B.S. Kerner and V.V. Osipov: *Autosolitons.* A New Approach to Problems of Self-Organization and Turbulence. 1994 ISBN 0-7923-2816-7
62. G.R. Heidbreder (ed.): *Maximum Entropy and Bayesian Methods.* Proceedings of the 13th International Workshop (Santa Barbara, USA, 1993) 1996 ISBN 0-7923-2851-5
63. J. Peřina, Z. Hradil and B. Jurčo: *Quantum Optics and Fundamentals of Physics.* 1994 ISBN 0-7923-3000-5
64. M. Evans and J.-P. Vigier: *The Enigmatic Photon.* Volume 1: The Field $\boldsymbol{B}^{(3)}$. 1994 ISBN 0-7923-3049-8
65. C.K. Raju: *Time: Towards a Constistent Theory.* 1994 ISBN 0-7923-3103-6
66. A.K.T. Assis: *Weber's Electrodynamics.* 1994 ISBN 0-7923-3137-0
67. Yu. L. Klimontovich: *Statistical Theory of Open Systems.* Volume 1: A Unified Approach to Kinetic Description of Processes in Active Systems. 1995 ISBN 0-7923-3199-0; Pb: ISBN 0-7923-3242-3
68. M. Evans and J.-P. Vigier: *The Enigmatic Photon.* Volume 2: Non-Abelian Electrodynamics. 1995 ISBN 0-7923-3288-1
69. G. Esposito: *Complex General Relativity.* 1995 ISBN 0-7923-3340-3

Fundamental Theories of Physics

70. J. Skilling and S. Sibisi (eds.): *Maximum Entropy and Bayesian Methods.* Proceedings of the Fourteenth International Workshop on Maximum Entropy and Bayesian Methods. 1996 ISBN 0-7923-3452-3
71. C. Garola and A. Rossi (eds.): *The Foundations of Quantum Mechanics Historical Analysis and Open Questions.* 1995 ISBN 0-7923-3480-9
72. A. Peres: *Quantum Theory: Concepts and Methods.* 1995 (see for hardback edition, Vol. 57) ISBN Pb 0-7923-3632-1
73. M. Ferrero and A. van der Merwe (eds.): *Fundamental Problems in Quantum Physics.* 1995 ISBN 0-7923-3670-4
74. F.E. Schroeck, Jr.: *Quantum Mechanics on Phase Space.* 1996 ISBN 0-7923-3794-8
75. L. de la Peña and A.M. Cetto: *The Quantum Dice.* An Introduction to Stochastic Electrodynamics. 1996 ISBN 0-7923-3818-9
76. P.L. Antonelli and R. Miron (eds.): *Lagrange and Finsler Geometry.* Applications to Physics and Biology. 1996 ISBN 0-7923-3873-1
77. M.W. Evans, J.-P. Vigier, S. Roy and S. Jeffers: *The Enigmatic Photon.* Volume 3: Theory and Practice of the $\boldsymbol{B}^{(3)}$ Field. 1996 ISBN 0-7923-4044-2
78. W.G.V. Rosser: *Interpretation of Classical Electromagnetism.* 1996 ISBN 0-7923-4187-2
79. K.M. Hanson and R.N. Silver (eds.): *Maximum Entropy and Bayesian Methods.* 1996 ISBN 0-7923-4311-5
80. S. Jeffers, S. Roy, J.-P. Vigier and G. Hunter (eds.): *The Present Status of the Quantum Theory of Light.* Proceedings of a Symposium in Honour of Jean-Pierre Vigier. 1997 ISBN 0-7923-4337-9
81. M. Ferrero and A. van der Merwe (eds.): *New Developments on Fundamental Problems in Quantum Physics.* 1997 ISBN 0-7923-4374-3
82. R. Miron: *The Geometry of Higher-Order Lagrange Spaces.* Applications to Mechanics and Physics. 1997 ISBN 0-7923-4393-X
83. T. Hakioğlu and A.S. Shumovsky (eds.): *Quantum Optics and the Spectroscopy of Solids.* Concepts and Advances. 1997 ISBN 0-7923-4414-6
84. A. Sitenko and V. Tartakovskii: *Theory of Nucleus.* Nuclear Structure and Nuclear Interaction. 1997 ISBN 0-7923-4423-5
85. G. Esposito, A.Yu. Kamenshchik and G. Pollifrone: *Euclidean Quantum Gravity on Manifolds with Boundary.* 1997 ISBN 0-7923-4472-3
86. R.S. Ingarden, A. Kossakowski and M. Ohya: *Information Dynamics and Open Systems.* Classical and Quantum Approach. 1997 ISBN 0-7923-4473-1
87. K. Nakamura: *Quantum versus Chaos.* Questions Emerging from Mesoscopic Cosmos. 1997 ISBN 0-7923-4557-6
88. B.R. Iyer and C.V. Vishveshwara (eds.): *Geometry, Fields and Cosmology.* Techniques and Applications. 1997 ISBN 0-7923-4725-0
89. G.A. Martynov: *Classical Statistical Mechanics.* 1997 ISBN 0-7923-4774-9
90. M.W. Evans, J.-P. Vigier, S. Roy and G. Hunter (eds.): *The Enigmatic Photon.* Volume 4: New Directions. 1998 ISBN 0-7923-4826-5
91. M. Rédei: *Quantum Logic in Algebraic Approach.* 1998 ISBN 0-7923-4903-2
92. S. Roy: *Statistical Geometry and Applications to Microphysics and Cosmology.* 1998 ISBN 0-7923-4907-5
93. B.C. Eu: *Nonequilibrium Statistical Mechanics.* Ensembled Method. 1998 ISBN 0-7923-4980-6

Fundamental Theories of Physics

94. V. Dietrich, K. Habetha and G. Jank (eds.): *Clifford Algebras and Their Application in Mathematical Physics.* Aachen 1996. 1998 ISBN 0-7923-5037-5
95. J.P. Blaizot, X. Campi and M. Ploszajczak (eds.): *Nuclear Matter in Different Phases and Transitions.* 1999 ISBN 0-7923-5660-8
96. V.P. Frolov and I.D. Novikov: *Black Hole Physics.* Basic Concepts and New Developments. 1998 ISBN 0-7923-5145-2; Pb 0-7923-5146
97. G. Hunter, S. Jeffers and J-P. Vigier (eds.): *Causality and Locality in Modern Physics.* 1998 ISBN 0-7923-5227-0
98. G.J. Erickson, J.T. Rychert and C.R. Smith (eds.): *Maximum Entropy and Bayesian Methods.* 1998 ISBN 0-7923-5047-2
99. D. Hestenes: *New Foundations for Classical Mechanics (Second Edition).* 1999 ISBN 0-7923-5302-1; Pb ISBN 0-7923-5514-8
100. B.R. Iyer and B. Bhawal (eds.): *Black Holes, Gravitational Radiation and the Universe.* Essays in Honor of C. V. Vishveshwara. 1999 ISBN 0-7923-5308-0
101. P.L. Antonelli and T.J. Zastawniak: *Fundamentals of Finslerian Diffusion with Applications.* 1998 ISBN 0-7923-5511-3
102. H. Atmanspacher, A. Amann and U. Müller-Herold: *On Quanta, Mind and Matter Hans Primas in Context.* 1999 ISBN 0-7923-5696-9
103. M.A. Trump and W.C. Schieve: *Classical Relativistic Many-Body Dynamics.* 1999 ISBN 0-7923-5737-X
104. A.I. Maimistov and A.M. Basharov: *Nonlinear Optical Waves.* 1999 ISBN 0-7923-5752-3
105. W. von der Linden, V. Dose, R. Fischer and R. Preuss (eds.): *Maximum Entropy and Bayesian Methods Garching, Germany 1998.* 1999 ISBN 0-7923-5766-3
106. M.W. Evans: *The Enigmatic Photon Volume 5: O(3) Electrodynamics.* 1999 ISBN 0-7923-5792-2
107. G.N. Afanasiev: *Topological Effects in Quantum Mecvhanics.* 1999 ISBN 0-7923-5800-7
108. V. Devanathan: *Angular Momentum Techniques in Quantum Mechanics.* 1999 ISBN 0-7923-5866-X
109. P.L. Antonelli (ed.): *Finslerian Geometries A Meeting of Minds.* 1999 ISBN 0-7923-6115-6
110. M.B. Mensky: *Quantum Measurements and Decoherence Models and Phenomenology.* 2000 ISBN 0-7923-6227-6
111. B. Coecke, D. Moore and A. Wilce (eds.): *Current Research in Operation Quantum Logic.* Algebras, Categories, Languages. 2000 ISBN 0-7923-6258-6
112. G. Jumarie: *Maximum Entropy, Information Without Probability and Complex Fractals.* Classical and Quantum Approach. 2000 ISBN 0-7923-6330-2
113. B. Fain: *Irreversibilities in Quantum Mechanics.* 2000 ISBN 0-7923-6581-X
114. T. Borne, G. Lochak and H. Stumpf: *Nonperturbative Quantum Field Theory and the Structure of Matter.* 2001 ISBN 0-7923-6803-7
115. J. Keller: *Theory of the Electron.* A Theory of Matter from START. 2001 ISBN 0-7923-6819-3
116. M. Rivas: *Kinematical Theory of Spinning Particles.* Classical and Quantum Mechanical Formalism of Elementary Particles. 2001 ISBN 0-7923-6824-X
117. A.A. Ungar: *Beyond the Einstein Addition Law and its Gyroscopic Thomas Precession.* The Theory of Gyrogroups and Gyrovector Spaces. 2001 ISBN 0-7923-6909-2
118. R. Miron, D. Hrimiuc, H. Shimada and S.V. Sabau: *The Geometry of Hamilton and Lagrange Spaces.* 2001 ISBN 0-7923-6926-2

Fundamental Theories of Physics

119. M. Pavšič: *The Landscape of Theoretical Physics: A Global View*. From Point Particles to the Brane World and Beyond in Search of a Unifying Principle. 2001 ISBN 0-7923-7006-6
120. R.M. Santilli: *Foundations of Hadronic Chemistry*. With Applications to New Clean Energies and Fuels. 2001 ISBN 1-4020-0087-1
121. S. Fujita and S. Godoy: *Theory of High Temperature Superconductivity*. 2001 ISBN 1-4020-0149-5
122. R. Luzzi, A.R. Vasconcellos and J. Galvão Ramos: *Predictive Statitical Mechanics*. A Nonequilibrium Ensemble Formalism. 2002 ISBN 1-4020-0482-6
123. V.V. Kulish: *Hierarchical Methods*. Hierarchy and Hierarchical Asymptotic Methods in Electrodynamics, Volume 1. 2002 ISBN 1-4020-0757-4; Set: 1-4020-0758-2
124. B.C. Eu: *Generalized Thermodynamics*. Thermodynamics of Irreversible Processes and Generalized Hydrodynamics. 2002 ISBN 1-4020-0788-4
125. A. Mourachkine: *High-Temperature Superconductivity in Cuprates*. The Nonlinear Mechanism and Tunneling Measurements. 2002 ISBN 1-4020-0810-4
126. R.L. Amoroso, G. Hunter, M. Kafatos and J.-P. Vigier (eds.): *Gravitation and Cosmology: From the Hubble Radius to the Planck Scale*. Proceedings of a Symposium in Honour of the 80th Birthday of Jean-Pierre Vigier. 2002 ISBN 1-4020-0885-6
127. W.M. de Muynck: *Foundations of Quantum Mechanics, an Empiricist Approach*. 2002 ISBN 1-4020-0932-1
128. V.V. Kulish: *Hierarchical Methods*. Undulative Electrodynamical Systems, Volume 2. 2002 ISBN 1-4020-0968-2; Set: 1-4020-0758-2
129. M. Mugur-Schächter and A. van der Merwe (eds.): *Quantum Mechanics, Mathematics, Cognition and Action*. Proposals for a Formalized Epistemology. 2002 ISBN 1-4020-1120-2
130. P. Bandyopadhyay: *Geometry, Topology and Quantum Field Theory*. 2003 ISBN 1-4020-1414-7
131. V. Garzó and A. Santos: *Kinetic Theory of Gases in Shear Flows*. Nonlinear Transport. 2003 ISBN 1-4020-1436-8
132. R. Miron: *The Geometry of Higher-Order Hamilton Spaces*. Applications to Hamiltonian Mechanics. 2003 ISBN 1-4020-1574-7
133. S. Esposito, E. Majorana Jr., A. van der Merwe and E. Recami (eds.): *Ettore Majorana: Notes on Theoretical Physics*. 2003 ISBN 1-4020-1649-2
134. J. Hamhalter. *Quantum Measure Theory*. 2003 ISBN 1-4020-1714-6
135. G. Rizzi and M.L. Ruggiero: *Relativity in Rotating Frames*. Relativistic Physics in Rotating Reference Frames. 2004 ISBN 1-4020-1805-3
136. L. Kantorovich: *Quantum Theory of the Solid State: an Introduction*. 2004 ISBN 1-4020-1821-5
137. A. Ghatak and S. Lokanathan: *Quantum Mechanics: Theory and Applications*. 2004 ISBN 1-4020-1850-9
138. A. Khrennikov: *Information Dynamics in Cognitive, Psychological, Social, and Anomalous Phenomena*. 2004 ISBN 1-4020-1868-1
139. V. Faraoni: *Cosmology in Scalar-Tensor Gravity*. 2004 ISBN 1-4020-1988-2
140. P.P. Teodorescu and N.-A. P. Nicorovici: *Applications of the Theory of Groups in Mechanics and Physics*. 2004 ISBN 1-4020-2046-5

KLUWER ACADEMIC PUBLISHERS – DORDRECHT / BOSTON / LONDON